Autonomic Nervous System

Daniel Pedro Cardinali

Autonomic Nervous System

Basic and Clinical Aspects

Springer

Daniel Pedro Cardinali
Facultad de Ciencias Médicas
Pontificia Universidad Católica Argentina
Buenos Aires, Capital Federal
Argentina

ISBN 978-3-319-57570-4 ISBN 978-3-319-57571-1 (eBook)
DOI 10.1007/978-3-319-57571-1

Library of Congress Control Number: 2017949889

Printed on acid-free paper

This Springer imprint is published by Springer Nature
The registered company is Springer International Publishing AG
The registered company address is: Gewerbestrasse 11, 6330 Cham, Switzerland

Prologue

The autonomic nervous system (ANS) is an important component of the nervous system consisting of a complex set of neurons and neural pathways that control the function of the various visceral organ systems. The overall function of the ANS is to maintain the body homeostasis and to react adaptively to changes in the external and internal milieu.

The ANS innervates the heart, the smooth muscle in all the organs, the abdominal viscera, the exocrine and endocrine glands, and the immune system. Thus, the ANS participates in the regulation of breathing, circulation, digestion, metabolism, and the internal milieu, exocrine and endocrine gland secretion, immune responses, body temperature, and reproduction [1, 2]. Unfortunately, such medical importance tends to be underscored in many books on physiology or neural sciences in which the subject takes up much less space than that accorded, for example, to somatosensory or cognitive functions.

The basic structure and operation of the ANS were defined at the beginning of the last century, primarily by Gaskell and Langley, who recognized its two main divisions: the sympathetic and the parasympathetic [3]. Furthermore, Langley designated the enteric nervous system as a third division based on the submucous plexus of Meissner and the myenteric plexus of Auerbach located in the wall of the gastrointestinal tract, albeit controlled by the sympathetic and parasympathetic divisions. Overcoming the classical concept of a purely efferent system, it is presently accepted that the ANS is composed of visceral afferents, integration centers, particularly in the brainstem, hypothalamus, and limbic cortex, and visceral sympathetic and parasympathetic efferents; thus, the ANS extends both into the central nervous system (CNS) and to the periphery.

Conceptually, the bio-psycho-social-ecological nature of the individual is truly expressed by the function of his or her ANS. Its name is misleading because none of the components shows "autonomy" in an integrated body. Nor are they solely "passive" or generated "without elaboration by mind." All body systems are dependent and affected by the action of others in a multicellular organization.

These dynamic relationships are the core of homeostasis, a key concept in physiology. "Homeostasis" is used today to define not only the strategies that allow the body's proper response to changes in the environment (reactive homeostasis), but also the remarkably developed, temporal mechanisms that allow the body to predict the timing of environmental stimuli (predictive homeostasis based on biological rhythms).

Autonomic reflexes are mediated by neural pathways in the brainstem and spinal cord and generally regulate organ and system performance very rapidly (in milliseconds). Autonomic control is also mediated by specific brain regions, such as the hypothalamus, which is responsible for medium-term (minutes) and long-term (hours/days) regulation of internal organ systems. Importantly, autonomic reflexes are dynamic, where adaptations can alter rapid homeostatic control over longer time scales [4].

This book discusses the ANS from both an enlarged and a timed perspective. First, it presents how the organization of the ANS is built in four different hierarchical levels. Next, it discusses how the ANS function changes in the three body configurations (wakefulness, slow-wave sleep, and rapid eye movement, REM, sleep) found during a 24-h cycle. Finally, the most important clinical implications for this enlarged and timed vision of the ANS are discussed.

The Autonomic Nervous System – Basic and Clinical Aspects is designed as a comprehensive textbook for advanced medical students and health professionals. It primes for a detailed and complete understanding of the neuroscience behind the ANS and a proper clinical applicability of this knowledge. ANS dysfunction and clinical manifestations involve multiple variables, which are often undervalued in clinical practice. However, symptoms and signs of ANS disturbances should always be considered according to their diagnostic implication, their impact on the quality of life of patients, and their prognostic value for life expectancy.

References

1. Saper CB. The central autonomic nervous system: conscious visceral perception and autonomic pattern generation. Annu Rev Neurosci. 2002;25:433–69.
2. Benarroch EE. The central autonomic network: functional organization, dysfunction, and perspective. Mayo Clin Proc. 1993;68:988–1001.
3. Oakes PC, Fisahn C, Iwanaga J, DiLorenzo D, Oskouian RJ, Tubbs RS. A history of the autonomic nervous system: part II: from Reil to the modern era. Childs Nerv Syst. 2016;32:2309–15.
4. McDougall SJ, Munzberg H, Derbenev AV, Zsombok A. Central control of autonomic functions in health and disease. Front Neurosci. 2014;8:440.

Contents

Abbreviations

3V	Third ventricle
5-HT	Serotonin
ACh	Acetylcholine
ACTH	Adrenocorticotropin
AD	Alzheimer's disease
AgRP	Agouti protein-related peptide
AMPK	AMP-activated protein kinase
AN	Ambiguous nucleus
ANP	Atrial natriuretic peptide
ANS	Autonomic nervous system
AP	Area postrema
APP	Amyloid precursor protein
ARC	Arcuate nucleus
AV3V	Anterior ventral region of the third ventricle
AVP	Arginine vasopressin
AVPV	Anteroventral periventricular
Aβ	Amyloid β
BAT	Brown adipose tissue
BCRs	B cell receptors
BF	Basal forebrain
BMD	Bone mineral density
BNP	Brain natriuretic peptide
BP	Blood pressure
BRAC	Basic rest–activity cycle
BZD	Benzodiazepine
CALB	Calbindin
CALR	Calretinin
CART	Cocaine and amphetamine-regulated transcript
CCG	Controlled clock genes
CCK	Cholecystokinin
CGRP	Calcitonin gene-related peptide
CNP	C-type natriuretic peptide
CNS	Central nervous system
CO	Carbon monoxide

COMT	Catechol-*O*-methyl transferase
CRH	Corticotropin-releasing hormone
CSF	Cerebrospinal fluid
CVC	Cutaneous vasoconstriction
DA	Dopamine
DDR	DNA damage response
DLMO	Dim light melatonin onset
DMH	Dorsomedial hypothalamus
DRN	Dorsal raphe nucleus
DSM	Diagnostic and Statistical Manual of Mental Disorders
E	Epinephrine
ECL	Enterochromaffin-like cells
ECN	External carotid nerve
EEG	Electroencephalographic
EGFR	Epidermal growth factor receptor
EKG	Electrocardiogram
EMA	European Medicines Agency
EPSP	Excitatory postsynaptic potential
FDA	Food and Drug Administration
FEO	Food-entrained oscillators
fMRI	Functional magnetic resonance imaging
FSH	Follicle-stimulating hormone
GABA	γ-Aminobutyric acid
GH	Growth hormone
GHRH	Growth hormone-releasing hormone
GLP-1	Glucagon-like peptide-1
Glu	Glutamate
Gly	Glycine
GnRH	Gonadotropin-releasing hormone
GRP	Gastrin-releasing peptide
His	Histamine
HRV	Heart rate variability
ICN	Internal carotid nerve
ICSD	International Classification of Sleep Disorders
IFN	Interferon
Ig	Immunoglobulin
IGF1	Insulin-like growth factor 1
IL	Interleukin
IPSP	Inhibitory postsynaptic potential
IRS	Insulin R substrate
IRS-1	Insulin receptor substrate 1
Kiss1 R	Kisspeptin receptor
Kiss1	Kisspeptin
LAN	Light at night
LC	Locus coeruleus

L-Dopa	L-Dihydroxyphenylalanine
LDT	Laterodorsal tegmental nucleus
LH	Luteinizing hormone
LHA	Lateral hypothalamic area
LPB	Lateral parabrachial nucleus
MAO	Monoamine oxidase
MCH	Melanin concentrating hormone
MCI	Mild cognitive impairment
MMC	Migrating motor complex
MnPO	Median preoptic area
MS	Metabolic syndrome
MSH	Melanocyte-stimulating hormone
MT	Melatonin receptor
NE	Norepinephrine
NF	Nuclear factor
NK	Natural killer
NMDA	*N*-methyl-D-aspartate
NO	Nitric oxide
NOS	NO synthase
NPY	Neuropeptide Y
NREM	Non-REM
NTS	Nucleus tractus solitarius
OC	Optic chiasm
O-GlcN-Ac	*O*-β-D-*N*-acetylglucosamine
OVLT	Organum vasculosum laminae terminalis
PACAP	Pituitary adenylate cyclase-activating peptide
PARP	Poly ADP-ribose polymerase
PB/PC	Parabrachial/precoeruleus
PBN	Parabrachial nucleus
PET	Positron emission tomography
PG	Prostaglandin
PGO	Ponto-geniculo-occipital
PH	Posterior hypothalamus
POMC	Pro-opiomelanocortin
PPARs	Peroxisome proliferator-activated receptors
PPT	Pedunculopontine tegmentum nucleus
PPT/LDT	Pedunculopontine and laterodorsal tegmenti
PRL	Prolactin
PSG	Polysomnography
PTH	Parathyroid hormone
PVH	Paraventricular hypothalamus
PVN	Paraventricular nucleus
PZ	Parafacial zone
REM	Rapid eye movement
RHT	Retino-hypothalamic tract

RNS	Reactive nitrogen species
ROS	Reactive oxygen species
rRPa	Rostral raphe pallidus
SCG	Superior cervical ganglion
SCGx	Superior cervical ganglionectomy
SCN	Suprachiasmatic nuclei
SFO	Subfornical organ
SIF	Small, intensely fluorescent
SLD	Sublaterodorsal
SN	Substantia nigra
SON	Supraoptic nucleus
sPVz	Sub PVN zone
TCRs	T cell receptors
Th	T helper
TMN	Tuberomammillary nucleus
TNF	Tumor necrosis factor
T-reg	T regulatory cells
TRH	Thyrotropin-releasing hormone
TRP	Transient receptor potential
TSH	Thyroid-stimulating hormone
VIP	Vasoactive intestinal peptide
VLPO	Ventrolateral preoptic nucleus
VMH	Ventromedial hypothalamus
VMN	Ventromedial nucleus
VTA	Ventral tegmental area
WAT	White adipose tissue
WHO	World Health Organization

The Enlarged Autonomic Nervous System

1

Abstract

The traditional view of the autonomic nervous system (ANS) considers only its peripheral part, the sympathetic and parasympathetic systems, to which sometimes the enteric ANS is added as an independent entity. However, this view is insufficient to understand the most important function of the ANS: the maintenance of homeostasis. This Chapter describes the hierarchical organization of the autonomic motor pathways and their similarities with the somatic motor organization. It analyzes the types of neurons, neuronal circuits, and electric potentials identified in autonomic neurons.

Keywords

Acetylcholine • Autonomic posture • Basic neuronal circuits • Circadian clock • Fight or flight reaction • Golgi I and Golgi II neurons • Hierarchical organization of the ANS • Homeostasis • Norepinephrine • Somatic and visceral motor neurons

Objectives

After studying this chapter, you should be able to:

- Describe the control of reactive and predictive homeostasis as the most important function of the ANS
- Describe the hierarchical organization of the autonomic and somatic motor pathways and their similarities
- Name the components of the autonomic posture in comparison with the mechanisms of body posture
- Describe the two types of neurons and the three neuronal circuits that give rise to the basic neuronal organization of ANS
- Name the different types of electrical potentials identified in autonomic neurons

D.P. Cardinali, *Autonomic Nervous System*, DOI 10.1007/978-3-319-57571-1_1

The Control of Homeostasis Is the Most Important Function of the ANS

The traditional view of the autonomic nervous system (ANS) considers only its peripheral part, the sympathetic and parasympathetic systems, to which sometimes the enteric ANS is added as an independent entity. However, this view is insufficient to understand the most important function of the ANS: the maintenance of homeostasis.

The term "homeostasis" was coined by the distinguished American physiologist Walter B. Cannon as a comprehensive concept to describe the physiological factors that maintain the equilibrium state of the body, and therefore, life [1]. Following Claude Bernard, Cannon perfected the idea of the constancy of the internal environment, giving it its present meaning. As explained by Cannon, he chose as a prefix homeo- ("like") instead of homo- ("the same") to take into account the normal variations that any physiological variable exhibits, thus avoiding the misleading consideration of a "constant (fixed) internal milieu."

The development of chronobiology as an independent discipline added another aspect to homeostasis. This term is used today to define not only the strategies that allow the organism's proper response to changes in the environment ("reactive homeostasis"), but also the time-related mechanisms, remarkably developed, that allow the body to predict the occurrence of a given environmental stimulus to initiate the appropriate corrective responses beforehand ("predictive homeostasis") [2]. Predictive homeostasis comprises the anticipatory mechanisms that precede a regularly occurring environmental phenomenon, thus facilitating a better physiological adaptation to it. For example, the increase in plasma cortisol that precedes waking anticipates the energy demands of a changing posture; the increased gastrointestinal secretion preceding the usual lunch time anticipates the changes in the content of the digestive tract.

Let us suppose that a diurnally active mammal in search of food finds the food at a place about 2 h from the shelter it uses to escape its predator at night. This situation makes it essential for the animal to predict nightfall about 2 h in advance. It do this using the position of the sun or other environmental variables, such as temperature or humidity. However, the existence of clouds or a large daily variation in ambient temperature make the value of such reference parameters unreliable. It is then extremely useful for survival to possess a system of time control integrated into its own organism that allows a temporal prediction without having to depend on the reading of external signals. This was probably the strategy selected as an advantage by evolution.

The circadian clock is ideal to fulfill such a function: we could have a sufficiently accurate idea of the time of day by simply analyzing our biological structure periodically and without consulting our wristwatch. That is, a "day" and "night" have been created within the organism to allow it to optimize adaptation [3].

Generally, the nervous system can produce a small number of actions: (a) the contraction of a muscle or of a group of striated or smooth muscles; (b) the exocrine, endocrine, or paracrine secretion of a cell or cell group; (c) changes in cellular permeability. This is achieved via motor neurons, which can be classified per their targets into three categories:

- Somatic motor neurons, which originate in the central nervous system (CNS), project their axons to the skeletal muscles (such as the muscles of the limbs, abdominal, and intercostal muscles) that are involved in locomotion.
- Special visceral motor neurons, also called branchial motor neurons, which directly innervate the branchial muscles (that motorize the gills in fish and the face and neck in land vertebrates).
- Visceral preganglionic neurons, which indirectly innervate the heart, the smooth muscle in all organs, the abdominal viscera, the exocrine and endocrine glands, and the immune system. They synapse onto neurons located in the autonomic ganglia of the ANS (sympathetic and parasympathetic postganglionic neurons).

As a consequence: (a) the motor command of skeletal and branchial muscles is monosynaptic (involving only one motor neuron, somatic and branchial respectively, which synapses onto the muscle); (b) the command of the viscera is disynaptic, involving two neurons, the visceral preganglionic neuron located in the CNS, which synapses onto a postganglionic neuron, which synapses onto the organ.

The ANS, which is in charge of the innervation of smooth muscles of all organs and innervation of the viscera, immune tissue, and exocrine and endocrine glands, can produce the three actions mentioned above (smooth muscle contraction; exocrine, endocrine, or paracrine secretion; permeability changes). It must be noted that for the command of visceral muscles, the ganglionic neuron, parasympathetic or sympathetic, is the real motor neuron, as it is the one that directly innervates the muscle, whereas the general visceral motor neuron is, strictly speaking, the preganglionic one [4].

All vertebrate motor neurons are cholinergic, that is, they release the neurotransmitter acetylcholine (ACh). Parasympathetic ganglionic neurons are also cholinergic, whereas most sympathetic ganglionic neurons are noradrenergic, that is, they release the neurotransmitter norepinephrine (NE).

It is not unusual to hear or read or to remain implicit that actions of the sympathetic and parasympathetic divisions promote opposite effects in the organs they innervate. However, for the precise aim of a homeostatic response requiring effectiveness, energy economy, and sometimes a fast response, the sympathetic and parasympathetic neurons must collaborate for the final net effect.

In general, the sympathetic activity is designed to place the individual in a situation of defense in the face of danger, real or potential. Sympathetic stimulation leads to variations in visceral functions designed to protect the integrity of the organism and to ensure survival. In fact, a sympathectomized animal hardly survives if left free and exposed in its natural environment. However, in addition to calm and resting states, our daily life involves many perturbations that induce active conditions such as locomotion, eating, and communication. Thus, the mobilization of physiological variables does not occur in an all-or-nothing manner and is not exclusive of extreme "fight or flight" conditions; rather, they can occur with graded intensities in ordinary daily situations. During such active periods, cardiovascular, respiratory, and body temperature regulation needs to be adjusted to situational demands, which differ from those during resting states, by modulating or resetting homeostatic points. For instance, if an increase in blood pressure (BP) is needed, a greater gain in the response can be obtained by the concomitant enhancement of sympathetic activity and inhibition of parasympathetic activity. This is obtained not only in the CNS areas, but also via reciprocal inhibitory connections at the level of the heart wall [5].

Autonomic nervous system activity adapts to body changes and supports somatic reactions. The autonomic response may be secondary to the somatic, parallel to, or, most commonly, before (the "autonomic posture") [3]. The ANS is designed to produce sustained actions, compared with those of the somatic nervous system. On the other hand, the ANS performs this multisystem control in a continuous and constant manner throughout the life cycles, both at rest and during activity in the three body configurations within a 24 h cycle discussed in Chap. 2.

One of the most surprising features of the ANS is the speed and intensity with which it modifies the visceral functions. In a few seconds, for example, the heart rate and BP can increase up to double the normal rates, there may be intense sweating, the urinary bladder may empty, or digestive motility and secretion may be activated.

The sympathetic stimulation causes general responses in the organism for two reasons. First, there is a high degree of irradiation of the connections between preganglionic and postganglionic neurons. The proportion of preganglionic fibers to postganglionic neurons in sympathetic ganglia is 1/4–1/30, and since the degree of axonal branching is high, it is estimated that each preganglionic axon contacts an average of 120 neurons. For this reason, the activity of a few preganglionic neurons is highly amplified numerically via their ganglionic connections. Second, sympathetic activity produces a stimulation of the secretion of catecholamines from the adrenal medulla. In circulation, almost all epinephrine (E) found is derived from the adrenal medulla, but only 20% of the NE is. The rest of the circulating NE comes from the peripheral sympathetic terminals, i.e., as the neurotransmitter that has escaped presynaptic reuptake.

A massive activation of the sympathetic system produces a set of reactions defined as an alarm response (fight or flight) [6]. The most obvious visceral phenomena of this massive activation are:

- Pupil dilation, to increase the visual field
- Piloerection, to simulate a larger body size
- Sweating, to lose heat that is produced by muscular activity
- Increased cardiac activity and BP, to provide greater blood flow to the muscles
- Bronchodilation, to increase the entrance of air into the lungs
- Hyperglycemia
- Inhibition of the digestive function
- Inhibition of the urinary and genital functions

In contrast, the activity of the parasympathetic system is related to protective and conservation functions, which favor the proper functioning of the different visceral organs. The functional components of the parasympathetic system do not act simultaneously under normal conditions, but participate in specific reflexes or in integrated reactions to promote a specific visceral function. Thus, stimulation of different nuclei of parasympathetic neurons promotes responses such as:

- Pupillary constriction, to protect the retina from excessive illumination
- Decreased heart rate, to avoid excessive activity
- Bronchoconstriction, to protect the lungs
- Increased motility and digestive secretions, to promote digestion
- Urinary activity and urination
- Genital activity (erection)

Therefore, the parasympathetic effects are, in consonance with their fundamental purpose, more localized. In the parasympathetic ganglia, each preganglionic neuron contacts a few postganglionary neurons, whose divergence is much smaller than that of the sympathetic system [6].

Both efferent systems exert a control of the function of the visceral organs in variable forms. Some effector organs receive innervation from a single system. For example, the smooth muscle of most blood vessels is only controlled by sympathetic vasoconstrictor innervation (except for vessels of genital organs, which also receive parasympathetic innervation, and cholinergic vasodilator fibers in the skeletal muscles and brain). In these cases, control of the visceral activity depends on the variations of the frequency of discharge of the impulses by sympathetic innervation. As the autonomic nerves have a basal tonic activity (of low pulse frequencies, 0.5–5 Hz), the activity may increase or decrease. In most vessels, an increase in sympathetic frequency determines vasoconstriction and a decrease, vasodilation [7].

Sympathetic and parasympathetic nerve fibers simultaneously innervate many effector organs. The activity of the organ depends on the interaction or balance between the signals of both systems, which exert antagonistic effects. This interaction can be developed by opposite actions on the same effector cells, as in the heart, where the sympathetic excites the nodal cells thus increasing the heart rate, whereas the parasympathetic inhibits the same cells, thus reducing their frequency.

Another possibility is the action on different cells, with opposite effects. In the iris, the sympathetic excites the meridian muscular fibers, which produces mydriasis, whereas the parasympathetic excites the circular fibers and produces miosis. In these cases, the functional balance is relatively complex. In normal circumstances, the two systems are reciprocally active, because the central and reflex signals excite one system and inhibit the other. When both sympathetic and parasympathetic systems innervate the same target cells, more complex interactions arise. For example, the simultaneous activation of sympathetic fibers may exaggerate the cardiovascular response to parasympathetic activity and vice versa. This phenomenon is mediated by mutual influences at the presynaptic and postsynaptic domains [7].

There Are Functional Similarities in the Hierarchical Organization of the Autonomic and Somatic Motor Pathways

To understand the hierarchical organization of the ANS, several concepts derived from the somatic motor system are useful [3]. Any movement, even the simplest one, involves an enormous amount of information processing and the participation of numerous neuronal groups. On the other hand, each movement of our body is based on a posture; thus, it is important to consider how the motor system provides global responses for both components, the posture needed, and the movement itself.

To achieve this, there are four hierarchical levels in which the somatic motor system is organized (Fig. 1.1): (a) spinal cord; (b) brainstem; (c) motor cerebral cortex; (d) premotor cortical areas.

The circuits of the three basic motor reflexes, i.e., the myotatic reflex, the tendon reflex, and the withdrawal reflex, are found in the spinal cord, the centers of regulation of the dorsolateral and ventromedial motor neuron systems are located in the brainstem, and the motor programs defined by the secondary motor areas (premotor, motor supplemental area, parietal cortex) are situated in the primary motor cortex.

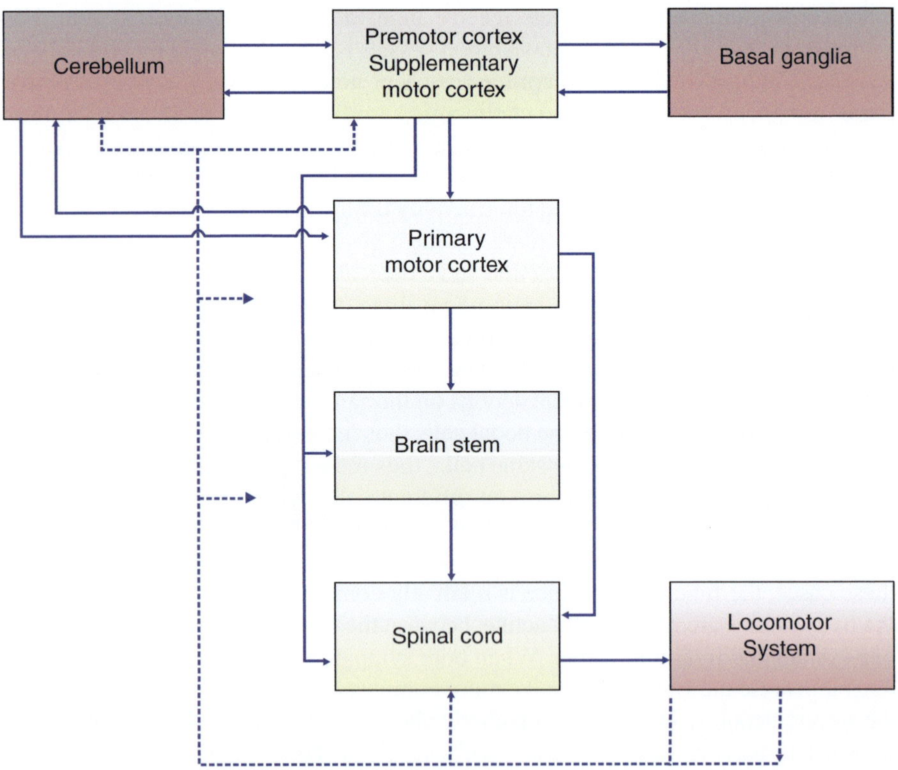

Fig. 1.1 Hierarchical organization of the somatic motor system. Modified with permission from Cardinali [3]

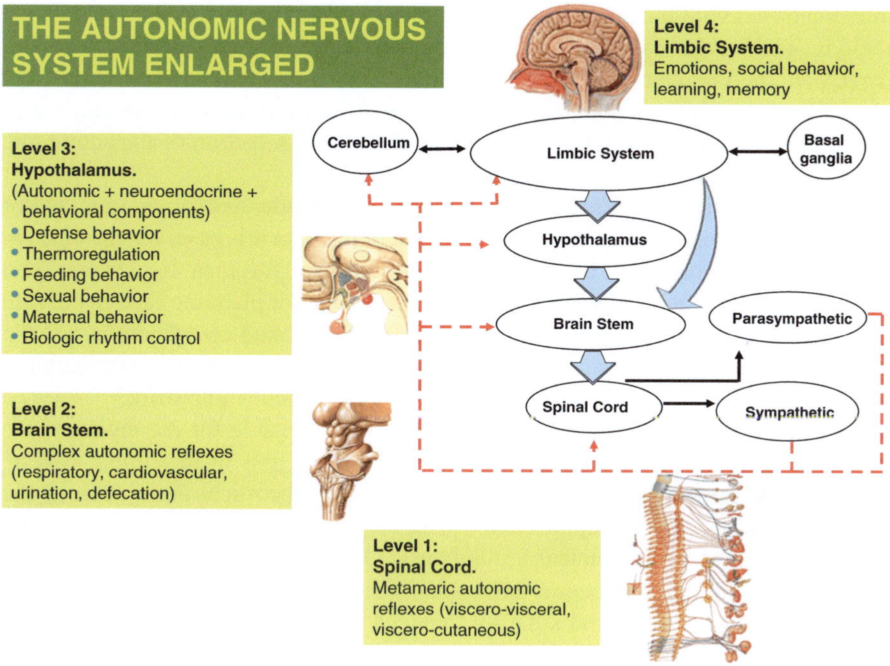

Fig. 1.2 Hierarchical organization of the autonomic nervous system (ANS). The figure was prepared in part using image vectors from Servier Medical Art (www.servier.com), licensed under the Creative Commons Attribution 3.0 Unported License (http://creativecommons.org/license/by/3.0/)

Also, two other brain areas, the cerebellum and the basal ganglia, have an essential regulatory function with regard to the somatic motor system [3].

In the ANS, a similar arrangement can be found (Fig. 1.2). The four major levels are: (a) spinal cord; (b) brainstem; (c) hypothalamus; (d) limbic system.

The spinal cord contains connections that mediate segmental autonomic reflexes involved in visceral function. These localized autonomic phenomena acquire intersegmental significance in the brainstem. In the brainstem, many complex autonomic responses, such as cardiovascular and respiratory regulation, are found.

The other two hierarchical levels of the ANS organization are the hypothalamus and the limbic system. In the hypothalamus, autonomic motor programs acquire their homeostatic nature. The cardiocirculatory and respiratory response to hemorrhage is completed by a neuroendocrine response, i.e., the secretion of arginine vasopressin (AVP) and adrenocorticotropin (ACTH), and by a behavioral response (thirst and fluid intake, etc.). These neuroendocrine mechanisms have strong similarities to those of the somatic posture needed for the execution of orders derived from the top level of the motor system.

The limbic system gives the homeostatic reaction its emotional tone and social significance [8]. The amygdala provides affective or emotional value to incoming sensory information and has multiple downstream targets that participate in the autonomic and neuroendocrine–immune responses. The anterior cingulate cortex is interconnected with the anterior insula and is subdivided into the ventral and dorsal regions. The insular cortex is the primary interoceptive cortex that integrates visceral, pain, and temperature sensations. This hierarchical level also comprises the cerebellum, which participates in the coordination of the executed autonomic programs, and the basal ganglia, which are relevant for the selection of the autonomic program best adapted to a given situation [9].

It is useful to compare the performance of somatic and autonomic motor responses with that of a building under construction [3]. In this case, we can observe three basic categories, in hierarchical order: masons, foremen (overseers), and architects. Schematically, the architects are responsible for planning (activity before the start of the work), the foremen for the management and coordination, and the masons for the construction itself. All three functions are indispensable for building and the failure of one of them will affect the success of the whole work.

The main "architects" of the ANS, which are responsible for the layout of the autonomic programs, are in cortical and subcortical regions of the limbic system. They include the limbic association cortex and the subcortical areas such as the amygdala, hippocampus, septal nuclei, olfactory bulb, and portions of the basal ganglia (ventral striatum, nucleus accumbens). The ventral portion of the basal ganglia are part of a loop that begins and ends in the limbic areas. The limbic association cortex projects to ventral striatum (nucleus accumbens) then to the thalamus, and finally back to the limbic cortex. The overall function of this loop is to select a sequence of autonomic actions (behaviors) while suppressing others.

The "overseers" or areas of execution in the motor system are linked to the movement itself (dorsolateral system: primary motor cortex, red nucleus) and the posture (ventromedial system: vestibular nuclei, inferior colliculus superior, reticular formation) needed for that movement. In the case of the ANS, the hypothalamus plays such a function. The main autonomic behaviors coordinated by the hypothalamus are:

- Defense behavior
- Nutritive or appetitive behavior
- Thermoregulatory behavior
- Maternal and sexual behavior

Typically, autonomic behaviors involve the coordinated expression of autonomic, neuroendocrine, somatic, and motivational mechanisms. Recent observations underlined the role of the cerebellum in the appropriate coordination of these components [4].

The "masons" of the somatic motor system, which represent the final common pathway of the system and are directly responsible for muscle contraction, are: (a) the motor units of the spinal cord; (b) the motor units in the nuclei of the cranial nerves.

The "masons" of the ANS include the motor neurons (postganglionic) of the sympathetic, parasympathetic, and enteric systems. These neurons are separate functional units (vasomotor muscular, cutaneous vasomotor, sudomotor, pilomotor, visceromotor) that exercise a specific and appropriate control over a target organ or cell.

From data derived from the somatic motor system, there are three important aspects to be considered for the physiological understanding of the hierarchical organization described above: (a) there is somatotopy, i.e., an orderly anatomical map at each level of organization; (b) at each level of the motor system, information from the periphery is received and modifies the descending order of command; (c) the upper levels have the capacity to control or suppress the information that reaches them (afferent control).

For other authors, the hierarchical organization of the integrated control of autonomic responses involved in homeostasis, adaptation, and emotional and goal-oriented behaviors includes the following levels: (a) spinal; (b) bulbopontine; (c) pontomesencephalic; (d) forebrain [10]. The bulbopontine (lower brainstem) level is involved in the reflex control of circulation, respiration, gastrointestinal function, and micturition. The pontomesencephalic (upper brainstem) level integrates autonomic control with pain modulation and responses to stress. The brainstem autonomic areas include the periaqueductal gray, the parabrachial nucleus (PBN), the nucleus tractus solitarius (NTS), the ventrolateral medulla, and the medullary raphe. The forebrain regions involved in the control of autonomic functions include the insular cortex, the anterior cingulate cortex, the amygdala, and the hypothalamus.

However, such a view considers only partially the role of the hypothalamus as a central component in the ANS hierarchy. Maintaining the hypothalamus as an independent level has the advantage of recognizing its role in integrated autonomic, neuroendocrine–immune, and behavioral responses. Therefore, the areas of the central autonomic network: (a) are reciprocally interconnected; (b) receive converging visceral and somatosensory information; (c) generate stimulus-specific patterns of autonomic, endocrine, and motor responses; (d) are regulated according to the respective body configuration: wakefulness, slow wave or non-REM (NREM) sleep, REM sleep [3].

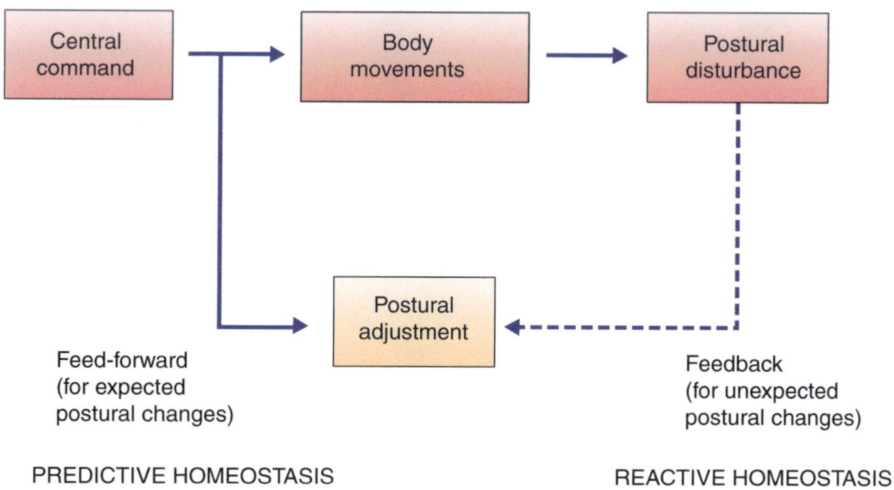

Fig. 1.3 Body posture is the position of the trunk relative to that of the limbs, and both, as a whole, in space. The postural reflexes are a set of antigravity reflexes, articulating with each other as a program. The postural adjustment includes anticipatory feed-forward programs and compensatory servo-assisted, feed-back mechanisms. In the same way, there is an "autonomic posture" that includes the mechanisms of anticipatory predictive homeostasis and the corrective mechanisms of reactive homeostasis. Modified with permission from Cardinali [3]

The Autonomic Posture

The adjustment program of body posture includes compensatory and anticipatory mechanisms. The spinal motoneurons are under the continuous influence of descending impulses from the upper regions and from the corresponding muscular and skin territories. One of the fundamental descending programs regulating the activity of motoneurons is that of posture, derived from neurons located in the brainstem [11].

The term "posture" defines the position on the trunk and the limbs. The postural reflexes are a set of antigravity reflexes, articulating each other as a program. This postural adjustment program includes feed-back and feed-forward compensatory mechanisms (Fig. 1.3) In the same way, there is an "autonomic posture" that includes the anticipatory mechanisms of predictive homeostasis and the corrective mechanisms of reactive homeostasis.

In the case of body posture, we anticipate with a proper body position the predictable changes given by muscle activity and the force of gravity, and correct this appropriate position with compensatory changes triggered by sensory information (Fig. 1.4). The position of the body in space varies accordingly to the movements performed. Therefore, we do not have a "single posture," but the correct one that is adapted to the movements performed. This maintenance of the postural equilibrium requires, in addition to the movements, advanced programming, and an on-line regulation of the process to adapt to the changes. For this, the integration of four sensorial modalities is indispensable: (a) vision; (b) position of the head; (c) proprioception;

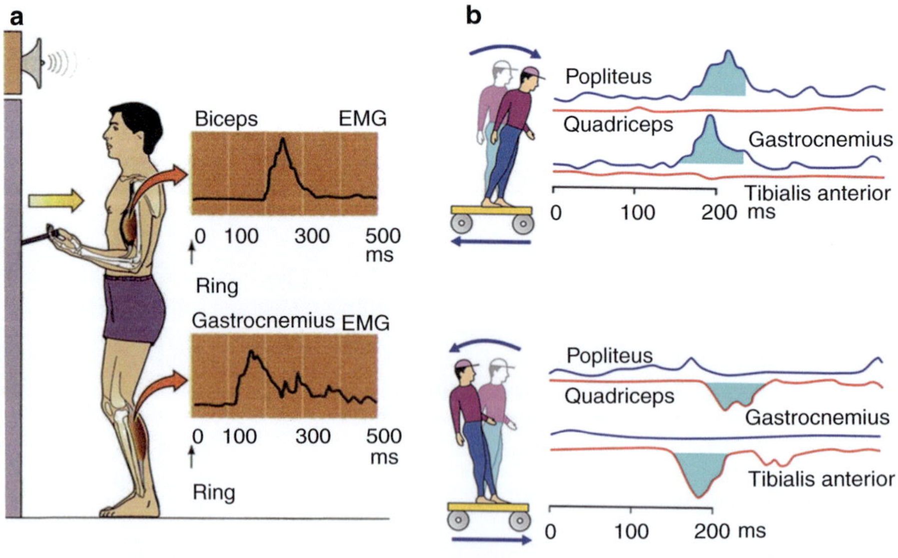

Fig. 1.4 With any motor plan, the proper posture program is executed first. Note that postural responses are always triggered before voluntary movements (**a**). If a normal individual is placed on a forward-leaning platform, the extension of the lower limb stabilizes the body, which causes the extensor (antigravitational) reflex to increase in the lower limbs as it is practiced (**b**). If the platform is tilted backward, the extension of the lower members, antigravitational in the previous case, now produces destabilization. In this second case, and after a few repetitions of the test, the extensor reflex diminishes, until it is completely extinguished. There is strong evidence for the role of the cerebellum in the mechanisms of both body posture and autonomic posture. Modified with permission from Cardinali [3]

(d) exteroception (touch). Based on these data, the nervous system produces an early postural program suitable for movement and provides a series of automatic adjustments if unanticipated problems occur (Fig. 1.4) [12].

In the case of the autonomic posture, and as discussed in Chap. 2, the circadian system generates a map of acrophases (maximal neurovegetative functions controlled by the ANS) that allows the adequate neuroendocrine–immune configuration for each of the three autonomic configurations of the body to be anticipated in a 24-h cycle (wakefulness, NREM sleep, REM sleep) [3]. In the face of unexpected demands, the modification of the predetermined neuroendocrine–immune configuration and the readjustment of the autonomic function occur (Fig. 1.5).

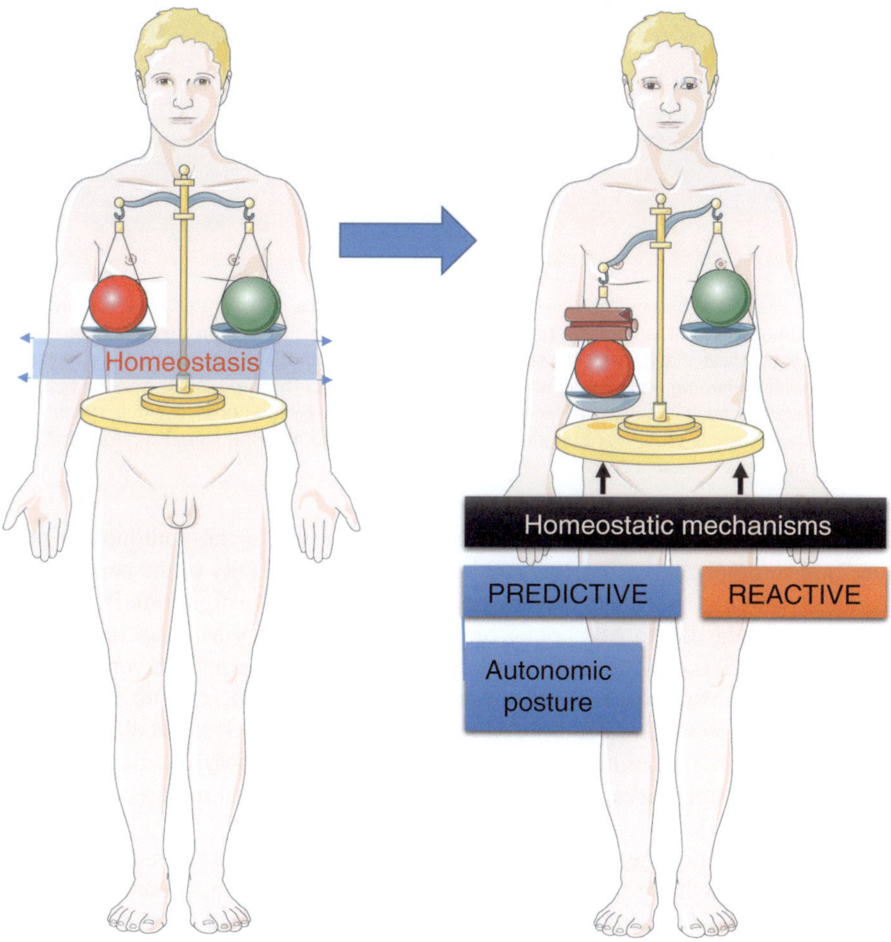

Fig. 1.5 In the case of autonomic posture, the circadian system generates a map of acrophases (maximum neurovegetative functions controlled by the ANS), which allows the adequate neuroendocrine–immune configuration for each autonomic configurations of the body (wakefulness, slow wave sleep, REM sleep) (predictive homeostasis) to be anticipated. Based on interoception, when unexpected demands arise, the modification of the predetermined neuroendocrine–immune configuration and the readjustment of the autonomic function arise (reactive homeostasis). The figure was prepared in part using image vectors from Servier Medical Art (www.servier.com), licensed under the Creative Commons Attribution 3.0 Unported License (http://creativecommons.org/license/by/3.0/)

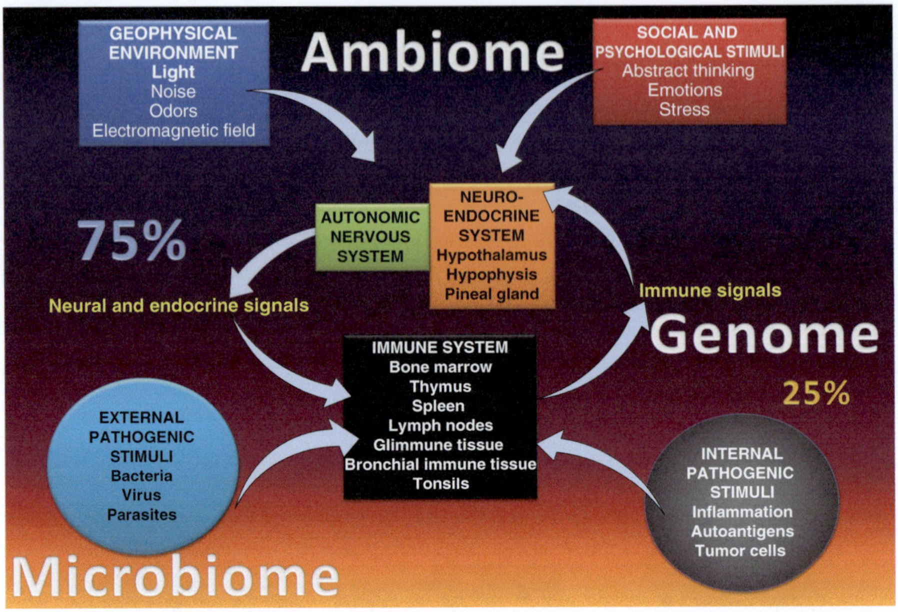

Fig. 1.6 Basis of the autonomic posture. The way in which the nervous system communicates with the immune system is twofold: (a) through the neuroendocrine system (hypothalamic–pituitary axis and pineal gland), via the secretion of pituitary, adrenocortical, thyroid and gonadal hormones and melatonin, thus modulating the immune response; (b) through the ANS, in both the parasympathetic and sympathetic divisions, which supplies the lymph nodes, thymus, spleen, and bone marrow. Various groups of hypothalamic neurons react to humoral signals (cytokines) produced by immunocompetent cells

The neuroendocrine–immune mechanisms involved in the "autonomic posture" are summarized in Fig. 1.6. The link between the activity of the nervous and immune systems has been the subject of numerous investigations in the last 50 years, giving rise to psychoneuroimmunoendocrinology as a discipline [13]. Multidirectional interactions among the immune, endocrine, and nervous systems have been demonstrated in humans and nonhuman animal models. Neuroendocrine–immune interactions can be conceptualized using a series of feedback loops, which culminate in distinct neuroendocrine–immune phenotypes. Behavior can exert profound influences on these phenotypes, which in turn reciprocally modulate behavior [13].

The way in which the nervous system communicates with the immune system is twofold: (a) through the neuroendocrine apparatus (hypothalamic–pituitary axis and pineal gland), via the secretion of pituitary, adrenocortical, thyroid, and gonadal hormones, and melatonin, all of which have a modulatory effect on the immune response; (b) through the ANS, both the sympathetic and parasympathetic divisions, which innervates the lymph nodes, thymus, spleen, and bone marrow (Fig. 1.6). Both pathways carry the link among the limbic, motivational, and immune response. Galen (second century AD), in his writings "On tumors against nature," had sensed this association, stating that breast cancer appeared in women whose menstruation was either abnormal or nonexistent because of the accumulation of "waste of black bile" (melancholia). A depressed patient is prone to developing inadequate immune responses; in contrast, a normal emotional balance contributes to a normal immune defense.

On the other hand, and because of the immune reaction, important changes in neuronal activity are verifiable. Several groups of central neurons react to humoral signals produced by immunocompetent cells (cytokines), such as interleukin (IL) 1 and 6, tumor necrosis factor-α (TNF-α), or interferon-γ (IFN-γ) [13]. These cytokines give rise to signs and symptoms that accompany acute or chronic infection (loss of appetite or anorexia, depressed motor activity, loss of interest in daily activities), in addition to activating the adrenal pituitary axis and producing thermogenesis (disease behavior).

The ambiome is defined as the set of nongenetic, changing elements that surround the individual and that contribute to the development and building of the human being, and therefore the state of health or the appearance of disease. It is part of the biopsycho-social–ecological reality of the individual from which the microbiome has been extracted as being very important in recent years. The microbiome defines the set of microorganisms that are normally located in different places in the human body, in particular the digestive tract [14]. The microbiome is in a commensal symbiotic relationship with the host. These microbial components aid in the digestion of food, produce vitamins, and protect against the colonization of other microorganisms that may be pathogenic. There are few physiological and immunological parameters that are not deeply affected by the presence and nature of the microbiome, with host resistance to infections being one of the most prominent factors. The gut microbiome is highly dynamic, exhibiting daily cyclic fluctuations that have repercussions for the host metabolism and provide evidence for the cross-regulation of prokaryotic and eukaryotic circadian rhythms [14]. We could see the microbiome forming part of our internal environment.

Genetic factors explain only some (<30%) of the changes linked to health and disease, as revealed by studies in twins. The ambiome and the microbiome correspond to the rest of the changes, which are essentially epigenetic. The biopsychosocial–ecological nature of the individual changes in the three autonomic body configurations, wakefulness, slow-wave sleep, and REM sleep is discussed in Chap. 2. Therefore, the autonomic posture is an essential and prior program for the homeostatic responses of organs and systems [15].

Hence, rather than being a mere top–down or reflex regulation, signals from the organs influence the functioning of the brain. For example, the reflex regulation of BP and the heart rate is not only subject to modulation by ascending information from the body, but also by descending information from several areas in the hypothalamus and cortex. The CNS has the capacity to control its output via the ANS using an amazing differentiation. For example, not only do the biological clock and prefrontal cortex contain neurons that influence the parasympathetic or sympathetic motor neurons, they also contain different neurons that project to diverse body compartments.

In the end, this leads to integrated responses whereby visceral sensory information reaches higher centers in the CNS via vagal or spinal sensory pathways, causing a reaction that considers factors such as the time of day, the season, the reproductive status, or the mood. Based on all this information, the brain sets the balance of the different parts of the ANS, causing its output to change its emphasis as per the situation. A disturbed balance, either as a result of behavior or of disease of any of the organs, leads to pathological conditions affecting the functioning of the entire individual [15].

Basic Neuronal Organization of ANS

Although dozens of types of neurons have been described in the nervous system by their morphological characteristics, when the length of the axon (indicative of the function they play) is considered, only two types of neurons are distinguished (Fig. 1.7) [3]:

- Golgi type I neurons with an identifiable axon, which are involved in the transfer of information between brain regions or in providing a basal tone of excitation to widespread brain areas. The difference between the two subsets of Golgi I neurons is the degree of axonal branching. In projection neurons, ramifications are limited to one or a few brain areas, whereas in widely distributed neurons ("spider web" neurons), axonal arborization ends in many areas or in some cases most of the cerebral cortex.
- Golgi type II neurons, with no identifiable or poor developed axon, that fulfill the function of interneurons in local circuits.

These two neuronal types (Golgi I and Golgi II) generate the three basic circuits:

- Local circuits, consisting of interneurons.
- Projection circuits or "point to point" connections, which relate to distant local circuits between them.

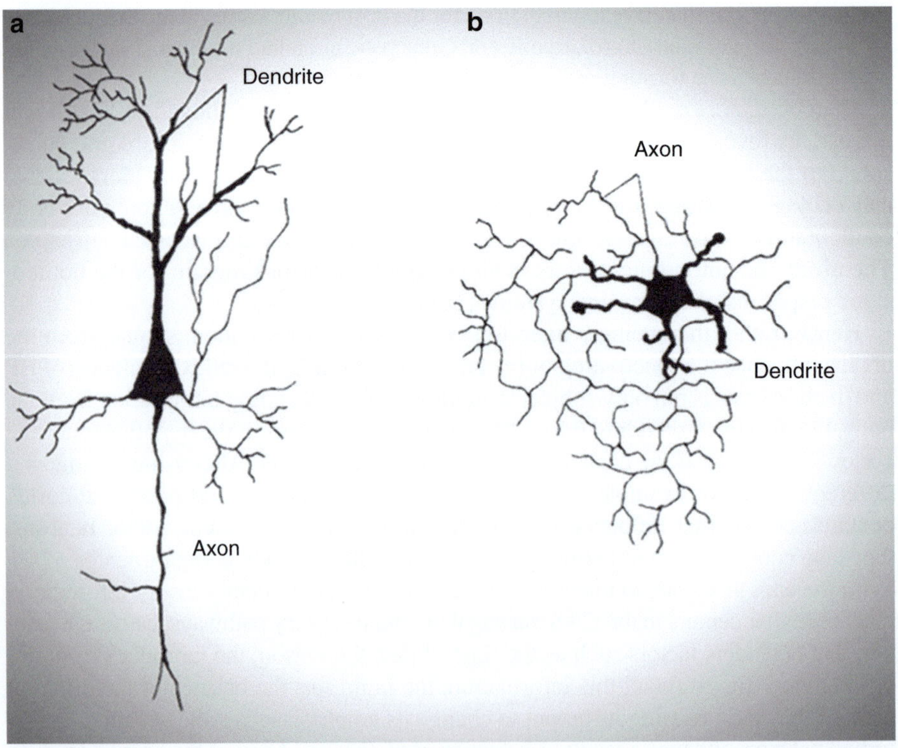

Fig. 1.7 Golgi type I (**a**) and Golgi type II neurons (**b**). Modified with permission from Cardinali [3]

- "Spider web" circuits, by which local modifications of brainstem nuclei are transformed into global states of CNS, e.g., wakefulness or sleep. These circuits are of fundamental importance for understanding the homeostatic function of the ANS.

Synaptic potentials are how a neuron can modify the membrane potential of the cells with which it connects. For this, the presynaptic neuron releases a chemical transmitter or, less frequently, the transmission is performed by an electrical mechanism [11].

In chemical transmission, the neurotransmitter interacts with receptors on the surface of the postsynaptic membrane resulting in the generation of synaptic potentials, which may be inhibitory – inhibitory postsynaptic potential (IPSP; i.e., of a hyperpolarizing nature) – or excitatory – excitatory postsynaptic potential (EPSP; i.e., of a depolarizing nature). The duration of synaptic potentials varies from a few milliseconds to, in some cases, seconds, or minutes (Fig. 1.8). These potentials are local and summable.

The integration signal (Fig. 1.8) is observed in the "trigger area" of the neuronal membrane, where various local potentials, propagated electrotonically, are summed, giving rise to the action potential. Generally, but not always, the "trigger zone" is in the axonal cone. This area is characterized by a high concentration of Na^+- and K^+-dependent voltage channels and constitutes the lowest threshold portion of the entire cell membrane. If the sum of the synaptic potentials reaches the threshold, an action potential is generated; hence, the signal produced is called "integrative" [11].

The driving signal is the action potential (Fig. 1.8). Whereas receptor synaptic potentials only passively propagate with sharp decreases in amplitude as a function

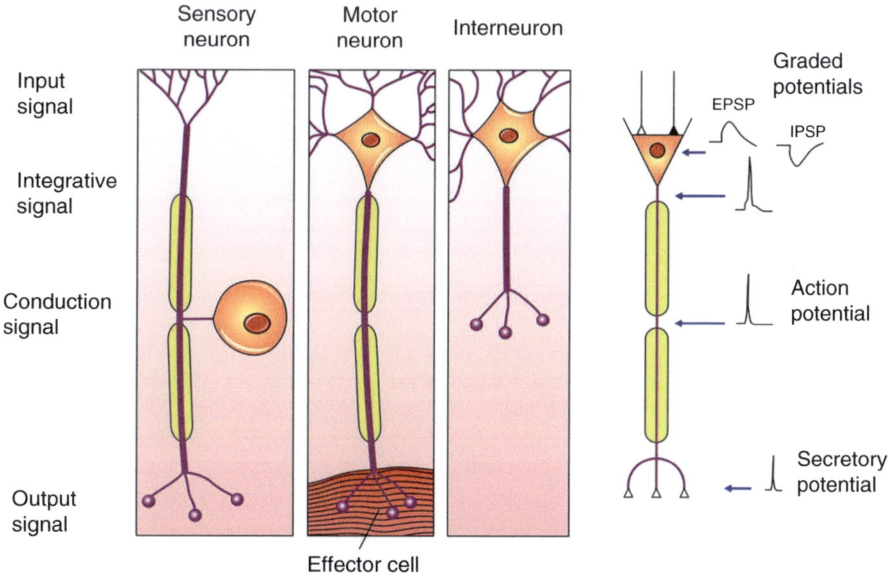

Fig. 1.8 The different signals of reception, integration, conduction, and secretion in neurons (*left*).The different electric potentials found in each segment (*right*). Modified from Cardinali [3]

of distance, the action potential (or "spike potential") has the following properties: (a) it actively propagates along the axons or in certain cases, as the pyramidal neurons of the cerebral cortex, also in dendrites; (b) it does not diminish its intensity as a function of distance; (c) it is of an "all or nothing" nature; (d) it is similar in all neurons, regardless of the neuron's function. The action potential amplitude is approximately 100 mV and the duration potential is 0.5–2 ms.

Although the Na^+-dependent action potential is the fastest method of signal conduction in the CNS, in the dendrites of the central neurons there are Ca^{2+} voltage channels displaying most of the properties of the Na^+ action potential. The main difference is the amplitude, i.e., a few mV for Ca^{2+} action potential vs 100 mV of Na^+ action potential.

The output signal (Fig. 1.8) is observed in synaptic axon terminals, where depolarization causes the release of neurotransmitter (chemical type synapses) or disturbs the resting potential of the postsynaptic neuron (electric type synapses) owing to the apposition of the membranes. In the case of chemical synapses, transmitter release depends on the entry of Ca^{2+} and involves the generation of a secretory potential [11].

Based on their conduction velocity (Fig. 1.9), nerve fibers are generally classified into:

- A fibers, myelinated, 2–20 mm thick with a velocity of 15–120 m/s. They are the sensory or motor fibers found in the somatic nerves. They comprise four subgroups, from highest to lowest speed: α, β, γ, δ. In the ANS the A fibers found are Aδ.

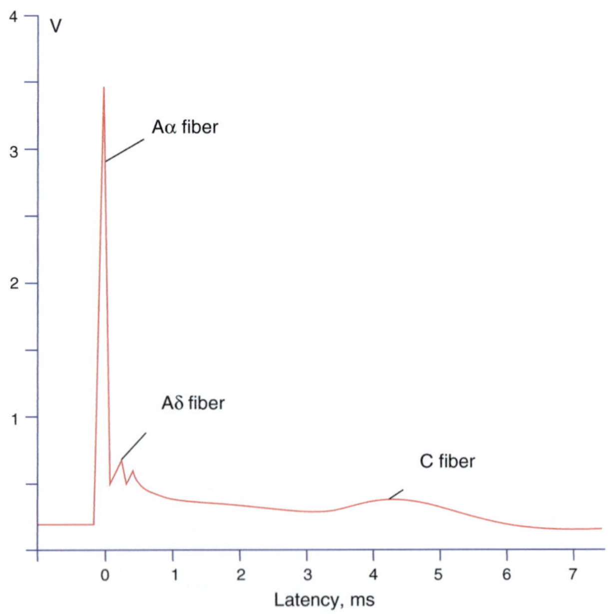

Fig. 1.9 Compound action potential in a peripheral nerve with peaks generated by different types of nerve fibers. Modified from Cardinali [3]

- B fibers, myelinated, 1–3 mm thick with a velocity of 3–15 m/s. They constitute the white communicating branches (preganglionic afferents) of the sympathetic chain.
- C fibers, unmyelinated, <1 mm thick, with a velocity of <2 m/s. They are the afferent amyelinic fibers of the visceral nerves and the sympathetic postganglionic nerves.

References

1. Cannon WB. Organization for physiological homeostasis. Physiol Rev. 1929;9:399–431.
2. Moore-Ede MC, Czeisler CA, Richardson GS. Circadian timekeeping in health and disease. Part 1. Basic properties of circadian pacemakers. N Engl J Med. 309:469–76.
3. Cardinali DP. Neurociencia aplicada. Sus fundamentos. Buenos Aires: Editorial Médica Panamericana; 2007.
4. Saper CB. The central autonomic nervous system: conscious visceral perception and autonomic pattern generation. Annu Rev Neurosci. 2002;25:433–69.
5. Hoit BD, Walsh RA. Chapter 5. Normal physiology of the cardiovascular system. In: Fuster V, Walsh RA, Harrington RA, editors. Hurst's the heart. 13th ed. New York: McGraw-Hill; 2011.
6. Waxman SG. Chapter 20. The autonomic nervous system. In: Clinical neuroanatomy. 27th ed. New York: McGraw-Hill; 2013.
7. Barrett KE, Barman SM, Boitano S, Brooks HL. Autonomic nervous system. In: Ganongs review of medical physiology. 25th ed. New York: McGraw-Hill Education; 2016.
8. Waxman SG. Chapter 19. The limbic system. In: Clinical neuroanatomy. 27th ed. New York: McGraw-Hill; 2013.
9. Strata P. The emotional cerebellum. Cerebellum. 2015;14:570–7.
10. Benarroch EE. Central autonomic network. Functional organization and clinical correlations. Leander: Futura Publishing; 1997.
11. Kibble JD, Halsey CR. Neurophysiology. In: Medical physiology: the big picture. New York: McGraw-Hill Education; 2015.
12. Waxman SG. Chapter 13. Control of movement. In: Clinical neuroanatomy. 27th ed. New York: McGraw-Hill; 2013.
13. Ashley NT, Demas GE. Neuroendocrine-immune circuits, phenotypes, and interactions. Horm Behav. 87:25–34.
14. Erny D, Hrabe de Angelis AL, Prinz M. Communicating systems in the body: how microbiota and microglia cooperate. Immunology. 2017;150:7–15.
15. Buijs RM. Chapter 1. The autonomic nervous system: a balancing act. In: Buijs RM, Swaab D, editors. Handbook of clinical neurology. Autonomic nervous system. New York: Elsevier; 2013, p. 1–11.

The Timed Autonomic Nervous System

2

Abstract

As the Earth rotates on its axis, it has two distinct environments: light and darkness. As the axis of rotation of the Earth is tilted, the relative duration of the periods of light and darkness changes systematically during the year. Because of the process of evolution, living beings have responded to these two situations by developing specific mechanisms for prediction and have successfully adapted to the time of day and the seasons of the year. The biological rhythms are the basis of predictive homeostasis. This Chapter describes the sleep/wake cycle as the major 24-hour rhythm as well as the components of the three different ANS physiological programs that occur during it. The significance of the glymphatic system and its link to sleep are discussed.

Keywords

Clock genes • Dreaming • γ-Aminobutyric acid • Glymphatic system • Melatonin • Non-REM sleep • Orexin • Phase maps • Polysomnography • REM sleep • Seasonality • Suprachiasmatic nucleus • Ventrolateral preoptic area • Zeitgeber

Objectives

After studying this chapter, you should be able to:

- Explain why the biological rhythms are the basis of predictive homeostasis.
- Summarize the behavioral and electroencephalographic (EEG) characteristics of each of the stages of NREM and REM sleep and the mechanisms responsible.
- Describe the pattern of normal night-time sleep in adults and the variations in this pattern from birth to old age.
- Describe the interplay between brainstem neurons that contain norepinephrine (NE), serotonin (5-HT), and ACh in addition to γ-aminobutyric acid (GABA) and histamine (His) in mediating transitions between sleep and wakefulness.
- Discuss the sleep/wake cycle as the major 24-h rhythm and the role of the suprachiasmatic nuclei (SCN) in its regulation.

D.P. Cardinali, *Autonomic Nervous System*, DOI 10.1007/978-3-319-57571-1_2

- Explain the components of the three different ANS physiological programs ("body configurations") that occur in a normal 24-h day/night cycle.
- Explain the meaning and mechanisms of dreaming.
- Describe the significance of the glymphatic system and its link to sleep.

Biological Rhythms Are the Basis of Predictive Homeostasis

As the Earth rotates on its axis, it has two distinct environments: light and darkness. As the axis of rotation of the Earth is tilted, the relative duration of the periods of light and darkness changes systematically during the year. Because of the process of evolution, living beings have responded to these two situations by developing specific mechanisms for prediction and have successfully adapted to the time of day and the seasons of the year.

The brain pacemaker creates a "day" and "night" in the body, as an approximate mirror of the outside world. We wake up every day at about the same time, relatively independently of the previous time devoted to sleep. We have a greater tendency to carry out certain tasks (physical or mental) at certain times of the day or night depending on what chronotype we have (an early chronotype, "larks," a late chronotype, "owls," or an intermediate one).

We perceive the seasons in our emotionality, physical strength, or ability to lose weight. The disturbances originating in a traveler of transmeridian flights, the emotional imbalances that often accompany the onset of winter and the troubles experienced by those workers who must comply with rotating shifts are proof of the existence of biological clocks and calendars, in conjunction with the geophysical cycles.

Although every physiological response exhibits a 24-h rhythm, there are differences between these rhythms at the time when a peak occurs. The "phase maps" are the graphic description of these maxima for many physiological periodic changes (Fig. 2.1). Such a sequence and spacing of the maximum values of daily rhythms reveal the ordered cause–effect relationships in bodily processes of all kinds, from genomic to behavioral, and their normality is what we could define as the quintessence of health [1].

These maps may experience temporary disruptions when the body is forced to make a quick adjustment phase, as happens after a transmeridian flight [2]. In such circumstances, the different rhythmic functions do not resynchronize with the same speed and the normal temporal relationships between phases are lost. Full resynchronization requires a few days (about 1 day for every hour of phase shift), and during this period the "jet-lag" syndrome is seen. Phase maps are also distorted in chronic or acute diseases, even mild ones [3, 4]. As we discuss in Chap. 8, full recovery is achieved by controlling for both the underlying disease and the accompanying chronobiological disruption.

It is now established that clock mechanisms are genomic. Since life originated about 4,500 million years ago, in an environment where day and night already existed, successful species reproduced in their genome such geophysical reality. Thus, the day and night have left an indelible mark, which is as universal as the genetic code in all forms of life. In every living cell, a cyclic mechanism of interaction between transcription factors, genes, and proteins exists close to 24-h periods (in humans, slightly longer than 24 h; Fig. 2.2).

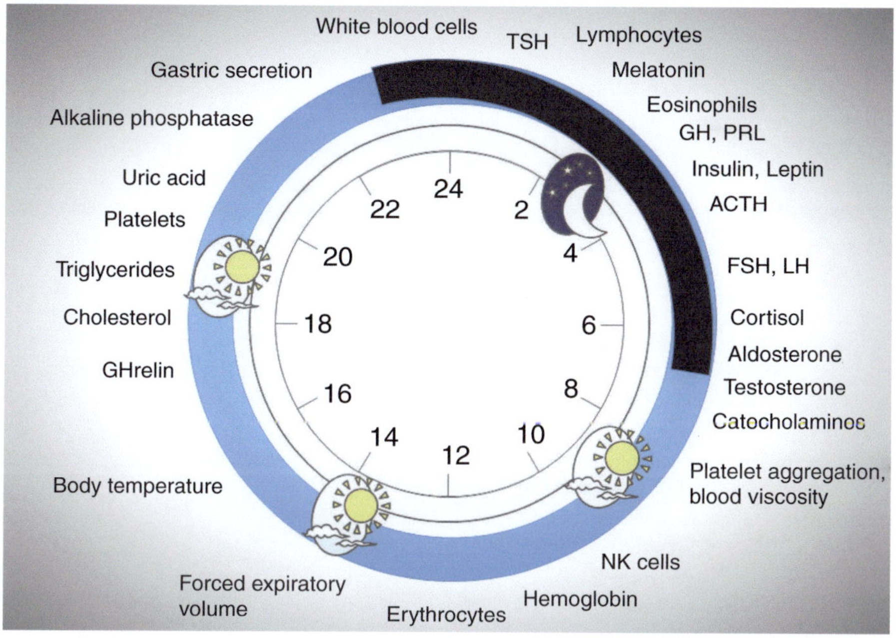

Fig. 2.1 Phase maps of various circadian rhythms in man. As the individual is in normal light–dark conditions, each of the rhythms shown is a period of exactly 24 h. What varies for each one of them is the time in which the daily maximum (or "acrophase") of the rhythm, shown in the diagram) is presented. A synchronized phase map characterizes normality. Reproduced with permission from Cardinali [1]

These intertwined feedback loops involve a small number of clock genes [5]. The positive arm of the daily clock consists of transcription factors *Bmal1* and *Clock*. The protein products of these genes form heterodimeric complexes that control the transcription of other clock genes, in particular, *Per* (*Per1, Per2, Per3*) and *cryptochrome* (*Cry1, Cry2*), which in turn provide the negative signal feedback that inhibits *Bmal1* and *Clock* to complete the circadian cycle. Other clock genes (*Rev-erbα*, *Rorα*, *NR1D1*, *timeless*) provide additional force to the translation/transcription loops. The expression of clock genes is cyclic and is regulated in part by phosphorylation of proteins, thus controlling protein stability, nuclear reentry, and transcription complex formation (Fig. 2.2). Collectively, the molecular circadian clock operates in every cell of the body to regulate at least half the genome [6].

To generate physiological and behavioral responses consistently, the phases of these myriads of cellular clocks must be orchestrated by a pacemaker. One of paramount importance resides in the SCN of the anterior hypothalamus [7]. This core master clock is a key regulator of 24-h cycles in many body functions, including sleep and wakefulness, thermoregulation, glucose homeostasis, and fat metabolism [8]. SCN integrity is necessary for the generation and maintenance of the 24-h rhythms, and for their synchronization by light–dark cycles. Although the complex behaviors such as sleep, waking or feeding involve many brain areas working in a network, in the case of circadian rhythms, the participating brain region is single and has minimal volume.

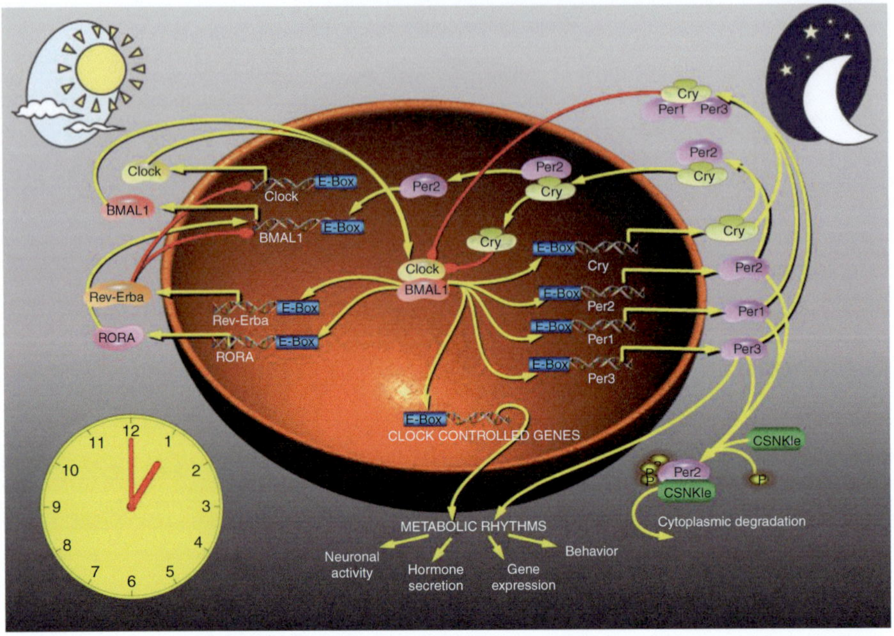

Fig. 2.2 Simplified model of the circadian intracellular mechanisms. The process begins when the CLOCK and BMAL1 transcription factors are dimerized and trigger the transcription of the *Per* (*Per1, Per2*, and *Per3*) and *Cry* (*Cry1* and *Cry2*) genes. *Per* and *Cry* are translated into their respective proteins, which increase throughout the day. When the PER and CRY proteins reach a certain level, they form heterodimers that enter the nucleus and regulate the CLOCK-mediated BMAL1 transcription of their own genes. This process takes about 24 h. There are also accessory mechanisms to the clock (e.g., *Rev.-ERBA* and *RORA*). Almost 50% of the transcriptome shows a 24-h variation

The circadian apparatus includes [7]: (a) a hypothalamic pacemaker, the SCN; (b) a series of physiological outputs under the control of SCN; (c) molecular clocks present in cells of all tissues and organs. The SCN has hypothalamic (endocrine) and extrahypothalamic projections of a behavioral type. The circadian effects of SCN are exerted: (a) on neuroendocrine neurons of the hypothalamic paraventricular nucleus (PVN); (b) on autonomic PVN neurons; (c) on hypothalamic structures associated with sleep generation (e.g., the ventrolateral preoptic area, VLPO); (d) other hypothalamic areas (sub-PVN zone, sPVz; dorsomedial hypothalamus, DMH; median preoptic area, MnPO), intermediate between the SCN and autonomic and neuroendocrine neurons; on extrahypothalamic structures (lateral geniculate body, paraventricular thalamic nucleus) for the synchronization of hypothalamic conducts and locomotor activity (Fig. 2.3) [9].

The SCN contains local projection neurons that communicate with one another and with other hypothalamic structures [10]. The axons of many SCN neurons terminate within the nucleus itself, thus forming local circuit connections and/or collaterals from longer range projections. The SCN core projects densely to the SCN shell, which projects only sparsely back to the core. Neuronal cell bodies in the SCN are small (~10 μm), have simple dendritic arbors, and are closely apposed.

Neurons in the SCN core and shell regions are distinguished by their neurochemical content. The neuropeptide vasoactive intestinal peptide (VIP) is found in

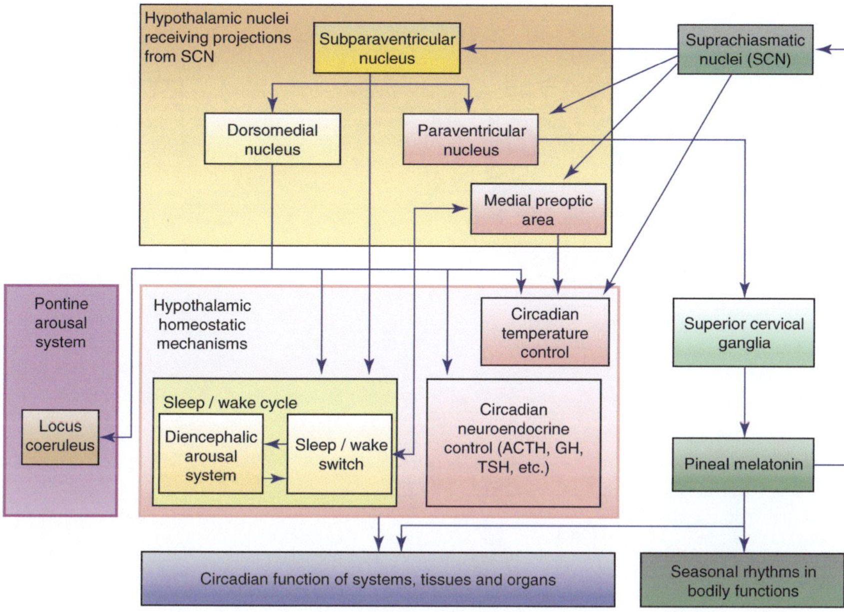

Fig. 2.3 Transmission of circadian information from the oscillator to the hypothalamic systems that control circadian rhythms, including the sleep–wake rhythm. The key steps identified include multisynaptic transmission from the suprachiasmatic nuclei (SCN) to the hypothalamic control systems through adjacent nuclei of the anterior hypothalamus, multisynaptic transmission to the pineal gland controlling the secretion of melatonin, direct SCN pathways to the sleep-promoting and awakening regions, and the integration of reactive and predictive homeostasis in relation to the sleep–wake rhythm in the medial preoptic area. Modified with permission from Cardinali [8]

about 10% of all SCN cells, whereas AVP is present in 20% of all cells [11]. The VIP-positive neurons are mainly located in the ventral and central parts of the SCN (core). In humans, the volume of the VIP core subdivision is 0.03 mm^3 and contains about 1,700 VIP-immunoreactive neurons, with a mean density of about 63,000 neurons/mm^3 [12]. Besides neurons containing VIP, substance P, gastrin-releasing peptide (GRP), calretinin- and calbindin-containing neurons are also found in the SCN core (Fig. 2.4). Most of the AVP-positive neurons are in the dorsomedial part of the SCN (the shell). In humans, the volume of the AVP subdivision is 0.2 mm^3 and contains about 6,900 AVP-immunoreactive neurons, with a mean density of 29,000 neurons/mm^3. In this region, neurons containing cholecystokinin (CCK) and prokineticin 2 are found in addition to AVP neurons (Fig. 2.4).

In most SCN neurons, neuropeptides are colocalized with GABA, and almost all synapses among SCN neurons are GABAergic. Electrophysiologically, it has been shown that glutamate (Glu) is also a transmitter in the efferent pathways of the SCN.

The increased electrical activity of SCN during the day occurs in both nocturnal mammals such as the hamster or the rat and in daytime species such as the primates. In primates, however, the secretion of corticosteroids and the onset of activity and phase of sympathetic predominance and temperature rise occur at the beginning of

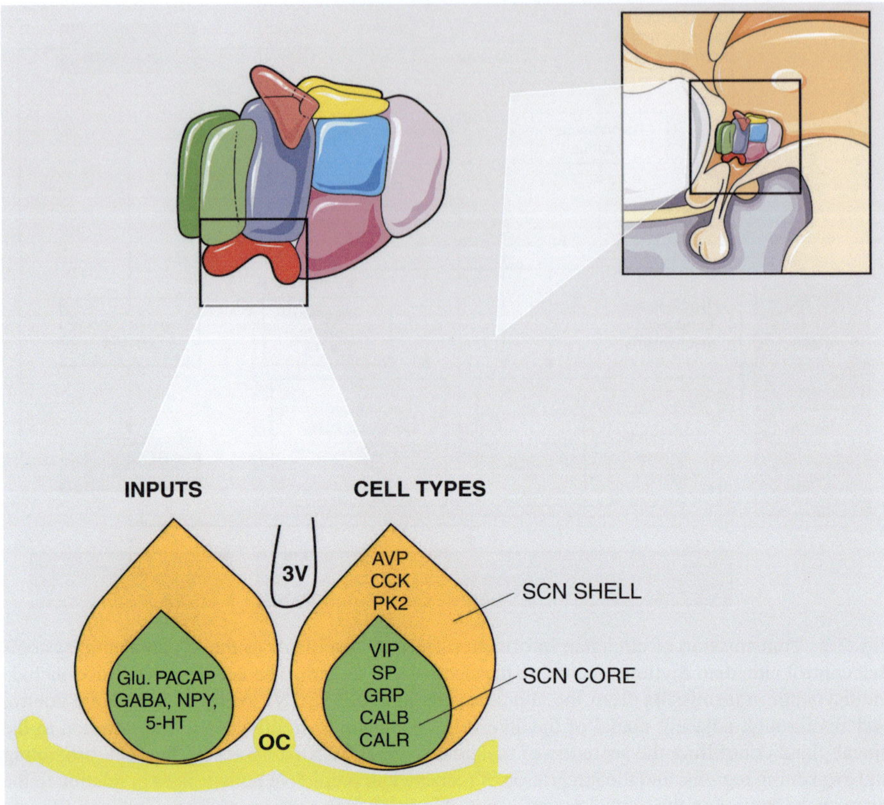

Fig. 2.4 Suprachiasmatic nuclei organization illustrating the compartmentalization of inputs and cell types and in a coronal plane. On the *left*, inputs originating from melanopsin-containing retinal ganglion cells (Glu; PACAP, pituitary adenylate cyclase-activating polypeptide), intergeniculate leaflet of the thalamus (NPY, neuropeptide Y; GABA) and dorsal raphe (5-HT). On the *right* the different cell populations described, containing AVP, *CCK* cholecystokinin, *PK2* prokineticin 2; VIP; *SP* substance P, *GRP* gastrin-releasing peptide, *CALB* calbindin, *CALR* calretinin. *3V* third ventricle, *OC* optic chiasm. The figure was prepared in part using image vectors from Servier Medical Art (www.servier.com), licensed under the Creative Commons Attribution 3.0 Unported License (http://creativecommons.org/license/by/3.0/)

the light phase, and not during the dark phase, as in the rat. That is, the signal produced by the SCN on the different effectors mentioned is interpreted in different ways in diurnal and nocturnal species.

Research into animals and humans has shown that only a few key environmental periodic clues, relevantly the light–dark cycle, are effective at synchronizing the internal clocks (with periods slightly longer than 24 h) to exactly 24 h [2]. A synchronizer agent can “reset” or modify the phase of the biological clock (*Zeitgeber*, time-giver in German). The variability of the response is always predictable, depending on when the synchronizer stimulus is applied, circadian rhythms are phase-advanced, phase-delayed, or remain unchanged (Fig. 2.5).

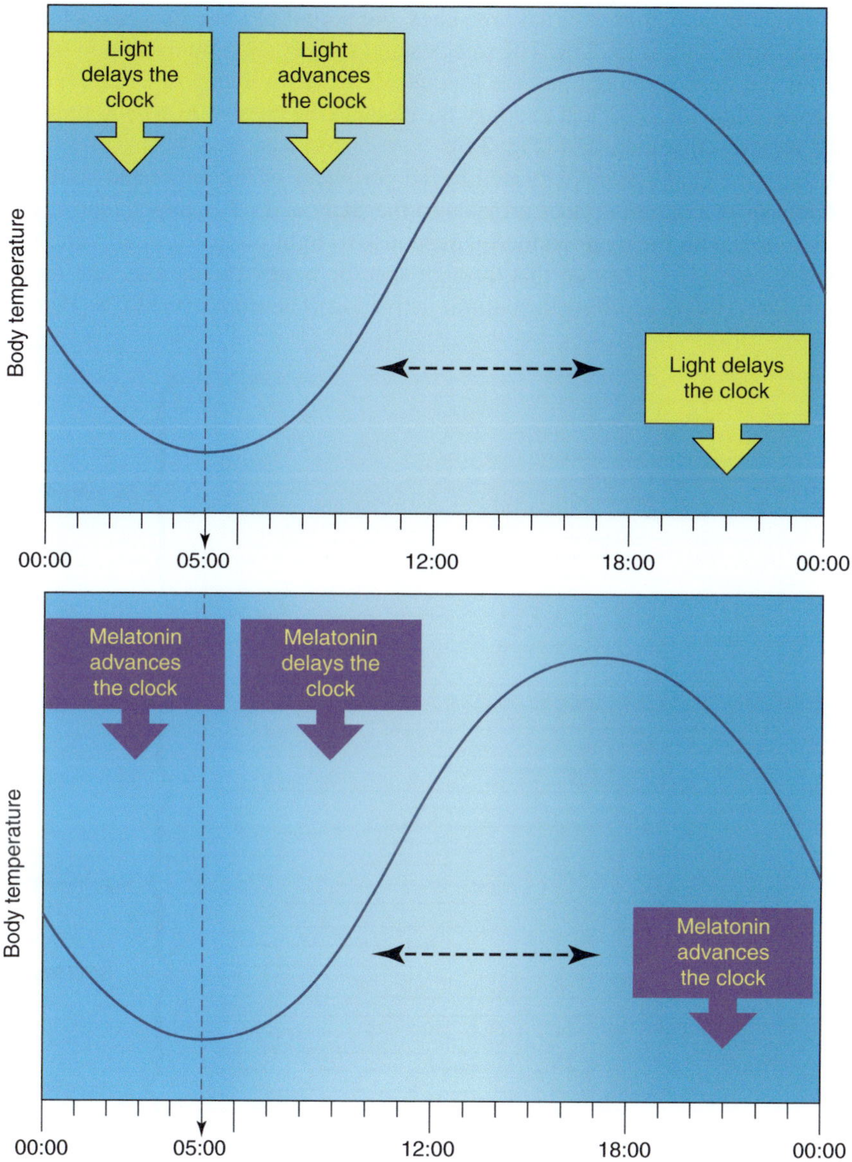

Fig. 2.5 Central body temperature response to the application of a light pulse or administration of melatonin (3 mg). A pulse of light during the evening or the early part of the night, slows the clock and sleep begins later on subsequent days. A light pulse during the second half of the night and early morning advances the clock and sleep begins earlier on subsequent days. Melatonin has the opposite effect. Neither light nor melatonin changed the clock if applied during the day (*dotted line*). Reproduced with permission from Cardinali [1]

Without the action of external time cues, the period of these oscillators tends to be longer than 24 h (Fig. 2.6). The rate is set at exactly 24 h by the action of light, which is the main (although not the unique) Zeitgeber in humans. Brief exposures to morning light are sufficient to adjust the clock to the precise 24-h solar time. This action requires an intact SCN (Fig. 2.6).

A group of ganglion cells located in the periphery of the retina and containing melanopsin as a photopigment projects to the SCN and other hypothalamic areas and is linked with the neuroendocrine response to light observed in all vertebrates including man [13]. They do this through specific neural projections (the retinal–hypothalamic tract), resulting in genomic activation of neurons in the SCN (Fig. 2.7).

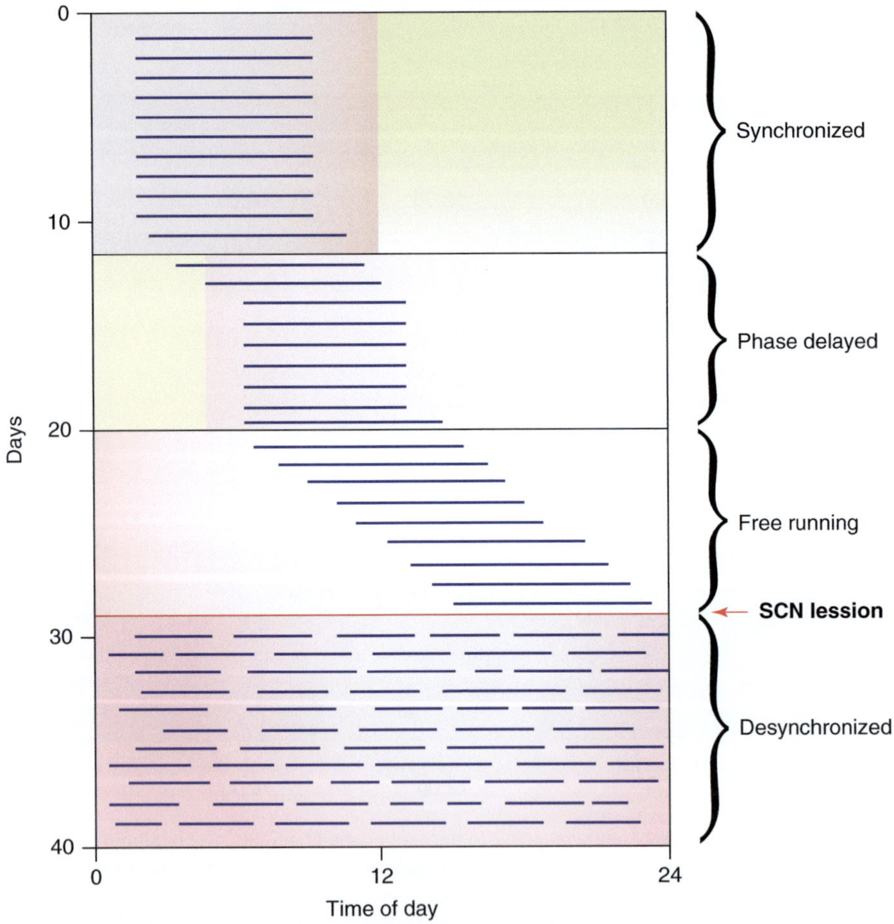

Fig. 2.6 Locomotor activity rhythm in the hamster. The activity on successive days (*blue bars*) coincides with the dark phase in the synchronized situation. If a few hours' delay is made at the beginning of the night, the animal adapts with a delay (equivalent to "jet-lag"). In permanent darkness, the locomotor activity rhythm follows the endogenous period (greater than 24 h). After SCN lesion, an abolition of rhythm is observed. Modified with permission from Cardinali [8]

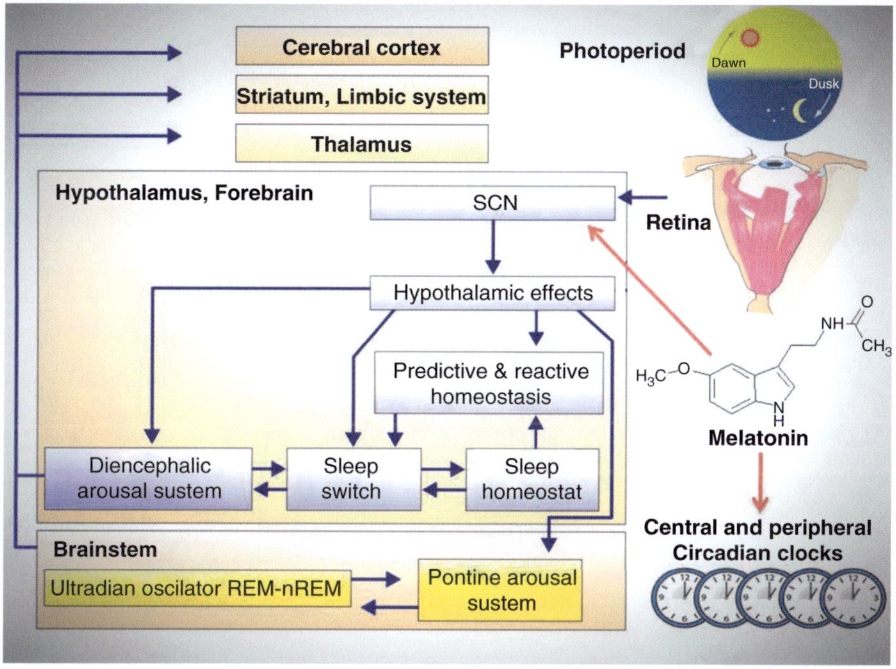

Fig. 2.7 The mechanisms triggering and maintaining sleep are in the hypothalamus and basal forebrain. The ultradian rhythm slow-wave sleep/REM sleep depends on mechanisms of the brainstem. Melatonin has effects on both central and peripheral oscillators. Reproduced with permission from Cardinali [1]

The information generated in these nuclei is transmitted to specific areas of the basal hypothalamus, which control the two major channels of body communication: the endocrine system and the ANS.

In humans, light during the first part of the night delays the clock and that during the second part of the night and early morning accelerates the clock (Fig. 2.5). At other times of the day, exposure to light exerts no appreciable effect to advance or delay the phase of the circadian rhythms. This mechanism explains why exposure to artificial light fastens the endogenous clock in the morning, whereas during the first part of the night it tends to perpetuate and aggravate sleep deprivation by causing a phase delay [2].

A major synchronizer of the SCN clockwork is pineal melatonin (Figs. 2.7 and 2.8) [1]. Melatonin synchronizes the human circadian system as per a phase response curve that is about 12-h out of phase with the phase response curve produced by light (Chap. 8). Projections of the SCN driving the daily melatonin rhythm inhibit the firing of neurons in the sPVz of the hypothalamus. From this zone, a multisynaptic pathway begins that includes the medial forebrain bundle, reticular formation, and the intermediolateral cell column of the cervical spinal cord, the superior cervical ganglia (SCG), and the postganglionic sympathetic fibers that end near the pineal cells to stimulate melatonin synthesis (Figs. 2.7 and 2.8).

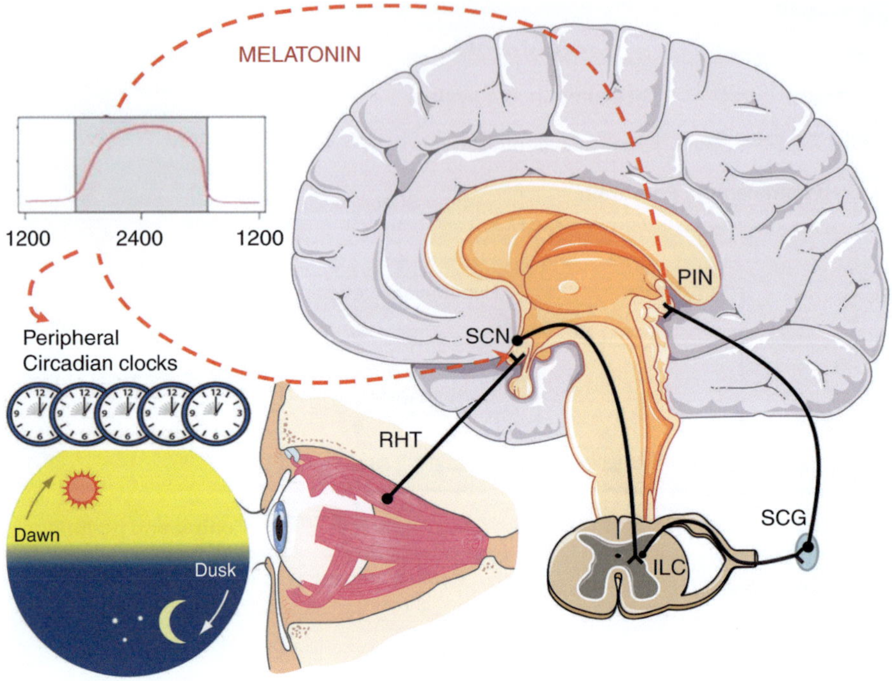

Fig. 2.8 Control of melatonin synthesis by environmental light. Retinal ganglion cell receptors project via the retino-hypothalamic tract (RHT) to the SCN. From here, and through a multisynaptic pathway including the cervical sympathetics, pineal melatonin release is modulated. Melatonin both feedback at the SCN and affects peripheral clocks. Reproduced with permission from Cardinali [1]

Melatonin phase-shifts circadian rhythms in the SCN by acting on MT_1 and MT_2 melatonin receptors in SCN neurons [14]. The phase-and amplitude-altering effect of melatonin is caused by its direct influence on the electrical and the metabolic activity of the SCN. The circadian rhythm of melatonin secretion has been shown to be responsible for sleep rhythm in both normal and blind subjects (i.e., in the absence of the synchronizing effect of light). If an individual who usually falls asleep after midnight wishes to advance his or her sleep schedule to rise early for work, the indication is the administration of melatonin at 18:00–19:00 h to achieve the desired phase advance of the circadian clock (Fig. 2.5) [1].

The SCN communicates day–night cycle information to the rest of the body through neural and humoral signals, including the ANS and the neuroendocrine system [7]. By this information, the peripheral cellular circadian clocks become synchronized to exactly 24 h. As is discussed in Chap. 4, the clocks at the periphery are also able to respond to other environmental cues, such as food, altering their phases to these cues accordingly.

Both melatonin and cortisol secretion are controlled by the circadian clock and in turn are essential for synchronizing peripheral circadian rhythms. The secretion of melatonin is very consistent from day to day for a given period of life and shows

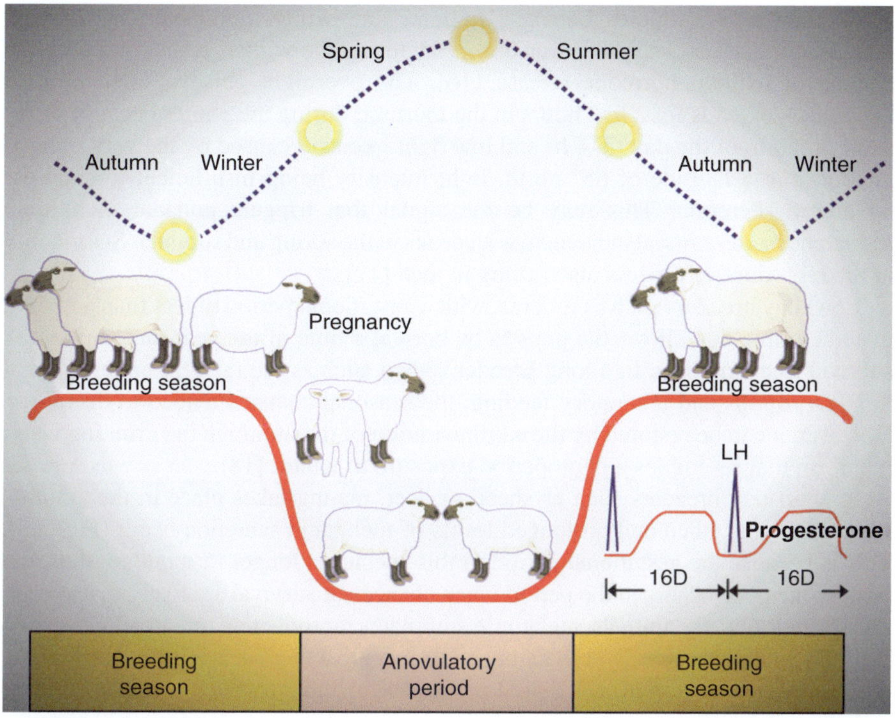

Fig. 2.9 Reproductive seasonal rhythm in the sheep (a short-day breeder). Modified from Cardinali [8]

far fewer contingent changes because of the stress that cortisol shows. Thus, melatonin reflects more accurately the circadian signal given by the SCN [1].

From an evolutionary point of view, it is not difficult to imagine that a successful adaptation to the environment in which animals compete for nutrients that are always scarce needs to optimize processes of high-energy consumption, such as reproduction [15]. Thus, almost all species undertake seasonal mating in the wild (Fig. 2.9). The most appropriate signal to the circadian system to transmit information about the season is the duration of photoperiod. Other environmental signals (temperature, humidity, etc.) do not have the same degree of reproducibility year after year as the length of the photoperiod.

In humans, there is seasonality in reproduction. The statistics on births indicate the seasonality in the reproductive process, with maximum activity during the summer. If multiple births are computed, which are indicative of ovarian overstimulation and are therefore independent of social factors such as sex behavior, the seasonal differences are even more significant. These studies have been conducted in populations in the northern hemisphere in periarctic zones (Scandinavia, Labrador Peninsula). In these countries, the activity of the pituitary–ovarian axis and the incidence of conception in human populations decline during the dark months of the year.

In comparative studies between summer and winter conducted in Finland, increased melatonin secretion was observed during the winter, coinciding with the decline of ovarian hormone release [16]. The overnight pulse of melatonin in the plasma extends for a few hours in the morning during the winter because of the short duration of the day (3–4 h) and low light intensity caused by the very oblique incidence at a latitude of 65° north, light intensity being insufficient to suppress melatonin secretion. This may be the signal that triggers gonadal involution. Menarche has a seasonal incidence, with peaks in the spring and summer. Seasonality in the reproductive process also occurs in men [17].

Long day breeders such as rodents, with a gestational period of less than a month, mate at a time that allows the pups to be born at a time of the year that maximizes survival (the summer). In a long breeder rodent such as the rat kept under optimal light conditions and laboratory feeding, the seasonal changes tended to disappear. However, it can be restored by the administration of melatonin in the drinking water in a form that resembles a photoperiod exposure to winter [18].

In short day breeders such as sheep or deer, mating takes place in the autumn, when nights lengthen and prolonged levels of melatonin secretion occur (Fig. 2.9). This is because the gestational period in this species is longer (5 months); thus, the birth of the pups occurs in the period when chances of survival are likely to be greatest (spring). In these animals, melatonin stimulates reproductive function by decreasing the sensitivity of the hypothalamus to the negative feedback of gonadal steroids.

Another seasonal rhythm is that of mood [19]. Although seasonal trends of mood and emotions have been recognized in the medical literature for over 100 years, it was not until the 1980s that the distinguishing features of an affective disease involving the recurrent winter depression were established. This form of affective illness ("seasonal affective disorder") is now the subject of numerous investigations [19]. Every fall or winter, patients with this condition get tired easily, eat high-calorie carbohydrate diets, show weight increases, and have exaggerated anxiety or sadness. With the arrival of spring, patients emerge out of depression, and in certain circumstances, may show moderate manic symptoms.

Light therapy is useful in seasonal affective illness: the morning exposure to light (intensity of at least 2,500 lux, equivalent to the light intensity of a sunset) for 2 h daily is recommended [19].

One could see seasonal affective disorder as the human equivalent of hibernation. Decreased libido in the winter reduces the chances of birth the following winter (an inappropriate time of the year for newborn survival). In addition, food intake increase in the winter facilitates reproductive success because overweight mothers tend to have larger fetuses. In the winter, at high latitudes, nocturnal melatonin secretion in humans is about 40 min longer than in the summer [16]. In a hibernating animal species, the prolongation of 30 min in the secretion of melatonin is sufficient to signal winter.

Evidence has now accumulated that a seasonal change in thyroid hormone availability within the brain is a crucial element for seasonality [15]. This is mediated by local control of thyroid hormone metabolizing enzymes within ependymal cells lining the third ventricle of the hypothalamus. Within these cells, deiodinase type 2 enzyme is activated in response to the length of summer days, converting metabolically inactive thyroxine to tri-iodothyronine. The pars tuberalis of the pituitary gland plays an essential role. Specialized thyrotroph cells are regulated by the changing signal of day length, with long days activating TSH. In mammals, the pars tuberalis is regulated by the nocturnal melatonin signal [15].

The Sleep/Wake Cycle as the Major 24-h Rhythm

A fundamental circadian rhythm is that of sleep/wakefulness. The Greeks called sleep "the brother of death," because they thought that in the sleeping man, the soul temporarily abandoned the body and wandered through the world. Indeed, sleep is an active process, the metabolism of several brain regions being greater than in wakefulness. Moreover, sleep is complex, with two different electroencephalographically (EEG) defined stages, NREM and REM sleep. It is also endogenous and relatively independent of exogenous factors. During REM sleep, O_2 consumption in various brain areas of the limbic system exceeds that found in wakefulness [20].

In rats, total sleep deprivation (NREM plus REM sleep) causes death in about 20 days. Hair loss and discoloration, tail and leg skin lesions and increased food intake, together with 20% body weight loss and a doubling of energy expenditure, are seen in sleep-deprived rats. If the deprivation is only of REM sleep, rats die in about 40 days, with the same abnormalities. On average, humans cannot live without sleep for more than a few days (about 2 or 3 days). After sleep deprivation, humans recover approximately 33% of the total sleep time, 100% of the slow-wave sleep, and 30–50% of the REM sleep lost [20].

A reduction of daily sleep hours is common in contemporary society, which prioritizes a timeless organization, with continuous activity 24 h, 7 days a week ("24/7 society," Chap. 8). On average, we sleep about 2 h less every day than just 40 years ago, and this affects our cognitive performance, emotional stability, and health [1].

Electroencephalography recording and ocular musculature activity during sleep has allowed the study of the "basic architecture" of the neural process. Four stages of sleep are distinguished according to the type of EEG brain activity (Fig. 2.10). Stages N1–N3 correspond to the progressive slowing of the brain waves, from the α rhythm (8–13 Hz) to the δ rhythm (<3 Hz). A fourth stage, REM sleep, corresponds to a desynchronized EEG, similar to that of wakefulness [20].

Rapid eye movement (REM) sleep is accompanied by a profound loss of muscle tone (except for the diaphragm, the muscles of the middle ear, and the crico-esophageal sphincter, all of which have very few muscle spindles), ponto-geniculo-occipital (PGO) spikes on the EEG, eye movements synchronous with PGO spikes, and ANS signs, such as increased variability in BP and heart rate and the tendency toward poikilothermia (loss of control of body temperature). The more complex mechanisms of cardiocirculatory, respiratory, and thermal feedback temporarily cease to function during REM sleep, with only the autonomic spinal reflexes remaining.

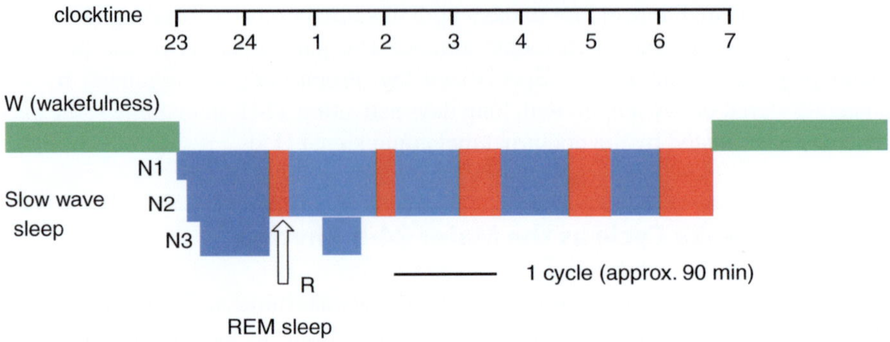

Fig. 2.10 Sleep stages based on electroencephalography (EEG). There is a slowing of the brain waves (N1–N3 stage) depending on the progression of sleep, reaching deep slow-wave sleep in about 30–45 min. At this time, a progressive acceleration of EEG occurs, running from stage N3 to N1 in about 30–45 min to reach the REM sleep stage (*R*), which lasts about 10–15 min. There are between 4 and 6 cycles per night. Reproduced with permission from Cardinali [1]

The normal polysomnographic register in humans indicates a slowing of the cerebral waves (stages N1–N3). An acceleration of the EEG rhythm then occurs, passing from stage N3–N1 in about 30–45 min (Fig. 2.10). Abruptly, a stage of REM sleep occurs that lasts about 10–15 min, beginning a new cycle of slowdown and subsequent acceleration of brain waves (Fig. 2.10). There are four to six REM sleep episodes per night, but a δ rhythm (N3 stage) is only detected during the first half of the night. In the second part of the night, the inter-REM periods show increased EEG frequency, but not passing the stage 2 or 3 in the final hours of sleep. In contrast, the REM periods are longer in the second part of the night. On average, normal adult polysomnographic recordings demonstrate 25% REM sleep and 75% NREM sleep (slow-wave sleep; 50% in stages N1 and N2, and 25% in N3). Video recordings detect a postural change of importance every 20 min, approximately, all of them corresponding to NREM sleep.

Three mechanisms have been identified as responsible for sleep (Fig. 2.11) [21]:

- A process called "S" (for sleep), determined by the previous individual history of sleep and wakefulness. The "S process" is manifested in the increased propensity to sleep after sleep deprivation. It is the accumulation of sleep debt, like the mechanism of an hourglass.
- A process called "C" (for circadian), controlled by the endogenous biological clock. It is independent of the previous history of sleep and wakefulness. The "C process" comprises the trend toward falling sleep with the decrease in body temperature (during the first part of the night) and the termination of sleep during the increase in body temperature (during the second part of the night). This is because of the "C process," that is, after a night spent awake one is sleepier at 04:00–05:00 h than at 07:00–08:00 h, regardless of the 2–3 h increase in sleep debt. Therefore, sleep is like a bank debt: it is impossible to pay the debt when the "bank window" (C process) is "closed." For a night worker who after a

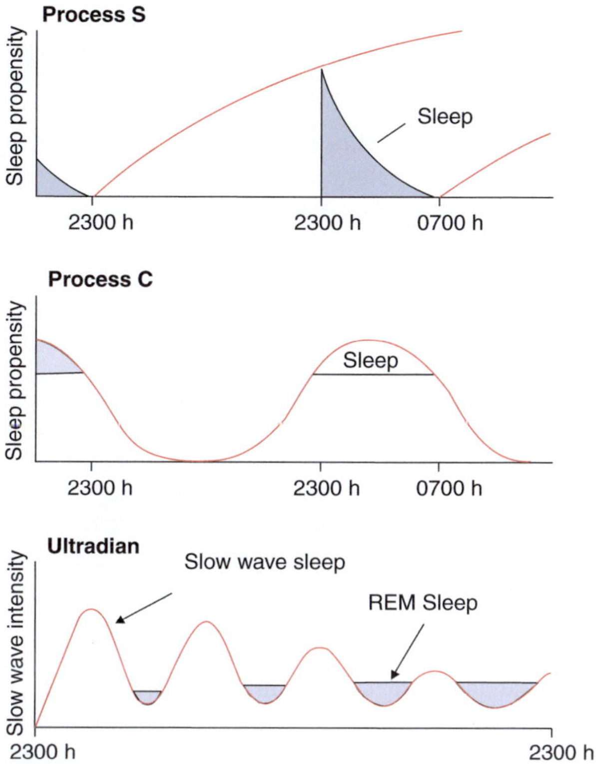

Fig. 2.11 Three interacting processes regulate the timing, duration, and depth, or intensity, of sleep: a homeostatic process that maintains the duration and intensity of sleep, a circadian rhythm that determines the timing of sleep, and an ultradian rhythm given by the NREM sleep–REM sleep sequence. Reproduced with permission from Cardinali [1]

sleepless night wants to sleep in the morning, he must wait until a more appropriate time (e.g., the siesta after lunch) to have a restful sleep.

- An ultradian component (frequency of about 90 min), perceptible both in sleep (slow-wave sleep and REM sleep alternation) and in wakefulness (periodicity of about 90 min with maximal and minimal attention, the so-called basic rest–activity cycle, BRAC) [22].

It is important to note that these three mechanisms apply not only to sleep, but also to the ultradian and circadian variations of most of the physiological phenomena in which they have been examined. Studies conducted in sighted individuals under normal sleep–wake cycles cannot establish whether the rhythm is endogenous and self-sustained (the definition of a circadian rhythm) or is driven by external factors such as sleep–wake or rest–activity cycles, the light–dark cycle, or feeding cycles. The constant routine protocol is a gold-standard method employed in circadian biology to differentiate whether or not a rhythm is intrinsically generated and sustained and, compared with measures taken under ambulatory baseline conditions, can measure the extent to which it is driven by external factors such as sleep, light, meal timing, or posture [2]. During a constant routine procedure, participants remain awake in bed for typically 30–50 h (1–2 circadian cycles) in a semirecumbent posture under dim light conditions and are fed hourly isocaloric meals. This procedure removes the direct impact of sleep, light, activity, and posture on rhythm expression and distributes calorie intake uniformly across the circadian cycle, hence removing or minimizing many invoked effects due to external factors that may mask the underlying endogenous circadian rhythm. An important point to consider in relation to the ultradian/circadian relation is that the ultradian period of 90 min is a harmonic of 24 h and that its progressive consolidation (rhythms of 1.5 h, 3 h, 6 h, 12 h, 24 h) is seen as the ontogenic development of the sleep/wake rhythm in human newborns.

The C process involves circadian changes in the promotion of wakefulness given by the SCN (Fig. 2.12) [7]. In primates that have lesions in the SCN, not only the desynchronization of circadian rhythms occurs, but a significant increase in the total time asleep (having removed a key area for maintaining wakefulness) is seen (Fig. 2.6). During the day, the electrical activity of the SCN increases and peaks toward the evening (around 18:00 h). SCN neurons help to counteract the increased pressure of sleep debt that accumulates during wakefulness [7].

Pineal melatonin begins to be released at the end of the afternoon, at about 18:00 h. Melatonin acts on specific receptors located at the SCN, reducing their electrical activity and therefore their ability to neutralize the pressure of the

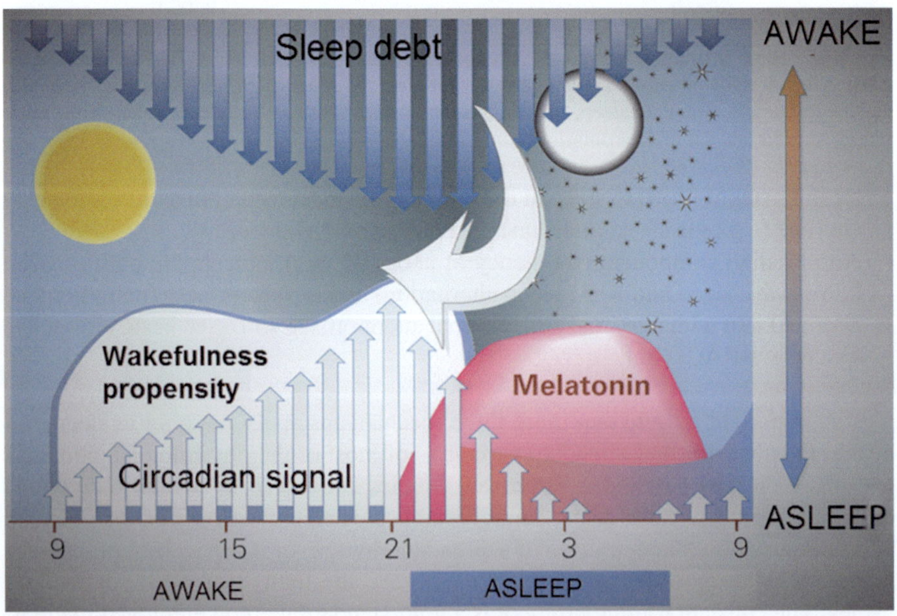

Fig. 2.12 The evening rise in melatonin feeds back to inhibit the wakefulness-promoting effect of the SCN. This is the trigger for consolidated sleep. Reproduced with permission from Cardinali [1]

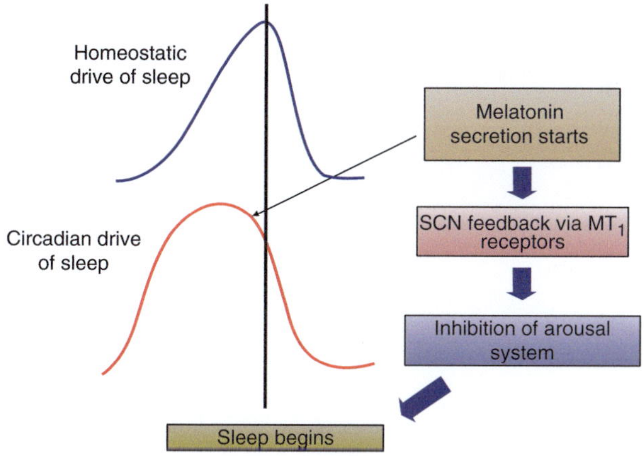

Fig. 2.13 The homeostatic sleep pressure increases during the day and is counteracted by a strong circadian promotion of wakefulness from the SCN. The secretion of melatonin inhibits SCN electrical activity, thus triggering sleep. Reproduced with permission from Cardinali [1]

S process [1]. There is evidence that this abrupt change in sleep propensity is crucial for sleep induction. Thus, melatonin is considered to be the signal that "opens the gates of sleep" (Fig. 2.13).

The best candidate in the search for the hypnogenic substance that mediates the homeostatic (S process) is an astrocyte-derived nucleoside, adenosine, acting on the sleep-active GABAergic VLPO and MnPO, and on the basal forebrain (BF; Fig. 2.14). Adenosine is a cellular product that accumulates because of metabolic activation in tissues and thus indicates the degree of activity. Caffeine, theobromine, and other xanthines are inhibitors of adenosine receptors, and coffee and tea may promote wakefulness by interfering with this mechanism [23].

The extracellular level of adenosine in the brain increases during prolonged waking. In several brain regions, the stimulation of presynaptic A1 adenosine receptors depresses Glu release and reduces the amplitude of excitatory postsynaptic currents. This is the mechanism by which increased adenosine levels after sleep deprivation influence the light responsiveness of the circadian clock. Indeed, after a 6-h sleep deprivation, the light response in the SCN was reduced compared with the control. Systemic injection of caffeine restored this attenuation of the SCN light response almost completely [24], supporting the interactive role of adenosine and Glu. The reduced response to light of SCN neuronal activity after sleep deprivation provides evidence that the pacemaker may be modified by sleep homeostatic pressure.

In addition, the cortical neuronal population expressing neuronal nitric oxide (NO) synthase has recently emerged as another candidate for involvement in the homeostatic physiological sleep response (S process) [23]. The accumulation of adenosine during the increase in IL-1 in infectious processes, given by the increase in prostaglandin (PG) D2 in the organum vasculosum laminae terminalis (OVLT) explains in part the symptomatology of the disease behavior (Chap. 4). IL-1 has also been linked to daily sleep debt (Fig. 2.14).

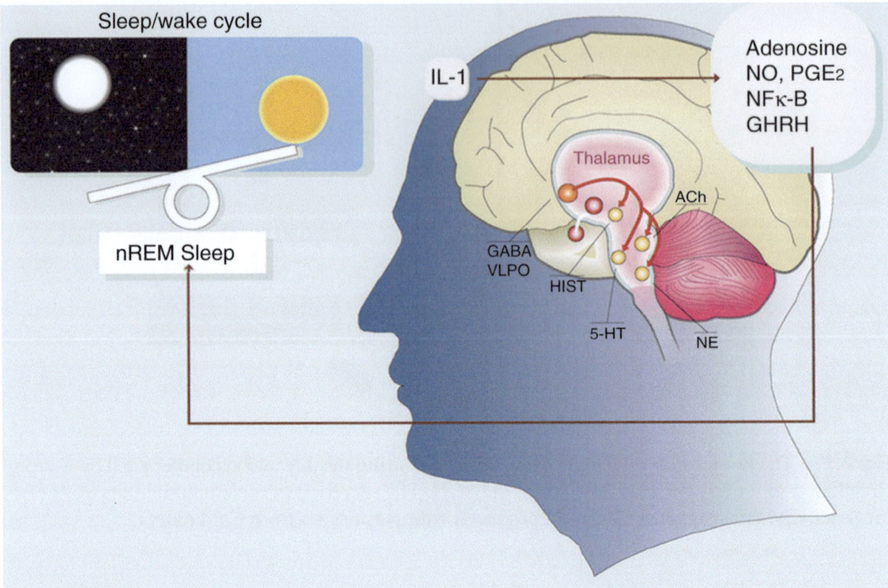

Fig. 2.14 Effect of humoral signals on slow-wave sleep. *NFκB* transcription factor kappa B

Neurophysiology of Sleep

The pioneering research of Bremer at the mid-1930s (Fig. 2.15) demonstrated that lesions at a low medullary level in cats did not modify the sleep–wake cycle ("*encephale isolé*"), whereas a cut between the pons and intercollicular midbrain produced chronic sleepiness ("*cerveau isolé*") [25]. Years later, other investigations demonstrated that a midpontine section suppressed REM sleep, an indication that the REM sleep generator was located below this level (Fig. 2.15).

Bremer's observations ruled out the initial concept that sleep was a merely passive phenomenon produced by the inactivation of sensory stimuli arriving at the diencephalic/telencephalic structures. Following Bremer's ideas, Moruzzi and Magoum demonstrated that the electrical stimulation of pontine reticular formation activated the cortex and produced awakenings, and concluded that the forebrain was kept alert by the tonic activity of the pontine reticular formation [26]. Thus, during those years, the overall framework of knowledge held that the reticular formation maintains an ascending activation from wakefulness to the thalamus and cortex, producing recordings of EEG activation or unambiguous wakefulness. It was hypothesized that a passive inactivation of the pontine reticular formation caused the reduction of sensory input and generated sleep. However, many observations did not fit with this theory. For example, thalamic stimulation produced sleep or wakefulness depending on the frequency of stimulation. Likewise, transection of the pons, rostral to the fifth cranial nerve, induced wakefulness, thus questioning the

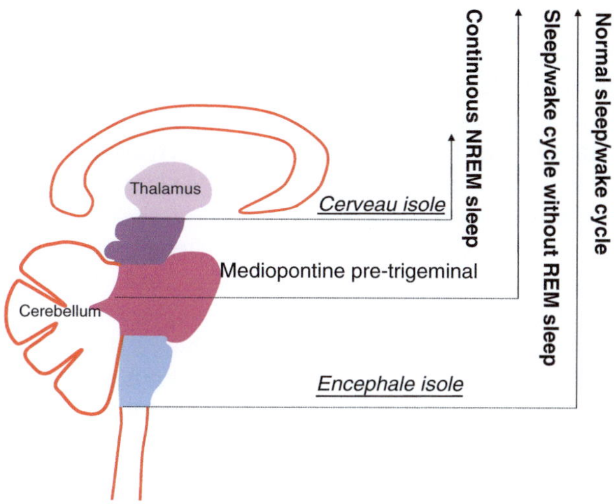

Fig. 2.15 In 1935, and working on sleep EEG in cats, Frédéric Bremer identified the changes produced by surgical cuts. He noticed that in the *cerveau isolé* there was permanent sleep, whereas in the *encéphale isolé* the wake/sleep rhythm was maintained. Medial pontine cuts eliminated REM sleep, leaving the slow sleep/wake rhythm intact. This located the origin of REM sleep in the structures of the brainstem

theory of a passive inactivation of pontine reticular formation. Indeed, the findings were compatible with the view that inputs from the lower pons or medulla inhibit a wakefulness center in the rostral pons or rostral sites (thalamus) to induce sleep. Thus, sleep was not merely a result of the deactivation of arousal centers, but an active state of the brain.

After World War I, a worldwide epidemic of influenza with neurotropic sequelae led the Rumanian neuropathologist von Economo to identify in patients three types of lesions associated with different premortem statuses of sleep and waking [27]. These were:

- Type 1: lesions from the posterior hypothalamus variably extending to mesencephalic reticular formation associated with premortem somnolence or coma.
- Type 2: lesions of the anterior hypothalamus (VLPO and MnPO) and nearby areas of basal forebrain associated with insomnia
- Type 3: lesions of the posterior–lateral hypothalamus, commonly in type 1 survivors of a somnolence/coma syndrome, associated with narcolepsy.

From these observations, von Economo concluded that the posterior hypothalamus contains promoters of wakefulness and the anterior hypothalamus promoters for sleep. These findings were confirmed by subsequent research demonstrating that lesions of the anterior hypothalamus or preoptic/BF (substantia innominata and horizontal limb of the diagonal band of Broca) reduced sleep and conversely their electrical stimulation produced sleep onset [23].

Histaminergic and orexin-containing neurons of the posterior–lateral hypothalamus constitute the posterior hypothalamic wake promotion center. Tuberomammillary nucleus (TMN) histaminergic neurons are the only source of His in the brain and these cells extensively project innervation to the forebrain and brainstem (Chap. 4). The TMN firing rate has a decreasing pattern from a continuum from wakefulness to NREM sleep and REM sleep, and plays a fundamental role in the generation of wakefulness [23].

The orexinergic neurons are found only in the lateral hypothalamic area (LHA) and project widely to the brain and spinal cord. The orexin A and B neuropeptides are excitatory, acting via OX1 and OX2 receptors and promote waking by activating forebrain and brainstem wake-active cell groups (Fig. 2.16). The most important evidence that orexins play a fundamental role in the regulation of wakefulness and sleep was the demonstration that in narcolepsy with cataplexy a loss of orexin signaling occurs. Orexin neurons act as a conductor of orchestration for vigilance states, behaviors, and autonomic functions. Body temperature regulation by orexin

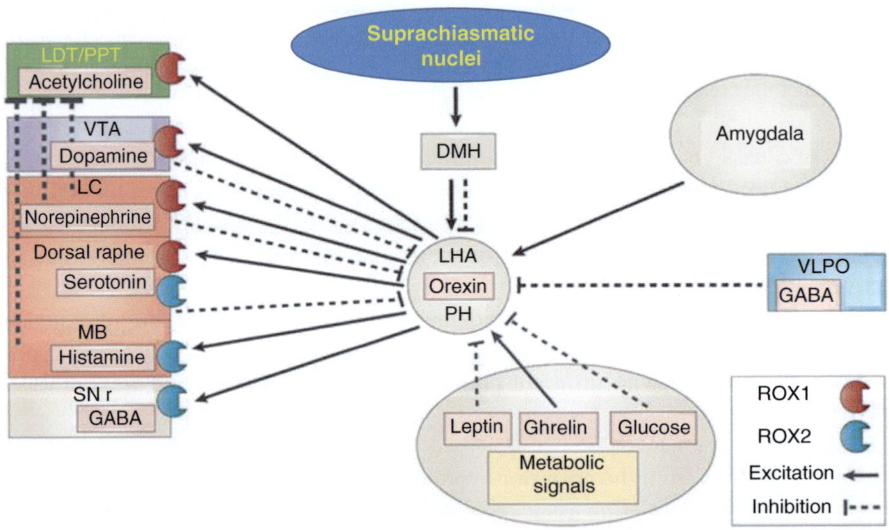

Fig. 2.16 Interactions of neurons containing orexin with other regions of the brain involved in sleep and wakefulness regulation. Orexinergic neurons in the lateral hypothalamic (LHA) and posterior hypothalamus (PH) are strategically located to serve as a link between the limbic system, the systems involved in energy and monoaminergic homeostasis, and cholinergic neurons in the brainstem. The SCN sends signals to orexin neurons through the dorsomedial hypothalamus (DMH). *VLPO* ventrolateral preoptic area, *NDR* dorsal raphe, *LC* locus coeruleus, *LDT* laterodorsal tegmental nucleus, *PPT* pedunculopontine tegmental nucleus, *SNR* substantia nigra pars reticulata, *TMN* tuberomammillary nucleus

neurons seems to be mediated by one of its cotransmitters, whereas cardiovascular and respiratory regulation are mediated by orexin itself [28].

The ascending reticular activating system involved in arousal and EEG activation, and consequently in wakefulness, is schematized in the left panel of Fig. 2.17. In addition to the posterior hypothalamic wake-promoting center described above, it is composed of the cholinergic laterodorsal tegmental (LDT) nucleus and pedunculopontine tegmentum (PPT) nucleus, the noradrenergic locus coeruleus (LC), 5-HT-containing dorsal raphe nucleus (DRN) neurons, and dopamine (DA)-containing neurons in the ventral tegmental area (VTA) and substantia nigra that

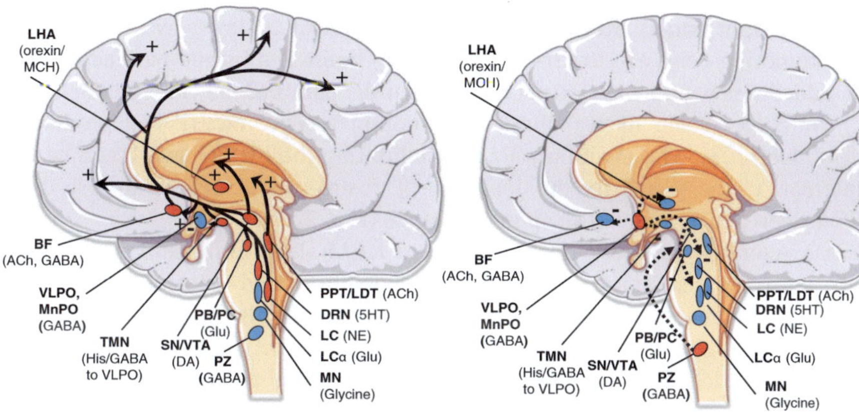

Fig. 2.17 *Left*: schematic representation of the neuronal pools promoting arousal to the forebrain. During wakefulness, the histaminergic neurons in the ventral tuberomammillary nucleus (TMN) at the bottom of the posterior hypothalamus provide a strong inhibitory influence on the VLPO/MnPO (median preoptic area). The components of the ascending reticular activating system further include the raphe nuclei (5HT neurons), the locus coeruleus (LC, noradrenergic neurons), the pedunculopontine and laterodorsal tegmenti (PPT/LDT) ACh-containing neurons, DA-containing neurons in the substantia nigra (SN) and the ventral tegmental area (VTA), Glu-containing neurons in the parabrachial/precoeruleus area (PB/PC) and the basal forebrain (BF), mainly cholinergic, but also containing a population of GABAergic neurons whose stimulation produces sustained wakefulness and EEG gamma activity [6]. Orexin-melanocyte concentrating hormone (MCH) neurons on the LHA provide stimulatory input to the wakefulness-promoting areas. *Right*: schematic drawing for a possible dual control, rostro-caudal and caudo-rostral, of sleep generation and maintenance. The rostral sleep-promoting pathway includes the VLPO/MnPO area, which is active during NREM sleep, and inhibits the activity of the arousal centers of the brainstem, hypothalamus, and cortex. The caudal sleep-promoting pathway includes the parafacial zone (PZ) area in the rostral medulla, which inhibits the PB/PC area, thus decreasing its wakefulness-promoting activity on BF. The LC-α and magnocellular nuclei (MN), participating in REM-induced atonia, are also depicted. *Red labels* denote activation, whereas *blue labels* represent inhibition. The figure was prepared in part using image vectors from Servier Medical Art (www.servier.com), licensed under the Creative Commons Attribution 3.0 Unported License (http://creativecommons.org/license/by/3.0/)

stimulate the cortex directly and indirectly via the thalamus, hypothalamus, and BF. Recent evidence indicates that the activation of a population of BF GABAergic neurons produce sustained wakefulness and EEG gamma activity, and that the pontomesencephalic PBN and precoeruleus area glutamatergic pathway to BF is very important for EEG activation and wakefulness, as their lesions produced a coma-like state (Fig. 2.17) [23].

The VLPO and MnPO play a key role in promoting sleep. The neurons in these nuclei contain GABA and the neuropeptide galanin and they innervate all the arousal-promoting regions, including the LDT/PPT, LC, VTA, substantia nigra, DRN, TMN, and the orexin neurons. Thus, the VLPO and MnPO promote sleep by coordinating the inhibition of arousal regions during NREM and REM sleep (Fig. 2.17, right panel). Recently, the identification of a slow-wave GABAergic sleep-promoting center in the rostral medullary brainstem, in the parafacial zone (PZ), added a new pathway. PZ neurons inhibit the parabrachial (PB)/precoeruleus (PC), thus decreasing its wakefulness-promoting activity on BF and producing NREM sleep [23]. The medullary Pz area inhibiting the PB/PC area could be the counterpart of the VLPO for the dual control, rostro-caudal and caudo-rostral, of NREM sleep generation and sleep maintenance (Fig. 2.17).

Since Aserinsky and Kleitman discovered REM sleep in humans in the mid-1950s, an impressive amount of research has demonstrated that the pons plays a key role in the generation of REM sleep. In the mid-1970s the proposal that the NREM/REM cycles arise from a reciprocal interaction between REM-off (monoaminergic) and REM-on cells (cholinergic) in the medial pons was made. A flip–flop model for REM-off and REM-on neurons was proposed (Fig. 2.18). Two neuronal pools of mutually inhibitory neurons in the upper pons form a switch for controlling transitions between NREM and REM sleep. GABAergic neurons in the ventrolateral periaqueductal gray and the adjacent lateropontine tegmental area fire during NREM states to inhibit entry into REM sleep. During REM sleep, they are inhibited by a population of GABAergic neurons in the sublaterodorsal region that fire during REM sleep. This mutually inhibitory relationship produces a REM–NREM flip–flop switch, promoting rapid and complete transitions between the two states [20].

The core REM switch is also modulated by other neurotransmitter systems. Noradrenergic neurons in the LC and serotoninergic neurons in the DRN inhibit REM sleep by actions on both sides of the flip–flop switch (exciting REM–off and inhibiting REM–on neurons), and during REM sleep they are silent. Cholinergic neurons from the PPT/LDT promote REM sleep by having opposite actions on the same two neuronal populations. The orexin neurons inhibit entry into REM sleep by exciting neurons in the REM-off population (and by presynaptic effects that excite monoaminergic terminals), whereas the VLPO neurons promote the entry into REM sleep by inhibiting this same target. During REM sleep, a separate population of glutamatergic neurons activates a series of inhibitory interneurons in the medulla and spinal cord, which inhibit motor neurons, thus producing the atonia of REM sleep. Withdrawal of tonic excitatory input from the REM-off regions may also contribute to the loss of muscle tone. At the same time, ascending projections from

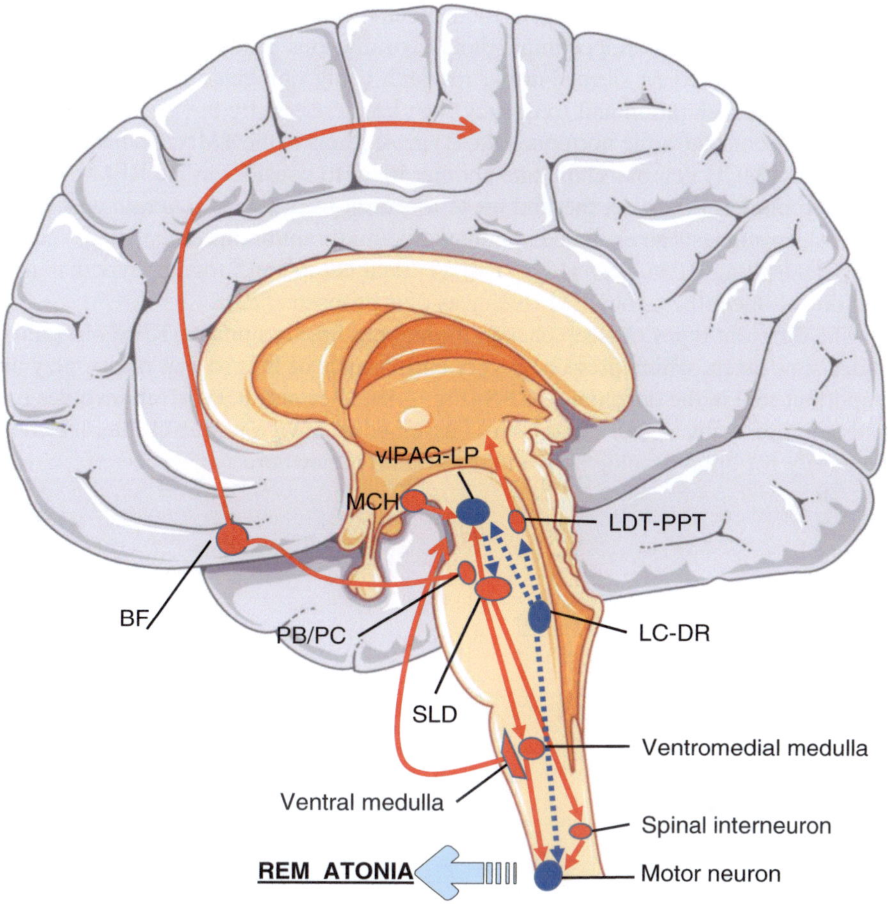

Fig. 2.18 Schematic drawing illustrating the hypothetical circuitry involved in REM sleep regulation. A flip–flop mutual inhibition between the sublaterodorsal tegmental area (SLD) REM-on neurons and ventrolateral periaqueductal gray matter/lateropontine tegmental areas (vlPAG/LPT) REM-off neurons are proposed to regulate transitions into and out of REM sleep. Pedunculopontine tegmental/laterodorsal tegmental area (PPT-LDT) REM-on neurons inhibit lateropontine tegmental areas/ventrolateral periaqueductal gray matter (LPT-vlPAG) REM-off neurons, but they are not mutually inhibited by the latter and thus they are not part of the REM flip–flop switch. The same unidirectional relationship occurs with the serotoninergic dorsal raphe nucleus and noradrenergic LC (DRN/LC) that activate REM-off neurons, but are not inhibited by the SLD REM-on neurons. During REM sleep, glutamatergic neurons from the PB/PC area project rostrally to the basal forebrain (BF) and regulate the EEG components of REM sleep and caudally to the ventromedial medulla and spinal cord to activate GABA/glycine interneurons inhibiting motor neurons to cause atonia. The inhibition of the SLD given by vlPAG/LPT is overcome during REM sleep by neurons containing MCH and other neurotransmitters. A population of GABAergic cells located caudally at the ventral medulla also participates in the REM sleep switch, indicating a "dual command," rostral–hypothalamic and dorsal–medullary, for REM sleep regulation. *Solid lines* denote pathways that are active during REM sleep; *dashed lines* pathways that are inactive during REM sleep. The figure was prepared in part using image vectors from Servier Medical Art (www.servier.com), licensed under the Creative Commons Attribution 3.0 Unported License (http://creativecommons.org/license/by/3.0/)

glutamatergic neurons in the PBN activate forebrain pathways that drive EEG desynchronization and hippocampal theta rhythms, thus producing the characteristic EEG signs of REM sleep. Further research using optogenic tools showed that REM episodes (duration and frequency) can be increased by the photostimulation of melanin-concentrating hormone (MCH) projections in the TMN and median septum. Thus, MCH neurons constitute another input to consider in the REM–NREM flip–flop model. Likewise, the finding of a ventral medulla GABAergic control of REM sleep suggests an extended hypothalamic/midbrain/brainstem, perhaps redundant, controlling REM sleep (Fig. 2.18). A "dual command," rostral–hypothalamic and dorsal–medullary, for REM sleep was thus proposed [29].

The different types of sleep change throughout life. In humans, REM sleep precedes slow sleep, which prevails in very early stages of life, so that it may play an important role in the developing CNS (Fig. 2.19). Because REM sleep involves the activation of many neural circuits, it is assumed to have a powerful internal drive necessary for brain development and maturation in newborns.

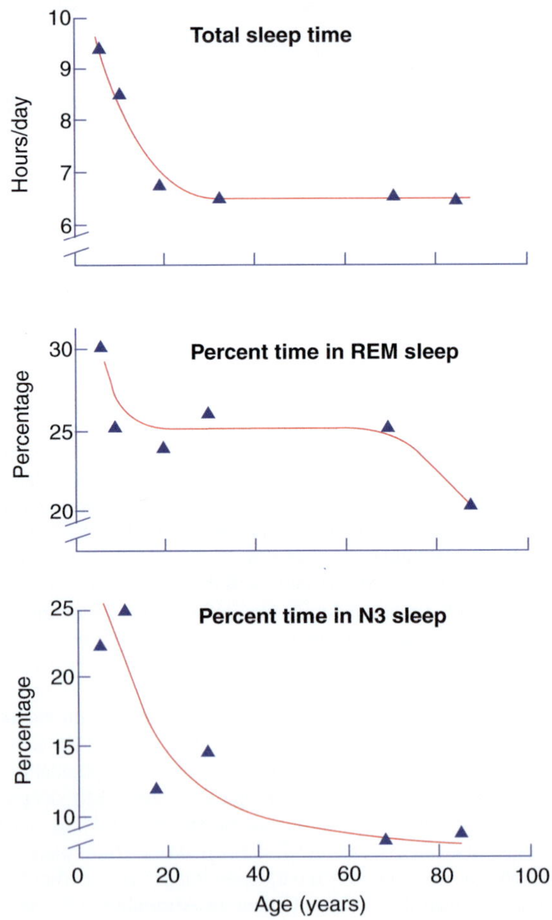

Fig. 2.19 Polysomnography recording throughout a lifetime. The depth of NREM sleep decreases with age. Reproduced with permission from Cardinali [1]

Slow sleep decreases exponentially with age and often disappears after 60 years of age [30]. This decrease in the depth of sleep causes frequent awakenings and a return to an ancestral pattern of interrupted sleep. The duration of slow-wave sleep decreases rapidly (almost 30 min per decade); thus, deep slow-wave sleep (stage N3) after 50 years of age is less than 10% of the total sleep period. This decrease in the depth of sleep is accompanied by an increase in the N1 and N2 stages, whereas the duration of REM sleep and total sleep time remains stable (Fig. 2.19). Many of the elderly often complain of sleep disturbances. In these cases, the physician should always investigate for the quality of wakefulness. If there is no daytime sleepiness, sleep is sufficient and productive, this is merely a complaint about sleep disruption resulting from the belief that sleep is normal if not interrupted [20].

In humans, the maximum threshold for awakening occurs in the N3 stage, whereas the minimum threshold for awakening corresponds to REM sleep. On average, 75% of normal individuals report dreaming when awakened during REM. Although this percentage is higher than those recorded in the NREM stages, what changes is the content of dreaming. As the time spent in REM sleep tends to be proportionally greater as the night progresses, there is a tendency to wake up toward the end of the REM sleep period, with more vivid and memorable dreams, at least for a short period.

There are other differences between slow-wave sleep and REM sleep. The slow-wave sleep period depends upon waking and is prevalent when the individual has not slept or slept little the previous night. In fact, the most accurate indicator of homeostatic "debt" is the time spent in slow-wave sleep. This link with previous waking does not exist for REM sleep. Both types of sleep also differ in their relationship to hypnotic drugs: alcohol and barbiturates decrease REM sleep, whereas benzodiazepines affect it less, although sufficiently to result in a nonphysiological sleep.

What are the bases of the physiological process by which stages of sleep and wakefulness are defined? The cortico-thalamic circuit is instrumental in defining wakefulness, slow-wave sleep, and REM sleep. Each of these stages is characterized by the level of thalamic "gate" to permit or not the passage of ascending sensory information. Wakefulness and REM sleep are characterized by an "open gate," which allows the arrival of sensory information (exteroceptive in waking, interoceptive in REM sleep) into the cortex. In NREM sleep the gate is closed and there is minimal information input into the cerebral cortex [30].

In Fig. 2.20, the physiological basis of this function is summarized. In thalamic relay nuclei, most synapses on the main cells are derived from the cerebral cortex (cortico-thalamic connections), brainstem, and inhibitory interneurons that control the various modes of activity of the main cells and thus how the information is sent from the thalamus to the cortex. The many inputs that operate on the thalamic relay cells use various neurotransmitters [20]. Retinal and cortical afferents are always excitatory and use the Glu as a neurotransmitter. Approximately 25% of the neurons of the relay nuclei (e.g., lateral geniculate body) are cells with a small soma whose dendritic tree extends along the sheet where they are located and do not project

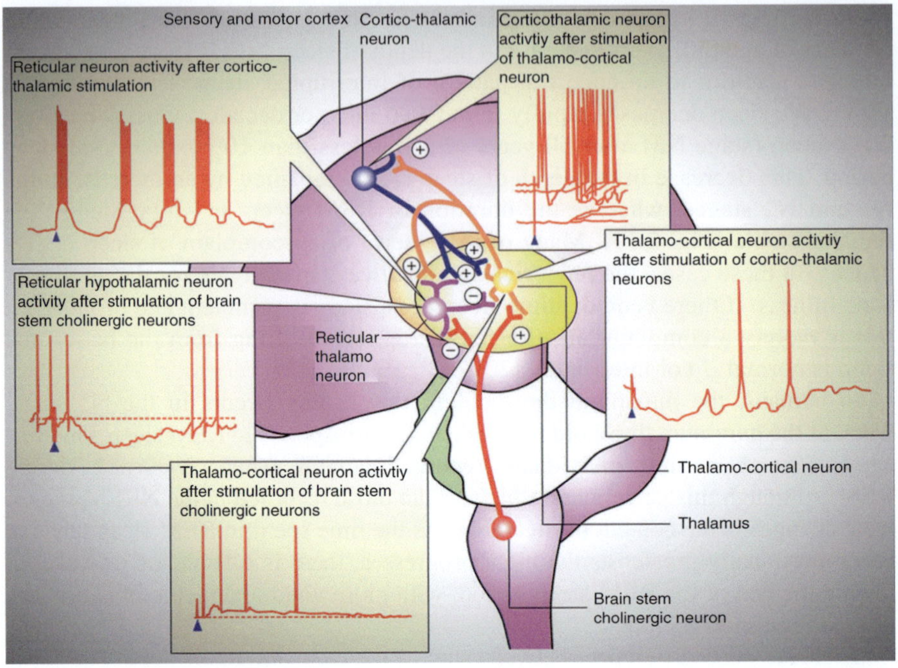

Fig. 2.20 Thalamic–cortical mechanisms. Represented neurons include pedunculopontine tegmental–laterodorsal tegmental cholinergic neurons, the GABAergic neurons of the reticular nucleus of the thalamus, the glutamatergic neurons of the thalamic nuclei, and the glutamatergic cortical neurons that project the thalamus. Different recordings, based on the transmission in the form of relay or of gate of the circuit are shown. Cholinergic projections depolarize relay neurons and hyperpolarize thalamic reticular neurons. The final result depends on the stimulation of thalamo-cortical and cortico-thalamic neurons on each one of them and the response characteristics in salvos (spindles) of the thalamic reticular neurons (see text). Modified with permission from Cardinali [8]

outside the core. They are interneurons with an inhibitory function and they use GABA as a neurotransmitter [30].

Cortico-thalamic glutamatergic afferents synapse with reticular nucleus cells, with inhibitory interneurons and with relay cells, so that the influence on core performance thalamic relay has both excitatory (direct) and inhibitory (through interneurons) components.

Afferents coming from the brainstem are heterogeneous. Mesencephalic reticular formation acts as a functional unit with the thalamus in what is defined as "gates." Important regulation of these gates is given by cholinergic influences, from the PPT/LDT area. DRN serotonergic neurons and LC noradrenergic neurons also affect the "gates." To a lesser extent, VTA dopaminergic projections are also involved. These inputs modulate globally the excitability of thalamic relay neurons, thereby influencing how much information is directed to the cortex.

There are thus two functional modes for thalamic relay nuclei cells (Fig. 2.20):

(a) "Open gate," in which the transmission of information is carried out with the triggering of action potentials of variable frequency only. Under these conditions, transmission through the thalamic relay nuclei reliably reflects the arrival of information from sensory pathways and the transfer rate approaches 1 (one PEPS produces one action potential).

(b) "Closed gate" when the thalamic relay cell is sufficiently hyperpolarized (membrane potential = −75 mV or more negative), it operates in a salvo mode, characterized by an initial depolarization with a variable number (2–7) of spikes linked to the Na^+ voltage-dependent channel. This is the basis of the "closed-gate mode" that appears in slow sleep, and coincides with a very characteristic EEG pattern, consisting of springs of waves at the frequency of 7–13 Hz, called sleep spindles. This mechanism is cyclic and hyperpolarization leaves the cell ready to start the process again.

Interestingly, in the "closed-gate" form, giant δ spikes are capable, as a grand mal epileptic focus, to co-opt the activity of most CNS neurons at their own pace. Indeed, it is as if a physiological "seizure focus" becomes the absolute regulator of brain activity. This state of being prone to epilepsy, has a clinical relevance: the EEG of epileptic patients during sleep records show a clear tendency toward the predominance of epileptic seizures in slow-wave sleep stage [30].

Three Different ANS Programs ("Body Configurations") Occur in a 24-h Day/Night Cycle

Sleep is not just a neurological phenomenon and a common mistake is to consider it an exclusive phenomenon of the CNS. Together with wakefulness, slow-wave sleep and REM sleep comprise three different ANS programs (Fig. 2.21 and Table 2.1) [1, 8].

Several physiological functions vary both in the passage from wakefulness to sleep and within each sleep stage; the sympathetic and parasympathetic systems, key regulators of the automatic functions of the body, are responsible for these changes. The sympathetic nervous system has evolved as predominant in wakefulness and in response to major threat to our species during evolution, i.e., physical trauma. It is thus linked to the consumption of energy (catabolism) to fight or flight at the threat and promote mechanisms to mitigate the consequences of trauma: vasoconstriction, increased blood coagulability, increased innate and humoral immunity (that keep wounds germ-free), etc. (Fig. 2.21 and Table 2.1). The sympathetic nervous system dominates wakefulness. This is a stereotypical hyperactivity, directed to place the individual in a situation of defense in the face of circumstantial danger, real or potential. Sympathetic overstimulation leads to variations in visceral functions designed to protect the integrity of the organism and to ensure survival. In fact, a sympathectomized animal hardly survives if left free in its natural environment.

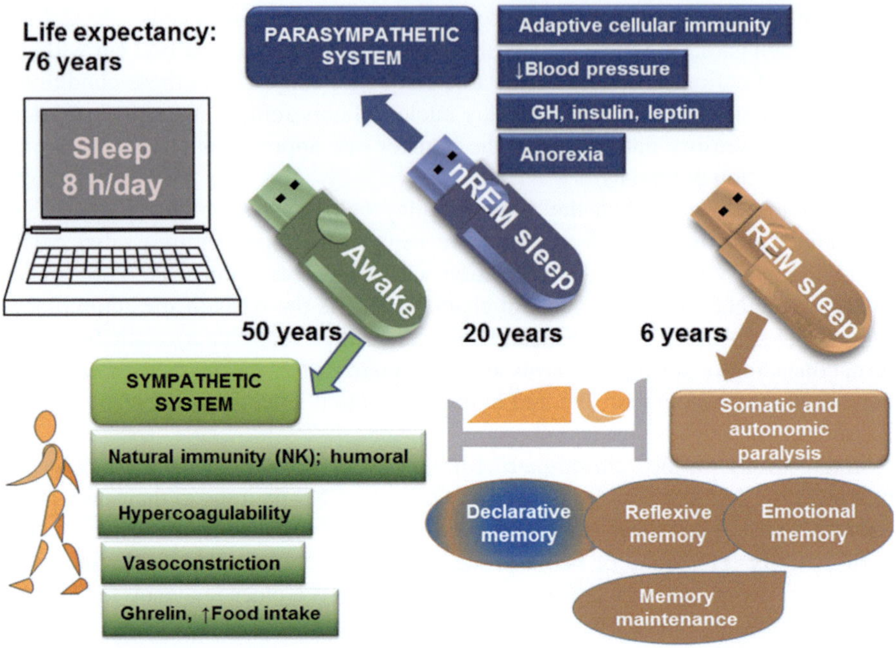

Fig. 2.21 The three different "bodies," wakefulness, slow-wave sleep (NREM sleep), and REM sleep, must necessarily follow each another harmoniously to ensure health. A 76-year-old man sleeping 8 h daily lives 50 years in the physiological state of wakefulness, 20 years in slow-wave sleep, and 6 years in REM sleep

Table 2.1 The three physiological states ("bodies") of our life

	Wakefulness	NREM sleep	REM sleep
	"Active brain in an active body"	"Inactive brain in an active body"	"Hallucinating brain in a paralyzed body"
Neurochemical "microclimate"	Tonic firing of neurons in the locus coeruleus (noradrenergic) and raphe nuclei (serotonergic) driven by orexinergic hypothalamic neurons. Phasic discharge of the pedunculo-pontine nucleus of the pontine tegmentum, PPT (cholinergic)	Inhibition by VLPO area of the arousal systems. Decreased aminergic activity in the face of a progressive increase in cholinergic activity (tonic firing of PPT neurons). Both responsible for decreased consciousness	Prevalent cholinergic activity (PPT nucleus) concomitant with extreme reduction of aminergic activity. REM sleep and wakefulness are states of cortical activation with different neuromodulating pattern (cholinergic vs noradrenergic) and different contents of consciousness

Table 2.1 (continued)

	Wakefulness	NREM sleep	REM sleep
Afferent	Actively functioning. Thalamocortical circuit in "open-gate fashion" so that sensory information can reach the cerebral cortex. Activated dorsolateral prefrontal cortex (working memory)	Thalamo-cortical circuit in "closed-gate fashion" (which prevents sensory information from reaching the cerebral cortex. A 25% decrease in cerebral blood flow and oxygen consumption. Synthesis of neurotrophins. Glymphatic flow increased.	Thalamic activity changes to operation "open-gate fashion" as during wakefulness
Efferent	Actively functioning	Episodic muscle activity, hypotonia	Skeletal muscle paralysis as a protective mechanism to prevent the locomotor correlates of a highly activated brain
Content awareness	Attention, logical thinking, memory	Disconnection, episodic memory	Dream activity characterized by vivid hallucinations, illogical thinking and intense emotion.
Perception	Externally generated	Absent	Generated internally, preferential activation of the pons and limbic system with deactivation of the dorsolateral prefrontal cortex
Physiological pattern in organs and systems	Predominance of sympathetic activity. Augmented plasma NE and cortisol	Parasympathetic hyperfunction in organs and systems. GH, prolactin and insulin secretion	Disconnection of autonomic regulatory system (this prevents expression of dreaming emotions). Antihomeostatic physiology

Reproduced with permission from ref. [1]

PPT pedunculopontine tegmentum, *NE* norepinephrine, *VLPO* ventrolateral preoptic area, *GH* growth hormone

Our species is programmed to eat only sporadically (every 2–3 days); thus, a mechanism that optimizes the maximal intake at the right time was selected for wakefulness. Thus, wakefulness is linked to an increased food intake given by the secretion of orexinergic hormones such as ghrelin. Note that the same chemical signal that increases appetite, orexin, is also a central neurotransmitter in maintaining alertness (Fig. 2.16).

For modern man, the trauma has become a minor factor and instead new diseases resulting from the prolongation of life and the type of diet and living conditions manifest. During evolution, endothelial injury and organ hypoxia were associated almost exclusively with trauma. Today, endothelial injury is precipitated by stressors such as hypertension, diabetes, or dyslipidemia. It is postulated that the pathophysiology of cardiovascular disease implies a prominent amplification of a triple response to trauma, (a) adrenergic response; (b) inflammation; (c) coagulation, which are aggravated by the sleep deprivation conditions described above. Selected components act to limit bleeding, defend infection of wounds, and initiate cell reconstruction. These mechanisms are highly conserved, as indicated by the phylogenetic age of the renin–angiotensin system [31].

The parasympathetic system in slow-wave sleep serves as the anabolic counterpart of the predominance of the catabolic sympathetic system during wakefulness. It promotes energy accumulation, adaptive immunity, and augmented secretion of anabolic hormones, such as growth hormone (GH), and of anorexic hormones, such as leptin and insulin. The eighteenth-century French clinicians believed that the parasympathetic system was the "master of sleep." Today, we know that this must be rephrased to point out that "the parasympathetic is the master of slow wave sleep", that is, of about 75% of the night (Fig. 2.21 and Table 2.1) [8].

Although the common view is that we humans are homeotherms (that is, we have a regulated body temperature, Chap. 5), in a substantial part of our life we lack that control. Wakefulness is characterized by a constant interaction of the hypothalamic (automatic) and behavioral mechanisms (facultative: I have cold and seek shelter) that control body temperature. In the passage to slow-wave sleep, the inactivation of behavioral control occurs, but the temperature is still regulated by the automatic processes discussed in Chap. 5. During REM sleep, the situation changes radically: at this stage both forms of temperature control are halted and there is no heat production to compensate for the cold. That is, during REM sleep, we acquire a similar state to amphibians and reptiles, whose body temperature depends on ambient temperature (poikilothermic animals).

Indeed, at REM sleep, most supraspinal autonomic reflex mechanisms are suppressed: the complex mechanisms of cardiovascular, respiratory, and thermal control temporarily stop working, with only the basic autonomic reflexes of the spinal cord persisting. Like a transoceanic flight in which most control mechanisms of aircraft become disconnected for 10–15 min and the risk of an accident is high, during REM sleep, there is a greater risk for strokes, heart attacks, and other acute episodes. As in the latter part of the night, there is a prevalence of REM sleep, such accidents tend to be higher late at night/early morning. This state of regional disconnection is the equivalent of leaving the body without its basic homeostatic mechanisms.

The impoverishment of slow-wave sleep and the consequent decrease in parasympathetic tone have strong effects on the neuroendocrine–immune network [32]. The observed immune changes include reduction of acquired immunity, particularly of cellular immunity, whereas innate and humoral immunity tends to increase. Many conditions that depend on an adequately controlled cellular immune response (viral diseases, oncology, autoimmunity) are aggravated by this imbalance. In turn, cancers and viral diseases are accompanied by a significant reduction in slow-wave sleep (and thus a greater parasympathetic withdrawal), either because they alter directly NREM sleep via the inhibition of the secretion of melatonin or because some of its symptoms trigger arousal (for example, coughing in lung disorders). However, it should not be forgotten that there is no absolute predominance of one system over the other, but a delicate interplay between the sympathetic and parasympathetic that is responsible for each of these body system configurations (Chap. 1).

For all that has been said up to now, it is obvious that we cannot skip over the slow-wave sleep repair period after spending several hours in the physiological setting of wakefulness (of sympathetic predominance of the catabolic type, with high energy consumption and potential damage to organs and tissues). Everything is prepared during NREM sleep for the anabolic recovery, with the release of hormones such as GH and typical responses of cellular immunity. This intricate and subtle mechanism is altered during sleep deprivation [32].

Finally, it should be noted that not only are the mechanisms of sleep neural, but that important humoral components also exist. The idea of a humoral origin of sleep dates from ancient times. In the pre-scientific stage, it was thought that "vapors" derived from the digestive system produced sleep. By the late nineteenth century, "fatigue substances" or hypnotoxins were postulated from experiments in which the cerebrospinal fluid (CSF) of sleep-deprived dogs produced sleep when injected into control dogs. The hypnotoxin theory lost strength after Von Economo's studies on sleeping sickness, which clearly showed that lesions of the anterior hypothalamus produce insomnia whereas lateral hypothalamic lesions lead to hypersomnia.

In the initial search for humoral factors inducing sleep, several substances were identified. In the cerebral venous effluent of sleeping rabbits a nonapeptide was isolated (δ sleep-inducing peptide). Three other substances isolated from the brain areas of sleeping animals were characterized as uridine pyrimidine nucleoside, oxidized glutathione tripeptide, and a "factor S," later identified as muramyl peptide. It is of interest that this peptide induces the synthesis of IL-1 by astrocytes.

The following criteria have been proposed to consider a humoral substance as a regulator of sleep [33]: (a) it must induce or maintain physiological sleep or induce sleep equivalent to that obtained after sleep deprivation; (b) the concentration, turnover or its receptors should vary with changes in sleep propensity; (c) normal sleep should be inhibited by the effect of the inhibiting substance; (d) it should promote sleep by acting on one or more parts of the neural circuitry shown to operate in sleep; (e) the conditions that promote or inhibit sleep must be accompanied by changes in the amount or metabolism of the substance; (f) the induced sleep should be rapidly reversible without significant physiological consequences.

More recently, several substances have satisfied most of the requirements stated above. An important one is the GH-releasing hormone (GHRH), a peptide of the secretin/glucagon family. Slow-wave sleep is reduced in transgenic mice deficient in receptors for GHRH or in GHRH production. GHRH action is exerted on the VLPO area on a subset of GABAergic neurons. These neurons also respond to IL-1, which raises the possibility that the mechanism by which IL-1 promotes sleep is by increasing the response of these neurons to GHRH [34].

Two cytokines are of great importance in inducing sleep, IL-1 and TNF-α. The following evidence indicates their importance: (a) central or peripheral administration of IL-1 or TNF-α increases slow-wave sleep and suppresses REM sleep; (b) the selective inhibition of IL-1 or TNF-α reduces NREM sleep; (c) knockout mice for receptor type I IL-1 or p55 TNF show less slow-wave sleep; (d) sleep deprivation increases mRNAs for IL-1 and TNF-α in the brain; (e) there is a parallel rate of increase in mRNA and protein levels of IL-1 and TNF and of slow-wave sleep in rodents.

The intracerebroventricular administration of TNF-α or IL-1 toward the latter part of the period of wakefulness increases NREM sleep. By contrast, if administered during the first part of the waking period TNF-α or IL-1 inhibits NREM sleep, presumably by the increased activity of the pituitary–adrenal axis. TNF seems to exert somnogenic effects by promoting the attraction of microglia and their processes to the vicinity of the dendrites and synapses [35]. Both cytokines induce fever, but this action is not linked to their somnogenic activity. Production of IL-1 and TNF-α by neurons, glial cells, and by endothelial cells in the CNS has been shown, and it varies with the state of sleep propensity.

It is noteworthy that the injection of TNF-α or IL-1 into the somatosensory cortex induces an increase in ipsilateral slow-wave sleep that is unobserved contralaterally. Conversely, the inhibition of cytokines produces local changes in EEG activity that are not observed contralaterally. These results indicate a local action of cytokines and are consistent with the idea that there is a "local sleep" in neural networks that depends on the local metabolic activity and release of cytokines and other paracrine or autocrine substances [36]. Some of the cytokines with effects on sleep are listed in Table 2.2.

Table 2.2 Cytokines with activity on sleep

Prosomnogenic	Antisomnogenic
IL-1	IL-4
Tumor necrosis factor (α and β)	IL-10
IL-2	IL-13
IL-6	TNF soluble receptor
IL-8	IL-1 soluble receptor
IL-15	Insulin-like growth factor-1
IL-18	
Epidermal growth factor	
Fibroblast growth factor	
Neural growth factor	
Brain-derived neurotrophic factor	
Neurotrophins 3 and 4	
Glial-derived neurotrophic factor	
Interferon (α and β)	
Granulocyte macrophage colony-stimulating factor	
Tumor growth factor	
Prokineticin 2	

The Meaning of Dreaming

Rapid eye movement sleep is widely distributed on the zoological scale. Therefore, it should be considered a process of great evolutionary significance. Indeed, the body is jeopardized in the phase of sleep by the major disconnection of the motor system and the ANS, a truly antihomeostatic state. One obvious conclusion is that this paralysis helps to preserve the body of motor and vegetative dream acting, which would be of great danger near predators. But why endanger homeostasis to preserve dreaming? The answer is probably the great importance of sleep in the process of learning and memory, both procedural and declarative (Chap. 6) [37].

Dreaming is a state characterized by illogical thinking, hallucinations, and emotional changes. Strictly speaking, the dream is a "delirium" as it presents hallucinations, disorientation, memory loss, and confabulation. For Sigmund Freud (*The Interpretation of Dreams*, 1900) there is a manifest content of the dream story with a latent or symbolic content. For Freud, dreams are the hallucinatory fulfillment of desire and have the biological function of protecting sleep (insomnia occurs when this mechanism fails).

Using a portable EEG recorder that distinguishes wakefulness, slow-wave sleep, and REM sleep during normal activity of the individual, several aspects of dreaming were examined [38, 39]. This led to the formulation of the activation–synthesis model of dreaming, according to which the physical substrate of dreaming is the cholinergic activation of an aminergically demodulated cortex during REM sleep. This situation leads to a synthesis of visual hallucinations, loss of reflective knowledge, emotional intensification, and memory loss [38, 39].

According to the model of reciprocal interaction, the REM phases and associated dreams are activated/deactivated through specific nerve connections at the brainstem. "REM-on" nerve cells (cholinergic) and "REM-off" (NE, 5-HT) nerve cells interact with each other. According to the activation–synthesis model, during sleep, brainstem neurons, and immediately after, areas of the cortex and the limbic system, are stimulated ("activation"). Dreaming content originates from random nerve impulses, produced by PGO (cholinergic) waves in REM sleep. With these random signals, the sleeping brain attempts to do the same thing it does in waking: integrate nerve impulses and give them meaning ("synthesis"). That is, according to Hobson's view, dreams are of an exclusively random nature [38, 39].

Freud believed that the amount of sleep was determined by the day's experience, which activates emergency-related memories. However, from experimentally obtained data it could be concluded that the dreams do not represent narrative or episodic memories, but rather the aggregation of discrete memory and incomplete narrative fragments to create a new synthetic script of dreaming [39].

However, there is way to link both scenarios, psychodynamic and neurocognitive. Dreaming is a side effect of neural activity generated by interoception (ANS). Interoceptive mechanisms are defined very early in life and are affected by the early emotional environment (Chap. 5). As it can be postulated that the biological representation of Freudian preconscious is given by interoception, dreaming may not be an aleatory event, but may depend heavily on the emotional history of the individual.

The absence of episodic memory in dreams reflects the relative inaccessibility of information from the hippocampus at the time (Chap. 6). Elevated levels of ACh suppress the flow of information from the hippocampal cortex both in wakefulness and in REM sleep. In individuals awakened during slow-wave sleep, there are elements of episodic memory (what happened during the day) that are characteristic of dreaming at this stage.

The forebrain structures activated in REM sleep (and paralimbic/limbic areas) give emotional and social content to dreaming. There is a very important activation of the basal ganglia in REM sleep, which is not surprising in view of the relevant participation of the limbic and cognitive circuitry of the basal ganglia in higher brain functions (Chap. 6). The cerebellar vermis is also stimulated, indicating the important role of the cerebellum in cognition and emotion [40].

Dreams have a visually predominant component. There are two main aspects in the visual analysis: (a) spatial vision in the medial temporal and parietal regions; (b) recognition of the object in the inferior temporal cortex [8]. It has long been known that these medial temporal cortex and occipital areas involved in visual processing are responsible for generating higher order visual imagery in dreams. These areas are selectively activated during REM sleep. The disabling of executive areas of the dorsolateral prefrontal cortex during slow-wave sleep and their failure to wake during REM explain the prominent executive deficiencies occurring in sleep (disorientation, lack of logic, less active memory, and amnesia).

The Glymphatic System and Sleep

The classic view of cerebrovascular physiology has been that blood flow and cerebral metabolism are tightly coupled under the influence of substances such as H^+, adenosine, NO, and K^+ that ensure a rapid and matched supply of blood. In part driven by the use of cerebral blood flow measurements by functional brain imaging, it has become clear that astrocytes also play a role in modulating functionally associated changes in cerebral blood flow. The autonomic innervation of the cranial circulation has both a sympathetic component that arises predominantly from the SCG and a cranial parasympathetic component that traverses the pterygopalatine (sphenopalatine) and otic ganglion [41]. Neuropeptide transmitters such as neuropeptide Y (NPY), VIP, and pituitary adenylate cyclase-activating peptide (PACAP) have each been identified in components of the system.

A previously unrecognized system that drains waste from the brain has recently been characterized as active in sleep [42]. This system acts as a glial cell-dependent pipe and has been called the "glymphatic system." This pathway consists of CSF influx into the brain parenchyma via para-arterial spaces, exchange of solutes

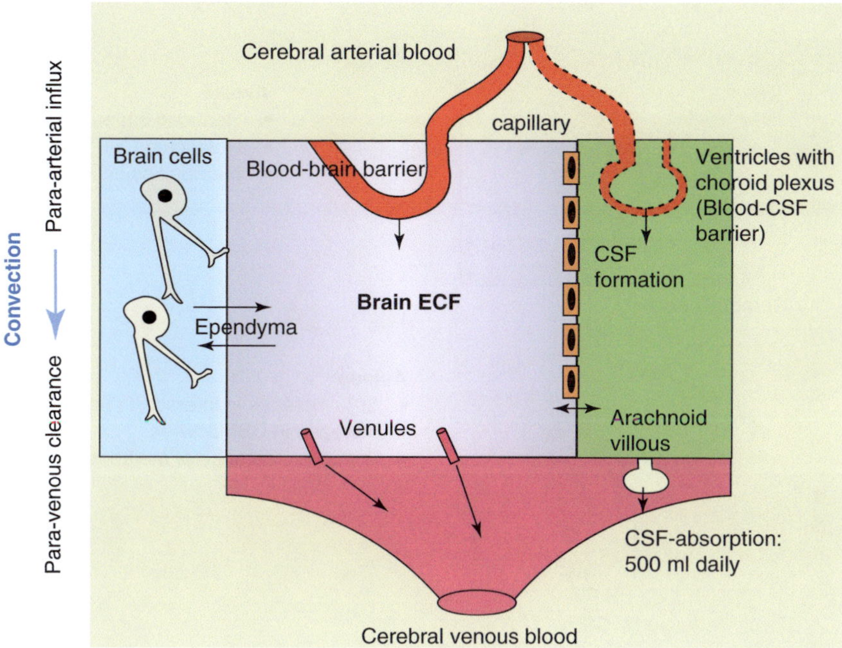

Fig. 2.22 Although the cerebrospinal fluid (CSF) influx is driven by arterial pulsation, the exchange of solutes with the interstitial fluid and fluid movement through the brain parenchyma are driven by convective bulk flow

(soluble proteins, waste products, metabolic wastes, and excess extracellular fluid) with the interstitial fluid and clearance along para-venous spaces (Fig. 2.22) [43, 44].

Although the CSF influx is driven by arterial pulsation, the exchange of solutes with the interstitial fluid and fluid movement through the parenchyma are driven by convective bulk flow rather than diffusion [43]. The exchange of solutes between the CSF and the interstitial fluid occurs mostly during NREM sleep, when the cortical interstitial space increases by more than 60% and provides a low-resistance path for the movement of CSF and interstitial fluid in the brain parenchyma (Fig. 2.23).

Brain water homeostasis is mediated by integral membrane pore proteins called aquaporins, which transport and regulate water movement in the brain. Aquaporin-4 is predominantly present in astrocytic endfeet near capillaries and in cells lining the ventricles, which are key sites for water movement among the cellular, vascular, and ventricular compartments [44]. The continuous aquaporin-4 expressing astrocytic endfeet lining the cerebral blood vessels create a low-resistance para-vascular channel for the movement of CSF. Postinjury reduction of aquaporin-4 expression has been associated with exacerbated glymphatic system dysfunction and aquaporin-4

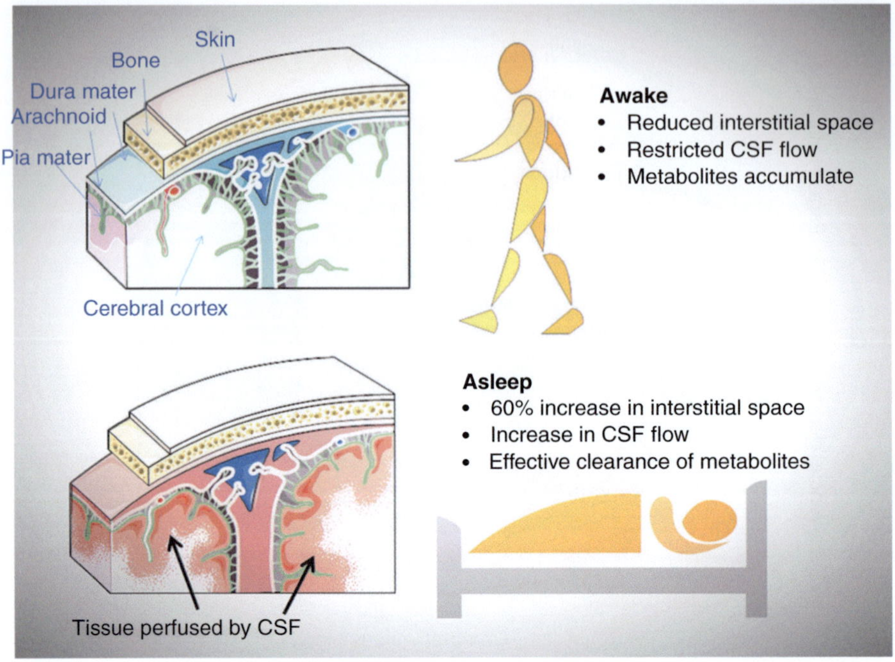

Fig. 2.23 The exchange of solutes between the CSF and the interstitial fluid occurs during NREM sleep, when the cortical interstitial space increases by more than 60% and provides a low-resistance path for the movement of CSF and interstitial fluid in the brain parenchyma. The figure was prepared in part using image vectors from Servier Medical Art (www.servier.com), licensed under the Creative Commons Attribution 3.0 Unported License (http://creativecommons.org/license/by/3.0/)

knockout mice exhibited slowed CSF influx and ~70% reduction in interstitial fluid solute clearance, indicating that the aquaporine-4 water channel mediates/facilitates the glymphatic pathway [44].

References

1. Cardinali DP. Ma Vie en Noir. Fifty years with melatonin and the stone of madness. Switzerland: Springer; 2016.
2. Golombek DA, Casiraghi LP, Agostino PV, Paladino N, Duhart JM, Plano SA, Chiesa JJ. The times they're a-changing: effects of circadian desynchronization on physiology and disease. J Physiol Paris. 2013;107:310–22.
3. Oldham MA, Lee HB, Desan PH. Circadian rhythm disruption in the critically ill: an opportunity for improving outcomes. Crit Care Med. 2016;44:207–17.
4. Agorastos A, Kellner M, Baker DG, Otte C. When time stands still: an integrative review on the role of chronodisruption in posttraumatic stress disorder. Curr Opin Psychiatry. 2014;27:385–92.
5. Hastings MH, Brancaccio M, Maywood ES. Circadian pacemaking in cells and circuits of the suprachiasmatic nucleus. J Neuroendocrinol. 2014;26:2–10.

6. Zhang R, Lahens NF, Ballance HI, Hughes ME, Hogenesch JB. A circadian gene expression atlas in mammals: implications for biology and medicine. Proc Natl Acad Sci U S A. 2014;111:16219–24.
7. Saper CB. The central circadian timing system. Curr Opin Neurobiol. 2013;23:747–51.
8. Cardinali DP. Neurociencia Aplicada. Sus Fundamentos. Editorial Médica Panamericana: Buenos Aires; 2007.
9. Cardinali DP, Pandi-Perumal SR, editors. Neuroendocrine correlates of sleep/wakefulness. New York: Springer; 2006.
10. Evans JA. Collective timekeeping among cells of the master circadian clock. J Endocrinol. 2016;230:R27–49.
11. Reghunandanan V, Reghunandanan R. Neurotransmitters of the suprachiasmatic nuclei. J Circadian Rhythms. 2006;4:2.
12. Hofman MA, Zhou JN, Swaab DF. Suprachiasmatic nucleus of the human brain: an immunocytochemical and morphometric analysis. Anat Rec. 1996;244:552–62.
13. Guido ME, Garbarino-Pico E, Contin MA, Valdez DJ, Nieto PS, Verra DM, Acosta-Rodriguez VA, de ZN, Rosenstein RE. Inner retinal circadian clocks and non-visual photoreceptors: novel players in the circadian system. Prog Neurobiol. 2010;92:484–504.
14. Hardeland R, Cardinali DP, Srinivasan V, Spence DW, Brown GM, Pandi-Perumal SR. Melatonin – a pleiotropic, orchestrating regulator molecule. Prog Neurobiol. 2011;93:350–84.
15. Wood S, Loudon A. Clocks for all seasons: unwinding the roles and mechanisms of circadian and interval timers in the hypothalamus and pituitary. J Endocrinol. 2016;228:X1.
16. Paakkonen T, Leppaluoto J, Makinen TM, Rintamaki H, Ruokonen A, Hassi J, Palinkas LA. Seasonal levels of melatonin, thyroid hormones, mood, and cognition near the Arctic Circle. Aviat Space Environ Med. 2008;79:695–9.
17. Levitas E, Lunenfeld E, Weisz N, Friger M, Har-Vardi I. Seasonal variations of human sperm cells among 6455 semen samples: a plausible explanation of a seasonal birth pattern. Am J Obstet Gynecol. 2013;208:406.
18. Jimenez-Ortega V, Cano P, Pagano ES, Fernández-Mateos P, Esquifino AI, Cardinali DP. Melatonin supplementation decreases prolactin synthesis and release in rat adenohypophysis. Correlation with anterior pituitary redox state and circadian clock mechanisms. Chronobiol Int. 2012;29:1021–35.
19. Melrose S. Seasonal affective disorder: an overview of a assessment and treatment approaches. Depress Res Treat. 2015;2015:178564.
20. Saper CB. The neurobiology of sleep. Continuum (Minneap Minn). 2013;19:19–31.
21. Borbely AA, Daan S, Wirz-Justice A, Deboer T. The two-process model of sleep regulation: a reappraisal. J Sleep Res. 2016;25:131–43.
22. Kleitman N. Basic rest-activity cycle – 22 years later. Sleep. 1982;5:311–7.
23. Garay A, Cardinali DP. New concepts in the neurophysiology of sleep and wakefulness. Physiol Minirev. 2016;9:26–36.
24. Van Diepen HC, Lucassen EA, Yasenkov R, Groenen I, Ijzerman AP, Meijer JH, Deboer T. Caffeine increases light responsiveness of the mouse circadian pacemaker. Eur J Neurosci. 2014;40:3504–11.
25. Bremer F. Preoptic hypnogenic area and reticular activating system. Arch Ital Biol. 1973;111:85–111.
26. Moruzzi G, Magoun HW. Brain stem reticular formation and activation of the EEG. Electroencephalogr Clin Neurophysiol. 1949;1:455–73.
27. Reid AH, McCall S, Henry JM, Taubenberger JK. Experimenting on the past: the enigma of von Economo's encephalitis lethargica. J Neuropathol Exp Neurol. 2001;60:663–70.
28. Kuwaki T. Thermoregulation under pressure: a role for orexin neurons. Temperature (Austin). 2015;2:379–91.
29. Peever J, Fuller PM. Neuroscience: a distributed neural network controls REM sleep. Curr Biol. 2016;26:R34–5.

30. De Andres I, Garzon M, Reinoso-Suarez F. Functional anatomy of non-REM sleep. Front Neurol. 2011;2:70.
31. Nishimura H. Renin-angiotensin system in vertebrates: phylogenetic view of structure and function. Anat Sci Int. 2017;92:215–47.
32. Pandi-Perumal SR, Cardinali DP, Chrousos G, editors. Neuroimmunology of sleep. New York: Springer Science+Business Media, LLC; 2007.
33. Krueger JM. The role of cytokines in sleep regulation. Curr Pharm Des. 2008;14:3408–16.
34. Geiger SS, Fagundes CT, Siegel RM. Chrono-immunology: progress and challenges in understanding links between the circadian and immune systems. Immunology. 2015;146:349–58.
35. Karrer M, Lopez MA, Meier D, Mikhail C, Ogunshola OO, Muller AF, Strauss L, Tafti M, Fontana A. Cytokine-induced sleep: neurons respond to TNF with production of chemokines and increased expression of Homer1a in vitro. Brain Behav Immun. 2015;47:186–92.
36. Mader EC Jr, Mader AC. Sleep as spatiotemporal integration of biological processes that evolved to periodically reinforce neurodynamic and metabolic homeostasis: the 2m3d paradigm of sleep. J Neurol Sci. 2016;367:63–80.
37. Stickgold R. Sleep-dependent memory consolidation. Nature. 2005;437:1272–8.
38. Hobson JA. REM sleep and dreaming: towards a theory of protoconsciousness. Nat Rev Neurosci. 2009;10:803–13.
39. Pace-Schott EF, Hobson JA. The neurobiology of sleep: genetics, cellular physiology and subcortical networks. Nat Neurosci Rev. 2002;3:591–605.
40. Strata P. The emotional cerebellum. Cerebellum. 2015;14:570–7.
41. Goadsby PJ. Chapter 16 – Autonomic nervous system control of the cerebral circulation. In: Buijs RM, Swabb D, editors. Handbook of clinical neurology. Autonomic nervous system. Elsevier; 2013, p. 193–201.
42. Jessen NA, Munk AS, Lundgaard I, Nedergaard M. The glymphatic system: a beginner's guide. Neurochem Res. 2015;40:2583–99.
43. Venkat P, Chopp M, Chen J. New insights into coupling and uncoupling of cerebral blood flow and metabolism in the brain. Croat Med J. 2016;57:223–8.
44. Tang G, Yang GY. Aquaporin-4: a potential therapeutic target for cerebral edema. Int J Mol Sci. 2016;17.

First Level: Peripheral Sympathetic and Parasympathetic Nervous System

3

Abstract

The periphery of the autonomic nervous system (ANS) comprises two parts: the sympathetic and the parasympathetic systems. The identity and functional feature of the ANS innervating the gastrointestinal is often considered the third section of the ANS, i.e., the enteric ANS. This Chapter discusses the neurochemical bases of the peripheral motor and sensory components of the ANS, including the neurotransmitters and receptors involved and their participation in spinal autonomic reflexes. The composition and functions of the enteric nervous system are also discussed.

Keywords

Adrenergic neurotransmission • Cholinergic neurotransmission • Co-transmission • Enteric nervous system • Ionotropic transmission • Metabotropic transmission • Parasympathetic nervous system • Prevertebral and paravertebral sympathetic ganglia • Sensory autonomic neurons • Spinal autonomic reflexes • Sympathetic nervous system

Objectives

After studying this chapter, you should be able to:

- Underline why cholinergic and adrenergic neurotransmission and peptidergic co-transmission are the basis of the peripheral motor constituents of the autonomic nervous system (ANS).
- Describe the location of the cell bodies and axonal trajectories of preganglionic and postganglionic sympathetic and parasympathetic neurons.
- Name the neurotransmitters that are released by preganglionic autonomic neurons, postganglionic sympathetic neurons, postganglionic parasympathetic neurons, and adrenal medullary cells.
- Name the types of receptors on autonomic ganglia and on various target organs and the processes involved in neurotransmission within the ANS.

D.P. Cardinali, *Autonomic Nervous System*, DOI 10.1007/978-3-319-57571-1_3

- Define sensory autonomic neurons, their anatomy, and how the basic sensorial dimensions are coded.
- Describe the homologies and differences between spinal motor reflexes and spinal autonomic reflexes.
- Describe the anatomical and physiological basis of genitourinary reflexes.
- Describe the anatomical and physiological basis of defecation.
- Describe the anatomical and physiological basis of pupillary reflexes.
- Describe the composition and functions of the enteric nervous system.
- Understand the meaning of local autonomic projections in neuroendocrine communication.
- Define how the ANS contributes to the maintenance of healthy bone tissue.

The Organization of the ANS at a Peripheral Level Comprises Two Neurons in a Series

The periphery of the autonomic nervous system (ANS) comprises two parts: the sympathetic and the parasympathetic systems (Figs. 3.1 and 3.2). The identity and functional features of the ANS innervating the gastrointestinal is often considered the third section of the ANS, i.e., the enteric ANS. Conceptually, however, the regulatory mechanisms at the digestive level, although complex, do not escape the general characteristics of the sympathetic and parasympathetic systems discussed below [1].

Moreover, the ANS has also been conceptualized as having five components: the sympathetic noradrenergic system, the sympathetic cholinergic system, the parasympathetic cholinergic system, the sympathetic adrenergic system, and the enteric nervous system [2]. The reason for the differential noradrenergic vs adrenergic responses is that sympathetic noradrenergic system activation dominates the responses to orthostasis, moderate exercise, and exposure to cold, whereas sympathetic adrenergic system activation dominates responses to glucoprivation and emotional distress [2].

Effectors of the sympathetic system are generally the smooth muscle, the heart, exocrine and endocrine glands, the adipose tissue, liver, lymphohematopoietic organs, and kidney. Most sympathetic ganglia are remotely located to the innervated organ, and therefore postganglionic axons (i.e., the axons of ganglion neurons) are long (Fig. 3.3). An exception is the short adrenergic neurons located on the wall of the genitourinary organs, which have a similar pattern to the parasympathetic array [3, 4].

Prevertebral ganglia (celiac, superior mesenteric, inferior mesenteric) give rise to postganglionic fibers, which through plexuses or nerves innervate organs of the abdominal and pelvic regions. The sympathetic preganglionic neurons have their bodies in the intermediolateral column of the spinal cord, between segments T1 and

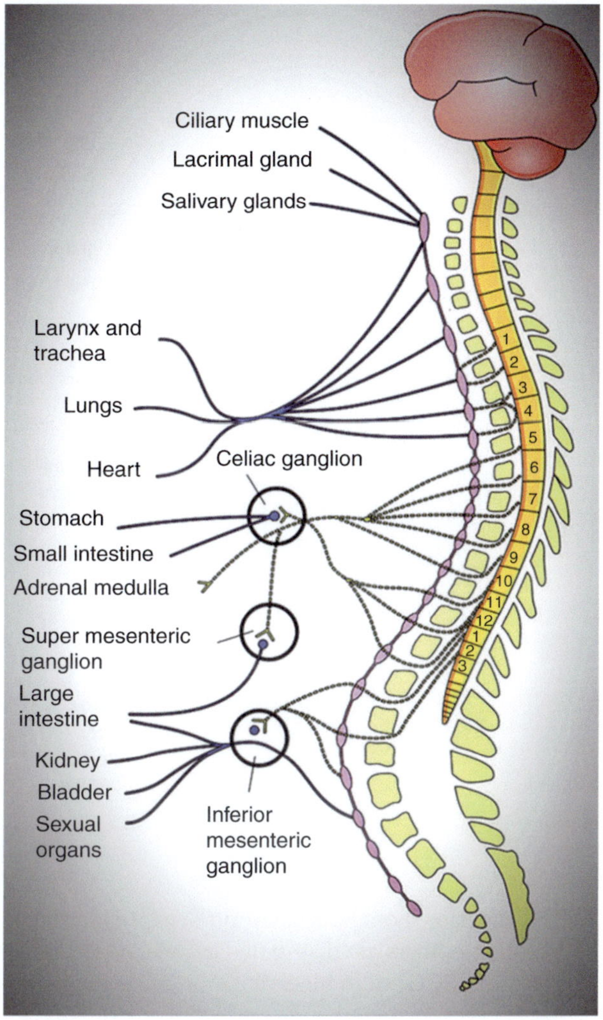

Fig. 3.1 The sympathetic branch of the autonomic nervous system. Modified with permission from Cardinali [1]

L2, so that the sympathetic system is also called thoracolumbar (Fig. 3.1). The preganglionic fibers exit through the corresponding anterior roots and follow the white communicating branches to the paravertebral sympathetic ganglion chain, from where four pathways can follow (Fig. 3.4) [3, 4]:

1. To synapse with ganglion neurons of the corresponding level or higher or lower levels. The postganglionic fibers of these ganglion neurons leave the ganglion through the gray communicating branch and join the somatic peripheral nerves to innervate blood vessels, exocrine or endocrine glands, immune tissue, and piloerector muscles in their territories. The postganglionic fibers present a variable segmental distribution, much more poorly delimited than that of the sensory and motor fibers.

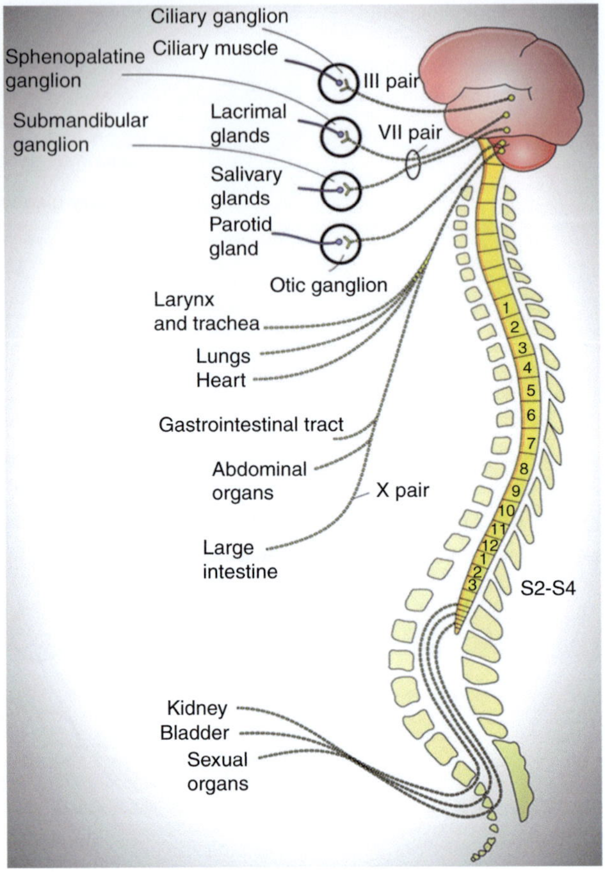

Fig. 3.2 The parasympathetic branch of the autonomic nervous system. Modified with permission from Cardinali [1]

2. The superior thoracic roots form synapses in the cervical ganglia, whose postganglionic fibers are distributed to innervate cranial structures (superior cervical ganglion, SCG) or to innervate thoracic organs via visceral, cardiac, and pulmonary nerves, by innervating these thoracic organs (middle cervical ganglion, stellate ganglion).
3. The preganglionic fibers that emerge from the spinal roots below the diaphragm pass through the paravertebral sympathetic chain and they give rise to the splanchnic nerves, to end up forming synapses in the prevertebral ganglia. From these, the postganglionic fibers form small visceral nerves that, in the form of plexuses, are distributed to the abdominal and pelvic viscera.
4. Finally, some preganglionic fibers also follow this path to form synapses directly with the chromaffin cells of the adrenal medulla, which represent the paravertebral sympathetic ganglia corresponding to the postganglionic cells.

These neurons form a bilateral chain parallel to the vertebral column, between the base of the skull and the coccyx. Normally, the anatomical distribution of each ganglionic chain consists of about 24 ganglia. In the cervical region, there are three

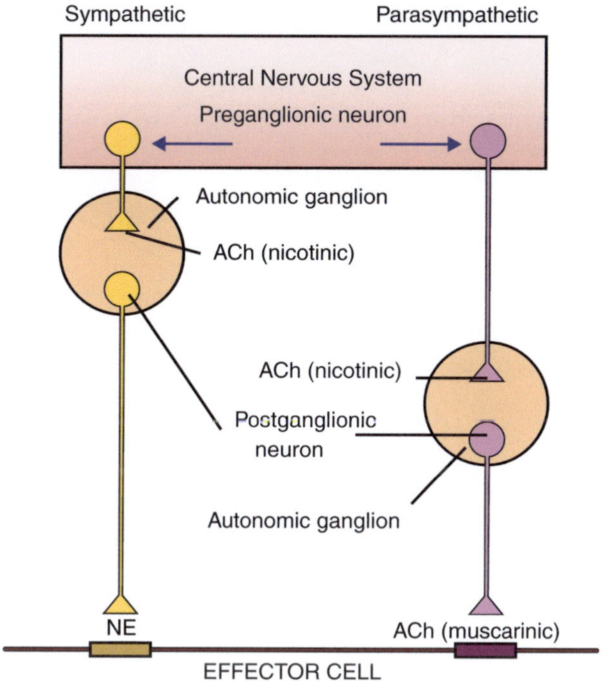

Fig. 3.3 Adrenergic and cholinergic neurotransmission in the autonomic nervous system. Modified with permission from Cardinali [1]

ganglia, superior, middle, and inferior, although the latter is often fused with the first thoracic ganglion forming the stellate ganglion. In the other regions, there are usually a pair of ganglia for each spinal metamere, 10 or 11 thoracic ganglia, 4 lumbar, 3 or 4 sacral, and a single coccygeal.

The sympathetic prevertebral ganglia are located around the branches of the abdominal aorta, forming the abdominal or solar plexus. The main prevertebral ganglia are celiac, superior mesenteric, inferior mesenteric, and aortorenal.

The preganglionic neurons of the parasympathetic system have their perikarya in nuclei of the brainstem or in the lateral column of the sacral spinal cord, which is why it receives the designation of craniosacral division (Fig. 3.2).

In the cranial portion the preganglionic neurons are in the visceral efferent nuclei, from where they distribute their fibers peripherally via the cranial nerves [5]:

- Edinger–Westphal nucleus: common ocular motor nerve (III).
- Superior and lacrimal salivary nucleus: facial nerve (VII).
- Lower salivary nucleus: glossopharyngeal nerve (IX).
- Dorsal nucleus of the vagus and ambiguous nucleus: vagus nerve (X).

These fibers are distributed to terminal ganglia, close to the effector organs, from which short postganglionic fibers are formed, giving rise to the innervation of the viscera of the head and neck, the thoracic cavity, and much of the abdominal cavity.

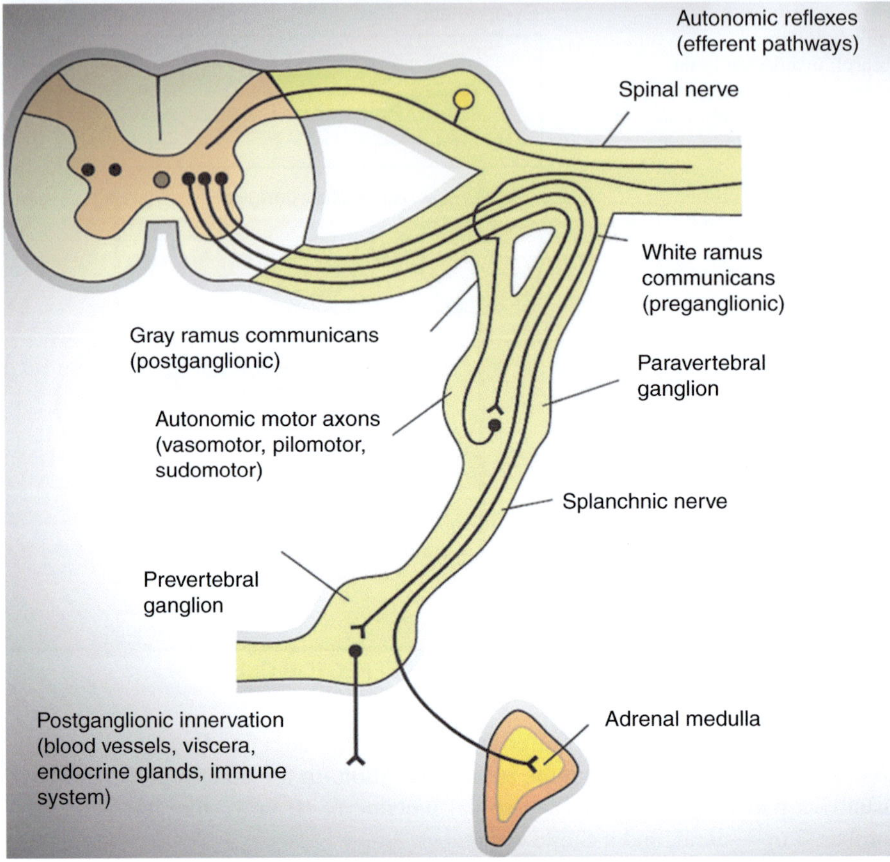

Fig. 3.4 Neurons of the autonomic reflex, integrated at the level of the intermediolateral column of the spinal cord. The visceral afferents enter through the spinal nerve. The sympathetic and prevertebral ganglia and the gray and white communicating branches are also shown. Modified with permission from Cardinali [1]

On the other hand, the preganglionic neurons of the sacral portion of the spinal cord, located in the intermediolateral columns between the S2 and S4 segments, send their fibers through the anterior roots to form the pelvic nerves and end in the terminal ganglia of the pelvic plexus, innervating the descending colon and the urogenital organs [6].

The parasympathetic terminal ganglia are formed by small clusters of neurons, located on or in the walls of the viscera, where the preganglionic fibers form synapses. In the cephalic region, there are four relatively large ganglia, associated with branches of the trigeminal nerve: the ciliary, sphenopalatine, optic, and submandibular ganglia.

In the cervical, thoracic, and pelvic regions, there are small ganglia that form visceral plexuses, whereas in the digestive tract they are in the myenteric plexus of Auerbach and the submucosal plexus of Meissner.

The cell bodies of the parasympathetic preganglionic neurons are located in the brainstem and sacral cord. Although some of their axons are myelinated, most are unmyelinated. Compared with the sympathetic system, the preganglionic axons are longer and their postganglionic fibers considerably shorter. This is because the parasympathetic ganglia are located in the immediate vicinity of the innervated organs [1].

Effectors of the parasympathetic system include smooth muscle and glands of the digestive tract, the excretory organs, the genital system, heart, and lung, lymphohematopoietic and endocrine organs and intraocular muscles. Except genital arteries and possibly the brain, there is no parasympathetic innervation of vascular smooth muscle. Nor is there a parasympathetic innervation of the skin [1]. By using different experimental models of autonomic hyper- or hypofunction, the conclusions listed in Table 3.1 can be reached [7].

Cholinergic and Adrenergic Neurotransmission and Peptidergic Co-transmission Are the Basis of the Peripheral Motor Constituents of the ANS

Transmitters identified, partially or totally, in neural pathways comprise five large families [1]:

- Biogenic amines: norepinehrine (NE), epinephrine (E), acetylcholine (ACh), 5-HT, dopamine (DA), His
- Amino acids: Glu, aspartate, GABA, Glycine (Gly), taurine and purine derivatives (adenosine, ATP, included in this group although they are not amino acids).
- Neuropeptides (over 70 different structures identified to date)
- Gases, such as NO or CO (carbon monoxide)
- Lipids, such as anandamide.

The criteria for a substance to be considered a neurotransmitter are as follows (Fig. 3.5):

- It must be synthesized by the presynaptic neuron and stored in synaptic vesicles (although exceptions are found, like the gases and lipids)
- The physiological neural stimulation must release it
- It must act on the postsynapsis in a similar manner to normal stimulus analyzed (action identity criterion)
- There should be effective mechanisms for terminating its action (reuptake in neural terminal, metabolism, passage to extrasynaptic space), to ensure the necessary speed and transience of transmitter action

Among them, the action identity criterion is the most important, as it involves the physiological effect itself [1].

Listed criteria have been completely fulfilled in the peripheral ANS, for the neurotransmitter nature of ACh or NE (various preganglionic and postganglionic

Table 3.1 Autonomic nervous system effects on effector organs

Organ, tissue	Parasympathetic stimulation	Sympathetic stimulation
Pupils		
Radial muscle of the iris	–	Contraction, α1 adrenoceptor mediated
Sphincter muscle of the iris	Contraction, M3, M2-receptor mediated	–
Ciliary muscle	Contraction, M3, M2-receptor mediated	Weak relaxation, β2 receptor mediated
Tracheobronchial muscle	Contraction, M2, M3-receptor mediated	Relaxation, β2 adrenoceptor mediated
Heart	Bradycardia, negative inotropism, M2 receptor mediated	Tachycardia, positive inotropism, β1 adrenoceptor mediated
Arteries		
Skin, mucosas	–	Constriction, α1 adrenoceptor mediated
Abdomen	–	Constriction, α1 adrenoceptor mediated Dilatation (blood epinephrine), β2 adrenoceptor mediated
Skeletal muscle	–	Dilation, M2 receptor mediated Dilatation (blood epinephrine), β2 adrenoceptor mediated
Coronary arteries	–	Constriction, α1 adrenoceptor mediated Dilation, β2 adrenoceptor mediated
CNS arteries	Dilation (?)	Constriction, α1 adrenoceptor mediated
Penis	Dilation, M2 receptor mediated	–
Kidney	–	Constriction, α1 adrenoceptor mediated
Veins	–	Constriction, α1 adrenoceptor mediated
Endothelium	Stimulation of NO synthase, M3 receptor mediated	
Bladder		
Detrusor muscle	Contraction, M3 receptor mediated	Relaxation, β2 adrenoceptor mediated
Internal sphincter	Relaxation, M2 receptor mediated	Constriction, α1 adrenoceptor mediated
Genital tract		
Seminal vesicles	–	Constriction, α1 adrenoceptor mediated
Vas deferens	–	Constriction, α1 adrenoceptor mediated
Intercourse	Erection, M3 receptor mediated	Ejaculation, α-1 adrenoceptor mediated
Uterus	–	Constriction, α1 adrenoceptor mediated Relaxation, β2 adrenoceptor mediated
Gastrointestinal tract		
Circular and longitudinal muscle	Contraction, M3 receptor mediated	Relaxation (direct) β adrenoceptor mediated Parasympathetic inhibition, α1 adrenoceptor mediated
Sphincters	Relaxation	Constriction, α adrenoceptor mediated
Exocrine glands		
Salivary	Secretion. M3 receptor mediated	

Table 3.1 (continued)

Organ, tissue	Parasympathetic stimulation	Sympathetic stimulation
Lacrimal	Secretion. M3 receptor mediated	–
Digestive	Secretion. M3 receptor mediated	Inhibition, α adrenoceptor mediated
Nasopharyngeal	Secretion. M3 receptor mediated	–
Sweat	–	Secretion. M3 receptor mediated
Bronchial	Secretion. M3 receptor mediated	Weak inhibition, α-1 adrenoceptor mediated. Weak stimulation, β adrenoceptor mediated
Adipose tissue	–	Lipolysis, β1 adrenoceptor mediated
Immune system		
Lymph nodes, spleen	Stimulation	Inhibition, α-1 adrenoceptor mediated
Splenic capsule	–	Constriction, α-1 adrenoceptor mediated
Endocrine system		
Juxtaglomerular apparatus	–	Renin secretion, β1 adrenoceptor mediated
Thyroid follicles	Thyroid growth, secretion of T4. M receptor mediated	Inhibition of T4 secretion, α1 mediated. Weak stimulus of T4 secretion, β adrenoceptor mediated
Thyroid C cells	Inhibition of CT secretion	Inhibition of CT secretion, α1 mediated. Weak stimulus of CT secretion, β adrenoceptor mediated
Parathyroid glands	Inhibition of PTH secretion	Inhibition of PTH secretion
Pineal gland	–	Melatonin release, β2 adrenoceptor mediated. Weak stimulus of melatonin release, α1 adrenoceptor mediated
Adrenal medulla		Preganglionic stimulation, M1 receptor mediated
Anterior pituitary		Postganglionic SCG sympathetic neurons inhibit FSH, LH, GH, PRL, and TSH secretion and augments ACTH secretion
Posterior hypophysis	–	Postganglionic SCG sympathetic neurons inhibit ADH secretion, β adrenoceptor mediated
Exocrine pancreatic glands	Secretion. M1 receptor mediated	Inhibition, α adrenoceptor mediated
Pancreatic islets, α cells	Secretion. M1 receptor mediated	Inhibition of glucagon secretion, α2 adrenoceptor mediated. Weak stimulation of glucagon secretion, β2 adrenoceptor mediated
Pancreatic islets, β cells	Secretion. M1 receptor mediated	Inhibition of insulin secretion, α-2 adrenoceptor mediated. Weak stimulation of insulin secretion, β2 adrenoceptor mediated
Liver	Glycogen synthesis (?)	Glycogenolysis, α1 adrenoceptor mediated Gluconeogenesis, β2 adrenoceptor mediated Lipolysis, β1 adrenoceptor mediated

NO nitric oxide, *CT* calcitonin, *PTH* parathyroid hormone, *SCG* superior cervical ganglion, *FSH* follicle-stimulating hormone, *LH* luteinizing hormone, *GH* growth hormone, *PRL* prolactin, *TSH* thyroid-stimulating hormone, *ACTH* adrenocorticotropic hormone, *ADH* antidiuretic hormone

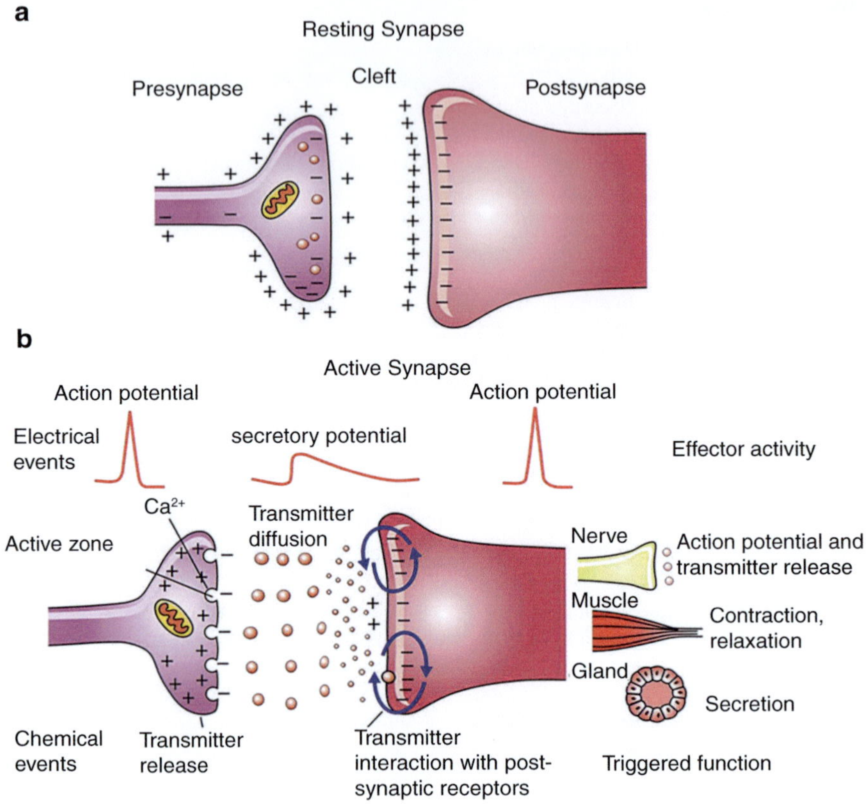

Fig. 3.5 Chemical synapse. (**a**) In the resting synapse the pre- and postsynaptic membranes are normally polarized. (**b**) In the active synapse, depolarization of the neural terminal (secretory potential) results in the release of the transmitter, which diffuses through the synaptic gap and produces local synaptic currents and potentials in the postsynaptic membrane, which initiate effector activity, neuronal transmission, neurotransmitter release, hormonal secretion, muscle contraction). Modified with permission from Cardinali [1]

territories). On the other hand, in central ANS neurons, the physiological characterization of neurotransmitters has been difficult, owing to the large number of synapses present, to the neuronal plurality of most brain regions, and to the very common coexistence of two or more neurotransmitters in the same synapse.

Biogenic amines participate in approximately 5% of the brain synapses, localizing in certain subcortical projection pathways to encephalic or descending spinal cord regions. The noradrenergic and serotonergic cortical, cerebellar, and subcortical innervation originates, almost exclusively, in brainstem nuclei that project in the form of a "spider web" to large cerebral areas, an anatomical disposition that speaks of its general modulatory function (Chap. 1). Something similar occurs for the central dopaminergic, cholinergic, and histaminergic systems. It is not surprising, then, that these monoaminergic systems have been linked to generalized alterations in brain function, such as psychiatric illness, emotionality, wakefulness, and sleep (Chap. 4).

In the ANS, ACh is the transmitter of all the preganglionic synapses, of all parasympathetic postganglionic neurons, and of some sympathetic postganglionic neurons (muscular vasodilator system, sweat glands). NE is the neurotransmitter of all the remaining postganglionic sympathetic neurons [3, 4].

For most chemical synapses, there is an exocytotic release from the presynapsis of the neurotransmitter substance contained in the synaptic vesicles. Exceptions to this rule are gases identified as neurotransmitters (NO, CO), which pass through membranes by simple diffusion, and neurotransmitter lipids such as anandamide, which are not stored in the vesicles.

In the case of NO, which is one of the most important local regulators of BP, three different NO synthase (NOS) isoforms have been described, two constitutively present (eNOS and nNOS), and one inducible (iNOS). In the CNS, both eNOS and nNOS are present. Centrally, NO functions mainly as a neuromodulator, having both sympathoinhibitory and sympathoexcitatory actions [8]. At the periphery, locally released NO from endothelial cells acts on adjacent vascular smooth muscle cells to produce vasodilatation. In addition, NO synthase activity has been demonstrated in preganglionic autonomic fibers innervating vascular smooth muscle. Furthermore, NO release from autonomic nitrergic nerves interferes with the release of NE. Both central and peripheral effects of NO under normal conditions are masked by the baroreflex [9] (Chap. 4).

In general, the action of these signals is exerted at the level of specific receptors in the postsynapsis. The gases are again the exception as they traverse the postsynaptic membrane and exert their action intracellularly. In the chemical synapse the synaptic message is unidirectional (it ranges from pre- to postsynapsis) and when it comes from synapses by exocytotic release of neurotransmitters present in vesicles, it shows a synaptic delay. This synaptic delay is largely due to the transmitter release process, and to a minimal extent by the passage of the transmitter through the synaptic gap. Its duration is about 0.5–1.0 ms [10].

In the ANS, the synaptic cleft is broad (synaptic varicosities or nondirected synapses) and thus differs from directed synapses, such as the neuromuscular plaque (Fig. 3.6). This allows the "transmission by volume" into the ANS, a name that is given to the wide diffusion of the autonomic transmitter to several postsynaptic cells.

Autonomic postganglionic fibers present a series of synaptic dilations or varicosities when they approach their targets, which have a length of approximately 1 μm and a diameter of between 0.5 and 2 μm. These varicosities contain a high density of mitochondria and synaptic vesicles. The synaptic space has a variable width; for example, in the innervation of the smooth muscle, it varies from 20 nm in the vas deferens to 1–2 μm in the large arteries. Transmitter release occurs in a passage, and propagates nerve impulses along the autonomous axon. The availability of varicosities and the wide synaptic space allow the released neurotransmitter to diffuse variable distances in the target organ and to activate multiple receptors, which expand the effect of autonomic activity (Fig. 3.6) [3, 4, 11].

This classic view of autonomic transmission is, however, a simplification, as it is now known that many of these neurons also use other types of molecules as neurotransmitters or as neuromodulators. Initially, these were encompassed under the

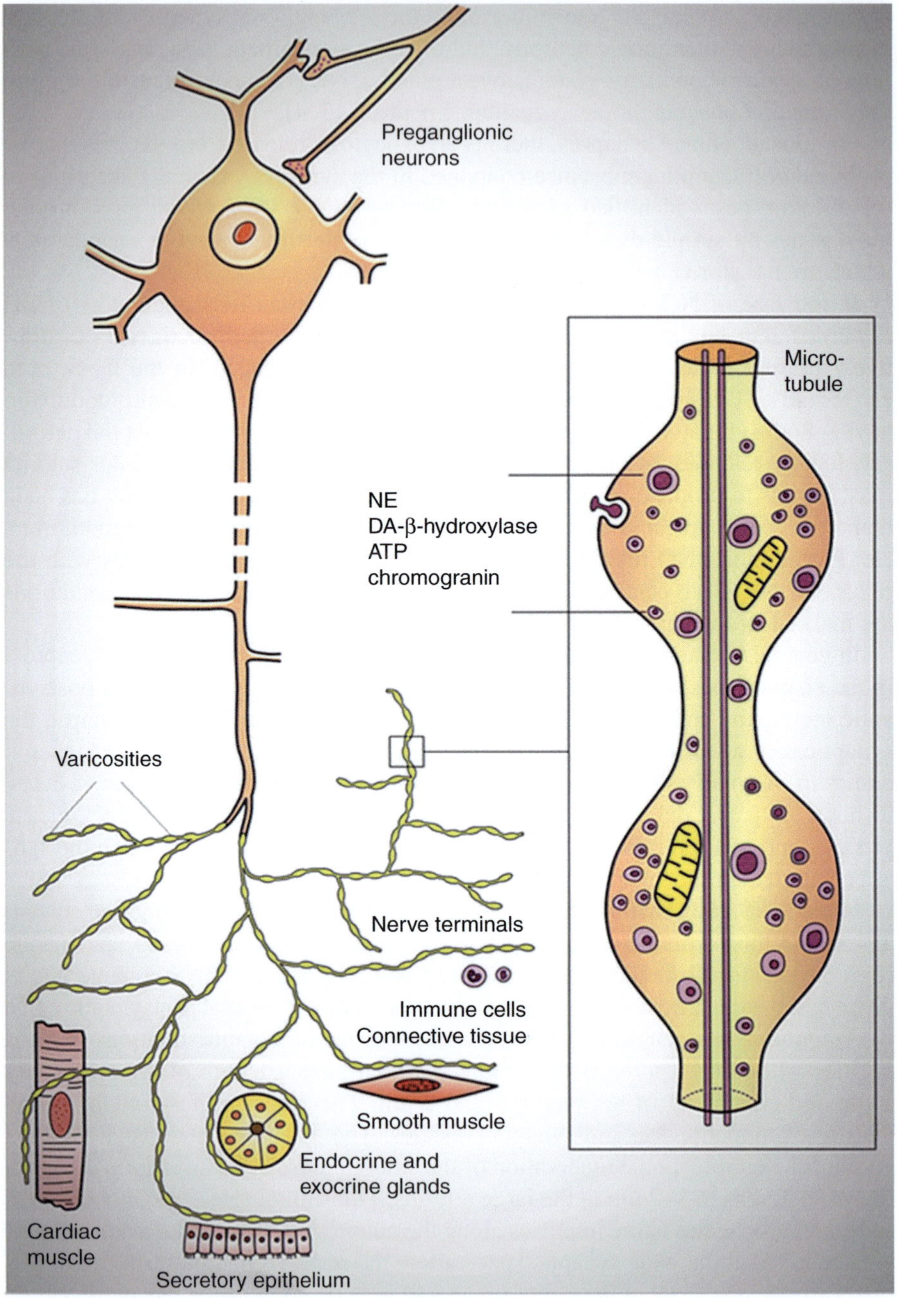

Fig. 3.6 Postganglionic noradrenergic neuron with its distribution in varicosities ("beads"). Modified with permission from Cardinali [1]

denomination of non-adrenergic–noncholinergic transmission, which mainly includes purinergic transmitters and neuropeptides. Purinergic nucleotides, such as ATP, are important transmitters in nerve fibers that innervate the intestinal smooth muscle, the urinary bladder, and the vas deferens, where they act directly as neurotransmitters or modulate the effects of NE or ACh. Several neuropeptides have been identified, widely distributed, in autonomic ganglia, the enteric system, and peripheral fibers. The most well-known include: enkephalin/endorphin, vasoactive intestinal peptide (VIP), substance P, calcitonin gene-related peptide (CGRP), neuropeptide-Y (NPY), somatostatin, bombesin, galanin, neurotensin, angiotensin, and cholecystokinin (CCK)/gastrin.

The coexistence of neuropeptides with classical transmitters in different autonomic neurons is the rule, for example, sympathetic and parasympathetic postganglionic neurons containing ACh and VIP, or adrenergic neurons containing NE and NPY. The classical neurotransmitter and the neuropeptide may be separately released under different excitation conditions (Fig. 3.7). It is postulated that neuropeptides might act as transmitters by themselves or as neuromodulators, altering the action of classical transmitters. Thus, stimulation of cholinergic fibers from the salivary or sweat glands causes release of ACh, which has a secretory effect, and VIP, which produces vasodilation [3].

The neuropeptides co-released with small molecule neurotransmitters in autonomic nerves do not usually act as co-transmitters, but rather as prejunctional neuromodulators or trophic factors. Autonomic co-transmission offers subtle, local variation in physiological control mechanisms, rather than the dominance of inflexible central control mechanisms envisaged earlier [12]. The variety of information imparted by a single neuron then greatly increases the sophistication and

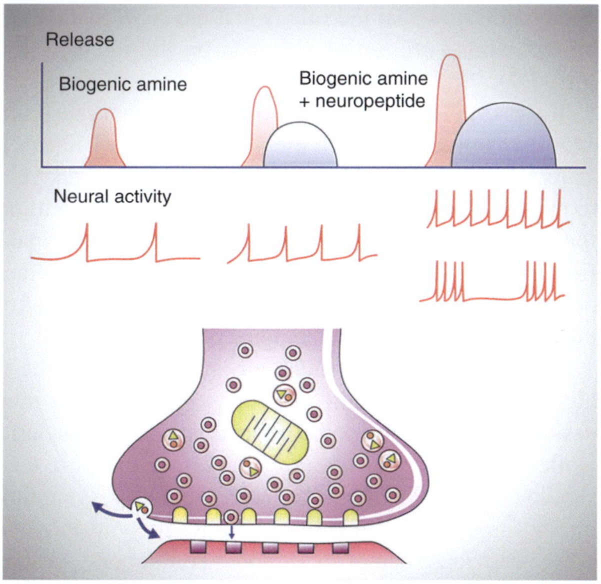

Fig. 3.7 Differential release of transmitters contained in small dense-cored vesicles (biogenic amine) and large dense-cored vesicles (neuropeptide). Modified with permission from Cardinali [1]

Table 3.2 Neurochemical code in different types of autonomic nerve fibers

Nerve fibers	Neurotransmitters
Vasoconstriction	NE, NPY
Vasodilation	ACh, VIP
Sudomotor	ACh, VIP, CGRP
Piloerector	NE, dynorphin
Sialomotor	CGRP, NPY
Intestinal ganglia	NE, somatostatin

NE norepinephrine, *NPY* neuropeptide-Y, *ACh* acetylcholine, *VIP* vasoactive intestinal peptide, *CGRP* calcitonin gene-related peptide

complexity of local control mechanisms. Co-transmitter composition shows considerable plasticity in development and aging, in pathophysiological conditions, and following trauma or surgery. For example, ATP appears to become a more prominent co-transmitter in inflammatory and stress conditions [12]. In situations such as congestive cardiac failure that are characterized by high levels of cardiac sympathetic drive, sympathetic co-transmitters such as NPY can be released in addition to NE. Even in the presence of β-adrenoceptor blockers, NPY is able to bind to its own receptors located on the cholinergic ganglia and ventricular myocytes, thus inhibiting ACh release during vagus nerve stimulation and limiting the subsequent bradycardia [13].

Most experimental evidence has shown that the autonomic sympathetic and parasympathetic control is organized in different functional units or groups of neurons that innervate specific target organs. Thus, in the sympathetic ganglia, subtypes of vasomotor, sudomotor, pilomotor, and visceromotor neurons are distinguished. Ganglion neurons of each subtype present a unique chemical code (Table 3.2). Today, we know that the rule is that the ANS releases a combination of neurotransmitters whose proportion depends on the intensity and frequency of stimulation (Fig. 3.7).

It is useful to pay attention to a differential aspect among the different families of neurotransmitters, namely, the mechanism of synthesis [1]. Biogenic amines, amino acids, gases, and lipids are synthesized by an enzymatic process at the synaptic terminals. The specific enzymes migrate to the terminal via axoplasmic transport, forming parts of vesicles, and catalyze at the terminal the synthesis of transmitter from specific precursors. Because of its catalytic nature, an enzyme molecule participates in the synthesis of thousands of molecules of the transmitter. This prevents the rapid reduction of the contents of the transmitter.

In contrast, neuropeptides are synthesized in the neuronal body as part of a higher molecular weight prepropeptide, which is incorporated into the synaptic vesicles and is processed (by acetylation, glycosylation, or hydrolysis reactions) while these vesicles migrate by axonal transport toward the neural terminal. It should be noted that both the axon and the presynaptic terminals are almost devoid of ribosomes, and hence of the ability to synthesize peptides or proteins (there is only a restricted synthesis of some components of the cytoskeleton). In this, they differ from the dendrites [3].

It is for this reason that after a prolonged neuronal stimulation there is a greater possibility of neuropeptide transmitter depletion than that of any co-transmitter (Fig. 3.7). The peptide transmitter deposits are depleted at terminals remote from the neuronal body, the terminals at close range being maintained for longer. This is the reason why, on certain occasions, terminals of the same neuron can release different combinations of transmitting substances [12]. Likewise, the coexistence of neurotransmitters does not mean that they are released with the same kinetics before presynaptic stimulation. As shown in Fig. 3.7, small vesicles with a dense center with a biogenic amine (e.g., NE) are released before the large vesicles (e.g., NPY).

Generally, the opening or closing of channels at the postsynaptic membrane by neurotransmitter in the autonomic synapses (Fig. 3.8) is produced by: (a) the direct association of neurotransmitter with the postsynaptic receptor coupled to a channel (ionotropic transmission); (b) by synthesizing intracellular second messengers, triggered by the association of the autonomic transmitter with its receptor, this second messenger being responsible for the modification of membrane conductance (metabotropic transmission) [7].

An example of ionotropic transmission is the transmission given by the nicotinic cholinergic receptor of the autonomic preganglionic synapse (Fig. 3.9). The receptor forms a constituent part of an ion channel that permeates Na^+ and to a lesser extent K^+, and that opens when ACh binds to the receptor.

An example of metabotropic transmission is the β action of NE, which occurs by increasing cAMP and the subsequent phosphorylation of K^+ channels to inactivate them. The result of this action is the postsynaptic depolarization that follows the increases in the concentration of intracellular K^+ (Fig. 3.10).

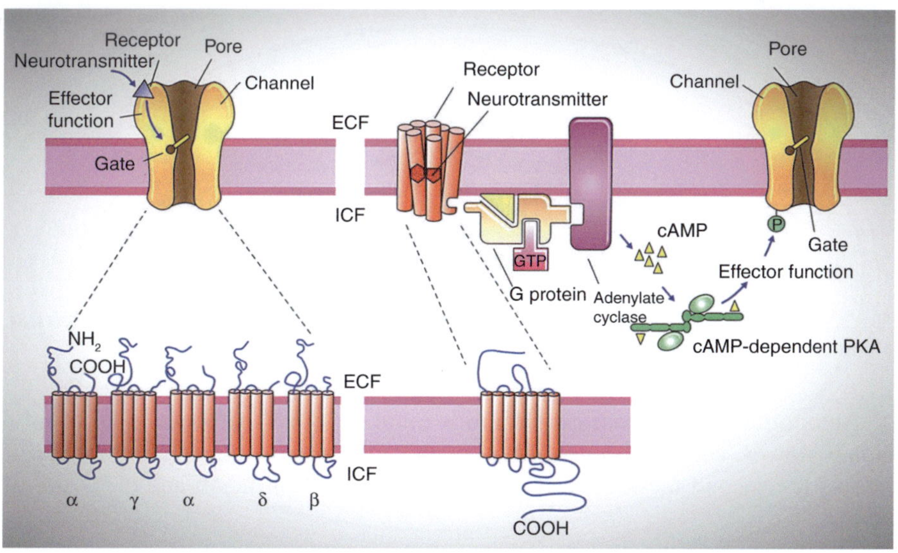

Fig. 3.8 Ionotropic and metabotropic effect of neurotransmitters. Modified with permission from Cardinali [1]

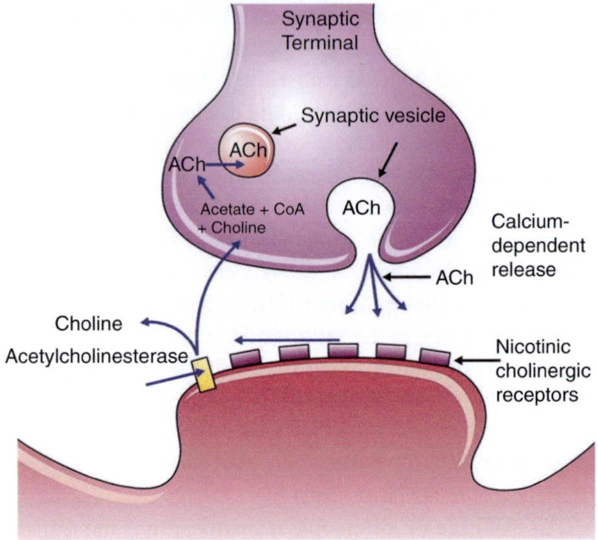

Fig. 3.9 Cholinergic nicotinic neurotransmission. Modified with permission from Cardinali [1]

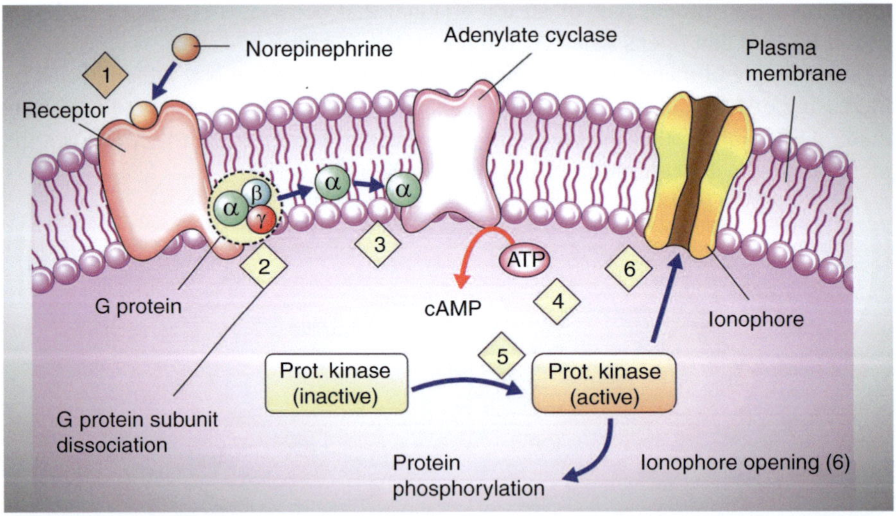

Fig. 3.10 Sequence of events produced by the interaction of NE with β-adrenoceptors. See text. Modified with permission from Cardinali [1]

Autonomic ionotropic transmission is rapid, whereas metabotropic transmission is slower. This fact is due to the time of synthesis and intracellular translocation of the second messenger synthesized from the cell membrane to its site of action.

The molecular biological studies on various neurotransmitter receptors indicate the existence of two structural superfamilies, which correspond to the two types of transmission (ionotropic and metabotropic) analyzed above [7]:

- Receptors associated with ionophores, whose quaternary structure is part of an ion channel. This is the case of the nicotinic cholinergic receptor, which comprises the following subtypes: muscle, ganglion, neuronal, neuronal CNS, neuronal α7 (more recently they have been molecularly defined as α1*-, α2*-, α3*-, α4*-, α6*-, α7*-, α8*-, and α9*-acceptors, nACh). This is also the case for the GABA receptor types A and C, the glycinergic receptor, the AMPA, kainate and *N*-methyl-D-aspartate (NMDA) receptors for Glu, the serotonergic receptor 5-HT3 subtype, and the ionotropic purinergic receptors. The structure consists of highly homologous protein sequence subunits that derive from a common gene. Each of the subunits contains 15–20 amino acid hydrophobic regions that constitute alpha-helices, which completely cross the plasma membrane.
- Receptors associated with G-proteins: the receptor structure constituting a single subunit with seven hydrophobic protein portions spanning the cell membrane. Each of, the α-helices has 15–20 amino acids, which are not arranged in the form of an ionic channel as in the previous case. The ion channel is located distant to the receptor and is not part of it; it can instead be used by several receptors (Fig. 3.10).
- This is a very large family of receptors and comprises, among others, α1 adrenergic receptors (whose subtypes are identified as α1A, α1B, and α1D), α2 (subtypes α2A, α2B, α2C), β (subtypes β1, β2, β3, and β4); muscarinic cholinergic receptors (subtypes: M1, M2, M3, M4, and M5); dopaminergic receptors (subtypes D1, D2, D3, D4, and D5); serotonergic (5-HT1A, 5-HT1A, 5-HT2A, 5-HT2B, 5-HT2C, 5-HT4A, 5-HT4A, 5-HT5A, 5-HT6, 5-HT7) subtypes; histaminergic (subtypes H1, H2, H3, H4); metabotropic Glu receptors (mGlu1 to mGlu8 subtypes), GABA type B receptor, melatonin receptor (MT1 MT2 subtypes), conopsins, and receptors of different types of neuropeptides.

There is intense “cross-talking” between the second messenger pathways, as in many cases individual enzymes, channels or cytoskeletal proteins can be modified at more than one site of the molecule by different systems of second intracellular messengers. It is estimated that the metabotropic pathway is about 10,000 times slower than the ionotropic pathway [11].

The activation of G proteins does not always lead to the synthesis of a second messenger mediating the effect. Effects of the G proteins are directly exerted on channels, without the participation of the phosphorylation. For example, the hyperpolarization produced by ACh in the heart is caused, in a first-stage effect, by a G protein that opens K^+ channels without the mediation of a second messenger. There are other similar actions described on Ca^{2+} channels. Finally, there are situations in which the channel is quickly affected by the G protein, and much later modified by the second messenger triggered by the same G protein (this is the case of the muscarinic M2 receptor type in the heart) [7].

In addition to the early and rapid effects of autonomic neurotransmitters, there are other late effects on neuronal function. In general, the action of the transmitter triggers modifications in gene expression, with lasting changes in cellular function until long after the synaptic action is over. One of the first known late effects was the trans-synaptic induction of tyrosine hydroxylase, the limiting enzyme in NE synthesis, in postganglionic sympathetic neurons because of nicotinic presynaptic ganglionic activation.

Regarding ACh synthesis, it comprises the acetylation of choline catalyzed by the enzyme choline acetyltransferase. ACh metabolism is accomplished in the synaptic gap and involves the reversal of the synthesis reaction, i.e., a hydrolysis catalyzed by the enzyme acetylcholinesterase. In muscarinic cholinergic synapses this hydrolysis is rapid enough to ensure complete inhibition of the synaptic effect of ACh, a fact that does not occur in nicotinic synapses, such as the preganglionic one.

There are two types of receptors for ACh: muscarinic and nicotinic.

Muscarinic cholinergic receptors mediate the effects of ACh in postganglionic synapses on smooth muscle, endocrine or exocrine glands, immune cells, and the heart. They also mediate some of the effects of ACh in the CNS and autonomic ganglia. Five types of muscarinic receptors (associated with G protein) are known to exist [11]:

- M1: associated with a decrease in K^+ conductance (and therefore excitatory), present in the cerebral cortex and autonomic ganglia.
- M2: associated with increased K^+ conductance (and therefore it is inhibitory), present in the heart, and presynaptically on several parasympathetic territories.
- M3 and M5: similar to M1, they are present in smooth muscle and glandular cells.
- M4: similar to M2.

As in the case of other G protein-associated receptors, the action of cholinergic agonists can be exerted through two mechanisms, one quick, direct, via the protein G effect on the channel, and a slower one through the second intracellular messenger. M1, M3, and M5 receptors mediate cholinergic responses accompanying postsynaptic stimulation such as smooth muscle contraction of the bronchi, bowel, or bladder. M2 and M4 receptors mediate inhibitory responses, bradycardia, and negative inotropism verifiable in the heart after vagal stimulation or inhibition of presynaptic transmitter release.

Nicotinic cholinergic receptors are in the autonomic ganglia, CNS, and muscle plate. In all cases, they are associated with ionotropic responses involving the opening of Na + channels (which, although to a lesser extent, also permeate K^+), and consequently depolarization (Fig. 3.9). These channels are composed by pentamers resulting from combinations of 17 subunits. Nicotinic receptors are broadly classified into two subtypes based on their primary sites of expression: muscle-type nicotinic receptors and neuronal-type nicotinic receptors. In both muscle-type and neuronal-type receptors, the subunits are somewhat similar, especially in the hydrophobic regions. In the autonomic ganglia, hexamethonium is the specific nicotinic blocker, whereas in the muscle plate, curare (or its active ingredient, tubocurarine) is the blocker [7, 11].

In relation to NE, their synthesis is started by tyrosine hydroxylation to L-dopa (L-dihydroxyphenylalanine) catalyzed by the enzyme tyrosine hydroxylase (Fig. 3.11). This is followed by the decarboxylation of levodopa to DA and by

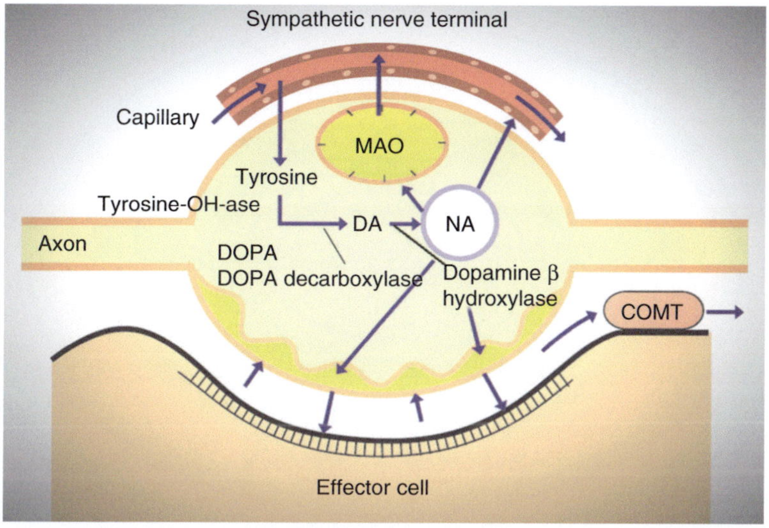

Fig. 3.11 Norepinephrine (NE) synthesis. Tyrosine is converted to L-DOPA by tyrosine hydroxylase. DOPA is then decarboxylated to DA by L-aromatic amino acid decarboxylase. Finally, DA is β-hydroxylated to NE by DA β-hydroxylase. NE is metabolized by either monoamine oxidase (MAO) or catechol-*O*-methyltransferase (COMT)

β hydroxylation of DA to NE. The limiting step in this sequence is the first one, i.e., the hydroxylation of tyrosine (Fig. 3.11).

Catabolism of NE is carried out by oxidative deamination by the action of monoamine oxidase enzyme (MAO), or by O methylation enzyme by catechol-*O*-methyl transferase (COMT; Fig. 3.12). However, under normal conditions, the most widespread form of termination of the action of NE is the presynaptic reuptake in an intact, unmetabolized form. The time at which this reuptake mechanism becomes saturated, metabolism by MAO or COMT ensues.

There are two main types of adrenergic receptors (or adrenoceptors): (a) α-adrenoceptors; (b) β-adrenoceptors.

As in the case of ACh, these receptor types were identified using adrenergic agonists and antagonists, and more recently, various subtypes have been cloned and identified. They belong to the superfamily of receptors associated with G proteins.

The α-adrenergic effect is one that shows the following sequence of activity for agonists: NE = E > > isoproterenol (a synthetic adrenergic agonist).

The β-adrenergic effect is one that shows the following sequence of activity for agonists: isoproterenol > E = NE.

In turn, each type of adrenoceptor is subdivided into α1 (whose subtypes identified are α1A, α1B, and α1D), α2 (subtypes α2A, α2B, and α2C), and β1, β2, β3, and β4 [7, 11].

The subcellular mechanism of action of the α1 adrenoceptor is the increase in conductance to Ca^{2+} and the activation of the turnover of inositol phospholipids. The subcellular mechanism of α2 adrenoceptor is the inhibition of adenylate cyclase.

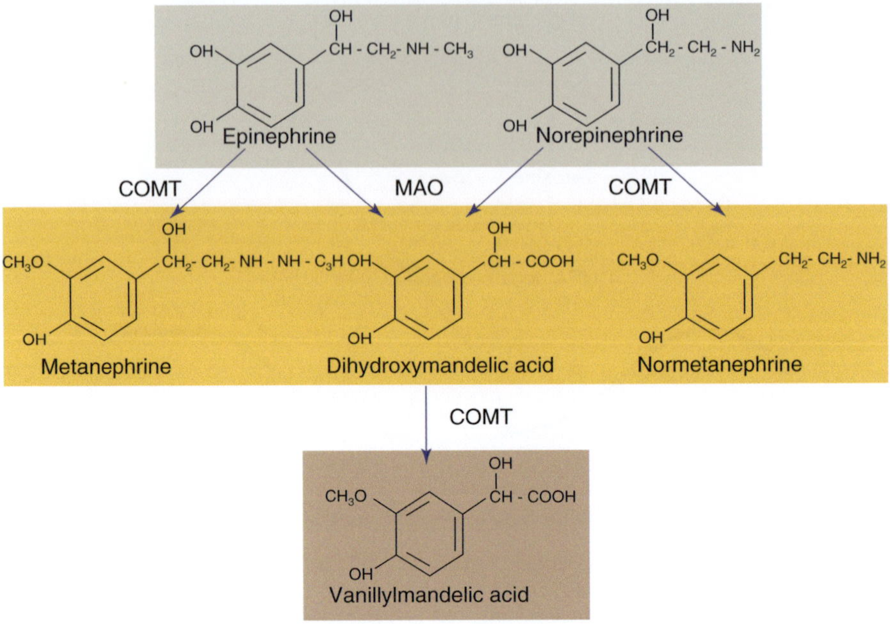

Fig. 3.12 Metabolism of NE by monoamine oxidase (MAO) and catechol-*O*-methyltransferase (COMT). Modified with permission from Cardinali [1]

The subcellular mechanism of action of β1, β2, β3, and β4 receptors is the activation of adenylate cyclase; α2 adrenergic receptors were initially described as having a presynaptic location and with negative modulation function of NE release. However, in the CNS, there are also α2 adrenoceptors at a postsynaptic location. The β1, β2, β3, and β4 adrenergic receptors have both pre- and postsynaptic localization.

The postsynaptic consequence of the mentioned autonomic ionotropic or metabotropic mechanisms is:

- Depolarization, which can be reached by a spatial or temporal sum to trigger an action potential. The depolarizing synaptic potential is called the excitatory post-synaptic potential, or EPSP (Fig. 1.8).
- Hyperpolarization with decreased postsynaptic excitability. The hyperpolarizing synaptic potential is called the inhibitory postsynaptic potential, or IPSP (Fig. 1.8).

The arrival of the action potential to the synaptic terminal produces its depolarization ("secretory potential"). Ca^{2+} voltage channels opened by depolarization are instrumental for a sharp rise in cytoplasmic Ca^{2+} concentration that causes membrane fusion of synaptic vesicles with the cell membrane, the opening of synaptic vesicles, and the exocytotic emptying of its contents into the synaptic cleft (Fig. 3.5). As the emptying of each vesicle is total, the amount of autonomic transmitter released in this manner is always a multiple of the unit concentration present in each vesicle. This is called the quantum release of transmitter.

In view of the importance of Ca^{2+} in neurotransmission, it is not surprising that the regulation of autonomic transmitter release is mainly performed at the level of voltage-dependent Ca^{2+} channels at the synaptic terminal [1]. This regulation is of two types:

- Intrinsic to the neuron, through changes in the membrane potential at rest following the previous neural activity.
- Extrinsic to the neuron by signals originating outside the cell. These signals may be the neurotransmitter itself or its precursors, another transmitter, postsynaptic metabolites, or hormones.

Examples of extrinsic regulation are [1]:

- Self-regulation, given by the same transmitter, which by interacting with the presynaptic autoreceptors that recognize it, modulates its own release. As examples, in sympathetic noradrenergic synapses, the presynaptic α2 adrenergic receptors are of inhibitory nature for NE release, whereas β presynaptic adrenergic receptors are excitatory of NE release (Fig. 3.13).
- Trans-synaptic regulation. This involves signals released by the postsynapses as a result of the action of the transmitter, which across the synaptic cleft, modify transmitter release. An example is given by various arachidonic acid metabolites, such as PGs, produced in the postsynapses by the action of NE and that are inhibitory for transmitter release (Fig. 3.13).
- Heterosynaptic regulation mediated via receptors for different neurotransmitters in the synaptic terminals. The regulation is exerted by nearby synapses that use a different type of neurotransmitter. An example of this phenomenon is presynaptic inhibition in primary sensory neuron nociception (transmitter: substance P, co-transmitter: neoendorphin) caused by enkephalinergic interneurons in the dorsal horn of the spinal cord (Fig. 3.13)
- Hormonal regulation, based on the various central and peripheral neuroendocrine phenomena. For example, the increased plasma levels of estradiol released by growing ovarian follicles cause the activation of neural systems that regulate the release of gonadotropin-releasing hormone (GnRH) and consequently the release of luteinizing hormone (LH; Fig. 3.13).

The phenomenon of neurotransmission is fleeting. This transience is achieved by effective mechanisms of the termination of neurotransmitter action, involving one of the following three mechanisms.

The first mechanism operable to biogenic amines and amino acids, is the active reuptake process by the terminal and synaptic vesicles. The passage through the presynaptic membrane is a process associated with a Na/K ATPase symporter, and is independent of presynaptic receptors. This uptake mechanism, which allows the reuse of most of the transmitter released, is predominant for NE, DA, and 5-HT; in the case of ACh, the choline precursor underwent reuptake. Genes are known, and the sequence has been elucidated, for various presynaptic transporters. Most of them

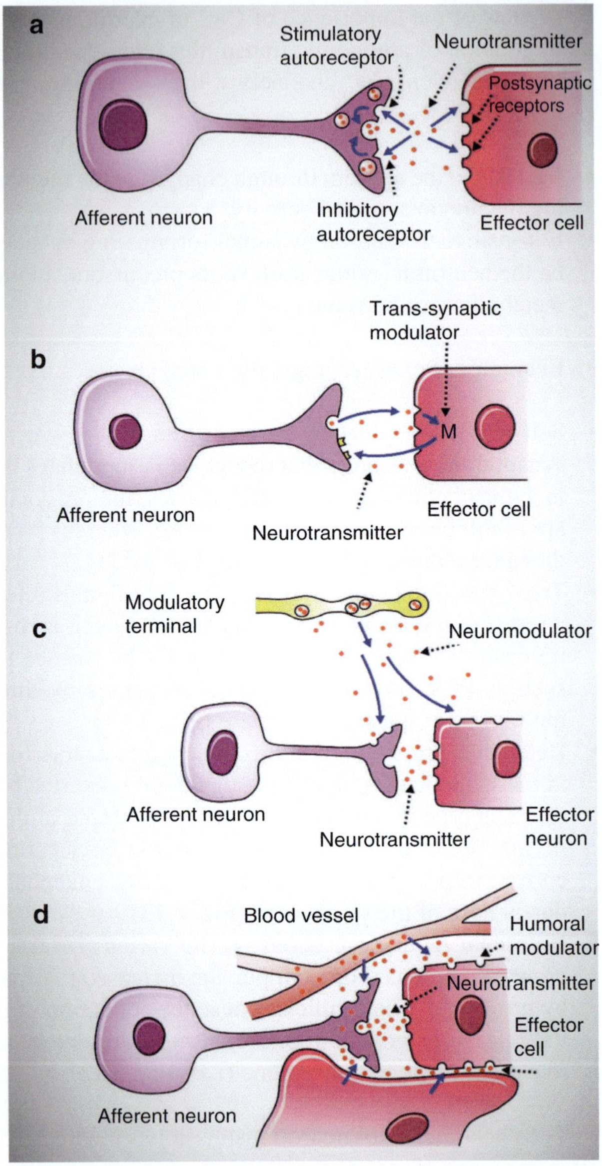

Fig. 3.13 Modulation of transmitter release. (**a**) Self-regulation. (**b**) Trans-synaptic regulation. (**c**) Heterosynaptic or interneuronal regulation. (**d**) Humoral regulation. Modified with permission from Cardinali [1]

comprise proteins with 12 hydrophobic transmembrane portions of 15–20 amino acids (α helices). Rapid regulation is achieved through changes in transporter plasma membrane expression and by intrinsic transport activity, mediated via transporter phosphorylation and the regulated associations of accessory proteins, including protein phosphatases and scaffolding proteins. Thus, carrier clearance capacity is likely to involve multiple regulatory proteins that localize, stabilize, and vary the activation state [14]. Interest in these carriers is that they are the sites of action of many drugs such as antidepressants and cocaine. In addition, genetic studies indicated gene polymorphisms for monoamine transporters in various psychiatric disorders.

The second inactivation mechanism for transmitters is metabolism (Fig. 3.12). This is the primary mechanism for ACh, which is inactivated by the acetylcholinesterase present in the postsynaptic membrane. Because of this inactivation, choline is produced and is taken up by the cholinergic terminal (Fig. 3.9).

The third mechanism of inactivation is the passage of transmitter from the synaptic cleft into the extracellular fluid or general circulation. Although this process occurs for all types of transmitter, it is the predominant one for neuropeptides, for which no presynaptic reuptake exists and the metabolism by extracellular peptidases is slow. The gases are also diluted by diffusion.

In Fig. 3.14, many current concepts of chemical transmission in the ANS are summarized. They have evolved from the simple consideration of the synapse as being mediated by a single transmitter, through a single postsynaptic receptor, to the present one, which includes a multiplicity of signals (M1–M3) acting through various pre- and postsynaptic receptors. The main interactions shown in the figure are:

- Inhibition of the release of a neuropeptide (M2) by a conventional transmitter (e.g., ACh) through presynaptic action (Rp'1, Fig. 3.14).
- Interaction between transmitters (M1 and M2) at a postsynaptic receptor (R'b, Fig. 3.14).
- Facilitation or inhibition of transmitter release or presynaptic electrical activity by peptide (M3) through a presynaptic receptor (Rp", Fig. 3.14).

Sensory Autonomic Neurons

According to a classical definition of senses [15], the following categories occur: teloreceptive (vision and hearing), proprioceptive (limb position), exteroceptive (touch, including temperature and pain), chemoreceptive (smell and taste), and interoceptive (visceral) modalities. For modern neuroanatomy, the existence of an A sensory system, including teloreception, exteroception/proprioception, and a B sensory system, including interoception/nociception, further classify the senses [16]. The development of small-diameter interoceptive afferents originating from small (B) cells is coordinated with the development of lamina I and II cells in the dorsal horn, clearly differing from the large-diameter exteroceptive afferents originating from large (A) cells that project to the deep dorsal horn and do not connect with lamina I neurons.

The main sensory input to the ANS comes from small-diameter sensory fibers by way of lamina I neurons in the superficial dorsal horn (B cells). This pathway supports organotopic homeostatic control of the body's condition, but also human feelings from the body, such as temperature, pain, itching, affective touch, muscle ache, vascular flush, etc. Once homeostatic information from the tissues is decoded, it is conveyed up to the anterior insula after making synaptic relays at different levels (spinal cord – lamina I and II, brainstem – homeostatic regions, thalamus) [16, 17].

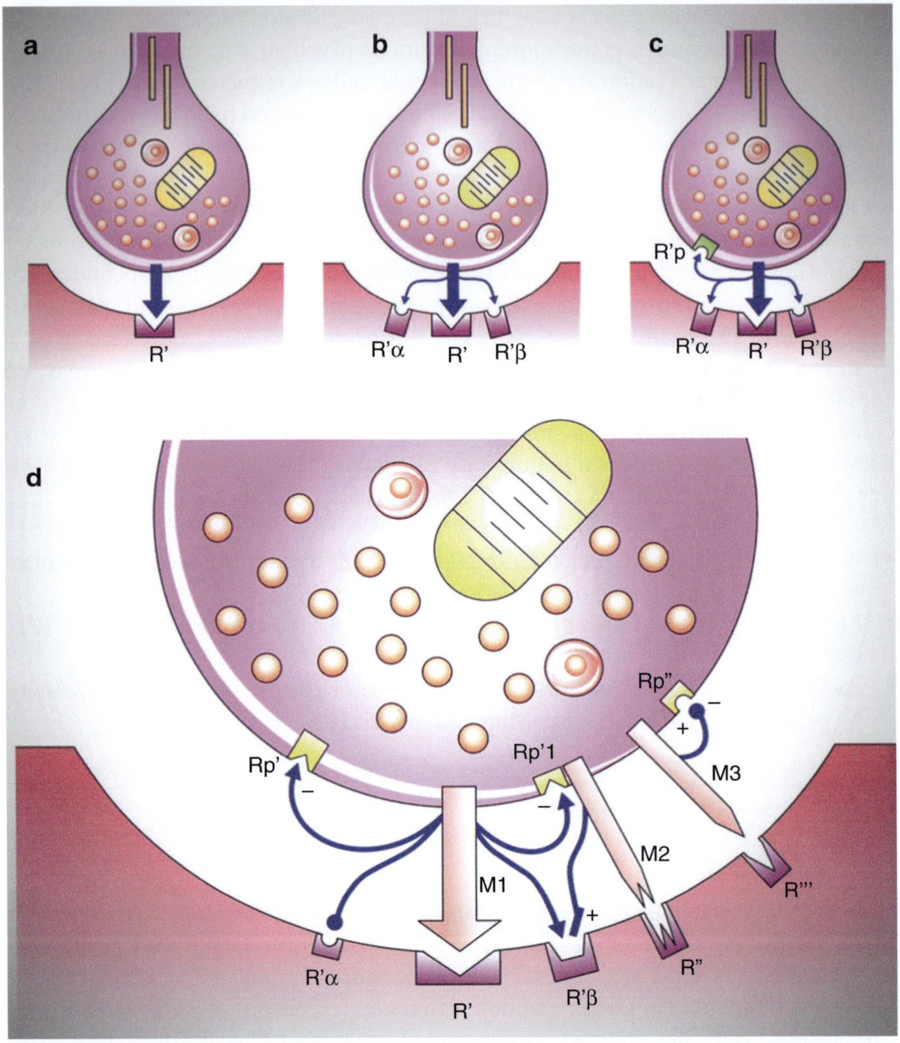

Fig. 3.14 Chemical neurotransmission. A transmitter acts on (**a**) a single or (**b**) several postsynaptic receptors, and (**c**) presynaptic receptors may also exist. In (**d**), the current concept of cotransmission (see text). Modified with permission from Cardinali [1]

Afferents of receptors in internal organs (interoceptors) are known collectively as "visceral afferents" (Fig. 3.15) [18]. This group includes afferents from thoracic, abdominal, and pelvic organs. A list of the principal unconscious sensory modalities measured includes arterial BP, by stretch receptors in the carotid sinus and aortic arch; central venous pressure, by stretch receptors in the walls of the great veins and atria; inflation of the lung by stretch receptors in the lung parenchyma; temperature of the blood in the head by neurons in the hypothalamus; arterial PO_2 by glomus cells in the carotid and aortic bodies; pH of CSF by receptors on the ventral

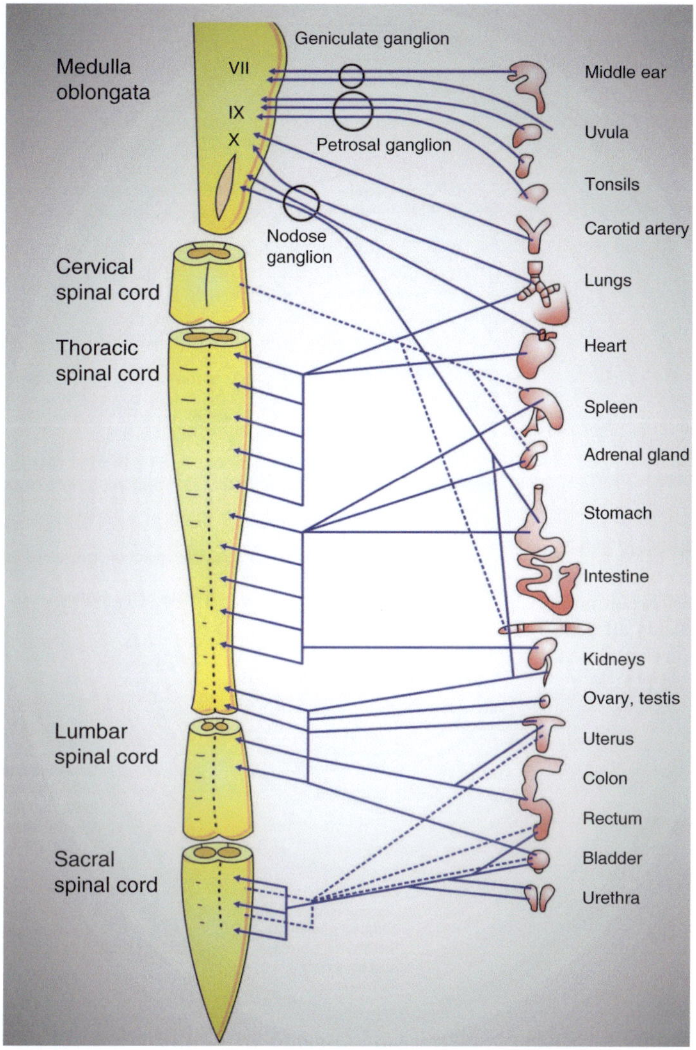

Fig. 3.15 Visceral afferents. Modified with permission from Cardinali [1]

surface of the medulla oblongata; osmotic pressure of plasma by cells in the organum vasculosum of the lamina terminalis and other circumventricular organs; arteriovenous blood glucose difference by cells in the hypothalamus and at the periphery (glucostats). Additionally, hormones and cytokines are sensed in many central and peripheral organs. Collectively, these unconscious sensory modalities constitute the afferent pathways of the autonomic reflexes (Fig. 3.16).

Some of these afferents enter the spinal cord and autonomic somatic pathways, and their cell bodies are in the spinal ganglia. Others travel through the X pair (approximately 80% of the vagal fibers are sensory; Fig. 3.17). It is noteworthy that, through the interoceptive afferents, the ANS builds a kind of "structural

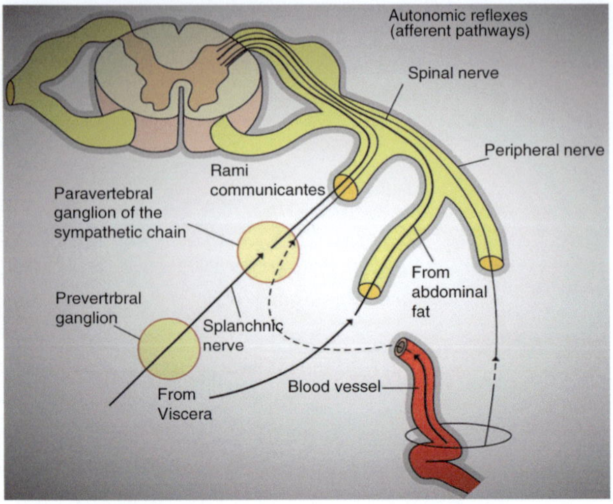

Fig. 3.16 Visceral afferent pathway as exemplified by the afferents of a blood vessel. Modified with permission from Cardinali [1]

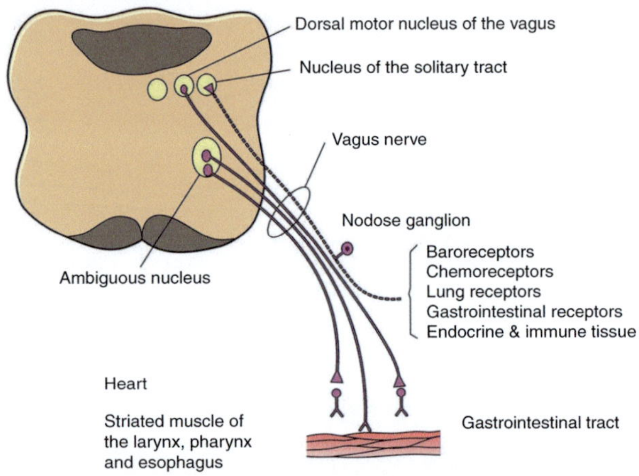

Fig. 3.17 Afferent and efferent fibers of the vagal nerve. 80% of X pair nerve fibers are sensory afferents. Modified with permission from Cardinali [1]

configuration of the interior space," analogous to the contribution of proprioception to the external configuration of the body image. The important projection of this sensorial information to the limbic system provides the basis for the hypothesis that they might constitute a kind of physiological correlate of unconsciousness. These fibers are also important in visceral nociception.

Visceral afferents present in the sympathetic (spinal) and vagus pathways transmit sensory inputs to the CNS about the physiological and pathological changes in the local environment of visceral organ systems. Visceral afferent fibers responsive to mechanical stimuli are either high or low threshold, with the high-threshold endings frequently serving as nociceptors. Many high-threshold mechanosensitive endings are triggered by chemical events and hence are bimodal in their sensitivity. Chemical stimuli activating these endings depend on the organ in which they are situated and the condition imposed [18].

Activation of these visceral afferents by mechanical and chemical stimuli elicits important ANS reflex responses. For example, visceral ischemia represents a pathophysiological condition associated with cardiovascular disease, which leads to the production and release of many metabolites including protons, bradykinin, 5-HT, His, endothelin, thromboxane and other cyclo-oxygenase products, and reactive oxygen species (ROS), etc. These chemical mediators, both individually and in combination, are involved in the activation of visceral afferents during ischemia and reperfusion.

There is very precise information in the CNS with regard to what happens both outside of and inside our body. Understanding this from a physiological point of view is important because it helps to explain, for example, how changes in emotionality can be detected as an early consequence of an organic disease such as cancer. The CNS has information on the functional status of systems that have long been considered independent of neural control, such as the immune system [1].

Most of the internal organs receive dual sensory innervation via sympathetic and parasympathetic nerves. Visceral pain information is conveyed mainly through sympathetic nerves, whereas reflex regulatory functions, for example, in the gastrointestinal tract, involved afferent fibers in parasympathetic nerves.

Autonomic sensory fibers travel through the visceral or peripheral nerves, pass through the autonomic ganglia without interruption and have their neuronal bodies in the spinal ganglia of the dorsal roots, through which they reach the spinal dorsal horn. Afferent autonomic pathways are organized into two patterns of connections:

- Reflex circuits in the spinal cord and especially in the brainstem, which allow adaptive responses of the visceral organs, for example, afferent baroreceptor or chemoreceptor neurons connect with the medulla oblongata, from where efferent fibers that control the corresponding effector responses are found (Chap. 4).
- Complex circuits, with upward projections via the anterolateral spinal cord and brainstem nuclei to the hypothalamus and the limbic system, where information is integrated. The responses affect multiple systems, autonomic, and emotional and immunoendocrine, for example, the set of connections that control eating behavior (Chap. 5).

Based on data obtained mainly in the somatosensory system, several general principles of sensory organization can be established. The responses of sensory systems have four basic dimensions: spatiality, temporality, modality, and intensity. All are deducible by conscious sensation and in the case of visceral perception, inferred from neuroanatomical and functional neuroimaging studies.

Spatiality and temporality of sensation or perception relate to the real world or to the internal medium itself. When something touches my skin, I can locate its position on my body (spatiality) and identify the beginning and end of the stimulus (temporality). Various studies indicate that visceral perception has a similar degree of accuracy and representation in specific brain areas [16].

Modality defines the type of sensation, that is, we do not have the experience of our environment as a whole, but through discrete elements produced by the interaction of appropriate stimuli with their specific sensory receptors. In the case of interoception, free nerve endings are the most common type of neural receptors.

The intensity, quantitative expression of feeling correlates with the amplitude of the receptor potential and with the discharge frequency of action potentials in the sensory nerve. In the case of conscious perception, the intensity is determinable both objectively and subjectively. In the case of interoception (unconscious), it can only be measured objectively. In general, it can be said that the intensity of a sensory stimulus is encoded through two common mechanisms: (a) frequency code (frequency of action potentials); (b) population code (number of sensory fibers stimulated).

A second form of extraction of properties from a stimulus depends on the phenomenon of adaptation or accommodation of sensory receptors. Invariably, a constant stimulus stops to stimulate the sensory receptor. Rapidly adapting receptors extract the dynamic characteristics of the stimulus (the time at which it is applied and when it varies). Instead, slowly adapting receptors extract the static properties of the stimulus (i.e., the time at which the stimulus is present).

The neural conduction velocity is a third form of abstraction stimulus. The different sensory modalities are conducted by nerve fibers of varying diameter (Fig. 1.9).

It should be noted that the pattern of action potentials coming from an afferent is not sufficient to fully sense the quality of a stimulus. For the brain to know that a stimulus is a change in osmolarity coming from a given portion of the intestine, a labeling of which afferent type has been activated must exist. Differing from the internet, in which messages travel down a shared common line, sensory neurons give each type of sensor its own private labeled line (Fig. 3.18). This implies the need for a large number of lines in the spinal cord.

In general, one of the common characteristics of the ordered representation of the sensory surface (somatotopic in the case of the somatosensory system, retinotopic for the visual system, tonotopic for hearing, viscerotopic for interoception) is the center–periphery antagonism of the peripheral fields. This center–periphery antagonism is a result of the existence of lateral inhibition in synaptic circuits and of the downward control of neural information, and is exerted at the upper levels of the sensory pathway. The organization of receptive fields in a center and a periphery with opposing characteristics increases the discriminative capacity of the stimulus and improves its contrast [1].

Descending control by the upper structures reduces the entry of irrelevant information. This downward, or input, control of sensory information describes the modulation at different levels of a neural pathway exerted by a structure located more centrally in the hierarchical level. Because of this top–down control, only a part of the information generated in the lower levels of the neural pathway reaches the higher levels.

This is the origin of central analgesia. Together with the lateral inhibition of circuits, downward control contributes to the center–periphery antagonism. Strictly speaking, it is an element of the hierarchical information processing.

The anterolateral spinal cord system consists of several ascending pathways, which together play an important role in the perception of pain and temperature in

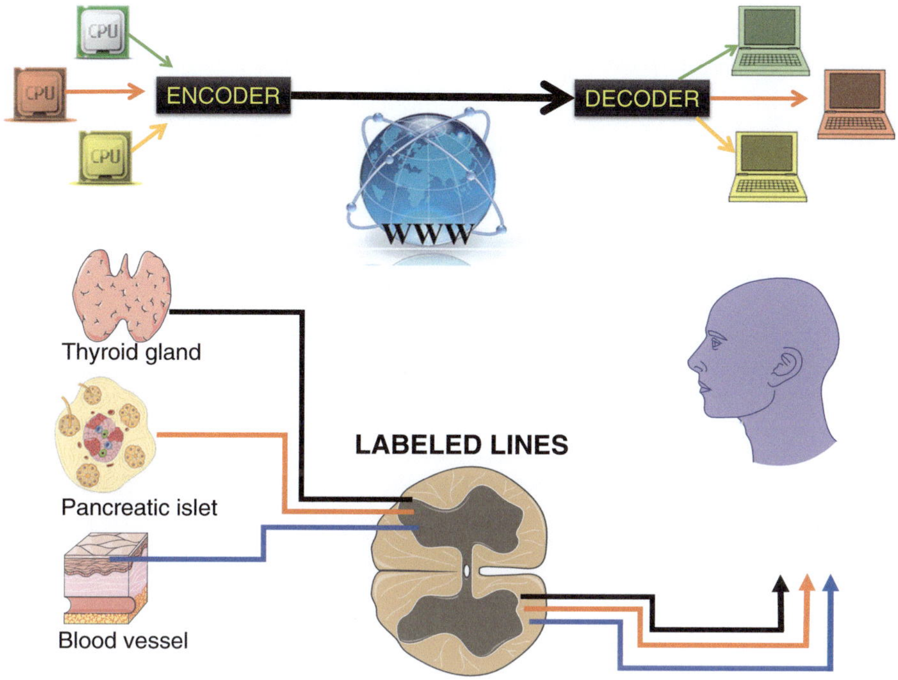

Fig. 3.18 Sensory information is reaching the brain via labeled lines, i.e., by giving each type of sensor its own private line. This differs from the internet, where the messages travel down a shared common line. To separate an individual message from the others, each packet of information is given a tag or label. The figure was prepared in part using image vectors from Servier Medical Art (www.servier.com), licensed under the Creative Commons Attribution 3.0 Unported License (http://creativecommons.org/license/by/3.0/)

interoception and constitute a secondary pathway for tactile sensitivity. The anterolateral system ends not only in the thalamus, but also in several regions of the brainstem (Fig. 3.19). Based on their termination site, three components of the anterolateral system can be identified: (a) spinothalamic (or neospinothalamic); (b) spinoreticular (or paleospinothalamic); (c) spinotectal. The last two are linked to interoception.

The analysis of the mechanisms of pain is useful in understanding the phenomena involved in interoception [19, 20]. As interoception, pain is transmitted by specific neural pathways that start at the level of the superficial and deep free nerve endings. These receptors respond to specific submodalities or are polymodal. They are in part chemoreceptors that respond to chemicals produced by the cells. The neurotransmitter in neurons of the myelinated fibers is the neuropeptide substance P. It has been demonstrated that there is release of substance P and other neuropeptides, such as neoendorphin, somatostatin or CGRP, from these fibers.

Two populations of second-order neurons for pain transmission and interoception are found in the dorsal horn of the spinal cord: (a) relay neurons, which send their axons to the thalamus or brainstem; (b) interneurons that connect to other, similar neurons in the dorsal horn, or to relay neurons [19, 20].

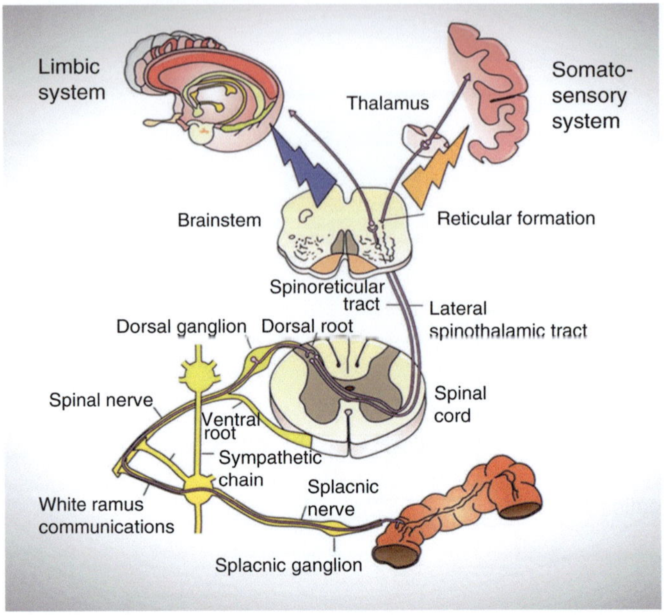

Fig. 3.19 The somatosensory and limbic pathways for abdominal pain. These connections underline the link between emotionality and digestive diseases. Modified from Cardinali [1]

It should be noted that a minority population of the anterolateral cord ascending fibers ends in the thalamus. Most end diffusely in the reticular formation of the brainstem, particularly in the area around the aqueduct of Sylvius (periaqueductal gray), which is closely linked to the limbic system, hence the capacity of these fibers to produce painful alertness and diffuse autonomic responses, such as stimuli of respiratory and cardiovascular changes.

In the sympathetic and parasympathetic ganglia, important neural information processing occurs, provided by similar synaptic circuits to those located at more complex neural levels. Also, various hormones (steroids, neurotrophins, pituitary and thyroid hormones) and cytokines affect the activity of the autonomic ganglia.

Let us take, for example, the SCG. In it, as in all other components of the chain and prevertebral ganglia, the electrical response to preganglionic stimulation includes the following aspects [4]:

- Quick depolarization, or rapid excitatory postsynaptic potential, responsible for the genesis of the action potential in the ganglion neurons. The transmitter involved in this rapid EPSP is ACh, acting through a nicotinic receptor, blocked by hexamethonium.
- Slow hyperpolarization, or slow IPSP, about 2 s in duration, produced by the release of DA from a group of catecholaminergic interneurons (small, intensely fluorescent, SIF, cells).
- Slow depolarization, or slow EPSP, about 30 s duration, produced by ACh acting through M1 receptors.

- A slow late hyperpolarization, or late slow IPSP, produced by the release of neuropeptides, in some cases, GnRH. Neurotransmitters such as GABA are also present in sympathetic ganglia.

In the author's laboratory, the SCG has been examined for the presence of hormone receptors and for the effects of various hormones on the neural mechanisms mentioned above (e.g., M1 cholinergic receptor, ACh and GABA release) [21, 22]. The participation of peripheral sympathetic innervation in neuroendocrine–immune regulation has been also examined in detail in the SCG territory (Table 3.1).

Spinal Autonomic Reflexes Have Homologies and Differences with Spinal Motor Reflexes

By analogy with the somatic motor system, in which the spinal α-motoneurons are the "final common pathway," the sympathetic and parasympathetic ganglion neurons are considered the final common pathway for the ANS. However, the homology between the two types of motor neurons, somatic and autonomic, is not complete.

Unlike somatic motoneurons, autonomic ganglion cells receive only a restricted intermetameric input. Although in the case of a sympathetic paravertebral chain, ganglion neurons can receive up- or downward influences from other metameres (via the sympathetic trunk), it is in the intermediolateral column of the spinal cord, and not in the sympathetic ganglia, that the integration of segmental spinal afferents with downward influences from a higher level is performed. The cell bodies of preganglionic autonomic neurons of the intermediolateral spinal column are smaller and more numerous than those of the α-motoneurons [3].

Synaptic connections between autonomic spinal afferents and efferents are referred to generically as the autonomic reflex arc. In the cutaneovisceral reflex the primary afferents come from the skin and the efferent sympathetic system innervates the viscera. When afferents come from viscera (viscerocutaneous and viscerosomatic reflexes), the efferent output can be both sympathetic (e.g., redness of the skin after visceral irritation) and somatic (e.g., reflex contraction of the abdominal muscles due to viscera inflammation). There are at least three synapses between afferent and efferent autonomic neurons. The existence of intervening synapses facilitates modulatory influences on autonomic reflexes [3, 4].

The spinal organization of the sympathetic system tends to be segmental or metameric. The preganglionic neurons of a particular spinal segment are in contact with the visceral afferent entering at that level. In certain organs, this feature is very pronounced: afferent heart, or excretory organs, make synaptic contacts at the segmental level with preganglionic sympathetic and parasympathetic neurons that innervate the same organs (intestinal–intestinal reflexes, cardio-cardiac reflexes, bladder evacuation, etc.).

Such segmental organization of visceral reflexes can be tested clinically. In pathological conditions of clinical significance, as in the case of inflammation of the gallbladder or appendicitis, the voluntary muscles of the affected metameres are contracted, and the corresponding dermatomes are bloodshot. This situation is explained by an inhibitory action of the visceral afferents originating from the affected body on the vasoconstrictor efferent pathway of the same spinal segment (vasodilation, reddening of the skin), and an excitatory action of these afferents on segmental α-motoneurons ("defense" reflex of the abdominal muscles). Reciprocally, stimulation of thermoreceptors in the skin produces a reflex inhibition of motility viscera innervated by the same spinal segment through the corresponding sympathetic neurons. This explains the therapeutic effect of applying heat in visceral pain [10].

After a complete spinal cord transection in humans, autonomic reflexes in segments below the section disappear for about 2–4 months. During the first phase of this paralysis, the skin is dry and pink, because the sympathetic activity of the fibers that innervate the sweat glands and vessels is very low. This hypoactivity gradually reverses, to become hyperreflexia. In this phase, the skin stimulation produces intense perspiration in areas innervated by deafferented spinal cord.

During the hyperexcitability after spinal cord section, exteroceptive stimuli, such as nipping the inner thigh or manually dilating the external anal sphincter, or interoceptive stimuli, such as contraction of the bladder muscles or dilation of the intestinal muscles, trigger a mass reflex. During this period, other phenomena, such as secretion of adrenal catecholamines, hypertension, piloerection, or profuse sweating, are observed [10].

After recovering from spinal shock, the isolated injured spinal cord is able to maintain a series of simple autonomic reflexes. For example, heating the skin produces heat loss by vasodilation and sweating. The transition from supine to upright, or blood loss, produces vasomotor compensatory segmental reflexes. Stimulation of the skin at the corresponding metameric level triggers defecation or urination.

Another comparative aspect with motor reflexes is useful in understanding some aspects of vegetative semiology. By losing neurogenic control, visceral effector organs usually acquire a greater level of intrinsic activity. For example, after the section of the sympathetic nerves, vascular smooth muscle is first paralyzed and vasodilation ensues. However, within a few hours, the myogenic tone increases and, depending on the internal pressure of the vascular filling, vasoconstriction occurs. When denervated, most effector organs are hypersensitive to neurotransmitters that reach them from the circulation. This explains why the miosis injury caused by postganglionic sympathetic fibers of the pupil decreases in states of emotional arousal, by action of plasma E, an effect not seen in the preganglionic injury (the postganglionic neurons remain intact). The explanation for this denervation hypersensitivity is that in the absence of direct innervation, enzyme systems responsible for the catabolism of neurotransmitters disappear and there is an increase in receptors in postsynaptic membranes, making it possible for bloodborne transmitters to act on the denervated synapse. Denervated sweat glands are a notable exception to this principle because they do not secrete sweat to cholinergic stimulation.

Denervation supersensitivity denervation is also observable in the CNS because of injury or prolonged pharmacological blockade. For example, after prolonged D2 blocker neuroleptic treatment, the tardive dyskinesia observed is attributed to plastic changes in the postsynaptic membranes [23].

Postsynaptic supersensitivity comprises pre- and postsynaptic mechanisms. Presynaptically, the denervated autonomic synapses lose the ability to reuptake the neurotransmitter, so that larger amounts of agonist reach receptor sites, thereby increasing the effect (presynaptic supersensitivity). Postsynaptically, after a few hours of denervation, cells show increased synthesis of receptors that translocate and disperse throughout the cell surface (postsynaptic supersensitivity). As part of postsynaptic changes after denervation, there is also an increased sensitivity to various nonspecific stimuli.

In the author's laboratory, two strategies were used to examine the neuroendocrine and immune consequences of manipulation of sympathetic ganglia (in this case, the SCG). The first consisted of the "deprivation experiment," by examining endocrine and immune sequelae of superior cervical ganglionectomy (SCGx) 1–4 weeks after surgery. The second strategy was to determine the effect of transient postsynaptic activation that occurs during the early phase of anterograde degeneration ("Wallerian") from sympathetic nerve endings in SCGx animals [24]. This process of postsynaptic activation observed during Wallerian degeneration is called "degeneration reaction." With a latency of 1–3 days, latency that depends on the distance between the SCG and the territory studied (i.e., the length of the axonal stump), nerve section is followed by transient hyperactivity in the denervated territory following the supraliminal release of the neurotransmitter NE from degenerating nerve endings, which continues for about 48 h before attaining a final and irreversible paralysis. One advantage of this experimental model is that the entire period of Wallerian degeneration in the path of the SCG can be easily controlled in the conscious animal by the degree of eyelid retraction (due to activity in the degeneration level of the periorbital muscles). It should be noted, however, that the continuous transmitter release found during degeneration is only a rough indication of the effect of the physiological release, which is phasic and dependent on the stimulation of the afferent pathway. Therefore, the degeneration reaction only indicates the potential role of the postsynaptic activation, but not its physiological details (Table 3.1). However, although it proved to be a model of great heuristic potential, it is unfortunately largely unnoticed today [24].

Urination, Defecation, and Pupillary and Sexual Responses Are Examples of Spinal Autonomic Reflexes

The function of the urinary bladder is the periodic storage and full voiding of urine produced continuously by the kidney. This function is based on the myogenic activity of bladder smooth muscle and on autonomic and somatic neural mechanisms. In controlling the bladder, prolonged phases of urine collection alternate with short expulsive periods (urination) [6].

During urine collection, neural activity prevents emptying of the bladder. The bladder is filled at a rate of about 50 ml/h. The plasticity of the bladder smooth musculature ensures that intravesical pressure is increased only mildly during filling.

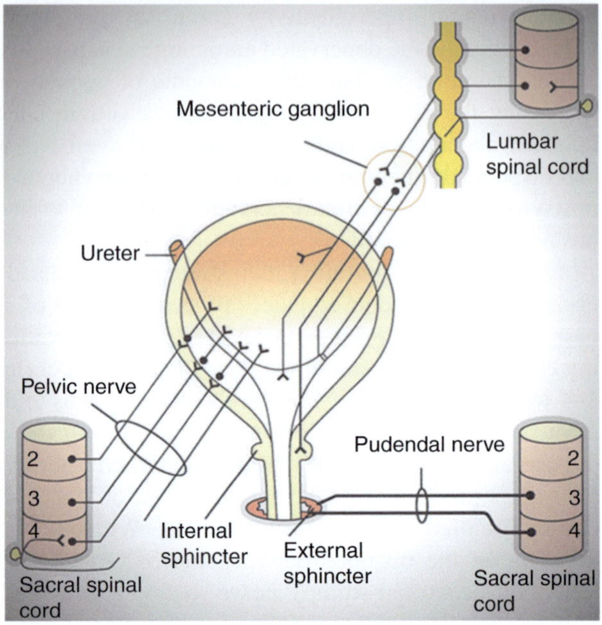

Fig. 3.20 Bladder innervation. Parasympathetic innervation stimulating the detrusor derives from S2–S4, whereas sympathetic (mesenteric ganglion) innervation inhibits the detrusor and stimulates the trigone. The external sphincter receives somatic innervation through the pudendal nerve. Modified with permission from Cardinali [1]

When the bladder is filled up to 150–250 ml, a perceived urgency to urinate occurs, a phenomenon triggered by brief increases in intravesical pressure. Upon reaching a content of 250–500 ml, volunteer continence mechanisms are overcome and the evacuation phase of urination is triggered [6].

In Fig. 3.20, the different neural and muscular components participating in micturition are depicted. Bladder muscle (detrusor muscle) is innervated by parasympathetic fibers from the pelvic nerve. This innervation is essential for normal bladder emptying. The sympathetic system inhibits the detrusor muscle and stimulates the bladder internal sphincter (trigone). Sympathetic preganglionic neurons are located in the intermediolateral column of the upper lumbar cord, and reach the bladder via the inferior mesenteric plexus [25].

The somatic motor system supplies the external sphincter through the pudendal nerves, which carry the axons of motor neurons located in segments S2–S4. The filling level of the urinary bladder is detected by stretch mechanoreceptors located in the bladder wall, whose afferent fibers run via the pelvic nerve, and the nerve cell bodies are in the dorsal root ganglia of the corresponding sacral segments.

Sensory information, particularly nociceptive, originates from the trigone and is carried by fibers traveling parallel to the sympathetic fibers and having the neuronal bodies in the dorsal ganglia of the upper lumbar region. This afference is important in the processes of bladder inflammation (pollakiuria, or pain when urinating) [25].

Up to certain limits, increased bladder volume induces relaxation of the detrusor muscle; thus, intravesical pressure does not increase and urinary continence can be maintained. This reflex action is mediated by sympathetic fibers, which inhibit the detrusor muscle via β-adrenoceptors, while stimulating the smooth muscle of the

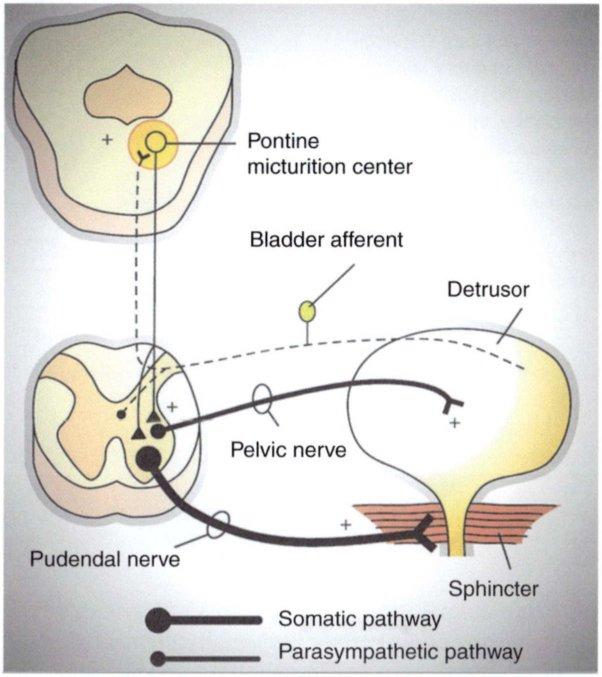

Fig. 3.21 Spinal and pontine reflex arches controlling urination. Modified with permission from Cardinali [1]

trigone and the urethra through α-adrenoceptors. The initial contraction of the bladder causes a greater excitation of the mechanoreceptors and, by reflex, a new increase in the contraction, so that the voiding reflex is self-reinforcing.

Electrical stimulation of the anterior brainstem region triggers the micturition reflex (Fig. 3.21). Once the bladder has begun to empty, the process accelerates exponentially ("positive feedback") because of: (a) increased activation of the mechanoreceptors of the bladder wall, this time by contraction of the detrusor; (b) activation of trigone afferents due to the passage of urine; (c) blocking supraspinal inhibitory influences of the reflex at the spinal cord level; (d) inhibition of sacral α-motoneurons that control the external bladder sphincter [6, 26].

After spinal cord section, and when the reflex restarts after the spinal shock, this is due exclusively to the lower arc operation shown in Fig. 3.21. This is called "automatic bladder," triggering the reflex by stimulation of the corresponding dermatome. The lower reflex arch is the single acting pathway in the newborn, and only in later stages does the upper reflex arc develop [6].

Defecation is under the control of the enteric intrinsic system, sacral parasympathetic innervation, and somatomotor mechanisms (Fig. 3.22) [26]. The role of the sympathetic system is only minor. Two sphincters close the distal end of the rectum: (a) the internal anal sphincter, composed of smooth muscle without voluntary control; (b) the external anal sphincter, composed of striated muscle innervated by motor neurons of the segments S2 and S4, whose axons travel via the pudendal nerve (Fig. 3.22). Usually, both anal sphincters are closed. The tonic contraction of

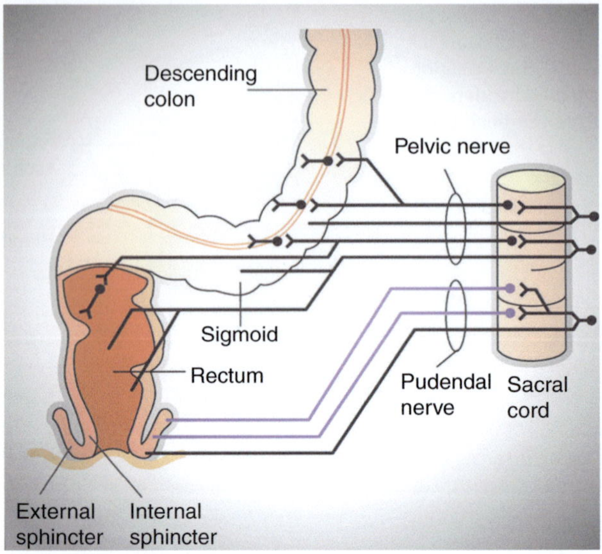

Fig. 3.22 Defecation. The external anal sphincter is under voluntary control, whereas the internal anal sphincter is under parasympathetic control. The internal sphincter is relaxed reflexively by rectal distension. Modified with permission from Cardinali [1]

the external anal sphincter is due to a reflex with afferences from the perianal muscles and surrounding tissue, especially the anal skin.

When the rectum is filled with intestinal content, peristaltic contractions of the descending colon relax the internal sphincter, and reflexively increase the contraction of the external sphincter. The relaxation of the internal sphincter as a reflex originated in the enteric autonomic system, whereas contraction of the external sphincter is a reflex that involves somatic afferents that run through the pelvic nerve to the sacral spinal cord (Fig. 3.22).

This sequence of phenomena is accompanied by the urgency to defecate, a conscious sensation triggered by the stretching of the rectal and colonic wall. After about 30 s to 1 min, the relaxation of the internal sphincter disappears simultaneously, and because of the rectal plasticity, the rectus musculature adapts to the new content. The result is that the need for defecation disappears.

Through these neural mechanisms, a healthy individual can maintain fecal continence up to a volume of rectal content of approximately 2 L. Cortical mechanisms participate in fecal continence: (a) through the excitation of α-motoneurons that innervate the external sphincter; (b) through the inhibition of the parasympathetic reflex (Fig. 3.22).

Defecation is initiated by a voluntary effort (increased intra-abdominal pressure), the simultaneous supraspinal facilitation of parasympathetic pathways, and the relaxation of both sphincters. Lesions of the sacral spinal cord eliminate the defecation reflex. The spinal cord section at the thoracolumbar level causes elimination of the supraspinal control with maintenance of the reflex, which can be excited in the paraplegic patient through other mechanisms (e.g., manual dilation of the anal sphincter), thus ensuring periodic bowel movement [27].

The ANS influences numerous ocular functions [28]. It does this by way of parasympathetic innervation from postganglionic fibers that originate from neurons in

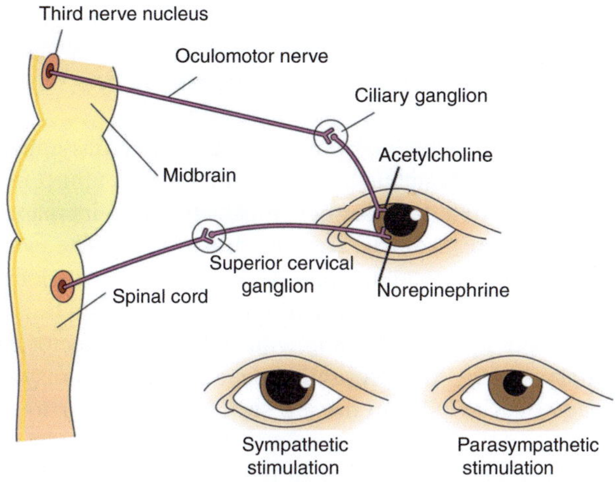

Fig. 3.23 Autonomic control of pupillary diameter. Modified with permission from Cardinali [1]

the ciliary and pterygopalatine ganglia, and by way of sympathetic innervation from postganglionic fibers that originate from neurons in the SCG.

Ciliary ganglion neurons project to the ciliary body and the pupillary sphincter muscle of the iris to control ocular accommodation and pupil constriction respectively. SCG neurons project to the pupillary dilator muscle of the iris to control pupil dilation (Fig. 3.23). Fibers of circular arrangement and others of radial disposition constitute the smooth muscle of the iris. The contraction of the first fibers, whose innervation is parasympathetic, provokes miosis, and that of the second fibers, innervated by the sympathetic system, causes mydriasis. The parasympathetic innervation of the iris originates in the Edinger–Westphal nucleus. The preganglionic fibers leave the nucleus, accompanying the third cranial nerve, and follow the path of the nerve until reaching the ciliary ganglion located behind the eyeball. From the ciliary ganglion, the postganglionic fibers innervate the ciliary muscle, which is responsible for the accommodation, and the constrictor muscle of the iris. Neurotransmission at this level is cholinergic [28].

Ocular blood flow is controlled both via direct autonomic influences on the vasculature of the optic nerve, choroid, ciliary body, and iris, and via indirect influences on retinal blood flow. In mammals, this vasculature is innervated by vasodilatory fibers from the pterygopalatine ganglion, and by vasoconstrictor fibers from the SCG. Intraocular pressure is regulated primarily through the balance of aqueous humor formation and outflow. Autonomic regulation of ciliary body blood vessels and the ciliary epithelium is an important determinant of aqueous humor formation; autonomic regulation of the trabecular meshwork and episcleral blood vessels is an important determinant of aqueous humor outflow. These tissues are all innervated by fibers from the pterygopalatine and the SCG. In addition to these classical autonomic pathways, trigeminal sensory fibers exert local, intrinsic influences on many of these regions of the eye, and on some neurons within the ciliary and pterygopalatine ganglia. Regarding sympathetic innervation, a first group of central fibers originates in the hypothalamus and projects to the intermediolateral horn of the first thoracic segments of the spinal cord. Here, they synapse with preganglionic neurons that send fibers to the SCG, where postganglionic fibers going to the eyeball originate. These adrenergic fibers innervate the iris dilator and the levator muscle (Müller muscle) [28].

From the functional point of view, the parasympathetic system constricts the pupil and the sympathetic system dilates the pupil (Fig. 3.23). Thus, a miotic pupil can be due to a decrease in sympathetic or an increase in parasympathetic activity and the opposite is true for a mydriatic pupil. The evaluation function can be performed by observing the pupillary responses to the local application of drugs.

Parasympatholytic (anticholinergic) drugs such as atropine block parasympathetic activity by competing with ACh at the effector cells of the iris sphincter and ciliary muscle, thus preventing depolarization. Parasympathomimetic (cholinergic) drugs such as pilocarpine are structurally similar to ACh and can depolarize the effector cell, thus causing miosis. Sympathomimetic (adrenergic) drugs such as E stimulate the receptor sites of the dilator muscle cells. A defect in the sympathetic pathway affects the pupillary dilator muscle and results in Horner's syndrome (Chap. 7). A defect in the parasympathetic pathway affects the pupillary sphincter muscle and results in a larger pupil. Defects in both pathways affect the pupillary dilator and sphincter muscles [29].

Instillation of cocaine, a sympathomimetic agent, produces mydriasis, whose absence indicates a lesion at some segment of the sympathetic chain, as in Horner's syndrome. Hydroxyamphetamine, however, only causes mydriasis in the case of preganglionic sympathetic injury because it acts by causing NE depletion from intact postganglionic synaptic terminals. An exaggerated mydriatic response to that of the contralateral eye, after instillation of phenylephrine, suggests the existence of postganglionic sympathetic denervation. Instead, an exaggerated miotic response to pilocarpine indicates postganglionic parasympathetic denervation.

In humans, the sympathetic, parasympathetic, and somatic nervous divisions participate in the control of sexual responses of erection, glandular secretion, emission, and ejaculation [30]. Penile erection can be caused by supraspinal centers, in response to visual, auditory, or psychological stimuli (psychogenic erection), or by a spinal reflex (reflex erection). In the latter, the impulses evoked by cutaneous stimulation of the genital area travel through the pudendal nerves to the sacral spinal segments S2–S4. The efferent pathway originates in the same segments, and continues through the parasympathetic pelvic nerves, to produce vasodilation of the arteries and closure of the penile arteriovenous shunts, thereby increasing the blood flow of the corpora cavernosa to allow erection. Not surprisingly, the vasculature, epithelia, and smooth muscle of all urogenital organs receive adrenergic innervation. These nerves contain non-adrenergic, noncholinergic neurotransmitters such as ATP and NPY. Cholinergic nerves increase motility in most urogenital organs. The major non-adrenergic, noncholinergic transmitters found to influence urogenital organs include those containing VIP/PACAP, galanin, and NO [31].

The glandular secretion of seminal vesicles, Cowper glands, and the prostate gland is controlled by the parasympathetic system. The emission of semen and glandular secretions to the urethra depends on the sympathetic activity, which causes contraction of the smooth muscle of the vas deferens and the excretory ducts. Sympathetic efferents are also responsible for closure of the bladder neck to prevent retrograde seminal flow into the bladder. Ejaculation is caused by a somatic reflex that causes rhythmic contractions of the bulbocavernosus and ischiocavernosus muscles innervated by the pudendal nerves. The repeated stimulation of the penis releases sacral centers from superior inhibitory signals. In situations of anxiety, in which sympathetic tone is high, premature ejaculation can occur [30].

In women, the physical expression of sexual arousal is related to parasympathetic activity, which, through the pelvic nerves, produces congestion of the clitoris and vulva and vaginal lubrication, and to somatic activity, via the pudendal nerves, which causes contraction of the vaginal sphincter and pelvic floor muscles during orgasm. Genital vasodilation seems mainly mediated by VIP. On the other hand, sympathetic stimulation induces contractions in the smooth musculature of the fallopian tubes and uterus. In both sexes, afferent somatic impulses are part of the reflex arcs, and their transmission to higher centers through the spinal cord is essential for the conscious perception of sexual phenomena and their regulation.

Penile erection or tumescence occurs during the REM sleep stage, although in adolescents it is not only confined to this stage. This has been proven in humans from 3 to 79 years of age. During the active sex life, it coincides with LH release pulses. Although its functional role remains unknown, the presence or absence of erection during sleep is used for the differential diagnosis between organic and psychogenic impotence. Similarly, clitoral erections and increased vaginal blood flow are seen in women during REM sleep [30].

The Enteric ANS as an Individual Entity

The enteric ANS is sometimes called the "second brain" because of the diversity of neuronal cell types and complex, integrated circuits that permit the enteric ANS to autonomously regulate many processes in the bowel [32]. In humans, the enteric ANS contains about 500 million neurons, far more than the number of neurons in the remainder of the peripheral ANS. It is a complex network of neurons and glia that resides in the myenteric and submucosal plexus of the bowel. The myenteric (Auerbach) plexus, located between longitudinal and circular muscle, primarily controls muscle contraction and relaxation. The submucosal (Meissner) plexus, found between circular muscle and bowel mucosa, regulates fluid secretion and absorption, modulates blood flow, and responds to stimuli from epithelium and lumen to support bowel function (Fig. 3.24).

The enteric ANS includes sensory neurons, interneurons, and motor neurons. Full arches are reflected in the enteric nervous system. The extrinsic innervation of the gastrointestinal tract given by the sympathetic and parasympathetic nerves acts on these components [33].

The extrinsic ANS is critically involved in modulating the local reflexes and secretion, absorption, digestion, motility, and sensation in the gastrointestinal tract, both during fasting and postprandially. In general, the craniosacral parasympathetic

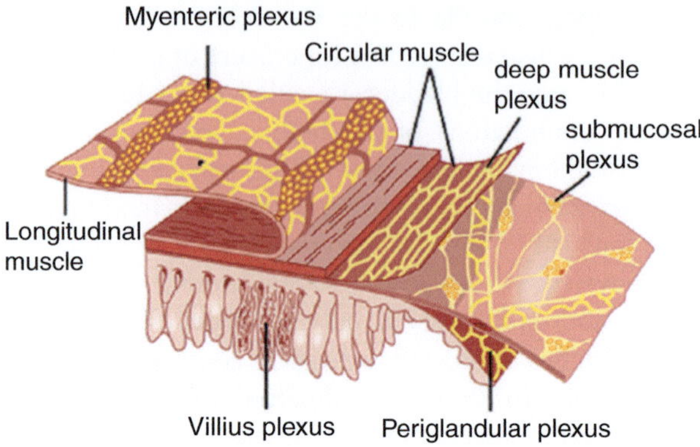

Fig. 3.24 Enteric neurons are organized into two nerve plexuses, the myenteric plexus and the submucosal plexus, with other minor side plexuses including muscle, periglandular layers, and the villi. The enteric nervous system includes sensory neurons, interneurons, and motor neurons. There are full arches reflected in the enteric nervous system

nervous system is excitatory (e.g., secretion, peristalsis) and the thoracolumbar sympathetic supply is mostly inhibitory (e.g., proabsorptive and inhibition of motility with stimulation of the sphincters). Although intrinsic neural and local paracrine control can sustain most gastrointestinal functions, the extrinsic autonomic nervous system provides the highly integrated digestive system with additional output [34].

Enteric glia are important components of the enteric nervous system and form an extensive network in the mucosa of the gastrointestinal tract. Initially regarded as passive support cells, it is now clear that they are actively involved as cellular integrators in the control of motility and epithelial barrier function [35]. Enteric glia form a cellular and molecular bridge among enteric nerves, enteroendocrine cells, immune cells, and epithelial cells, depending on their location.

The gut wall consists of several concentric layers, an inner mucosa, surrounded by several layers: submucosa, muscle and serous (Fig. 3.24). Motility of the digestive tract is characterized by the synchronized contraction of the inner and outer muscle layers, resulting in two types of movement: (a) peristalsis, which sets in motion the intestinal contents and ensures a cephalo-caudal flow; (b) segmenting, contributing to mixing food with digestive secretions. This is a myogenic contractile activity that has its origin in the gut wall itself, particularly in the nerve plexus. Myogenic origin explains the persistence of rhythmic patterns and motility in isolated denervated digestive segments [36].

The mucosa has three components: specialized epithelial cells lining the lumen; the underlying lamina propria, a layer of connective tissue containing small blood and lymphatic vessels, immune cells, and nerve fibers; and the muscularis mucosa, a thin layer of muscle cells. The submucosa is a layer of connective tissue directly beneath the mucosa, containing the largest blood vessels and the submucosal plexus (Meissner). This nerve plexus is particularly important for the control of secretion. In some areas, submucosal glands and lymphoid tissue are present. The external muscle is composed of an inner circular and an outer longitudinal layer of smooth muscle and is responsible for the motility of the gastrointestinal tract. Among these muscle layers are the myenteric nerve plexus (Auerbach), a division of the enteric nervous system that regulates motility. The serosa is an outer sheath of squamous mesothelial cells and connective tissues, where nerves and larger blood vessels travel in a bed of connective and adipose tissue [36].

Like any neural arch, there are sensory enteric neurons, interneurons, and motor neurons. Sensory information in the enteric nervous system comes from changes in luminal volume or environment. The endocrine and paracrine cells function as auxiliary sensors [10].

Reflexes regulate colon epithelial responses through cholinergic interneurons. Two categories of epithelial and submucosal motor neurons innervate cholinergic and VIPergic cells and each uses additional neuroactive substances. The sympathetic and parasympathetic tone modulates ion transport, with a cholinergic basal secretory influence. The loss of the regulatory mechanisms of sympathetic nerves in diabetic autonomic neuropathy is associated with the development of a "diabetic diarrhea" and can be corrected by the administration of $\alpha 2$ adrenoceptor agonists [33].

Peptidergic neurons release a specific combination of mediators (e.g., VIP, CCK, bombesin). These mediators can act either as classical neurotransmitters or alternatively as neuromodulators, with fine tuning of neural circuits in presynaptic sites of origin neurons or from other neurons. Individual neurotransmitters may have biphasic effects at different concentrations.

Figure 3.25 summarizes the regulation of gastric acid secretion by nerves and hormones. Secretion of gastric acid between meals is low. The cephalic phase of secretion (~30% response) is initiated by the sight, smell, taste, and swallowing of food and these stimuli activate the dorsal motor nucleus of the vagus nerve. In the body of the stomach, postganglionic nerves release ACh, which activates parietal cells directly by M3 receptors. ACh also induces His release from enterochromaffin-like cells (ECL), which stimulates the secretion of H^+ ions by the parietal cells. In the gastric antrum, vagal stimulation induces the release of gastrin-releasing peptide, from postganglionic fibers, and the release of gastrin, thus indirectly stimulating the secretion of H^+ ions. ACh also inhibits the release of somatostatin D cells in the corpus and pylorus [33].

The gastric phase (~70% response) secretion is induced by stimuli within the stomach. Vagal sensory nerves detect gastric distention by food and cause a vagovagal reflex whereby ACh is released and promotes the release of H^+. Partially digested proteins and amino acids stimulate gastrin release from G cells in the pylorus. Both the G and D cells (which release somatostatin) are open-type endocrine cells that directly detect stomach content. Gastrin stimulates further acid secretion. Acidification of pylorus stimulates somatostatin release, which inhibits acid secretion by negative feedback [36].

During the intestinal phase of digestion, products of protein entering the small intestine stimulate gastrin release from the G cells in the duodenum. Many substances, notably fatty acid, stimulate the secretion of hormones in the small intestine (secretin, CCK) which inhibits gastric acid secretion.

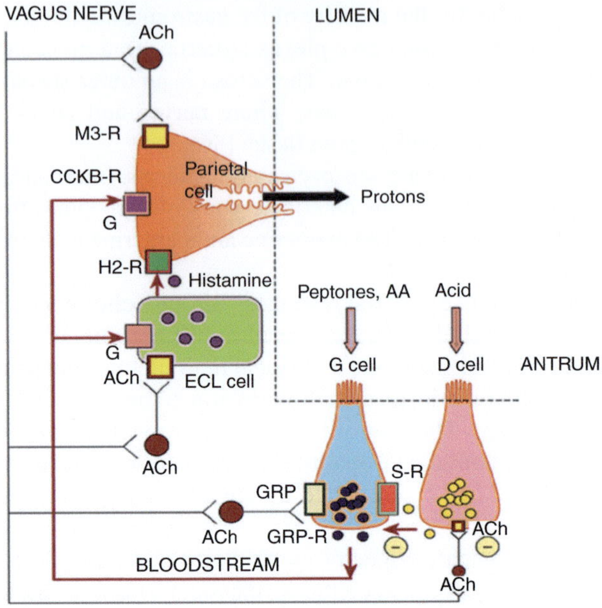

Fig. 3.25 Regulation of gastric acid secretion by nerves and hormones. During the cephalic phase of digestion, vagal cholinergic nerves stimulate directly the parietal cells and induce histamine release from enterochromaffin-like cells (ECLs), which also stimulates parietal cells. Vagal fibers also release gastrin-releasing peptide (GRP) in the antrum to induce gastrin (G cells), which through the bloodstream induces histamine release and stimulates the parietal cells. During the gastric phase of digestion, food in the stomach causes local reflexes that stimulate the secretion of gastrin. Acidification of the gastric antrum stimulates somatostatin release (D cells), which inhibits the release of gastrin and thus acid secretion. Vagal stimulation inhibits somatostatin release

The responses are exerted through the enteric ANS, the enteroendocrine cells, the CNS via the extrinsic innervation of the gastrointestinal tract given by the sympathetic and parasympathetic nerves, and the gut immune and tissue defense systems. It is apparent that the control of the digestive organs is an integrated function of all these effectors. The enteroendocrine cell release about 20 different hormones, together making the gut endocrine system one of the largest endocrine organs in the body. Influenced functions include satiety, mixing and propulsive activity, release of digestive enzymes, induction of nutrient transporters, fluid transport, local blood flow, gastric acid secretion, evacuation, and immune responses [36].

Gut content receptors, including free fatty acid, peptide, and phytochemical receptors, are primarily located on enteroendocrine cells. Hormones released by enteroendocrine cells act via both the enteric ANS and the CNS to optimize digestion. Toxic chemicals and pathogens are sensed and then avoided, expelled, or metabolized. These defensive activities also involve the enteroendocrine cells and signaling from enteroendocrine cells to the enteric nervous system and the CNS [37].

The various types of epithelial cells and smooth muscle cells are the primary effectors controlling gastrointestinal transit, absorption, and secretion. They are both directly and indirectly regulated by the enteric ANS. The intestinal epithelium is a single layer of cells that combines the potential for nutrient absorption, water, and electrolyte secretion (and reabsorption) with sensory and endocrine function and with its role as a physical barrier to prevent the uncontrolled entry of harmful substances and microbes into the body. Within the epithelium of the adult small intestine, many types of differentiated cells can be found: enterocytes, which play a key role in water and electrolyte secretion in the crypts and absorb various nutrients at the villus tips; enteroendocrine cells, which secrete many different types of endocrine and neurotransmitter-like signaling molecules; goblet cells, which secrete mucus into the gut lumen; and Paneth cells, which are of an immune nature. Enterocytes, enteroendocrine cells, and goblet cells are present in the villi, whereas Paneth cells reside in the crypts together with stem cells, which divide to replenish these differentiated cell types (Fig. 3.26) [26].

The colon lacks villi, but has crypts that house intestinal stem cells, differentiating mostly to form colonocytes (absorptive cells) and enteroendocrine cells. Enteric circuits controlling secretion and absorption are mostly located in the submucous plexus [36]. In the small intestine, together with systemic (sympathetic) pathways, local enteric reflexes act through the activation of secretomotor

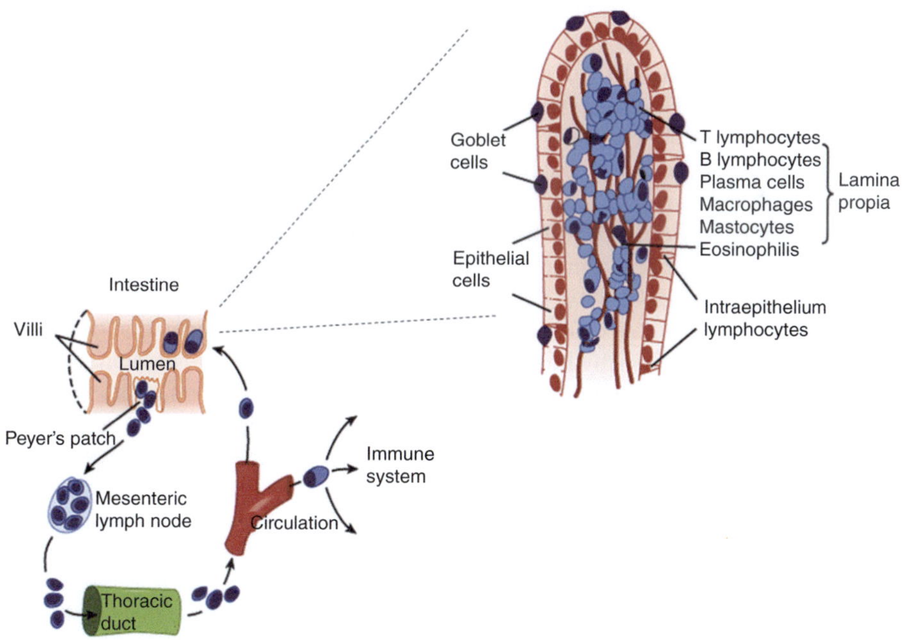

Fig. 3.26 The adaptive immune system of the mucosa or gut-associated lymphoid tissue monitors the contents of the intestinal lumen through a variety of mechanisms used by cells such as lymphoid and myeloid lineages. Lymphoid cell aggregates form Peyer's patches and lymphoid follicles located throughout the intestine. These lymphoid aggregates also play an important role in immune surveillance, guarding against pathogenic bacteria, viruses, and toxins, and allowing tolerance to dietary substances and potentially immunogenic bacteria

and vasodilator neurons to promote or inhibit secretion as part of digestion and whole body homeostasis.

As far as intestinal movements are concerned, different enteric ANS circuits direct various motor patterns depending on the region of the gut and the physiological circumstances [36]. Although esophageal and gastric motility are controlled via intrinsic and extrinsic innervation, propulsive and mixing contractile patterns in the small intestine and colon mostly rely on the enteric ANS. A basic circuit, including sensory neurons (intrinsic sensory neurons), orally directed interneurons and excitatory motor neurons, anally directed interneurons, and inhibitory motor neurons, mediates propulsion. A key element of this circuit is its polarity; stimuli that excite motor pattern generators activate an orally directed (ascending) pathway leading to smooth muscle contraction. The same stimuli excite an anally directed pathway that inhibits or relaxes the circular muscle. Both classes of motor neurons receive input from local sensory neurons and from relevant interneurons; they are the final common outputs of the monosynaptic and polysynaptic pathways [33].

In Fig. 3.27, the peristaltic reflex of the small intestine is schematized. Enteric sensory nerves detect the chemical or mechanical stimulation of the mucosa, or the stretching of the muscular layer. The signals are transmitted in an oral or anal direction by interneurons. Excitatory motor nerves release ACh and substance P, which cause muscle contraction on the oral side of the stimulus. Nervous motor inhibitors release VIP and NO, which cause muscle relaxation on the anal side of the stimulus [33].

The response to local stretching of the intestinal wall, for example by a bolus of partially digested food, results from the direct activation of mechanosensitive neurons in the myenteric plexus. The intrinsic sensory neurons have mechanosensitive elements in the myenteric plexus and probably in the smooth muscle. These neurons also project to the mucosa and respond to chemical stimuli indirectly via the release of 5-HT, ATP, and other mediators from EEC, giving them a polymodal stimulus–response profile. To control gut motility, enteric nerve circuits act in concert with myogenic control elements, notably interstitial cells of Cajal and fibroblast-like cells. By acting as intestinal pacemaker cells, interstitial cells of Cajal are responsible for the generation of intestinal slow waves and other rhythmic contractions of the smooth musculature and for excitatory and inhibitory transmission from enteric neurons to smooth muscle cells, and thereby integrating enteric ANS and slow-wave activity [33, 36].

Cells of the immune system and its wide range of products play an integral role in the regulation of fluids and are closely interrelated with the enteric nervous system and the endocrine paracrine network [37]. The lamina propria is a location for immunocompetent cells. Most of them (60%) are T lymphocytes with fewer B lymphocytes and plasma cells (25–30%), macrophages (8–10%), and mast and polymorphonuclear cells (2–5%). Inflammation causes an increase in the number of immune cells in the gut. Acute bacterial infections result in an increase in polymorphonuclear leukocytes, whereas lymphocytes in celiac disease characteristically increase. In inflammatory bowel disease, there is activation of all components of the immune system, with an increase in immunoglobulin (Ig) G-secreting cells. Therefore, the cause of the inflammatory reaction can determine the type of inflammatory cells recruited and the range of cytokines released.

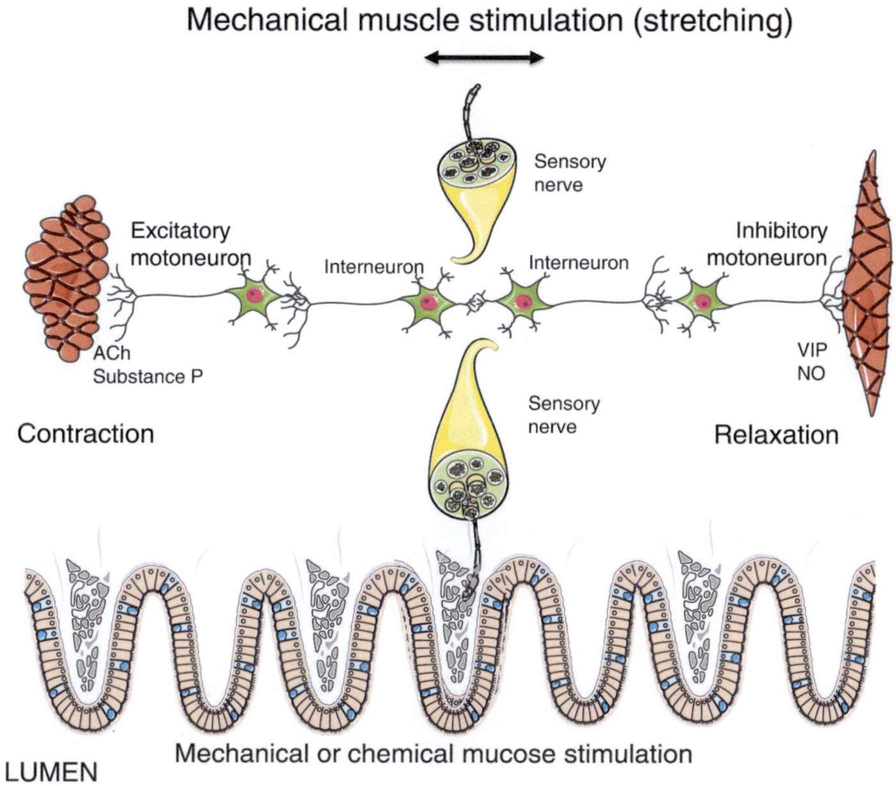

Fig. 3.27 Peristaltic reflex of the small intestine. Enteric sensory nerves detect the chemical or mechanical stimulation of the mucosa or the stretching of the muscular layer. The signals are transmitted in an oral or anal direction by interneurons. Excitatory motor nerves release ACh and substance P, which cause muscle contraction on the oral side of the stimulus. Inhibitory motor nerves release vasoactive intestinal peptide (VIP) and nitric oxide (NO), which cause relaxation of the muscle on the anal side of the stimulus. The figure was prepared in part using image vectors from Servier Medical Art (www.servier.com), licensed under the Creative Commons Attribution 3.0 Unported License (http://creativecommons.org/license/by/3.0/)

Cytokines, eicosanoids, and other peptide mediators interact with immune cells, neurons, and epithelial cells directly or alternately alter ion transport rates or gut barrier function. The adaptive immune system of the mucosa or gut associated lymphoid tissue monitors the contents of the intestinal lumen through a variety of mechanisms used by cells both as lymphoid myeloid lineages. Myeloid cells (dendritic cell populations, specific macrophages) extend through processes of the intestinal epithelial barrier, which are associated with the luminal environment. Lymphoid cell aggregates form Peyer's patches (larger aggregates in the distal small intestine) and lymphoid follicles located throughout the intestine (Fig. 3.26). These lymphoid aggregates also play an important role in immune surveillance, guarding against pathogenic bacteria, viruses, and toxins, and allowing tolerance to dietary substances and potentially immunogenic bacteria.

Specialized goblet cells secrete mucus in the intestine. The mucus forms a protective layer on the epithelial cells and antimicrobial peptides are secreted into the intestinal lumen. Paneth cells produce and secrete lysozyme and α-defensins that contribute to the defense. Clover peptides are also secreted in the lumen of the gastrointestinal tract. One of their many effects is promote the healing of mucosal lesions [33, 36].

In Chap. 1, we define microbiome as the set of microorganisms that are normally located in different places in the human body, in particular the digestive tract [38]. These microbial components aid in the digestion of food, produce vitamins, and protect against the colonization of other microorganisms that may be pathogenic. The gut microbiome is highly dynamic, exhibiting daily cyclic fluctuations, which have repercussions for the host metabolism and provide evidence for the cross-regulation of prokaryotic and eukaryotic circadian rhythms.

Figure 3.28 depicts the bases of the effect of the microbiome and the way in which they interact with the neuroimmune and bioenergetic afferents and with cognitive function. An essential pathway is the neural pathway through innervation of distant organs (e.g., intestine, spleen, liver) and neuroimmune signals that can be transmitted bidirectionally between the brain and the periphery [39]. Another important route is the general circulation, where circulating immune signals (cytokines, chemokines) and bioenergetic signals (e.g., endogenous metabolites or microbial origin) can exert a systemic impact on the brain microenvironment. Such immune/bioenergetic signals can have an impact on the blood–brain barrier signals thus spread to the brain, or they can cross the blood–brain barrier, exerting direct neuronal effects (Fig. 3.28).

The Overlooked Role of the Local Autonomic Projections in Neuroendocrine Communication

The dominant paradigm of endocrinology in the twentieth century endowed the innervation of endocrine glands with a merely secondary role compared with the corresponding trophic hormones. Indeed, both endocrine and immune tissues have three identifiable neural systems: (a) sympathetic noradrenergic neurons derived from para- or prevertebral chains in the corresponding metameres; (b) parasympathetic cholinergic neurons whose bodies are in the local parasympathetic ganglia; (c) peptidergic neurons, most of them located in the innervated organ. Concerning peptidergic innervation, there is an almost universal presence of substance P, a typical neurotransmitter of type C sensory fibers. That is, there is both sensory and motor innervation in endocrine and immune tissues. The sensory aspect is a strong evidence for the now recognized viscerotopic representation of endocrine and immune structures in the CNS, as discussed in this chapter.

The thyroid and parathyroid glands are a typical case of this threefold innervation (Fig. 3.29). These structures are innervated by: (a) noradrenergic neurons derived from the SCG, or from lower cervical nodes and reaching the thyroid through the SCG and the external carotid nerve; (b) parasympathetic cholinergic

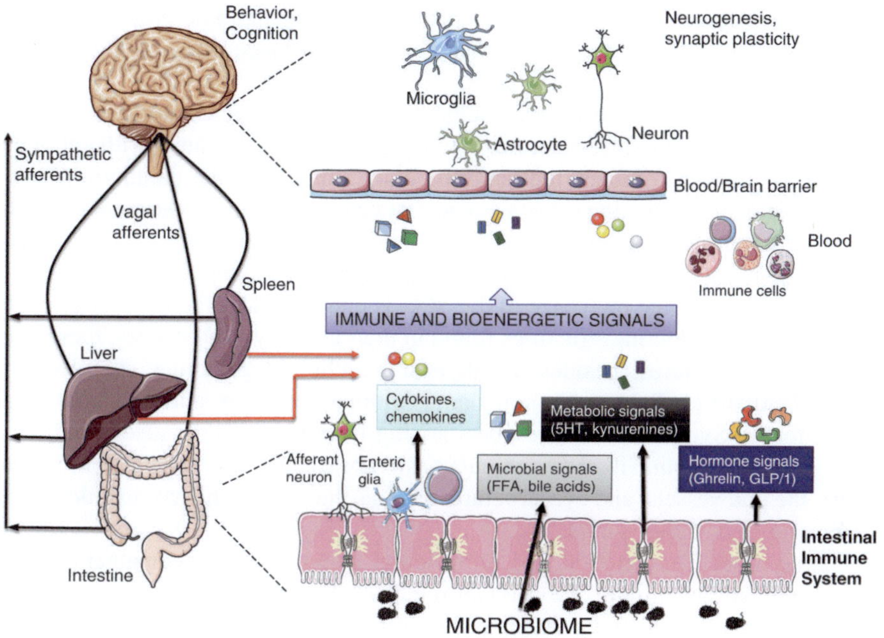

Fig. 3.28 The basis of the effect of the microbiome on brain function is shown. An essential pathway is the neural pathway, via innervation of distant organs (e.g., intestine, spleen, liver) and neuroimmune signals that can be transmitted bidirectionally between the brain and the periphery. Another important route is through the general circulation, where circulating immune signals (cytokines, chemokines) and bioenergetic signals (e.g., endogenous metabolites of microbial origin) can exert a systemic impact on the brain microenvironment. The figure was prepared in part using image vectors from Servier Medical Art (www.servier.com), licensed under the Creative Commons Attribution 3.0 Unported License (http://creativecommons.org/license/by/3.0/)

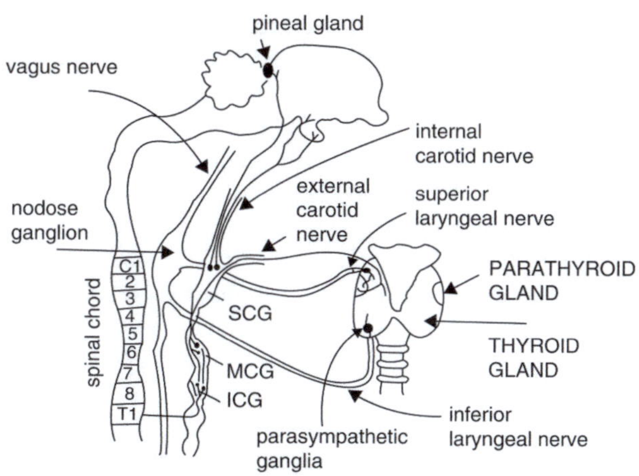

Fig. 3.29 Parasympathetic and sympathetic innervation of the thyroid/parathyroid glands in rats. Reproduced with permission from Cardinali [40]

neurons located in the local ganglia and receiving preganglionic projections from the vagus nerve; (c) peptidergic neurons, including substance P-containing neurons, which are present in local ganglia [24].

Sympathetic nerve fibers arriving at the thyroid/parathyroid territory are not only widely distributed in the vasculature, but also make physical contact with epithelial cells of the thyroid follicles, the parafollicular C-cells, and the parathyroid cells. Retrograde neuronal tracing demonstrated that neurons located in the caudal pole of the SCG that project via the external carotid nerve of the SCG are the only source of sympathetic innervation to the thyroid–parathyroid complex.

The neuroendocrine relevance of the SCG is underlined by the number of endocrine and neuroendocrine structures found in their territory [40]. Among these neuroendocrine structures, besides the thyroid–parathyroid glandular complex, the pineal gland, the medial basal hypothalamus/pituitary complex, and the carotid bodies are found. Sympathetic nerve fibers arising from the SCG are also widely distributed in the submaxillary/mandibular salivary glands, the oral and laryngeal cavity, the pial vessels, and eye structures such as the cornea, the iris, the nictitating membrane, and Müller's muscles.

Unlike other paravertebral and prevertebral ganglia of the sympathetic chain, the sympathetic ganglia of the cervical region lack communicating branches and, consequently, the preganglionic fibers reach the SCG from lower segments of the sympathetic chain. Postganglionic sympathetic fibers leave the SCG in two ways (Fig. 3.29). The internal carotid nerve pathway is followed by the postganglionic fibers innervating the intracranial structures, such as the pineal, the median eminence, the adeno- and neurohypophysis, and the choroid plexus. The external carotid nerve is the neural path by which the thyroid and the parathyroid glands and immune structures such as the submaxillary lymph nodes are innervated. In this case, some of the innervating neurons are located in the middle and/or lower cervical sympathetic ganglia and send their axons through the SCG and the external carotid nerve [24].

Concerning the hypothalamic–pituitary–thyroid axis, the relevance of SCG is best explained in terms of central and peripheral effects of the sympathetic nerve terminals, as thyrotropin (TSH) release and the thyroid response to exogenous TSH was inhibited during the increased release of NE from degenerating nerve terminals shortly after SCGx [24]. Therefore, in the presence of normal or elevated levels of TSH, NE release from local sympathetic nerves provides a negative signal for the release of thyroxine. The daily fluctuations in the concentration of thyroxine in serum and the thyroid content of catecholamines in rats were consistent with this modulatory function of peripheral NE in the secretion of thyroid acini.

In rats studied 2–4 weeks after SCGx, greater growth (goiter response) was identified. The effect of promoting goiter was ipsilateral to sympathetic denervation. The compensatory growth of the remaining lobe after a hemithyroidectomy persisted in the absence of the anterior pituitary gland, an effect that was completely blocked by an ipsilateral SCGx, indicating its essential neural nature [24].

Preganglionic parasympathetic input to local thyroid neurons derives from the dorsal motor nucleus of the vagus and, through the nodose ganglion, is conveyed via the inferior laryngeal nerves and the thyroid nerves (Fig. 3.29). Before entering the thyroid gland, the nerves display two types of ganglionic formation known as the laryngeal ganglion and the thyroid ganglion. Cholinergic and peptidergic nerve fibers arising from central nuclei and/or coming from local ganglia enter the thyroid, innervating blood vessels and endocrine structures. Indeed, parasympathetic fibers distributed in the thyroid and parathyroid gland were retrogradely traced to the medulla oblongata labeled the dorsal nucleus of the vagus [24].

In rats subjected to unilateral parasympathetic decentralization by ipsilateral inferior laryngeal nerve section, a low compensatory growth of the lobe after hemithyroidectomy was found. In hypophysectomized rats, section of the inferior laryngeal nerve produced an additional involution of the thyroid gland. Thus, the intact parasympathetic nerves are needed to maintain adequate trophism of the thyroid gland [24].

Concerning thyroid C and parathyroid cells, at the time of the increased NE release during nerve degeneration shortly after SCGx, both a decrease in the maximum release of calcitonin and a delay to reach that maximum after stimulation with an injection of calcium gluconate were observed. A similar inhibitory effect was seen as far as a hypocalcemic-stimulated parathyroid hormone (PTH) release. Blocking α-adrenoceptors suppressed the inhibition of C and parathyroid cell response during post-SCGx degeneration, whereas ß-adrenoceptor blockade did not affect the release of calcitonin or PTH, although it was effective in partially reversing the activity of α-adrenoceptor blockade. The results point to a significant inhibitory effect of sympathetic nerves on calcitonin and PTH release [24].

To study the effect of regional parasympathectomy on the release of calcitonin and PTH animals with inferior laryngeal or thyroid nerves, sections were carried out. After the injection of calcium chloride, a greater increase in serum calcitonin and lower blood calcium levels than in controls were seen. PTH levels after a hypocalcemia challenge were also higher in parasympathectomized rats. Thus, the thyroid parasympathetic innervation exerts an inhibitory influence on calcitonin secretion by C-cells. In summary, these few examples in the thyroid–parathyroid territory indicate that the nerves supplying the endocrine glands were alternative routes through which the brain communicates with them [24].

Similar phenomena occur for regional autonomic nerves in the adrenal glands, the gonads, and pancreatic islets. For example, extrapituitary mechanisms of adrenal cortical control, including sympathetic neural activity, have been implicated in controlling the amplitude of the cortisol awakening response, a diagnostic index of hypothalamic pituitary adrenal activity in humans. In addition, increases in sympathetic neural tone have been implicated in polycystic ovary syndrome, a leading cause of female infertility [41, 42]. In the ovary, the superior ovarian nerve inhibits ovarian estradiol secretion by activation of α2 adrenoceptors [43].

Intracellular glucose signaling pathways control the secretion of glucagon and insulin by pancreatic islet α- and β-cells respectively [44]. In addition, glucose also indirectly controls the secretion of these hormones through the regulation of the ANS that richly innervates this endocrine organ. Both parasympathetic and sympathetic nervous systems also have an impact on postnatal endocrine pancreas development and plasticity in adult animals. Defects in these autonomic regulations impair β-cell mass expansion during the weaning period and β-cell mass adaptation in adult life. Both branches of the ANS also regulate glucagon secretion. In type 2 diabetes mellitus, impaired glucose-dependent autonomic activity causes the loss of the cephalic and first phases of insulin secretion, and impaired suppression of glucagon secretion in the postabsorptive phase [45].

Collectively, these examples offer compelling evidence for the capability of autonomic neural activity to alter the functional sensitivity of endocrine glands. By using a dual viral tracing technique that enabled simultaneous exploration of the neural circuits of two organs, the following observations were made [46]:

(1) There are autonomic neurons exclusively innervating a given endocrine organ.
(2) A left-sided predominance in the supraspinal innervation of the endocrine glands (adrenal, ovary) studied so far.
(3) Viral co-infection of neurons, i.e., special neuronal populations coexist in different brain areas that are trans-synaptically connected with both paired endocrine and non-endocrine organs, endocrine glands and non-endocrine organs.

The number of common neurons seems to be related to the need to coordinate the action of different systems. The data on co-infection of neurons suggest that the CNS might have the capacity to coordinate different organ functions via common brain neurons, providing supraspinal innervation of the organs [46].

Therefore, the concept that nerves innervating endocrine glands constitute alternative pathways through which the brain communicates with the endocrine glands must be emphasized. The organization of these pathways, as for many other sensory and motor pathways of the central nervous system, is essentially hierarchical and in parallel (Fig. 3.30).

The ANS Contributes to the Maintenance of Healthy Bone Tissue

The skeleton is a specialized and dynamic organ that undergoes continuous regeneration. It consists of highly specialized cells, mineralized and nonmineralized connective tissue matrix, and spaces that include the bone marrow cavity, vascular canals, canaliculi, and lacunae. Removal of bone (resorption) is the task of osteoclasts, whereas formation of new bone is the task of osteoblasts. Tight molecular control of bone remodeling is vital for the maintenance of appropriate physiology and microarchitecture of the bone, providing homeostasis, also at the systemic level. The process of remodeling is regulated by the rich innervation of the skeleton, as it is the source of various growth factors, neurotransmitters, and hormones regulating the functions of the bone [47].

A ubiquitous autonomic innervation of all periosteal surfaces exists and its disruption may affect bone remodeling control and lead to various bone diseases (e.g., osteogenesis

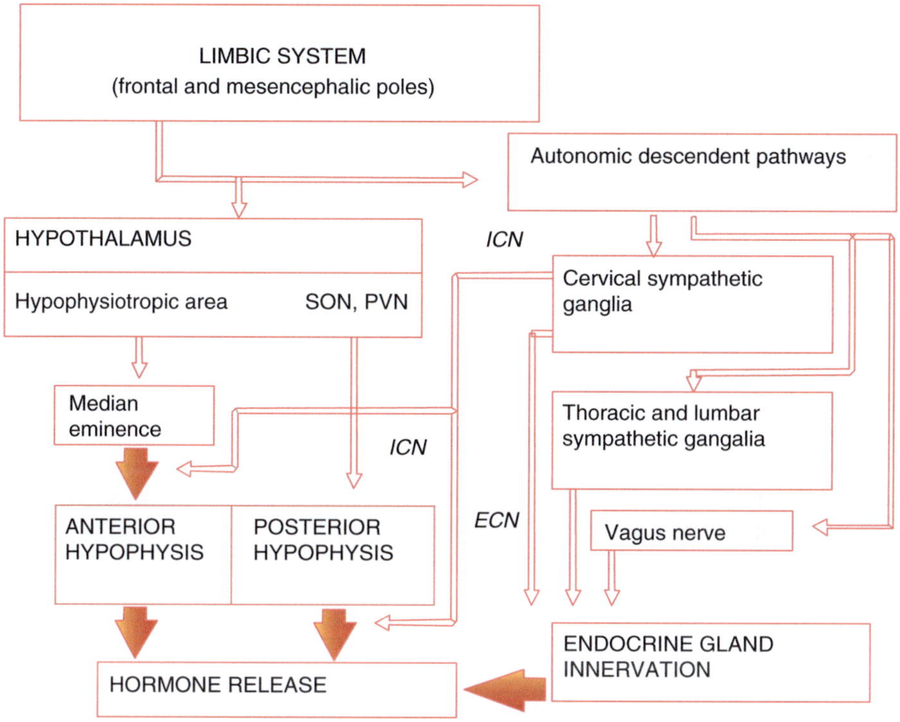

Fig. 3.30 Hierarchical and parallel organization of neurohormonal pathways under endocrine control. *ICN* internal carotid nerve, *ECN* external carotid nerve, *SON* supraoptic nucleus, *PVN* paraventricular nucleus. Reproduced with permission from Cardinali [40]

imperfecta). Patients with neurological disorders exhibit localized osteopenia and bone fragility, altered fracture healing, and excessive callus formation [48].

An intact ANS contributes to the maintenance of healthy bone tissue. The role of the ANS in abnormal bone formation and its association with clinical diseases has been proposed, for example, postmenopausal osteoporosis, adolescent idiopathic scoliosis, and depression-induced osteoporosis [49]. The long bones of the upper extremities receive nerve supply from the brachial plexus, which then branches to the median nerve to innervate the humerus and the ulnar and radian nerves, which supply the forearm bones. Sympathetic innervation of the lower limbs originates in the lumbar plexus, which supplies the femoral and deep saphenous nerves to the femur, and the tibial, medial, and popliteal nerves to the tibia and fibula. Basivertebral nerves in the spine supply intraosseous autonomic innervations of the vertebral bodies [50].

In the author's laboratory, studies were focused on results obtained in the field of projection of the SCG territory, the first sympathetic ganglion of the paravertebral chain, which includes the mandibular bone [51]. To assess in an anatomically specific way the effect of local sympathectomy on bone physiology, we examined the effect of unilateral SCGx on growth, bone mineral content, and bone mineral density of the ipsi- and contralateral hemimandibles. Total bone mineral content of the hemimandibular bones decreased on the side ipsilateral to the SCGx. Bone mineral density (i.e., the bone mineral content/bone mineral area ratio) was also significantly lower in the hemimandible ipsilateral to the SCGx.

The ANS is one of the factors modifying tooth eruption. Teeth are innervated by unmyelinated sympathetic axons originating in the ipsilateral SCG, and by unmyelinated and small myelinated sensory axons, most of them terminal branches of larger parent axons in the trigeminal nerve. The sympathetic nerve endings contain NE, whereas sensory dental axons contain SP-like immunoreactivity. Other neuropeptides are also present in dental nerves, such as VIP and NPY. Adrenergic nerves end at the odontoblast/predentin border, in the predentin adjacent to the odontoblast processes, and as free endings in the middle part of the predentin. In situations of degenerative autonomic neuropathy, an overall marked reduction in pulpal innervation, with an absence of large nerve bundles and the subodontoblastic plexus, has been reported in humans.

We examined the effect of a unilateral SCGx on the eruption rate of ipsilateral and contralateral rat incisors in two experimental situations: (a) without any further manipulation; (b) after the ipsi- or contralateral lower rat incisor had been cut out of

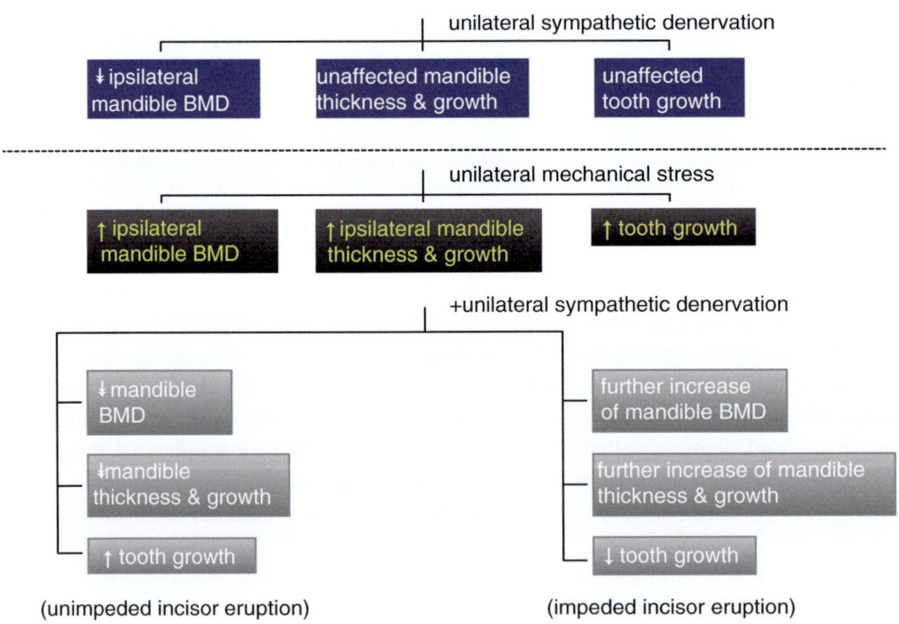

Fig. 3.31 Diagram summarizing the effect of unilateral sympathetic denervation on rat mandibular bone. The effect of a unilateral superior cervical ganglionectomy on bone mineral density (BMD), morphometric assessment of mandible thickness, and the growth and eruption rate of ipsilateral and contralateral incisors were analyzed in two experimental situations: (1) without any further manipulation; (2) after a unilateral mechanical stress given by cutting the ipsi- or contralateral lower incisor out of occlusion

occlusion (Fig. 3.31). In a first experiment, the eruption rate of ipsilaterally denervated incisors was similar to that of contralaterally innervated incisors, when assessed for up to 28 days after surgery. In a second experiment, under conditions of unilateral unimpeded eruption of incisors performed ipsilaterally or contralaterally to a unilateral SCGx, a significantly lower eruption rate of denervated incisors on the impeded eruption side and a significantly higher eruption rate of denervated incisors on the unimpeded side were observed (Fig. 3.31).

The results indicated that incisor eruption is not modified by local sympathetic denervation unless the contralateral lower rat incisor was cut out of occlusion. When this was done, sympathetically denervated incisors exhibited higher eruption rates on the unimpeded eruption side and lower eruption rates on the impeded side. Therefore, normal presynaptic neural activity, and not merely an augmented transmitter release, seems to be a prerequisite for maintaining a minimal normal unimpeded incisor eruption and for maintaining the unimpeded eruption to attain abnormally high velocities under conditions of stimulated incisor growth. This dual role of the ANS on organ activity is found in several other tissues, such as endocrine and immune tissues. For example, a normal sympathetic output maintains basal cell proliferation in lymph nodes and curtails overstimulation following an antigen challenge [52].

References

1. Cardinali DP. Neurociencia Aplicada. Sus Fundamentos. Editorial Médica Panamericana: Buenos Aires; 2007.
2. Goldstein DS. Chapter 2 – Differential responses of components of the autonomic nervous system. In: Buijs RM, Swaab D, editors. Handbook of clinical neurology, autonomic nervous system. New York: Elsevier; 2013. p. 13–22.
3. Waxman SG. Chapter 20. The autonomic nervous system. In: Clinical neuroanatomy. 27th ed. New York: McGraw-Hill; 2013.
4. Barrett KE, Barman SM, Boitano S, Brooks HL. Autonomic nervous system. In: Ganong's review of medical physiology. 25th ed. New York: McGraw-Hill Education; 2016.
5. Fernandez-Gil MA, Palacios-Bote R, Leo-Barahona M, Mora-Encinas JP. Anatomy of the brainstem: a gaze into the stem of life. Semin Ultrasound CT MR. 2010;31:196–219.
6. De Groat WC, Griffiths D, Yoshimura N. Neural control of the lower urinary tract. Compr Physiol. 2015;5:327–96.
7. Westfall TC, Westfall DP. Neurotransmission: the autonomic and somatic motor nervous systems. In: Brunton LL, Chabner BA, Knollmann BC, editors. Goodman & Gilman's: the pharmacological basis of therapeutics. 12th ed. New York: McGraw-Hill Education; 2011. p. 1–58.
8. Ray A, Chakraborti A, Gulati K. Current trends in nitric oxide research. Cell Mol Biol. 2007;53:3–14.
9. Wehrwein EA, Joyner MJ. Chapter 8 – Regulation of blood pressure by the arterial baroreflex and autonomic nervous system. In: Buijs RM, Swaab D, editors. Handbook of clinical neurology, autonomic nervous system. Elsevier; 2013, p 89–102.
10. LeBlond RF, Brown DD, Suneja M, Szot JF. The nervous system. In: DeGowins diagnostic examination. 10th ed. New York: McGraw-Hill Education; 2015.
11. Kibble JD, Halsey CR. Neurophysiology. In: Medical physiology: the big picture. New York: McGraw-Hill Education; 2015.

12. Burnstock G. Chapter 3 – Cotransmission in the autonomic nervous system. In: Buijs RM, Swabb D, editors. Handbook of clinical neurology. Autonomic nervous system. New York: Elsevier; 2013, p 23–35.
13. Herring N. Autonomic control of the heart: going beyond the classical neurotransmitters. Exp Physiol. 2015;100:354–8.
14. Hahn MK. Chapter 8 – Antidepressant-sensitive norepinephrine transporters: structure and regulation. In: Biaggioni I, Burnstock G, Low PA, Paton JFR, editors. Primer on the autonomic nervous system. Third ed. San Diego: Academic; 2012. p. 49–50.
15. Sherrington C. The integrative action of the nervous system. Cambridge: Cambridge University Press; 1906.
16. Craig AD. Chapter 9 – Cooling, pain, and other feelings from the body in relation to the autonomic nervous system. In: Buijs RM, Swabb D, editors. Handbook of clinical neurology, autonomic nervous system. New York: Elsevier; 2013. p. 103–9.
17. Benarroch EE. Central autonomic network. Functional organization and clinical correlations. Leander: Futura Publishing; 1997.
18. Longhurst JC, Fu LW. Chapter 35 – Cardiac and other visceral afferents. In: Biaggioni I, Burnstock G, Low PA, Paton JFR, editors. Primer on the autonomic nervous system. Third ed. San Diego: Academic; 2012. p. 171–6.
19. Zhuo M. Contribution of synaptic plasticity in the insular cortex to chronic pain. Neuroscience. 2016;338:220–9.
20. Verriotis M, Chang P, Fitzgerald M, Fabrizi L. The development of the nociceptive brain. Neuroscience. 2016;338:207–19.
21. Gonzalez Burgos G, Biali FI, Cardinali DP. Picrotoxin-sensitive receptors mediate γ-aminobutyric acid-induced modulation of synaptic plasticity in rat superior cervical ganglion. Brain Res. 1997;751:148–51.
22. Landa ME, Gonzalez Burgos G, Cardinali DP. In vitro effects of thyroxine on cholinergic neurotransmission in rat sympathetic superior cervical ganglion. Neuroendocrinology. 1991;54:552–8.
23. Shireen E. Experimental treatment of antipsychotic-induced movement disorders. J Exp Pharmacol. 2016;8:1–10.
24. Romeo HE, Cardinali DP. The autonomic nervous system of the cervical region as a channel of neuroendocrine communication. Front Neuroendocrinol 1991;12:278–97.
25. Seth JH, Panicker JN, Fowler CJ. Chapter 10 – The neurological organization of micturition. In: Buijs RM, Swabb D, editors. Handbook of clinical neurology, autonomic nervous system. New York: Elsevier; 2013. p. 111–7.
26. Ranson RN, Saffrey MJ. Neurogenic mechanisms in bladder and bowel ageing. Biogerontology. 2015;16:265–84.
27. Low PA, Engstrom JW. Disorders of the autonomic nervous system. In: Kasper D, Fauci A, Hauser S, Longo D, Jameson JL, Loscalzo J, editors. Harrison's principles of internal medicine. 19th ed. New York: McGraw-Hill Education; 2015.
28. McDougal DH, Gamlin PD. Autonomic control of the eye. Compr Physiol. 2015;5:439–73.
29. Joos KM, Melson MR. Chapter 49 – Control of the pupil. In: Biaggioni I, Burnstock G, Low PA, Paton JFR, editors. Primer on the autonomic nervous system. Third ed. San Diego: Academic; 2012. p. 239–42.
30. Clement P, Giuliano F. Anatomy and physiology of genital organs – men. Handb Clin Neurol. 2015;130:19–37.
31. Jobling P. Autonomic control of the urogenital tract. Auton Neurosci. 2011;165:113–26.
32. Avetisyan M, Schill EM, Heuckeroth RO. Building a second brain in the bowel. J Clin Invest. 2015;125:899–907.
33. Kibble JD, Halsey CR. Gastrointestinal physiology. In: Medical physiology: the big picture. New York: McGraw-Hill Education; 2015.
34. Camilleri M. Chapter 42 – Gastrointestinal function. In: Biaggioni I, Burnstock G, Low PA, Paton JFR, editors. Primer on the autonomic nervous system. Third ed. San Diego: Academic; 2012. p. 205–9.

35. Ochoa-Cortes F, Turco F, Linan-Rico A, Soghomonyan S, Whitaker E, Wehner S, Cuomo R, Christofi FL. Enteric glial cells: a new frontier in neurogastroenterology and clinical target for inflammatory bowel diseases. Inflamm Bowel Dis. 2016;22:433–49.
36. Barrett KE. Gastrointestinal physiology. Second ed. New York: McGraw-Hill Education; 2014.
37. Berg CJ, Kaunitz JD. Gut chemosensing: implications for disease pathogenesis. F1000Res. 2016;5:2424.
38. Erny D, Hrabe de Angelis AL, Prinz M. Communicating systems in the body: how microbiota and microglia cooperate. Immunology. 2017;150:7–15.
39. Mayer EA, Tillisch K, Gupta A. Gut/brain axis and the microbiota. J Clin Invest. 2015;125:926–38.
40. Cardinali DP. Ma Vie en Noir. Fifty years with melatonin and the stone of madness. Switzerland: Springer; 2016.
41. Lansdown A, Rees DA. The sympathetic nervous system in polycystic ovary syndrome: a novel therapeutic target? Clin Endocrinol. 2012;77:791–801.
42. Engeland WC. Chapter 4 – Sensitization of endocrine organs to anterior pituitary hormones by the autonomic nervous system. In: Buijs RM, Swaab D, editors. Handbook of clinical neurology, autonomic nervous system. New York: Elsevier; 2013. p. 37–44.
43. Uchida S, Kagitani F. Autonomic nervous regulation of ovarian function by noxious somatic afferent stimulation. J Physiol Sci. 2015;65:1–9.
44. Rodriguez-Diaz R, Caicedo A. Neural control of the endocrine pancreas. Best Pract Res Clin Endocrinol Metab. 2014;28:745–56.
45. Thorens B. Neural regulation of pancreatic islet cell mass and function. Diabetes Obes Metab. 2014;16(Suppl 1):87–95.
46. Gerendai I, Toth IE, Boldogkoi Z, Halasz B. Recent findings on the organization of central nervous system structures involved in the innervation of endocrine glands and other organs; observations obtained by the transneuronal viral double-labeling technique. Endocrine. 2009;36:179–88.
47. Niedzwiedzki T, Filipowska J. Bone remodeling in the context of cellular and systemic regulation: the role of osteocytes and the nervous system. J Mol Endocrinol. 2015;55:R23–36.
48. Binks S, Dobson R. Risk factors, epidemiology and treatment strategies for metabolic bone disease in patients with neurological disease. Curr Osteoporos Rep. 2016;14:199–210.
49. Ji-Ye H, Xin-Feng Z, Lei-Sheng J. Chapter 14 – Autonomic control of bone formation: its clinical relevance. In: Buijs RM, Swaab D, editors. Handbook of clinical neurology, autonomic nervous system. New York: Elsevier; 2013. p. 161–71.
50. Elefteriou F, Campbell JP. Chapter 53 – Autonomic innervation of the skeleton. In: Biaggioni I, Burnstock G, Low PA, Paton JFR, editors. Primer on the autonomic nervous system. Third ed. San Diego: Academic; 2012. p. 257–9.
51. Boggio V, Ladizesky MG, Cutrera RA, Cardinali DP. Autonomic neural signals in bone: physiological implications for mandible and dental growth. Life Sci. 2004;75:383–95.
52. Cardinali DP, Cutrera RA, Esquifino AI. Psychoimmune neuroendocrine integrative mechanisms revisited. Biol Signals Recept. 2000;9:215–30.

Second Level: The Brainstem

4

Abstract

The sympathetic and parasympathetic divisions of the ANS are integrated and regulated by a hierarchy of central structures. This central autonomic neural network has been identified and mapped in recent years and functional neuroimaging data largely support it. Such a network involves reciprocal connections, both direct and indirect, between the two efferent systems and with more cranial neural clusters located in the brainstem, hypothalamus, limbic system, and the insular cortex. As examples of that organization, this Chapter discusses the neural mechanisms that control arterial blood pressure, heart rate, breathing, immunity, and gastrointestinal function and how they vary in the three body configurations (wakefulness, NREM sleep, REM sleep).

Keywords

Brainstem • Cardiorespiratory homeostasis • Carotid body • Cerebellum • Gastrointestinal function • Immunity • Migrating motor complex • Neural regulation of cardiovascular function • Neural regulatory system of breathing • Nucleus tractus solitarium • Parabrachial nucleus • Reticular formation

Objectives

After studying this chapter, you should be able to:

- Describe the location of forebrain and brainstem neurons that are components of central autonomic pathways.
- Describe the neural mechanisms that control arterial BP and heart rate, including the receptors, afferent and efferent pathways, central integrating pathways, and the effector mechanisms involved, and how they vary in the three body configurations (wakefulness, NREM sleep, REM sleep) during a 24-h cycle.
- Identify the location and functions of the dorsal and ventral groups of respiratory neurons, the pneumotaxic center, and the apneustic center in

D.P. Cardinali, *Autonomic Nervous System*, DOI 10.1007/978-3-319-57571-1_4

the brainstem and how they vary in the three body configurations (wakefulness, NREM sleep, REM sleep) during a 24-h cycle.
- Describe the participation of the cerebellum in the autonomic posture.
- Identify the roles and mechanisms of innate, acquired, humoral, and cellular immunity and how they vary in the three body configurations (wakefulness, NREM sleep, REM sleep) during a 24-h cycle.
- Describe the functional significance of the gastrointestinal system, and its roles in nutrient assimilation, excretion, and immunity and how they vary in the three body configurations (wakefulness, NREM sleep, REM sleep) during a 24-h cycle.

At the Brainstem, Various Complex Autonomic Responses Are Coordinated

The sympathetic and parasympathetic divisions of the ANS are integrated and regulated by the hierarchy of central structures plotted in Fig. 1.2. This central autonomic neural network has been clearly identified and mapped in recent years and functional neuroimaging data largely support it. Such a network involves reciprocal connections, both direct and indirect, between the two efferent systems and with more cranial neural clusters located in the brainstem, hypothalamus, limbic system, and brain (e.g. insular cortex).

The brainstem is the region between the spinal cord and the diencephalon. It comprises the medulla oblongata, the pons, and the midbrain. Many sensory and motor nuclei and the different components of reticular formation are in the brainstem and lesions affecting this area have profound motor, sensory, and consciousness-related consequences, which is why a discussion of its organization is necessary for an understanding of the physiology of the ANS.

The brainstem is a site of somatic and autonomic motoneurons and different types of sensory neurons that are organized into anatomically identifiable columns corresponding to the cranial nerves [1]. The cranial nerves have three defined functions: (a) to provide somatosensory and motor innervation to the neck and head; (b) to provide innervation to the sense organs; (c) to provide parasympathetic preganglionic innervation to autonomic ganglia that control visceral function.

As in the spinal cord, somatic motor efferent neurons, autonomic efferent neurons (preganglionic) and second-order sensory afferent neurons coexist in the brainstem. Second-order sensory neurons receive afferent fibers from somatic or visceral primary sensory neurons located in ganglia outside the brainstem or in the sense organs (the only exception to this rule is the mesencephalic nucleus of the V pair, which contains primary sensory neurons).

What is particular in the brainstem is that, in the case of motor and sensory, somatic and visceral neurons of the cranial nerves are subdivided into anatomically segregated functional groups. This subdivision consists of two organizational principles:

1. There are three types of motor neurons in the brainstem: somatic motoneurons, special visceral motoneurons, and general visceral motoneurons. Somatic

motoneurons innervate the muscles of the face and neck derived from the myotome, with the same origin as the rest of the skeletal musculature. They send their fibers through the III, IV, VI, and XII cranial nerve pairs (voluntary ocular musculature and tongue). Special visceral motoneurons also innervate the striated musculature of the face and neck, but in this case, the muscles are derived from the gill arches (mastication, facial expression, larynx, pharynx). They send their fibers through V, VII, IX, X, and XI cranial nerve pairs. General visceral motoneurons provide parasympathetic preganglionic autonomic innervation. They send their fibers through the III, VII, IX and X cranial nerve pairs.

2. There are four types of second-order sensory neurons in the brainstem: general somatic sensitivity, special somatic sensitivity, general visceral sensitivity, and special visceral sensitivity respectively. General somatic sensitivity comprises the sensitivity of face skin and oral and pharyngeal mucous membranes, and proprioception (V, VII, IX and X cranial nerve pairs). The special somatic sensitivity comprises that of the inner ear (VIII cranial nerve pair). The general visceral sensitivity comprises that coming from thoracic and abdominal organs (IX and X cranial nerve pairs). The special visceral sensitivity originates in the gustatory corpuscles (VII, IX, X cranial nerve pairs).

These seven functionally differentiable neuronal groups are grouped in columns along the brainstem, as represented in Fig. 4.1:

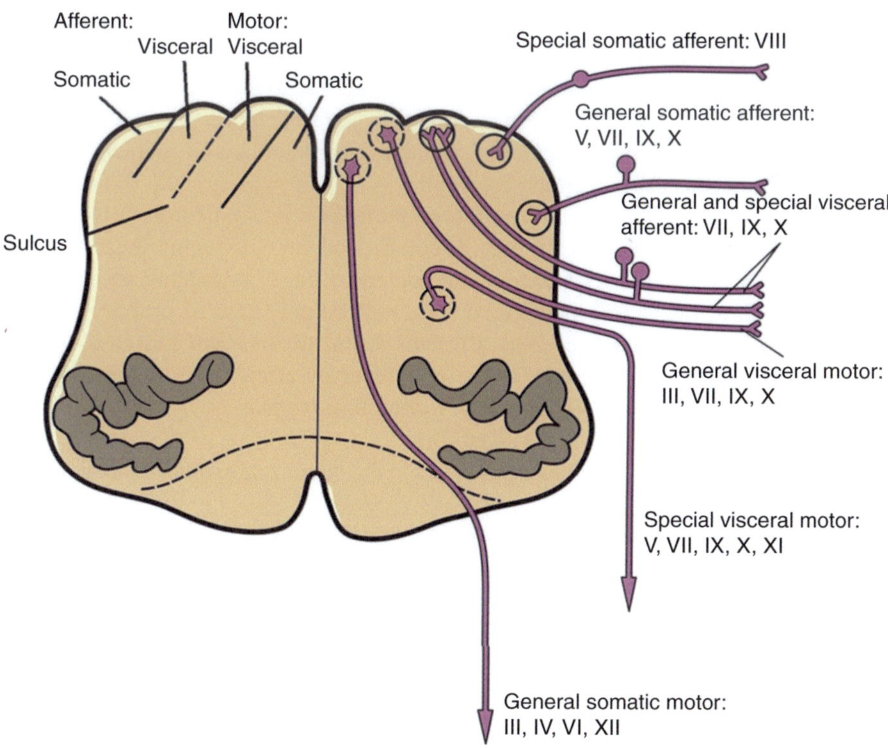

Fig. 4.1 Nuclei of cranial nerves and column organization of the brainstem. Modified with permission from Cardinali [3]

- The somatic motor column, which contains motor neurons that innervate the extraocular and tongue muscles (in rostrocaudal order).
- The special visceral motor column containing motor neurons that supply the muscles of the larynx, pharynx, face, and jaw. They are, in descending order: a) motor nucleus of the V cranial nerve pair (mastication); (b) motor nucleus of the VII cranial nerve pair (facial expression); (c) ambiguous nucleus (IX, X cranial nerve pairs; speech, swallowing); (d) nucleus of the spinal accessory nerve (XI cranial nerve pair).
- The general visceral motor column, which contains the neuronal bodies of parasympathetic preganglionic neurons: (a) Edinger–Westphal nucleus (III), preganglionic to ciliary ganglion; (b) upper salivary nucleus (VII), preganglionic to the sphenopalatine ganglion (lacrimal gland) and submaxillary ganglion (sublingual and submaxillary glands); (c) inferior salivary nucleus (IX), preganglionic to the otic ganglion (parotid gland); (d) dorsal motor nucleus of the X pair: parasympathetic preganglionic to the thoracic and abdominal organs.
- The general visceral afferent columns and the special visceral afferent columns form the nucleus tractus solitarium (NTS), which comprises two parts: (a) rostral, which is the relay site for taste and general visceral afferents from the digestive tract; (b) caudal, which receives the afferents of the carotid body and the general visceral afferent from the bronchi and lungs. The perikarya of these primary sensory neurons are in ganglia associated with the VII, IX and X cranial nerve pairs, and their central extensions form the solitary tract, ending in the secondary neurons that form the solitary nucleus. From here, the portion that mediates sensitivity projects to the thalamus, whereas that corresponding to cardiorespiratory regulation establishes direct contact with the reticular and indirect formation with the limbic system (through the parabrachial nucleus, PBN).
- The special somatic afferent column contains the vestibular and cochlear secondary relay neurons (VIII cranial nerve pair).
- The general somatic afferent column contains secondary neurons of oral and oropharyngeal somatic sensitivities. It consists of three divisions separated rostrocaudally into: (a) the mesencephalic nucleus of the V cranial nerve pair (proprioception of facial and mandibular muscles, the only case of primary sensory neurons in a brainstem) receives afferents from mucous and muscles; (b) the main nucleus of the V cranial nerve pair receives afferents from mucous and muscles; (c) the spinal nucleus of the V cranial nerve pair.

In summary, there are three basic principles in the organization of the cranial nerves:

1. Most motor nuclei are linked to individual cranial nerves.
2. Relay nuclei in the sensory pathways receive fibers from various cranial nerve pairs, e.g., the NTS receives taste information brought by the VII, IX, and X pairs, following the general principle that somatotopic sensory location ends in the same central area, regardless of the route it follows.
3. Neurons with similar functional properties occupy the same positions in the brainstem.

The NTS, located in the dorsal part of the medulla oblongata, and secondarily the NPB, constitute the main centers of central relay of visceral sensory information (Fig. 4.2). The main afferent inputs of the NTS comprise fibers from the cardiovascular, respiratory, gastrointestinal, and neuroendocrine–immune systems, along with gustatory collaterals and somatoesthetic sensitivity.

The NTS is the primary integrative center for cardiovascular control and other autonomic functions in the CNS. The NTS has long been identified as a site where the first synapse of the baroreceptor reflex is located. Therefore, the NTS, in addition to other key central nuclei in the hypothalamus and other forebrain regions, play important roles in mediating cardiovascular responses to acute stresses. The NTS not only integrates convergent information, but itself is the site of substantial modulation. Evidence demonstrated several neurotransmitters or neuromodulators such as Glu, NE, E, ACh, 5-HT, NO, angiotensin II, arginine vasopressin (AVP), β-endorphin, enkephalins, NPY, adenosine and insulin [2].

By means of neural circuitry mapping techniques a viscerotopic map was traced in the NTS and in the neuronal groups of the dorsal nucleus of the X cranial nerve pair. These data support the notion that vagal information in the gut–brain axis maintain specific pathways from the gastrointestinal tract through individualized vagal branches. The viscerotopic organization of the gut–brain afferent loop correlates with a parallel map in the efferent limb of the brain–gut axis forming a network

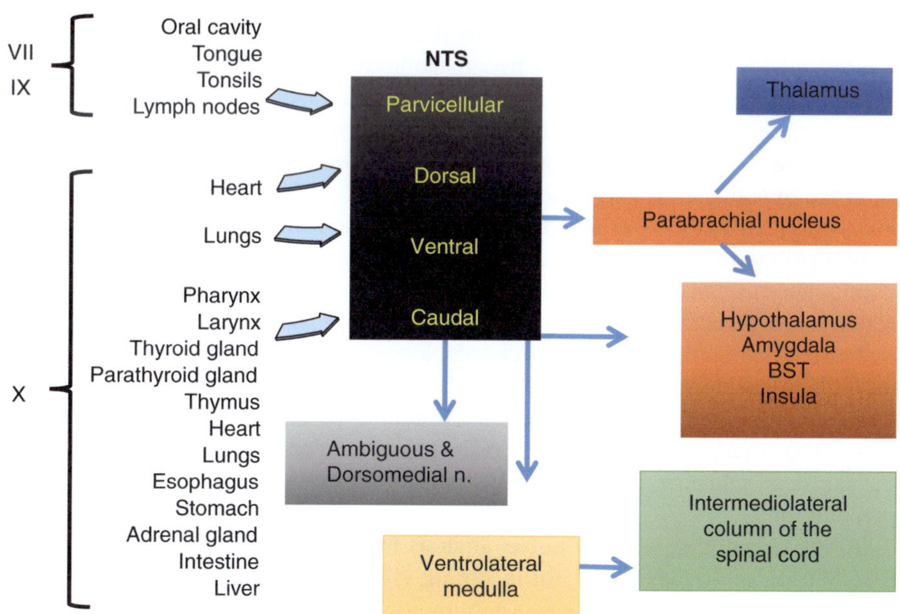

Fig. 4.2 Visceral sensory input to the nucleus tractus solitarius (NTS) brain relevant to the control of food intake and energy balance. All along the alimentary canal, various mechano- and chemosensors are located that transmit food- and nutrition-related signals via primary visceral afferents in the trigeminal (*V*), facial (*VII*), glossopharyngeal (*IX*), and vagus nerve (*X*) to the brainstem

ideally suited to mediate cephalic, gastric, hepatic, and intestinal reflexes accompanying the food intake.

The NTS is subdivided into several subnuclei, which play different functional roles according to the visceral afferents they receive (Fig. 4.2). Thus, cardiovascular afferences terminate predominantly in the dorsal subnucleus, pulmonary afferents in the ventral subnucleus, and gustatory input in the parvicellular subnucleus, whereas the caudal commissural subnucleus receives afferences from all visceral components. Projections from the NTS transmit a wide range of visceral information to higher nuclei, and establish circuits of reciprocal connection with regions of the reticular formation (particularly PBN and periaqueductal gray matter), hypothalamus, amygdala, limbic system, and insular cortex.

The descending efferent pathways of these circuits innervate, directly or through synapses in the NTS, the parasympathetic preganglionic neurons and the sympathetic intermediolateral columns. The descending autonomic pathways mainly travel ipsilaterally in the anterior portion of the lateral cord. Using neuroimaging (functional magnetic resonance imaging, fMRI; positron emission tomography, PET) the asymmetry and lateralization of the ANS function was verified.

The PBN plays a key role in the central autonomic network, as an intermediate between the brainstem reflex control and behavioral control systems at the diencephalic and telencephalic levels (Fig. 4.2). The most relevant parts of the PBN for controlling food intake and energy homeostasis are the medial and lateral subdivisions. The neurons located in the upper lateral nucleus of the lateral subdivision of PBN are particularly important for the control of food intake, and send dense projections to the ventromedial nucleus (VMN) of the hypothalamus related to the control of satiety. The upper lateral nucleus of the PBN has a large concentration of CCK-expressing neurons that are related to the induction of satiety and are activated by circulating leptin.

The central projections on the preganglionic autonomic neurons come mainly from the hypothalamus and various brainstem nuclei (dorsal raphe nucleus, DRN, locus coeruleus, LC, ventrolateral medulla reticular formation). The role of these central pathways is to maintain a tonic state of excitability of preganglionic autonomic neurons, modulate segmental reflexes, and generate organized patterns of activity in different functional groups of preganglionic neurons.

In the viscera, there are numerous receptors, such as osmoreceptors, baroreceptors, glucoreceptors, etc., which respond to changes in the internal environment. Their afferents participate in different autonomic reflexes of homeostatic importance. These receptors are the basis of interoception, and presumably they can be the physical substrate of what from a psychological point of view is called the preconscious.

A key question is whether nuclei such as the NTS or PBN are predominantly relay nuclei, or do they transform incoming visceral information? It is noteworthy that contrary to popular anatomical terminology, there is no need for relay nuclei because axon potentials do not need to be boosted by the synapse. A chemical synapse on its own simply adds an unavoidable delay.

Neurons form synapses in a nucleus to transform or change the incoming signal. The information obtained in the nuclei of the dorsal column (Goll and Burdach nuclei) in the somatosensory system indicates the existence of several transformations. On the one hand, there is differential convergence that results in a distorted viscerotopic representation of the body. In addition, because of the lateral inhibition, the discrimination of the stimuli is accentuated. As in other sensory territories, stimulus in the center activates NTS neurons whereas stimulus in the surround, through inhibitory feedback, inhibits the same NTS neurons. Another function is cortical gating: corollary discharge selectively gates input based on motor output [3].

The existence of the gating control exerted by superior mesencephalic, diencephalic, and cortical structures on the afferent information that arrives from receptors is exemplified by pain. The main function of this input control (central analgesia) is to suppress the irrelevant information originated in the periphery, allowing only that of sufficient intensity to have an adaptive meaning to pass. As seen in Figs. 4.3 and 4.4, the convergence of visceral and cutaneous afferents on the same second-order neural groups results in visceral pain being experienced on a portion of the cutaneous surface (referred pain). This referred pain is verified in the portion of the cutaneous surface of embryological origin similar to the affected viscera (dermatome) [4].

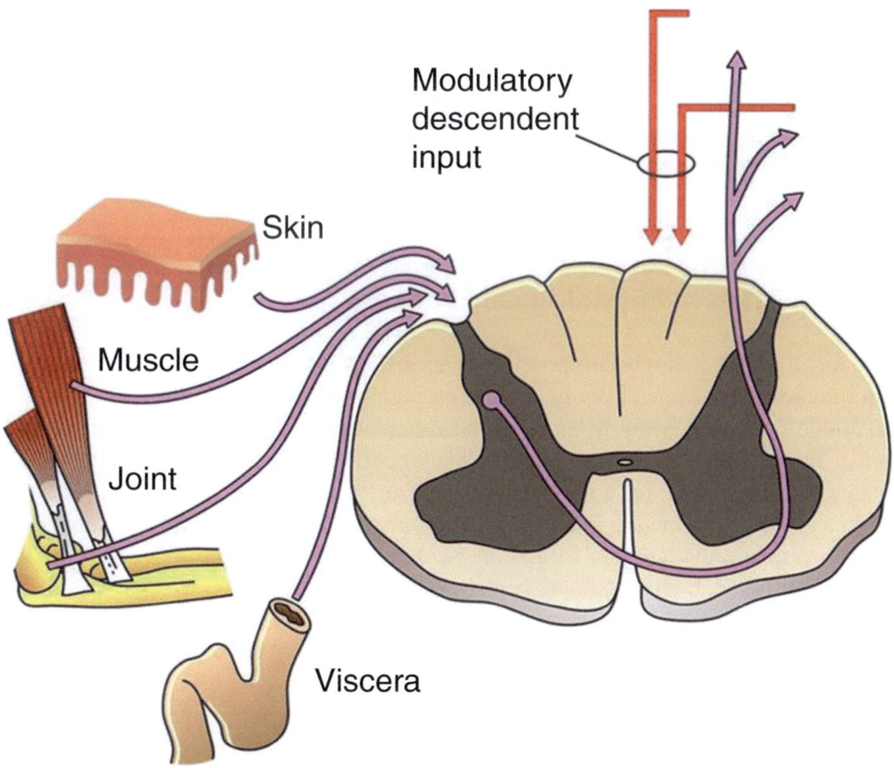

Fig. 4.3 The main function of this afferent control (central analgesia) is to suppress the irrelevant information originating at the periphery, only allowing that of sufficient intensity to have an adaptive meaning to pass. Modified with permission from Cardinali [3]

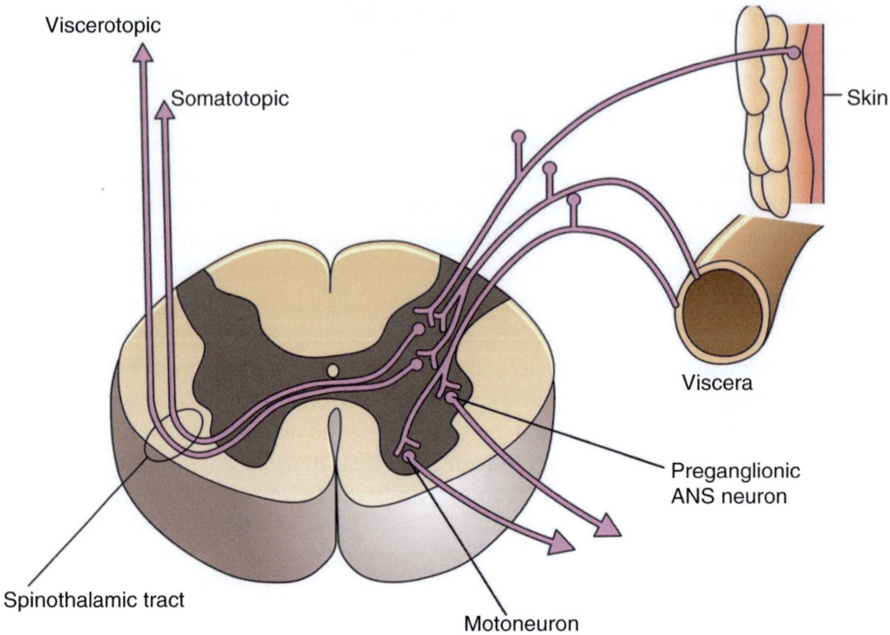

Fig. 4.4 The convergence of visceral and cutaneous afferents on the same second-order neural groups results in visceral pain being experienced on a portion of the cutaneous surface (referred pain). Modified with permission from Cardinali [3]

Monoaminergic Systems in the Brainstem Modulate 24-h Rhythms in Physiological Function

The reticular formation is composed of neurons that do not correspond to the different functional columns of the brainstem mentioned above. It is convenient from an anatomical point of view to analyze the reticular formation in the medial–lateral sense, thus distinguishing (a) the raphe nuclei at both sides of the medial line; (b) a magnocellular region; (c) a parvicellular region.

Pons reticular formation and midbrain reticular formation have different functions. Most neurons in the reticular formation display a great diversity of connections, diffusely distributed, both with upper centers and toward the spinal cord (distribution "in a spider web," Fig. 4.5). Based on the neurotransmitter employed, the following neural groups are distinguished, most of them with locations in the brainstem:

1. Noradrenergic system: located in the LC and contiguous nuclei, it provides innervation to the entire brain, including the spinal cord (Fig. 4.6). Except for a small intrahypothalamic group of neurons, this is the only source of central noradrenergic innervation. As discussed in Chap. 2, its role is fundamental in the maintenance of wakefulness and its alteration is linked to emotional illnesses such as depression, their spinal projection participating in muscular tone control and gait generation. The different NE receptors have been analyzed in Chap. 3.

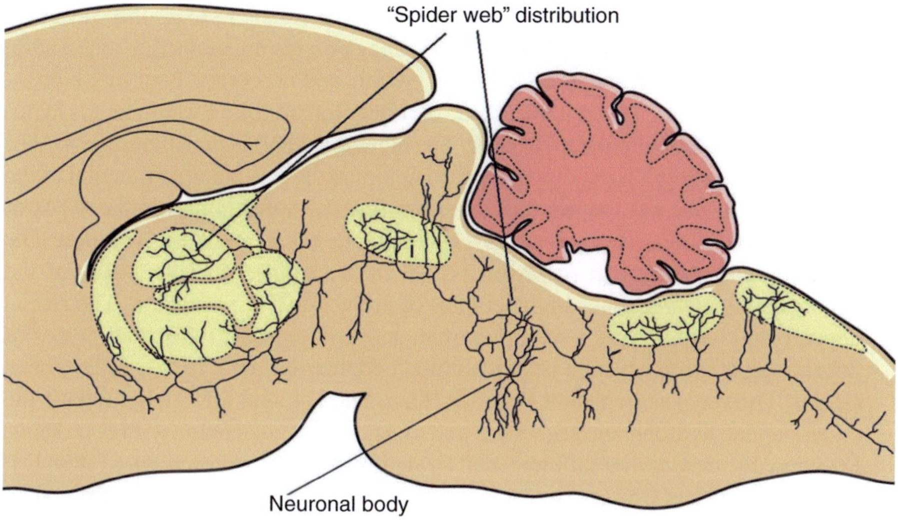

Fig. 4.5 "Spider web" neuron in a rat brainstem. Modified with permission from Cardinali [3]

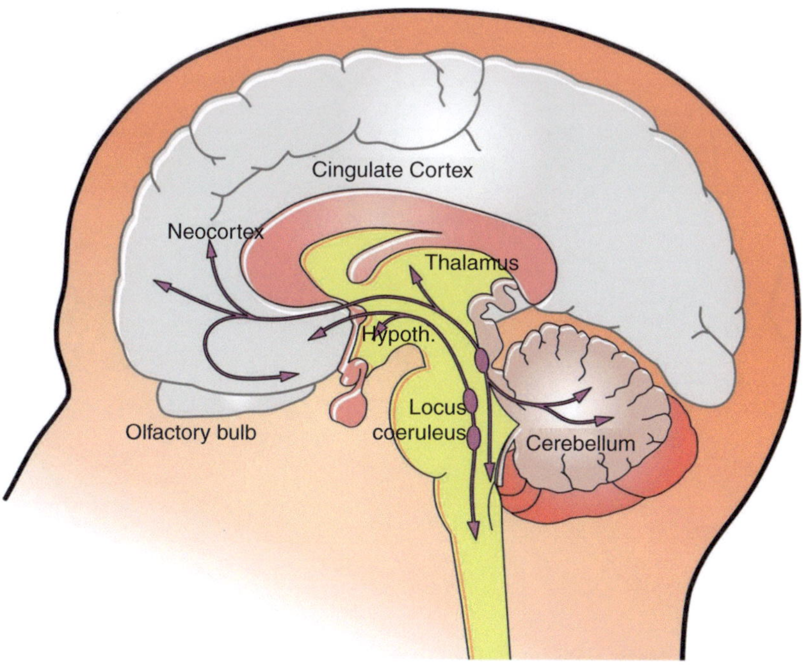

Fig. 4.6 Principal noradrenergic projections of the reticular formation. Modified with permission from Cardinali [3]

2. Dopaminergic system: several cell groups are in the midbrain (e.g., the substantia nigra pars compacta and the VTA, Fig. 4.7). They provide dopaminergic innervation to the basal ganglia, hypothalamus, limbic system, and neocortex. Four major dopaminergic systems are found in the CNS: (a) nigrostriatal; (b) mesolimbic; (c) hypothalamic tuberoinfundibular; (d) retinal. The nigrostriatal projection (neuronal bodies in the substantia nigra pars compacta) participates in the regulation of function of the basal ganglia, whereas the mesolimbic projection (neuronal bodies in the VTA, or A10) is involved in emotional states, psychiatric diseases, and drug addiction. The tuberoinfundibular system (neuronal bodies in the arcuate nucleus, ARC, of the hypothalamus), participates in the control of prolactin (PRL) secretion. The retinal dopaminergic neurons are a subgroup of amacrine interneurons. Six major types of dopaminergic receptors have been identified and their different isoforms have been cloned. These types are called D1, D2a, D2b, D3, D4, and D5. They all have the seven-peptide hydrophobic sequences that indicate their association with G proteins. Functionally, they can be differentiated by their nature into (a) excitatory (class D1, comprising D1 and D5) associated with a stimulating G protein and with the activation of an adenylate cyclase; (b) inhibitory (class D2, comprising D2, D3, and D4) associated with an inhibitory G protein and inhibition of an adenylate cyclase.

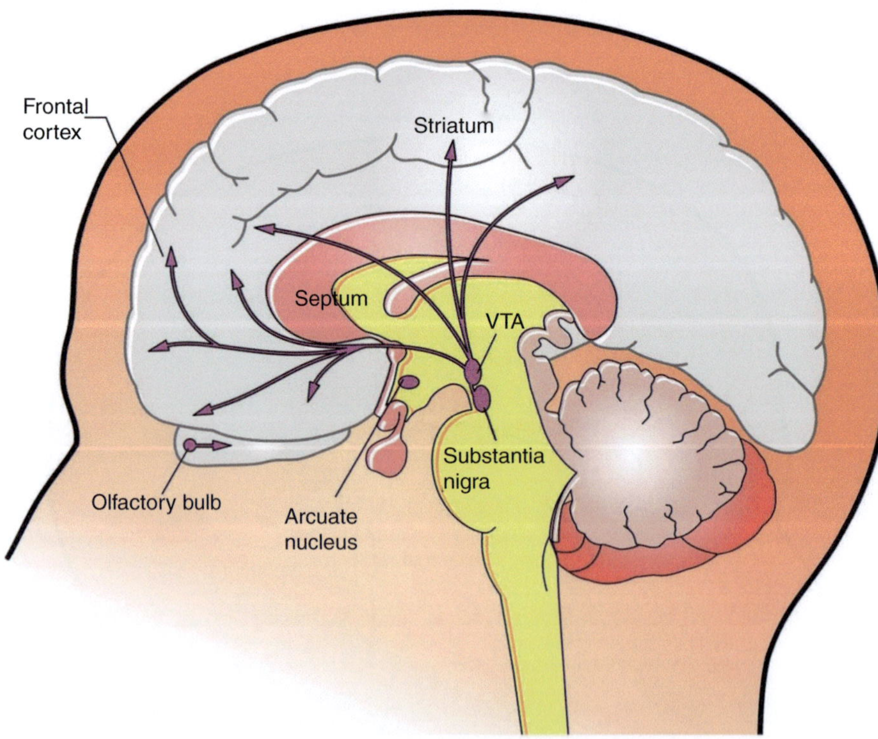

Fig. 4.7 Principal dopaminergic projections of the reticular formation. Modified with permission from Cardinali [3]

3. Serotoninergic system: comprising the DRN (a continuation of the periaqueductal gray matter; Fig. 4.8). DRN are the origin of almost all serotonergic innervation of the brain. The serotonergic receptors (5-HT-1, 5-HT-2, 5-HT-3, 5-HT-4, 5-HT-5, 5-HT-6, and 5-HT-7) are classified by the gene superfamily corresponding to:
   (a) The superfamily of receptors associated with G protein, including 5-HT1A, 5-HT1D, 5-HT1E, and 5-HT1F (all inhibit adenylate cyclase), the 5-HT2 receptor subfamily (5-HT2A, 5-HT2B, and 5-HT2C subtypes, all stimulate the synthesis of phosphoinositides and phospholipase C), and 5-HT-4, 5-HT-5, 5-HT-6, and 5HT-7 receptors (which stimulate adenylate cyclase). These receptors, especially 5-HT-1 (5-HT1D), are linked to the normal and pathological regulation of cerebral flow. 5-HT1A agonists have anxiolytic action
   (b) The superfamily of receptors that directly control channels, corresponding to the 5-HT 3 receptor.
4. Cholinergic system: there is a dense cholinergic innervation of the thalamus, striatum, limbic structures, and cerebral cortex (Fig. 4.9). Except for the striatum, which is mainly intrinsic (from local interneurons), in most other regions cholinergic innervation is extrinsic. The medial–septal area (with the diagonal band of Broca) is the origin of cholinergic innervation of the hippocampus (septohippocampal neurons). The nucleus basalis of Meynert is the origin of the cholinergic innervation of the neocortex and amygdala (Fig. 4.9). The group of cholinergic neurons of the

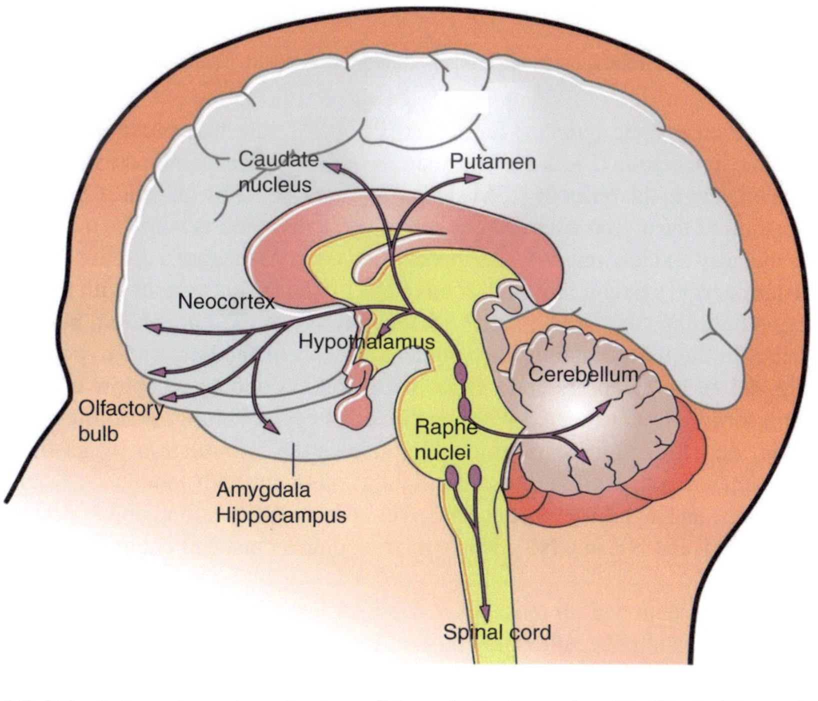

Fig. 4.8 Principal serotonergic projections of the reticular formation. Modified with permission from Cardinali [3]

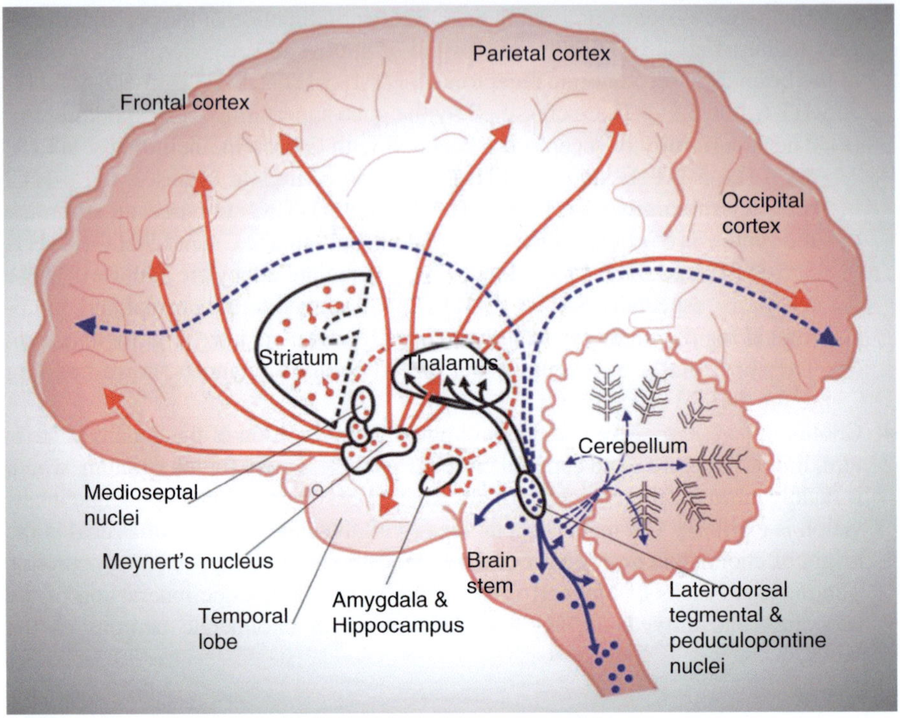

Fig. 4.9 Cholinergic projections in the human brain. Modified with permission from Cardinali [3]

pontomesencephalic reticular formation (PPT/LDT) are the source of cholinergic thalamic innervation (Fig. 4.9). The neurons of these nuclei participate in the waking reaction, and in the onset of REM sleep (they activate REM-on cells, Chap. 2). The alteration of the aforementioned cholinergic systems coexists with the alterations of the memory and learning, such as those observed in Alzheimer's disease (AD).

5. Histaminergic system: although it has been linked to this system with monoaminergic neurons "in a spider web," its location is not mesencephalic, but diencephalic, in the tuberomammillary nucleus (TMN) of the posterior hypothalamus (Fig. 4.10). Its receptors are a class of G protein-coupled receptors comprising four groups: H1 (associated with the circadian cycle, itching, systemic vasodilatation, and bronchoconstriction; H2 (tachycardia, stimulation of gastric acid secretion, smooth muscle relaxation, inhibition of antibody synthesis, T cell proliferation, and cytokine production); H3 (presynaptic autoreceptor, decreasing ACh, 5-HT and NE in CNS); H4 (which mediates mast cell chemotaxis).

As already mentioned, in most cases, these systems do not participate in point-to-point communication, but rather their diffuse distribution points to a general permissive role for other brain processes. This anatomical arrangement offers an ideal anatomical substrate for diffuse functions of alertness, emotionality, and neurovegetative control. In addition to the role discussed in Chap. 2 in the regulation of sleep/wake rhythm, the functions of reticular formation include:

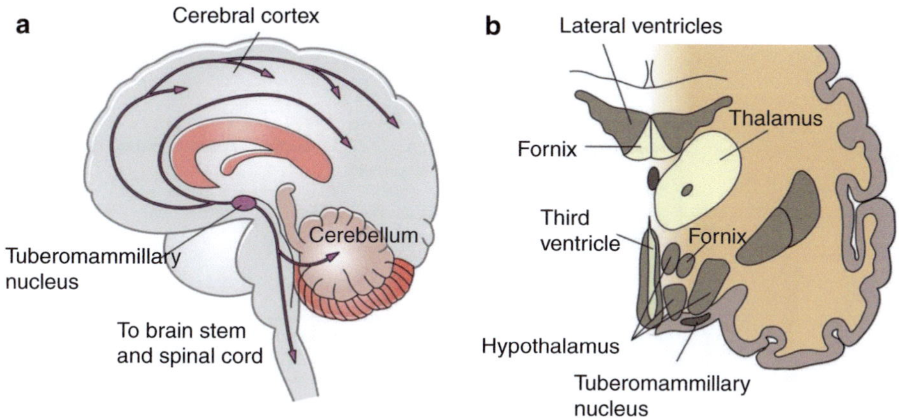

Fig. 4.10 Histaminergic projections originating in the tuberomammillary nucleus (TMN). (**a**) Distribution in a sagittal view. (**b**) Hypothalamic location of the TMN. Modified with permission from Cardinali [3]

- Activation of the CNS to trigger specific behaviors and control of alertness.
- Modulation of medullary reflexes and muscle tone, through the reticulospinal tracts.
- Modulation of locomotion (descending pathways of the LC).
- Modulation of the ventromedial and dorsolateral spinal motor neurons (LC, DRN).
- Participation in respiratory and cardiovascular control.
- Participation in the control of different visceral responses, such as urination, vomiting or defecation.
- Modulation of the flow of nociceptive information in the posterior horn of the medulla, through reticulospinal projections, mainly serotonergic.

Rather than an apparent reticular structure with little order, the monoaminergic areas of the brainstem represent a modular somatotopic organization like other brain areas, such as the neocortex. For example, the periaqueductal serotonergic DRN neurons have columns composed of afferent, output neurons and interneuronal circuits organized viscerotopically. Thus, important functions linked to these areas, such as defensive reactions, analgesia, and autonomic regulation, are integrated in overlapping longitudinal columns, and different types of aversive or painful stimuli trigger specific somatic, vegetative, and nociceptive programs.

24-h Rhythms in Cardiovascular Control

A schematic representation of the extrinsic and intrinsic innervation of the heart is depicted in Fig. 4.11. The neural regulation of the cardiovascular system, controlled by arterial mechanoreceptors, operates as a reflex arc (Fig. 4.12). Elevations or falls in BP cause a proportionally greater or lesser deformation of the arterial walls, which are encoded by baroceptors to a greater or lesser frequency of action potential firing. This information is carried to the CNS by afferent nerve fibers: (a) the aortic

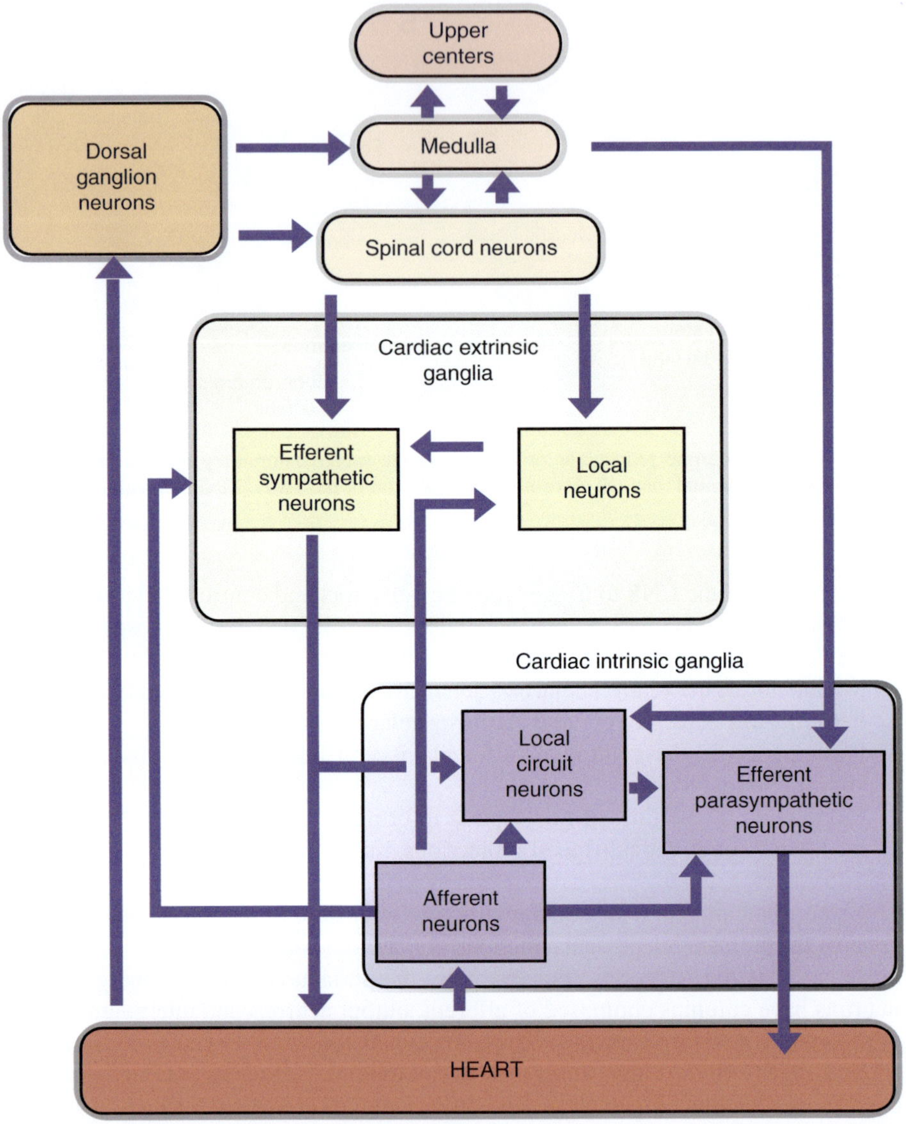

Fig. 4.11 Schematic representation of the extrinsic and intrinsic innervation of the heart

depressor nerve, which travels along the X cranial nerve pair; (b) the sinus nerve (Hering's nerve), which travels along the IX cranial nerve pair [5].

Primary afferent axons from the baroreceptors project to the caudal region of the NTS, where they synapse onto second-order neurons, which in turn send excitatory glutamatergic projections onto inhibitory neurons within the region of the caudal ventrolateral medulla. These inhibitory neurons synapse directly onto excitatory neurons within the rostral ventrolateral medulla and serve to inhibit the spontaneous

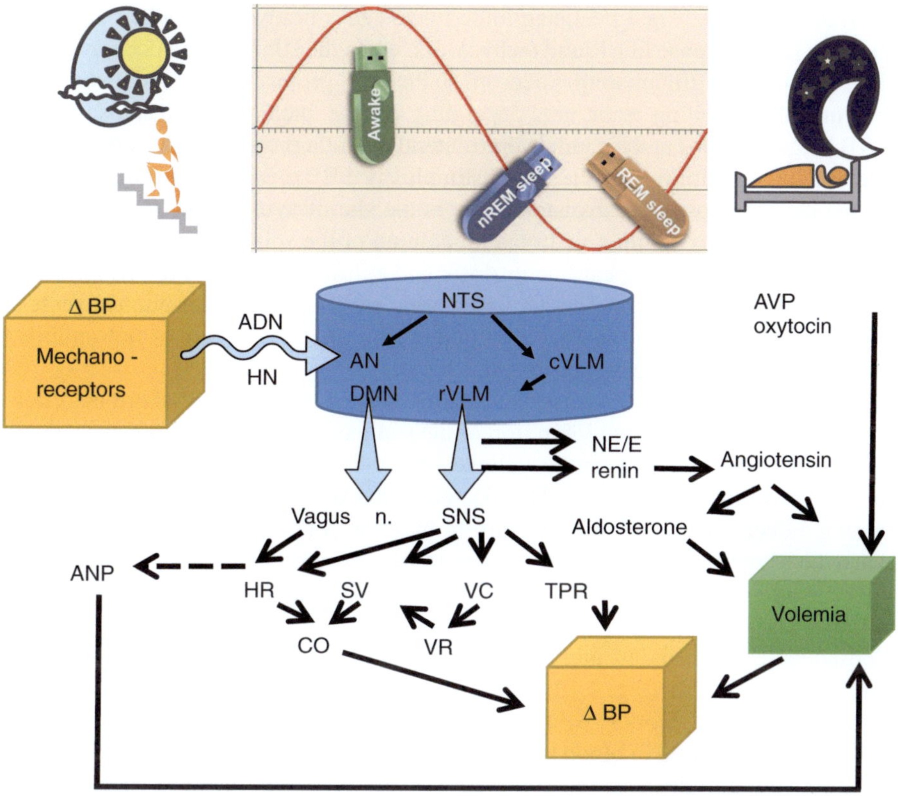

Fig. 4.12 Schematic representation of neurohumoral responses triggered by the stimulation of vascular mechanoreceptors by changes in blood pressure (BP) at the three body configurations in a 24-h cycle. *ADN* aortic depressor nerve, *HN* Hering's nerve, *rVLM* rostral ventrolateral medulla, *cVLM* caudal ventrolateral medulla, *AN* ambiguus nucleus. *DMN* dorsomedial nucleus of vagus, *SNS* sympathetic nervous system, *ANP* atrial natriuretic peptide, *HR* heart rate, *SV* systolic volume, *VC* venous capacitance, *TPR* total peripheral resistance, *CO* cardiac output, *VR* venous return. The figure was prepared in part using image vectors from Servier Medical Art (www.servier.com), licensed under the Creative Commons Attribution 3.0 Unported License (http://creativecommons.org/license/by/3.0/)

activity of premotor sympathetic neurons (Fig. 4.12). Glutamatergic neurons from the NTS also project to the X cranial nerve pair nuclei that in turn project to the heart via the X cranial nerve pair. The vagal output originates from preganglionic neurons located in the dorsal motor nucleus of the vagus and in the ventrolateral portion of the ambiguous nucleus (AN) in the medulla.

Studies in awake human subjects, in which fMRI was coupled with microelectrode recordings of muscle sympathetic nerve activity, has confirmed the operation of these medullary nuclei in humans [6]. Specifically, information on changes in the rostral ventrolateral medulla indicated that it is implicated in elevated sympathetic outflow associated with several cardiovascular diseases including hypertension and heart failure related to physical inactivity [7].

In the heart, the vagal postganglionic fibers cause bradycardia, when the initial stimulus is an increase in BP, and tachycardia, when the stimulus is a decrease in BP (Fig. 4.12). Other glutamatergic neurons in the NTS project to the caudal ventrolateral medulla, exciting it or not, depending on the increase or decrease in BP. GABAergic neurons at the caudal ventrolateral medulla project to the rostral ventrolateral area, inhibiting it (if the initial stimulus is a BP elevation) or not (when BP decreases). The rostral ventrolateral area is the site of sympathetic premotor neurons that project to the intermediolateral column of the spinal cord, the site of preganglionic neurons to the cardiac intrinsic and extrinsic ganglia (Fig. 4.11), and to the sympathetic peripheral ganglia innervating resistance and capacitance vessels. This sympathetic activity is thus inhibited during increases in BP or not inhibited during transitional decreases in BP [8].

Thus, when the triggering reflex stimulus is an increase in BP, there is, reflexively, a vagal activation and inhibition of the sympathetic output, with a consequent reduction in heart rate, systolic volume, and total peripheral resistance, and an increase in venous capacitance. This decreases the venous return to the heart, determining a decrease in BP and its return to baseline (Fig. 4.12). On the other hand, when the triggering stimulus is a decrease in BP, the vagus nerve is not activated and the sympathetic input is not inhibited. This causes an increase in heart rate, systolic volume, and total peripheral resistance, and a decrease in venous capacitance. Consequently, there is an increase in the venous return, which helps to increase cardiac output further (Fig. 4.12). These responses, triggered neurally, are extremely fast, correcting, in seconds, BP swings up or down the control values.

Orthostatic pooling of blood begins almost immediately upon the change from the supine to the upright posture. The main sensory receptors involved in orthostatic cardiovascular reflex adjustment are the arterial baroreceptors located in the carotid sinuses and aortic arch and mechanoreceptors located in the heart and lungs [9]. A decrease in BP, as occurs on assumption of the upright posture, removes this tonic inhibition with a resultant decrease in vagal outflow and an increase in sympathetic activity causing an increase in heart rate, cardiac contractility and vasomotor tone. Central modulation of vasomotor outflow is reinforced by local vasoconstrictor mechanisms, such as the veno-arteriolar axon reflex and a myogenic response. The veno-arteriolar axon reflex is triggered when venous pressure exceeds 25 mmHg, which results in vasoconstriction of the corresponding arteriole and is reported to elicit up to 30–45% of the total vasoconstriction in the legs in the upright posture [9].

The sympathetic and parasympathetic components of the ANS play a crucial role in maintaining cardiovascular homeostasis and enabling the body to respond to physiological stressors. Neurogenic mechanisms are not only essential for maintaining and regulating arterial BP, but also play a crucial role in regulating the distribution of blood flow between and within vascular beds. The sympathetic component of the ANS plays the predominant role in regulating vascular tone and whole-body hemodynamics via its effects on both resistance and capacitance vessels. By contrast, the overall contribution of the parasympathetic nervous system to the regulation of vascular tone and hemodynamics is small compared with its primary regulatory role in mediating negative chronotropic and inotropic effects on the heart [10].

The vascular endothelium plays an essential role in the regulation of blood vessel tone and cellular activity, helping to maintain a healthy vessel [11]. Endothelial cells produce several important vasoactive substances including NO, prostacyclin, endothelium-derived hyperpolarizing factor, endothelin, vasoactive prostanoids, and ROS. These factors, and other endothelium-derived substances, also modulate local thrombotic and inflammatory pathways influencing the progression of atherosclerosis and its complications. Exposure to risk factors for atherosclerosis and the presence of circulatory diseases, including atherosclerosis and heart failure, leads to endothelial activation, associated with reduced NO bioavailability and expression of proinflammatory cytokines, chemokines, selectins, and adhesion molecules. This enhances vasoconstrictor tone and promotes a proatherogenic milieu.

Several hormonal mechanisms are also involved in BP control, including the release of catecholamines, angiotensin II, aldosterone, AVP, oxytocin, and atrial natriuretic peptide (ANP), which act by supporting the maintenance of baseline BP, intensifying and prolonging the cardiovascular responses for minutes or even hours, making BP control more effective, especially in situations of prolonged elevations or falls of BP (hemorrhage, dehydration, drug reactions, etc.; Fig. 4.12). The actions of signaling molecules that are not classically viewed as such, e.g., cytokines and ROS, must also be considered [12].

Circadian clock genes are expressed in the heart and aorta and these genes and approximately 4–6% of the cardiac protein genes showed circadian rhythms in transcription [13]. Ex vivo experiments demonstrate that varied functions of the heart and aorta are dependent on the time in which tissues are collected. In murine knockout models of circadian genes, suppression of *Bmal1* in cardiomyocytes results in an abnormal electrocardiogram (ECG) with RR and prolonged QRS intervals. The hearts of *Bmal1* knockout mice were more susceptible to arrhythmia. Other studies have revealed that removal of *Bmal1* in endothelial cells or vascular smooth muscle cells alters the diurnal variation of BP. These findings are consistent with the presence and importance of circadian genes in the cardiovascular system [13, 14].

The alteration of normal day–night cycles, such as jet lag or shift work, leads to the desynchronization between the central and peripheral clocks and the deregulation of the clock genes (Chap. 8). Restoration of a normal daytime rhythm rescues from these changes, suggesting that maintaining a normal rhythm is crucial for cardiovascular health.

Cardiovascular function changes significantly in the three body configurations during the 24-h cycle [15]. BP decreases during NREM sleep and becomes variable in REM sleep. During REM sleep, transient BP increases of up to 40 mmHg occur that coincide with the phasic events of this stage of sleep in conjunction with vasoconstriction in the skeletal muscles. Pulmonary artery pressure remains stable. Variation of sleep-related BP can be described by a square wave function with changes in the onset and end of sleep and relatively constant values during sleep. This fall in BP during sleep is important for cardiovascular health [15].

The values of systolic pressure drop 15 mmHg or more during sleep and are heavily influenced by the S Process described in Chap. 2. That is, the BP drop accompanies the presence of sleep, regardless of the time of day at which it occurs. An initial fall in BP due to postural change and darkness (~ 7 mmHg) is followed by a period of instability when sleep is unstable (stage N1 of sleep) and by an abrupt drop once stable sleep is achieved (N2–N3 stages of sleep; ~ 7 mmHg). Within each sleep phase the BP is constant, in the NREM sleep, the values are lower than in wakefulness, and in REM sleep they are similar to those of relaxed wakefulness. Although there are transient increases in association with phasic events of REM sleep, the major disturbance of BP during sleep is in the awakening. BP at the end of sleep shows a rise largely due to postural changes. The magnitude of the changes is greater when awakening occurs at the N2 sleep stage compared with awakening in the REM stage [15].

In the case of the heart rate the closest approximation is to a sinusoidal pattern related to the central temperature and compatible with the influence of the C process (circadian), i.e., the changes occur regardless of sleep. The lowest point of the oscillation is in the middle of the night and remains, albeit attenuated, in individuals who are deprived of sleep. As for BP, at the beginning of sleep, there is a fall in the heart rate with two components (preparation for sleep; when sleep becomes stable). There is a close relationship between heart rate and metabolic heat production because of its circadian dependence. Heart rate is higher during the REM sleep phase, with transient tachycardia in relation to REM phasic events (Fig. 4.13).

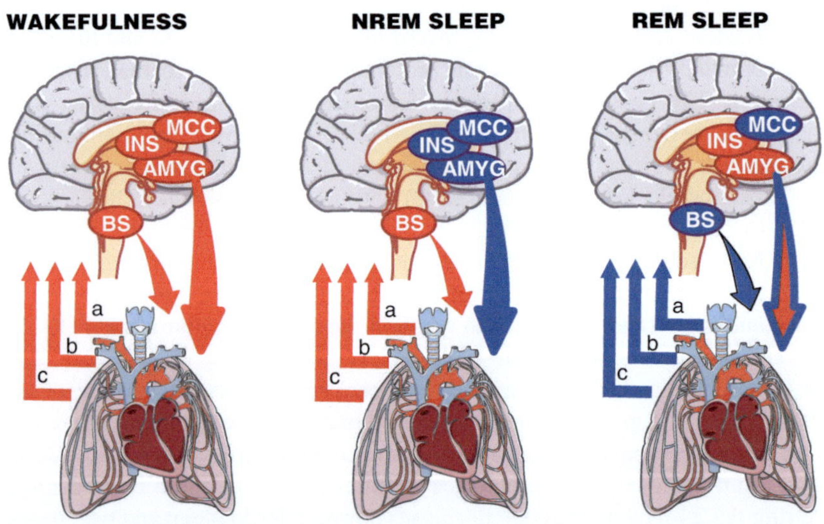

HR	+++	+	++
SNS	+++	+	++
PNS	+	+++	+

Fig. 4.13 Modulation of cardiac activity at the three body configurations in a 24-h cycle. *Letters* designate reflex loops (*a*: baroreflex; *b*: chemoreflex; *c*: respiration). The brain stem centers (BS) and central autonomic network including midcingulate cortex (MCC), insula (INS), amygdala (AMYG) are depicted. Relative changes in the heart rate (HR), sympathetic nervous system. In *red* increases and in *blue* decreases in activity are shown. Redrawn from Chouchou and Desseilles [16]. The figure was prepared in part using image vectors from Servier Medical Art (www.servier.com), licensed under the Creative Commons Attribution 3.0 Unported License (http://creativecommons.org/license/by/3.0/)

Heart rate variability (HRV) analysis combined with brain imaging has identified a close connectivity between autonomic cardiac modulation and activity in brain areas such as the amygdala and insular cortex during REM sleep, but no connectivity between the brain and cardiac activity during NREM sleep [16]. There is also evidence for an association between HRV and the intensity and emotionality of dreams.

Brief awakenings are a characteristic of normal sleep. They occur with high frequency (more than 15–20 per night) and are a normal situation in which an increase in heart rate (≥8 beats per minute), BP (<15 mmHg) and peripheral vasoconstriction are observed. Awakenings occur in both NREM sleep and REM sleep. There are two components of awakenings. The first component is a transient peak of heart rate and BP that occurs within 3–6 s of the event. The second component is dependent on the previous wake period. Most awakenings are brief and there is no such second component. For many normal individuals, the cardiovascular activation response occurs at a level of excitation below the occurrence of α changes in the EEG (cortical awakening). There are large individual differences in cardiovascular response to awakenings and interindividual variability of systolic BP response in an awakening can reach about 15 mmHg. Healthy individuals with large activating cardiovascular responses are at increased risk for cardiovascular disease because of frequent awakenings [17].

In NREM sleep, there is a period of relative autonomic stability with vagal predominance and increased gain of baroreceptors. A sinusoidal modulation of heart rate exists because of the coupling of cardiovascular and respiratory regulation centers, resulting in respiratory sinus arrhythmia. During inspiration, the frequency briefly accelerates to accommodate the increased venous return, resulting in increased cardiac output, whereas during expiration, a progressive decrease in heart rate occurs. This normal sinus heart rate variability, particularly during NREM sleep, is generally indicative of a state of heart health; thus, the absence of sinus arrhythmia has been associated with heart disease and advanced age [18].

Cardiac- and respiratory-related rhythms are observed in nerve activity supplying the heart and blood vessels [19]. These rhythms arise from phasic inputs related to cardiac/pulse or ventilation-related afferent activity and/or a common cardiorespiratory CNS network. Much of this heart rate fluctuation is linked to the phase of respiration (i.e., respiratory sinus arrhythmia) and its loss is a prognostic indicator of morbidity and mortality. Another possible origin of sympathetic rhythms is independent CNS oscillators. Separate oscillators, which are able to couple, may drive activity to different sympathetic nerves, and sympathetic neurons regulating the same target may be influenced by populations of weakly coupled or uncoupled oscillators [19]. Two major hypotheses have been proposed to account for cardiac- and respiratory-related rhythms in sympathetic discharges. The classic view holds that these rhythms are imposed upon sympathetic discharge by "external" inputs. The observation of a nonrespiratory and noncardiac-related sympathetic rhythm suggests that sympathetic rhythms might not arise exclusively from phasic inputs to tonic sympathetic tone generating networks [19]. Loss of vagal tone in cardiovascular diseases can be demonstrated by the diminished change in heart rate on administration of a vagolytic drug such as atropine and by the loss of respiratory sinus arrhythmia. The burst of cardiac vagal activity originates centrally at the level of the preganglionic neurons in the AN, which are inhibited during inspiration, but excited during postinspiration. The changes in BP during respiration are inversely related to the respiratory effort increases in BP, resulting in a decrease in respiratory rate. This effect is potentiated during NREM sleep in which small reductions in BP lead to increases in respiratory rate. These pauses and increases of respiratory rate serve as compensatory mechanisms to normalize BP and their disappearance in infants predisposes to a sudden death.

Sympathetic nerve activity is relatively stable during NREM sleep, and cardiovascular activity is reduced to more than half of that observed in the vigil at the N3 NREM sleep stage. NREM sleep, with relative hypotension, bradycardia, and reduced cardiac output and systemic vascular resistance, provides a "daily vacation" for heart activity. Bradycardia is mainly of a vagal origin, whereas hypotension is mainly attributable to a reduction in sympathetic vasomotor tone.

In the transition from NREM to REM sleep, abrupt vagal discharges can lead to pauses in cardiac rhythm or to asystolia. As discussed in Chap. 2, REM sleep occurs at 90-min intervals and several essential homeostatic mechanisms are disrupted during this period. The increase in limbic activity during REM sleep leads to significant sudden increases in cardiac sympathetic nerve activity at the level of the coronary vessels (Fig. 4.13). The baroreceptor gain is reduced and the heart rate fluctuates markedly, with marked episodes of tachycardia and bradycardia. Vagal tone is generally suppressed during REM sleep with very irregular breathing patterns that can lead to hypoxemia, particularly in patients with pulmonary or cardiac disease. Except for the diaphragm and cricopharyngeal sphincter there is hypotonia of the muscles of the airways. During sleep apnea, respiratory activity can be stopped for central or peripheral reasons several times each night, with adverse consequences for normal cardiorespiratory activity.

Changes in coronary blood flow occur at sleep states and during transitions between sleep states. In monkeys, nighttime coronary flow increases sporadically up to 100% during sleep. These periodic oscillations in blood flow are not associated with alterations in cardiac work or BP. Concomitant with sudden increases in heart rate during REM sleep, abrupt increases in coronary pressure and corresponding decreases in coronary vascular resistance were observed. These phenomena are seen predominantly during REM sleep in phase with POG events in EEG and ocular movements. There were no significant changes in mean BP. As the heart rate is elevated during sudden increases in coronary flow, increased cardiac metabolic activity was assumed to be the cause of coronary vasodilation. These sudden increases in coronary blood flow appear to be the result of increased adrenergic discharge as they were suppressed by removal of the corresponding sympathetic ganglia.

During severe stenosis of the coronary artery a decrease in coronary arterial blood flow is observed during REM sleep instead of the increases found in normal subjects. In these patients, there is a strong correlation between the magnitude of the increase in heart rate and the decrease in coronary flow. The link between REM sleep and the occurrence of myocardial ischemia in coronary patients is consistent.

24-h Rhythms in Respiratory Control

The spontaneous rhythm alternating between inspiration and expiration is produced by the automatic generation of a basal pattern in the brainstem. This spontaneous and stereotyped rhythm is modified by metabolic changes such as changes in pH or partial gas concentrations in the blood, or by mechanical changes (e.g., postural). Respiratory rate and tidal volume are modifiable to allow modifications of the internal environment. Interruptions of breathing (apnea) can be generated to allow phonation or swallowing.

The regulatory system of breathing includes: (a) a control center located in the CNS at the level of the brainstem, where the neural activity that triggers breathing starts; (b) effector muscles that produce ventilation (the respiratory muscles, especially the diaphragm); (c) a set of sensors located in the lungs and central and peripheral chemoreceptors that regulate respiratory activity [20].

The respiratory centers are divided into four major groups, two groups in the medulla and two in the pons (Fig. 4.14). The two groups in the medulla are the dorsal respiratory group and the ventral respiratory group. The two groups in the pons are the pneumotaxic center also known as the pontine respiratory group, and the apneustic center. The exact location of the apneustic center is not yet defined. The inspiratory center (dorsal respiratory group) is in the dorsal portion of medulla, in the NTS.

The expiratory center (ventral respiratory group) is in the AN, in the anterolateral part of the medulla (Fig. 4.14). The AN consists of a rostrally to caudally extending column of neurons expressing respiratory-related activity, with subregions containing motoneurons that innervate the muscles of the larynx and pharynx that are not considered part of the ventral respiratory group. It generally causes expiration, but can cause either expiration or inspiration depending upon which neuron in the group is stimulated.

The expiratory center sends an inhibitory impulse to the apneustic center, which is presumably located in the lower part of the pons and which releases stimulatory impulses to the inspiratory center causing inspiration. It receives an inhibitory impulse from the pneumotaxic center and from stretch receptors of the lung via the X cranial nerve pair and discharges inhibitory impulse to the expiratory center. The pneumotaxic center, located in the upper part of the pons, in the PBN, controls both rate and pattern of breathing and limits inspiration (Fig. 4.14).

Both the dorsal respiratory group and the other sub-nuclei of the NTS are the primary sites for vagal afferent projection of the lungs and afferents and chemoreceptors of the carotid and aortic baroreceptors. The NTS, including the dorsal respiratory group, is a key site integrating sensory information from the lung, in addition to information on the levels of arterial PCO_2, PO_2, and pH.

The ventral respiratory group and dorsal respiratory group contain both bulbospinal respiratory premotoneurons (i.e., neurons that project to spinal motoneurons, which in turn innervate the respective respiratory pump and abdominal muscles of breathing), and propriobulbar neurons (i.e., neurons that project to, and influence

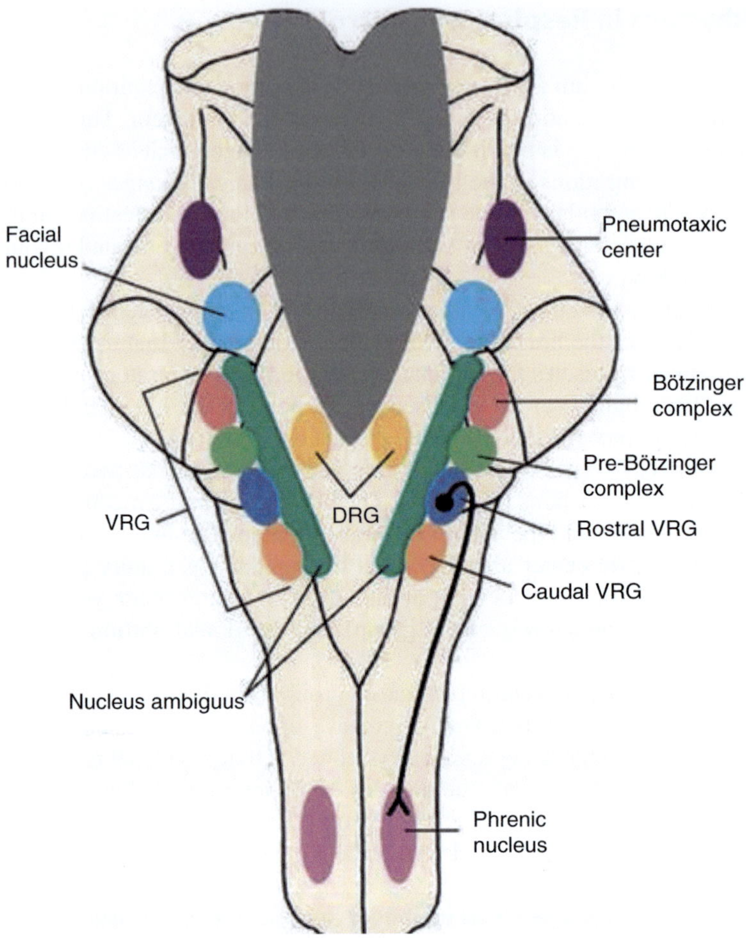

Fig. 4.14 The respiratory centers are divided into four major groups. The two groups in the medulla are the dorsal respiratory group (DRG) and the ventral respiratory group (VRG). The two groups in the pons are the pneumotaxic center and the apneustic center (whose exact location is not yet defined). The inspiratory center (DRG) is in the dorsal portion of the medulla, in the NTS. The expiratory center (VRG) is in the ambiguous nucleus. For other details, see the text

the activity of, other medullary respiratory neurons, but do not themselves project to motoneurons). The hypoglossal, trigeminal, and facial motor nuclei also innervate muscles important to pharyngeal motor control and the maintenance of upper airway patency.

In mammals, the Bötzinger complex is a group of neurons located in the rostral ventrolateral medulla and ventral respiratory column. In the medulla, this group is located caudally to the facial nucleus and ventrally to the AN. The Bötzinger complex plays an important role in controlling breathing and responding to hypoxia. It consists primarily of glycinergic neurons, which inhibit respiratory activity.

The Bötzinger complex has projections to the NTS, phrenic pre-motor neurons in the medulla, phrenic motor neurons in the cervical spinal cord, the dorsal respiratory group, and the ventral respiratory group (Fig. 4.14). Bötzinger complex neurons are intrinsic pacemakers that are important to the generation of the basic respiratory rhythm and the expression of rhythmic neuronal activity elsewhere in the respiratory network. Respiratory rhythm-generating pre-Bötzinger complex neurons coexpress μ-opioid and neurokinin-1 receptors (i.e., the receptors for substance P) that slow and increase the respiratory rate respectively. The presence of μ-opioid receptors in pre-Bötzinger complex neurons explains the respiratory rate depression that follows the administration of opioid drugs. During inspiration, the central respiratory drive potential is transmitted to phrenic and intercostal motoneurons via monosynaptic connections from inspiratory premotor neurons of the dorsal respiratory group.

Central sensors that detect changes in CO_2 are located on the ventral surface near the entrance of the VIII and XI cranial nerve pairs. These chemoreceptors respond to the local application of CO_2 or acids and are inhibited by anesthetics and local cold. These chemoreceptors are not in direct contact with blood, but are bathed in CSF and respond to changes in both arterial PCO_2 and CSF pH.

Carbon dioxide diffuses easily through the blood–brain barrier, but H^+ does not. Thus, the stimulus produced by increased ventilation is the increase in PCO_2, which, after crossing the blood–brain barrier, causes a fall in pH in brain tissue. Alterations of $PaCO_2$ are rapidly transmitted to the CSF, which has little CO_3H_2 as a buffer. After an acute change in arterial $PaCO_2$, there is an even greater change in PCO_2 in the CSF. The response time constant is about 60 s.

Peripheral sensors are located in the carotid and aortic bodies [21]. They measure PO_2, PCO_2, and arterial pH. They are sensitive to the decrease in PaO_2 that is measured directly and induces hyperventilation. Denervation of peripheral chemoreceptors leaves hypoxia without its fundamental homeostatic regulatory mechanism and because of hypoxia, CNS depression occurs. Increases in $PaCO_2$ and the decrease in arterial blood pH stimulate these receptors less, but amplify their response to hypoxemia.

The carotid body, a highly vascular tissue, receives afferent innervation from the carotid sinus nerve, which is a branch of the glossopharyngeal nerve. The increased sensory activity of the carotid body is maintained during the entire period of hypoxia with little adaptation. Thus, the exquisite sensitivity and the rapid response to a wide range of hypoxic intensities with little or no adaptation make the carotid body a unique oxygen-sensing organ in comparison with other tissues [22].

Autonomic nerves play an important role in regulating the functions of the airways, including airway smooth muscle tone, mucus secretion, and blood flow. Afferent nerves in the airway are important with regard to airway defenses (cough), inducing reflex effects, and through the release of neuropeptides (neurogenic inflammation). Cholinergic nerves are the major bronchoconstrictor pathway through the activation of muscarinic receptors on airway smooth muscle. By contrast, adrenergic nerves have little direct control of airway smooth muscle, circulating E being more important in adrenergic regulation. A neural bronchodilator pathway is mediated by release of NO. Several neuropeptides are expressed in airway nerves and play a co-transmitter role in concert with the classical autonomic transmitters [23]. Autonomic nerve function is regulated primarily through reflexes initiated upon bronchopulmonary vagal afferent nerves [24].

During NREM sleep, there is a decreased chemosensitivity and apneas at the onset of the sleep phase [25]. During REM sleep, the decrease and variability of chemosensitivity, with a greater propensity to apneas and the decrease in the neural discharge to the respiratory muscles (except the diaphragm), are accentuated.

During wakefulness, the respiratory control is exerted by three mechanisms:

- Metabolic mechanisms, which ensure arterial O_2 and CO_2 homeostasis by central and peripheral chemoreceptors.
- Voluntary control, which allows the adaptation of ventilation to phonation and other demands.
- Tonic depolarization of the spinal motor neurons of the respiratory muscles.

Respiratory changes during sleep reflect the inhibition of some of these control mechanisms (Fig. 4.15). Metabolic control (index of a predominant parasympathetic

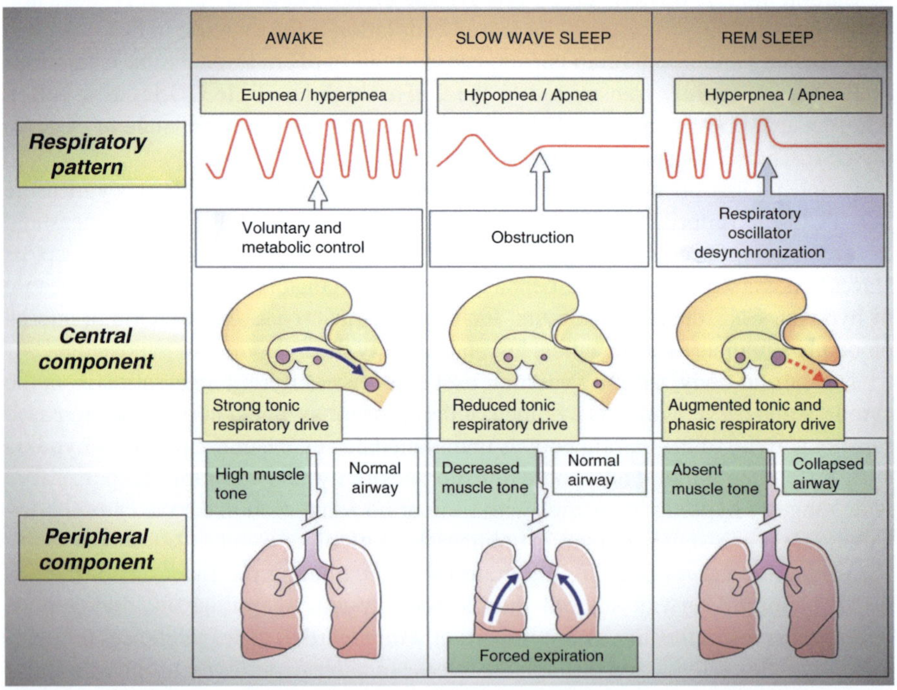

Fig. 4.15 Balance between the different components of respiratory control in the three body configurations in a 24-h cycle. When the voluntary control disappears and the peripheral component is reduced, there is a tendency toward hypopnea or apnea in sleep (slow-wave sleep). This is exacerbated in REM sleep because of the motor and autonomic disconnection. Hence, the apnea predominates in the latter part of the night. Modified with permission from Cardinali [3]

mechanism) prevails in slow-wave sleep and there is a decrease in this control during REM sleep. The predominance of REM sleep in the last stages of the night explains the higher incidence of episodes of apnea in the second part of the night [25].

The initiation of sleep and up to stage N2 of slow-wave sleep is accompanied by an unstable respiratory rhythm with successive episodes of hypo- and hyperventilation called "periodic ventilation." During stage N3, ventilation becomes regular in terms of respiratory rate and amplitude. Respiratory rate and depth are relatively constant, this being a stable period from the respiratory point of view.

Respiratory rhythm during REM sleep is characterized by being faster and mostly irregular, with apneic episodes and hypoventilation. The responsible mechanism is central, neural, to which is added the muscular hypotonia, which has a double influence: on the one hand, it diminishes the force of the expansion of the rib cage, and on the other hand it increases the resistance of the superior airway to the passage of the air. The diaphragm maintains irregular activity, but does not participate in the generalized atonia of REM sleep, because it lacks a significant number of muscle spindles.

Cardiorespiratory homeostasis involves the regulation of two motor systems, one that supplies the somatic (i.e., diaphragmatic, intercostal, abdominal, and upper airway) musculature and another the autonomic innervation of heart and vasculature. As seen, the activity of respiratory neurons varies greatly in the stages of sleep, as does the regularity of the heart rhythm. During REM sleep, there is tachycardia, polypnea, sweating, and dramatic elevations in arterial BP secondary to the intense autonomic activity that occurs in this period. Maintaining the perfusion of vital organs through the control of adequate BP is essential for homeostasis. Respiratory mechanisms are recruited to support cardiovascular action, helping the venous return and altering the heart rate by reflex.

The integration of the cardiorespiratory function during sleep requires the participation, in addition to the brainstem, of central areas of the autonomic hierarchy. This has been documented by imaging studies using PET. In REM sleep, the preferential activation of limbic and paralimbic regions of the anterior brain is demonstrated in comparison with wakefulness or NREM sleep. The serotonergic neurons of the DRN play an important role in vascular control. These neurons are damaged in heart failure, probably because of altered perfusion and hypoxia accompanying altered breathing.

The orbitofrontal cortex, parts of the formation of the hippocampus, the hypothalamus, and other structures of the CNS participate in regulation of the cardiorespiratory pattern [26]. The central nucleus of the amygdala is strategically positioned to regulate cardiac and respiratory functions in affective behavior, as it broadly projects to the brainstem (NTS, PBN, the dorsal motor nucleus of the vagus, periaqueductal gray). Many are involved in mediating the transient rise in arterial BP caused by the cold or the Valsalva maneuver. These structures are severely damaged, both in patients with heart failure and in obstructive sleep apneas.

The insular cortex deserves special attention among the cortical areas that express the regulatory action on cardiovascular control in the sleep and waking states

(Fig. 4.13). This area modulates the sympathetic activity (mainly the right insula) and the parasympathetic activity (mainly the left insula). There is marked damage of the insula under conditions of respiratory disorders and heart failure.

The Cerebellum and the Autonomic Posture

The cerebellum, the largest subcortical center for motor control, has been demonstrated in the last decade to be involved in the regulation of nonsomatic functions, such as respiration, feeding behavior, cognition, and working memory [27–29].

For example, activation of fastigial nuclear neurons predominantly increases ventilation via elevation of the respiratory frequency and/or tidal volume. Ablation of the fastigial nucleus did not significantly alter eupneic breathing, but did markedly attenuate the respiratory response to medium and severe hypercapnia, and hypoxia. The fastigial nucleus contains respiratory-modulated neurons and about 25% of these neurons do not show their respiratory-related phasic activity until exposed to hypercapnia. The fastigial nucleus also contains CO_2/H^+-chemosensitive sites that contributed to the respiratory response to hypercapnia. The involvement of the cerebellum in the control of cardiovascular and respiratory activity in the three body configurations of a 24-h cycle, i.e., wakefulness, NREM, and REM sleep, is thus significant.

Although it is not yet clear through which pathways such cerebellar nonsomatic functions are mediated, the direct bidirectional connections between the cerebellum and the hypothalamus are probably involved [30]. The direct hypothalamocerebellar projections originate from the widespread hypothalamic nuclei/areas and terminate in both the cerebellar cortex as multilayered fibers and the cerebellar nuclei. Immunohistochemistry studies have indicated that some of these projecting fibers are histaminergic. It has been suggested that through their excitatory effects on cerebellar cortical and nuclear cells mediated by metabotropic histamine H_2 and/or H_1 receptors, the hypothalamocerebellar histaminergic fibers participate in the cerebellar modulation of somatic motor and nonmotor responses. The histaminergic afferent system of the cerebellum, having been considered an essential component of the direct hypothalamocerebellar circuits, originates from the tuberomammillary nucleus in the hypothalamus [31].

The direct cerebellar–hypothalamic projections arise from all the three cerebellar nuclei, the fastigial nucleus, the interpositus nucleus, and the dentate nucleus (Fig. 4.16), and terminate to extensive regions of the hypothalamus, such as the lateral hypothalamic area (LHA) and the posterior and dorsal hypothalamic areas, in addition to the dorsal medial nucleus (DMN) and the paraventricular nucleus (PVN; Fig. 4.17) [32]. Neurophysiological and neuroimaging studies have demonstrated that these connections are involved in feeding, cardiovascular, osmotic, respiratory, micturition, immune, emotion, and other nonsomatic regulation. For example, electrophysiological data suggest that via the direct cerebellohypothalamic projections, the cerebellar outputs may reach, converge, and be integrated with some critical feeding signals, including gastric vagal afferents, CCK, leptin, and glycemia on single hypothalamic neurons [33]. Hypothalamic orexin neuronal projections to the cerebellum

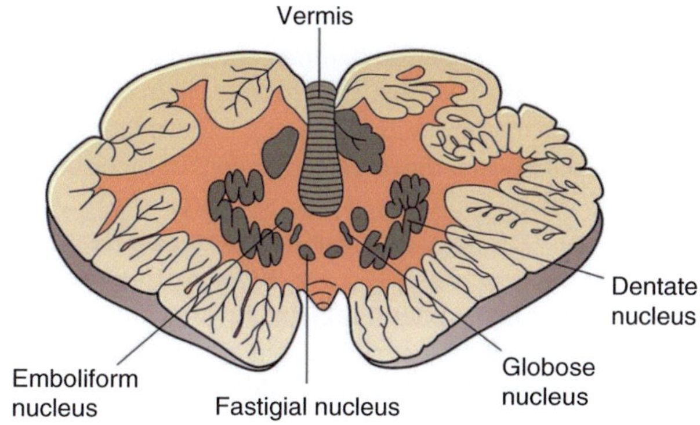

Fig. 4.16 Cerebellar nuclei. Modified with permission from Cardinali [3]

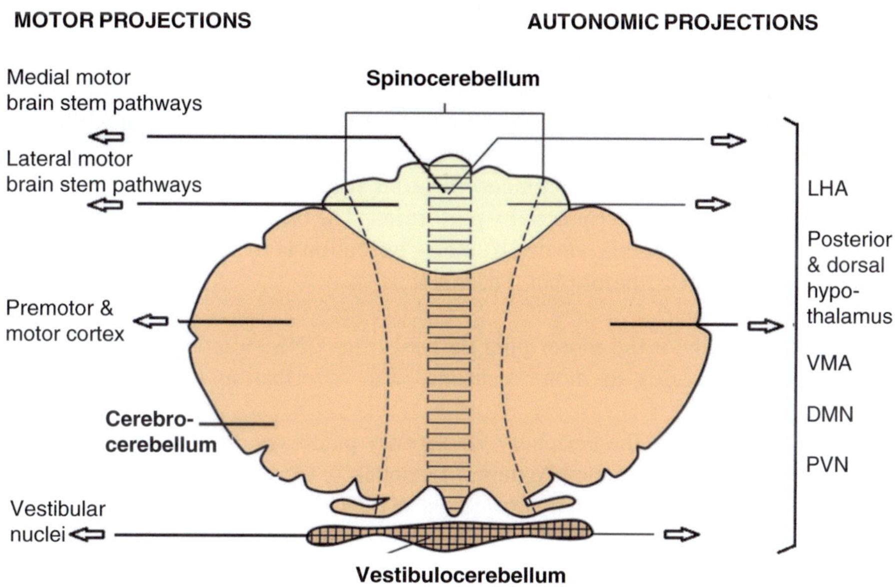

Fig. 4.17 Motor and autonomic projections of the cerebellum. Modified with permission from Cardinali [3]

co-ordinate vestibulo-cerebellar motor and autonomic functions associated with feeding behavior [34]. Furthermore, functional imaging studies provide substantial evidence that hunger, satiation, and thirst are accompanied by a cerebellar activation. Concerning the immune function, lesions of fastigial nuclei enhance cellular and humoral immunity, whereas lesions of the interpositus nucleus inhibit the immune function. The positive and negative immunoregulations by the two different cerebellar nuclei strongly show that the cerebellum involves an elaborate and critical modulation of immune homeostasis [35]. Collectively, these observations provide support for the hypothesis that the cerebellum is an essential modulator and coordinator for integrating motor, visceral, and behavioral responses participating in the autonomic posture.

In rodents, the reversible inactivation of the vermis during the consolidation or the reconsolidation period hampers the retention of the fear memory trace. This is a typical example of cerebellar participation in emotional learning conclusively demonstrated for motor control. Imaging experiments show that in humans the cerebellum is also activated during mental recall of emotional personal episodes and during learning of a conditioned or unconditioned association involving emotions (Fig. 4.18). The vermis participates in fear learning and memory mechanisms related to the expression of autonomic and motor responses of emotions. In humans, the cerebellar hemispheres are also involved at a higher emotional level.

To understand the cerebellar role in autonomic function it is necessary to briefly review the cerebellum function as derived from classical motor control experiments. Cerebellar neurons are distributed in the cortex, and in three pairs of nuclei located in the interior of the cerebellar hemispheres: (a) the fastigial nucleus; (b) the interpositus nucleus (globular plus emboliform nuclei); (c) the dentate nucleus (Fig. 4.16). Despite representing 10% of the brain weight, the cerebellum contains 50% of the brain neurons. The structure of the cerebellum is very systematized and is similar throughout the organ, suggesting a common basic function, modified by the type of information that each zone receives.

To perform its function of motor coordination, the cerebellum acts as a comparator of intention with the motor activity performed. The basic mechanism of function of the part of the cerebellum related to motor execution is to supervise the different executive stages of the motor plan. For this it uses:

(a) Information about the motor plan derived from CNS structures, from the primary motor cortex to motor neurons. This information is called "internal feedback."
(b) (b) Information on the periphery via sensory pathways originating in the skin, muscles, and joints (called "external feedback"). Correction of the plan is carried out by the cerebellum by projection to the neuronal groups constituting the descending motor systems.

This comparison of the "plan" with the "execution of the plan" allows the cerebellum to appreciate deviations and to correct them, not by direct action on the motoneurons, but by indirect influence through the descending motor pathways. An equivalent occurs for several autonomic behaviors. The cerebellum receives information from the limbic system on the autonomic strategy selected as adequate, and presents similar phenomena to the internal and external feedback indicated above. The cerebellar data are consistent with a dampening or coordinating role for the cerebellum in the presence of significant changes in BP that could be similar to motor coordination.

The function of the cerebellum is modified by experience, hence its importance in motor and autonomic learning. For this motor function, the cerebellum receives information from three sources: (a) the periphery; (b) the brainstem; (c) the cerebral cortex. The pathways that enter the cerebellum send collaterals to the cerebellar nuclei and the cerebellar cortex. For its autonomic function, the main afferent is that of the cortical and subcortical areas of the limbic system (Fig. 4.17). The exit of the cerebellar

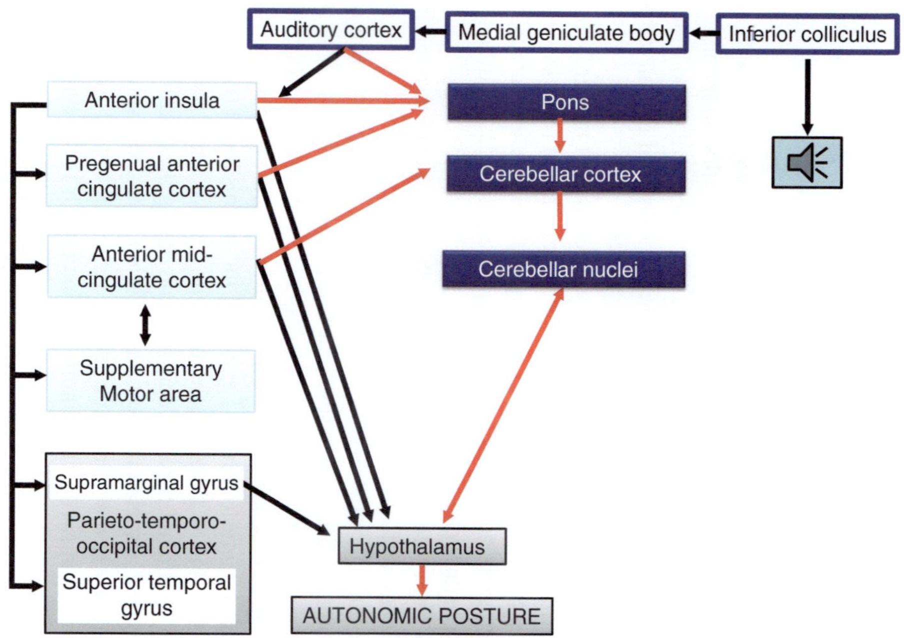

Fig. 4.18 A diagram describing the generation of autonomic responses following a loud tone presentation. Auditory signals influence auditory regions and the anterior insula, whose activity reflects stimulus saliency. Signal generated in the anterior insula modulate autonomic activity through direct projections to the hypothalamus and to other cortical areas within the salience network. Signals generated in the auditory cortex also influence cerebellar nuclei through pontine projections. The cerebellar cortex, in turn, influences autonomic activity through projections to the hypothalamus

cortex always passes through the cerebellar nuclei (except for that corresponding to the flocculonodular lobe, which is sent to the vestibular nuclei). This particularity allows the cerebellar nuclei to perform the integration of the input information with the elaboration of the cerebellar cortex. The cerebellar entrance and exit routes are projected by the three pairs of cerebellar peduncles: upper, middle, and lower.

Each portion of the cerebellar cortex projects to a given group of cerebellar nuclei (Fig. 4.16) following the distribution:

- The vermis projects to the fastigial nucleus
- The middle portion of the cerebellar hemisphere projects to the interposed nucleus
- The lateral portion of the cerebellar hemisphere projects to the dentate nucleus
- The flocculonodular lobe projects to the vestibular nuclei

In Fig. 4.17, the three functional divisions of the cerebellum and their motor and autonomic outputs are schematized: (a) vestibulocerebellum (flocculonodular lobe); (b) spinocerebellum, composed of the vermis and the intermediate portion of the cerebellar hemispheres; (c) the cerebrocerebellum, composed of the lateral portion of the cerebellar hemispheres.

The vestibulocerebellum is the phylogenetically oldest part of the cerebellum. Its function is the control of body posture and ocular movements. It receives vestibular information directly from the labyrinth, and through the vestibular nuclei. The exit is through the vestibular nuclei. Its autonomic participation is important in BP responses to rapid postural changes.

The spinocerebellum receives information from the spinal cord through the spinocerebellar bundles, and from the auditory, visual, vestibular, and ANS projections. These projections are organized somatotopically, indicating the comparative function of the motor plan and execution that fulfills the cerebellum. The projections that reach the spinocerebellum from the spinal cord are direct (through the ventral and dorsal spinocerebellar bundles) and indirect (from the mesencephalic relay nuclei, such as the inferior olive). Lesions of the vermis block the retention of the fear memory trace (Fig. 4.18) and in humans the cerebellum is activated during the mental recall of emotional personal episodes and during the learning of a conditioned or unconditioned association involving emotion. Thus, the spinocerebellum contributes substantially to the autonomic posture (Chap. 3; Figs. 1.5 and 4.18).

The cerebrocerebellum participates in motor planning. It is the center of a complex feedback system, which modulates cortical commands. The output of the cerebrocerebellum oversees the dentate nucleus, which projects to the lateral ventral nucleus of the thalamus, and from here to the premotor and motor cortex and to the LHA, DMN, and PVN areas of the hypothalamus.

The histological structure of the cerebellum is homogeneous, regardless of the function that the region fulfills. The cerebellar cortex is divided into three layers: molecular, Purkinje, and granule cells. The fundamental cellular elements present in the cerebellum are the GABAergic Purkinje cells, which are the only neurons to exit the cerebellar cortex and granule cells, their excitatory interneurons originating from the parallel fibers (Fig. 4.19). The different afferences that reach the cerebellum are made in one of two ways: (a) mossy fibers, which constitute the most important contribution and

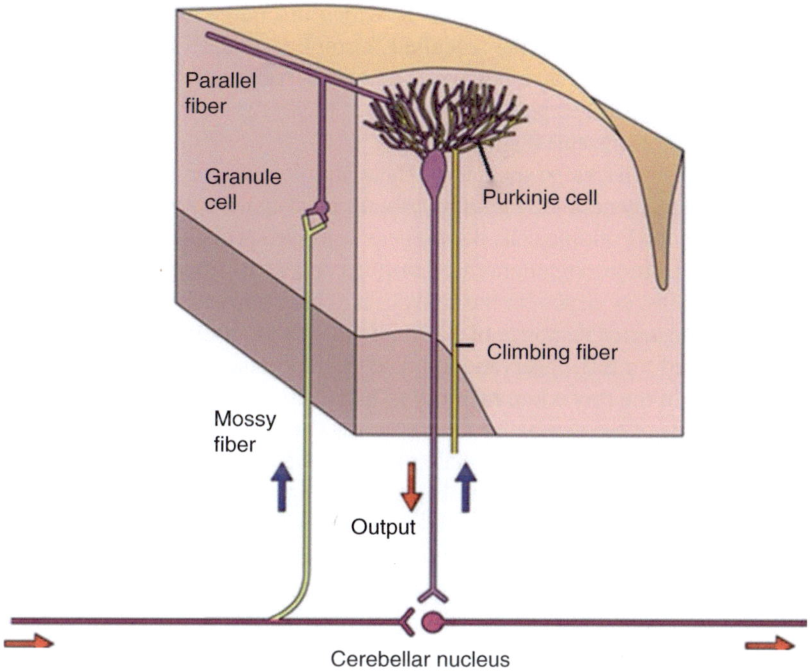

Fig. 4.19 The histological structure of the cerebellum is homogeneous regardless of the function that the region fulfills. The principal cell elements are: (a) the GABAergic Purkinje cells, which are the only projection neurons of the cerebellar cortex; (b) granule cells, which are excitatory interneurons that give rise to the parallel fibers. The different afferences that reach the cerebellum are made in one of two ways: (a) mossy fibers, which constitute the most important contribution and which comprise all the entrances to the cerebellum except for the inferior olive; (b) climbing fibers, originating in the lower olive. Both pathways send a stimulating collateral to the cerebellar nuclei before proceeding to the cerebellar cortex. Modified with permission from Cardinali [3]

which comprise all the entrances to the cerebellum except for the inferior olive; (b) climbing fibers, originating in the inferior olive. Both entry routes send a stimulatory collateral to the cerebellar nuclei before proceeding to the cerebellar cortex (Fig. 4.19).

Mossy fibers exert an indirect stimulatory influence on Purkinje cells through the granule cells, on which they synapse. The granule cells give rise to the parallel fibers, which activate the dendrites of the Purkinje cells, at the level of the molecular layer. The arrangement of the parallel fibers in relation to the dendritic trees of the Purkinje cells resembles that of old telephone wires in their contact with the posts that support them. Each Purkinje cell receives information from about 200,000 parallel fibers, and each parallel fiber contacts thousands of Purkinje cells aligned perpendicularly.

The climbing fibers, on the other hand, have a different distribution. They originate in the inferior olive, and ascend through one of the Purkinje cells, making several synaptic contacts with it. This link is almost individual, with only one single climbing fiber per Purkinje cell.

The synaptic connection between the climbing fiber and the Purkinje cell is the strongest detected in the CNS: only an action potential in the climbing fiber can produce a giant EPSP in the Purkinje cell, which discharge in response a salvo of action potentials, constituting the so-called "complex spikes." The terminals of the mossy fibers, on the other hand, behave as common excitatory synapses of the CNS, requiring the spatial and temporal summation of EPSP for the discharge of an action potential in the Purkinje cell ("simple spikes").

The mossy fibers discharge spontaneously at high frequencies (50–100/s). They are the main control element of the Purkinje cells, and during the movement or by sensory stimuli induce in the Purkinje cells discharges of simple spikes with a high frequency.

Only during the motor or autonomic learning process are complex spikes observed in the Purkinje cells, evidence for the activation of the climbing fibers. This discharge modulates the posterior response of Purkinje cells to the mossy fibers (post-tetanic inhibition). Based on this observation, a role of importance for climbing fibers has been postulated in the procedure memory (Chap. 6).

Other afferents of the cerebellar cortex are the "spider web" monoaminergic projections originating in the mesencephalon and diencephalon. For example, the histaminergic afferent system of the cerebellum, having been considered an essential component of the direct hypothalamocerebellar circuits, originates from the TMN in the hypothalamus. Unlike the mossy fibers and climbing fibers, the histaminergic afferent fibers, a third type of cerebellar afferents, extend fine varicose fibers throughout the cerebellar cortex and nuclei. Histamine directly excites the Purkinje cells and granule cells in the cerebellar cortex, and the cerebellar nuclear neurons. Therefore, the histaminergic afferents modulate these dominant components in parallel in the cerebellar circuitry and consequently influence the final output of the cerebellum. The hypothalamocerebellar histaminergic fibers/projections, bridging the autonomic centers to somatic structures, play a critical role in the somatic–ANS integration.

The use of PET and fMRI to identify brain structures in humans involved in cognitive functions indicated the important role of the cerebellum in language and cognition. An evolutionary fact of importance is that the cerebellum, along with the prefrontal cortex, has expanded significantly in humans relative to other hominids. Until recently, it was assumed that this development was linked with, and resulted in, motor skills of *Homo sapiens*. Many studies indicate that the human cerebellum is activated in the absence of movement, when the subject performs cognitive and verbal functions or experiences emotions.

In the evolution of the human cerebellum, the part that has developed most is the cerebrocerebellum, whose nucleus is the dentate nucleus. In the dentate nucleus, an older part is identifiable, present in the lower primates, in addition to a new portion, typical of humans. The new portions of the dentate are connected, via the thalamus, with the association areas of the cerebral cortex involved in language and the limbic system.

24-h Rhythms in the Immune Response

The way in which the nervous system communicates with the immune system is twofold: (a) via the endocrine system; (b) through the ANS, both sympathetic and parasympathetic divisions, which supplies innervation to the lymph nodes, thymus, spleen, and bone marrow. This interaction varies significantly in the three body configurations during a 24-h cycle (Fig. 4.20).

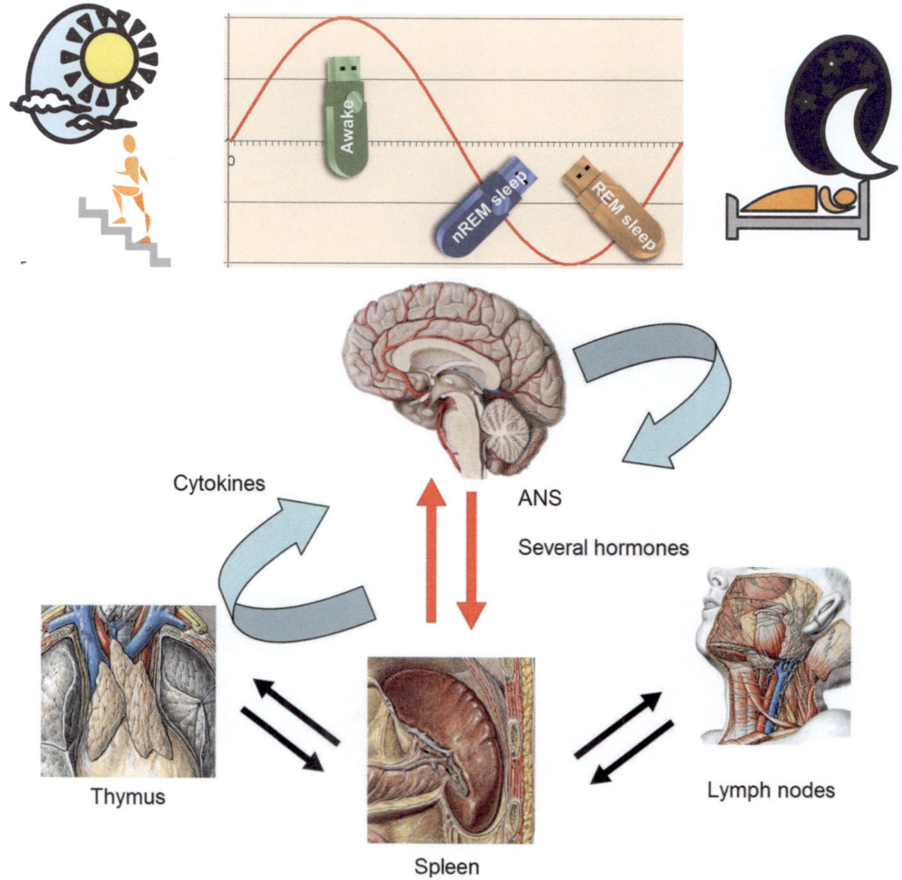

Fig. 4.20 Neuroendocrine–immune communication varies significantly among the three body configurations during a 24-h cycle. The figure was prepared in part using image vectors from Servier Medical Art (www.servier.com), licensed under the Creative Commons Attribution 3.0 Unported License (http://creative-commons.org/license/by/3.0/)

In turn, because of immune reaction significant changes are verifiable in neuronal activity. Various groups of hypothalamic neurons react to humoral signals (cytokines) produced by immunocompetent cells, such IL-1, IL-6, TNF-α, and IFN-γ. Besides participating in the normal humoral regulation of NREM sleep (Chap. 2), these signals produce the "illness behavior" accompanying the acute or chronic infection (loss of appetite or anorexia, depressed motor activity, loss of interest in daily activities), in addition to fever and the activation of the pituitary adrenal axis (Fig. 4.21).

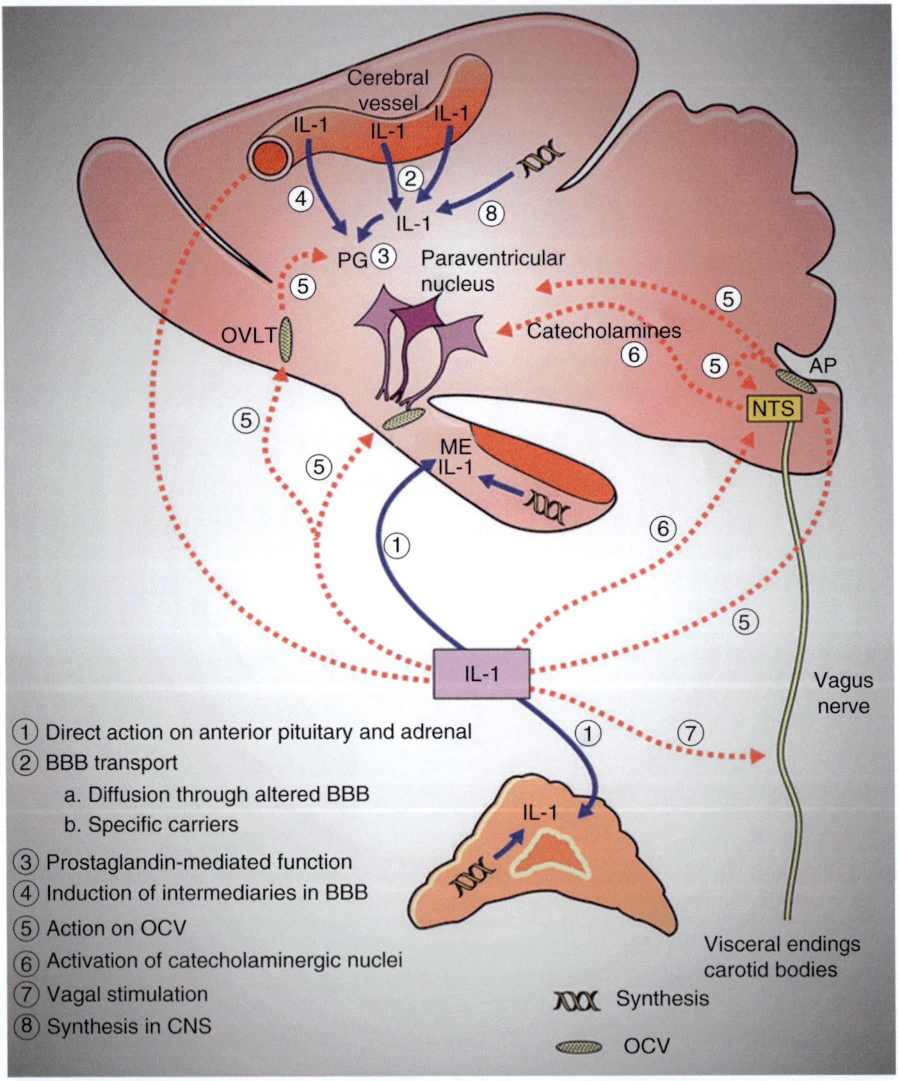

Fig. 4.21 The different mechanisms by which an increase in cytokines (e.g., IL-1) results in central changes (e.g., illness behavior, which includes increased central temperature). Prostaglandins are intermediates in the effects of IL-1 on the circumventricular organs. Modified with permission from Cardinali [3]

To deal with this subject, we first discuss briefly the bases of the immune response. Then, we analyze the data linking sleep with immune responsiveness. A definite immune configuration of wakefulness and NREM sleep can be achieved. However, no configuration is known for the REM sleep itself, because of the methodological problems of monitoring immune responses at short intervals.

The immune response consists of a first phase of antigen recognition, before the later cell activation to produce molecules to eliminate it. This activation, which constitutes the second phase of the immune response, is a set of processes that are finely regulated. Indeed, an uncontrolled activation of leukocytes because of the failure of the regulation could lead to the onset of illness or death of the individual. The third and final phase is the destruction of the non-self, and involves the generation of inflammation and oxidation that allows the elimination of the pathogen.

The immune response comprises innate or nonspecific mechanisms and acquired or specific mechanisms. Various pathogens, such as bacteria, viruses, fungi, and parasites, trigger an immune response. Pathogens are recognized by a pattern recognition receptor that gives rise to various signaling pathways to initiate activation of adaptive innate immune response.

The unspecific response develops and acts immediately to deal with any foreign agent that has managed to pass the natural barriers of the body. The innate response defends the body from external pathogens and against any of its own cells that have become dangerous, i.e., cancer cells. This response, which is fast because it is triggered within seconds and lasts a few hours, is carried out by several cells and soluble factors.

Innate immunity does not entail immune memory: it responds in the same form and intensity to subsequent infections. It recognizes groups of pathogens, not the subtle differences between them. Its cellular components are macrophages, neutrophils, basophils, eosinophils, natural killer (NK) cells, and cytokines (e.g., TNF-α, IL-1β, IL-6, IL-8).

Phagocytes ingest and destroy infectious agents and NK cells bind to tumor and virus-infected tissues to program them for destruction by apoptosis. In the innate response, immediate mechanisms of action (within minutes) are followed by other induced responses (lasting between 4 and 96 h). These mechanisms do not provide lasting immunity protection.

The acquired immunity is specific and with immunological memory, mediated by cells (T and B lymphocytes), cytokines and antibodies (Fig. 4.17). It comprises humoral immunity (antibodies that recognize extracellular pathogens or foreign molecules and make them sensitive to macrophage destruction) and cellular immunity (cytotoxic T lymphocytes, CD8+, which recognize and destroy infected cells). T helper (Th) lymphocytes (CD4+) coordinate the innate, cellular, and humoral responses by means of numerous cytokines (Figs. 4.22 and 4.23). The acquired and adaptive immunity possesses a system for self-regulation designed to avoid the response of activation against the antigens to extend undesirably in time and space.

Lymphocytes can recognize, thanks to their specific receptors, millions of different antigenic molecules, distinguishing even those that have a great structural similarity, and they do it with great specificity. When the lymphocytes (B-cells in the bone marrow and T-cells in the thymus) proliferate, a whole series of genetic

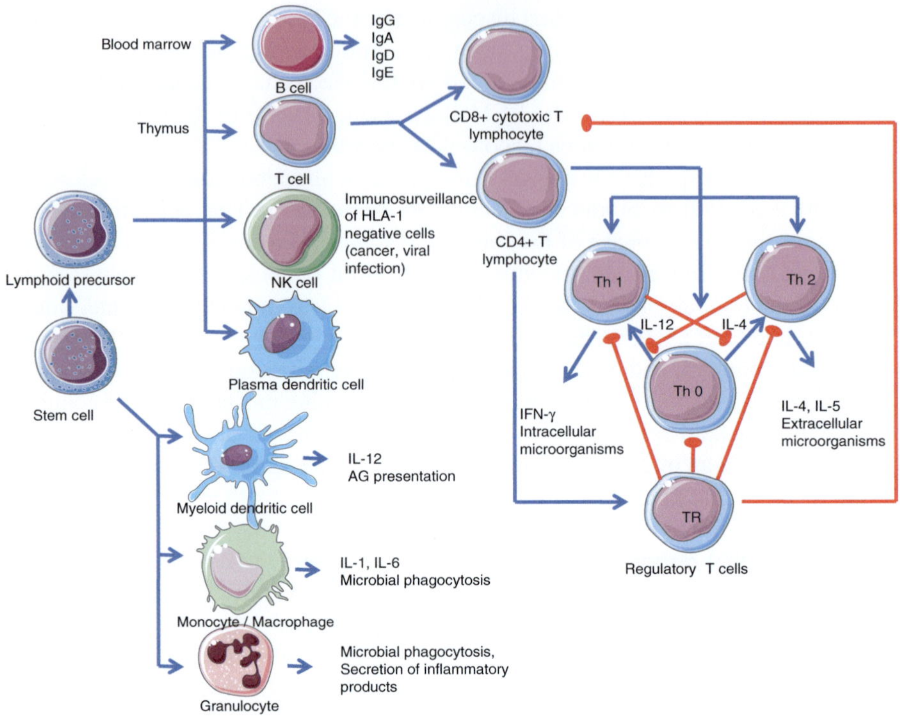

Fig. 4.22 Schematic model of intercellular interactions of immune cells in acquired immunity. Stem cells differentiate into T cells, dendritic antigen-presenting cells, NK cells, macrophages, granulocytes, or B cells. Foreign antigen is processed by the dendritic cells, and the peptide fragments are presented to T lymphocytes. Activation of CD4+ T cells leads to Th lymphocytes. Activation of CD8+ T cells leads to the induction of cytotoxic T lymphocytes. For the production of antibodies against the same antigen, the antigen binds within the receptor complex and induces B cell maturation into plasma B cells secreting Ig. The figure was prepared in part using image vectors from Servier Medical Art (www.servier.com), licensed under the Creative Commons Attribution 3.0 Unported License (http://creativecommons.org/license/by/3.0/)

combinations allow the expression of millions of possible different receptors on their membrane. These receptors (T cell receptors, TCRs, in T lymphocytes; Fig. 4.23) and B cell receptors (BCR) in B lymphocytes (which are class M immunoglobulins) allow recognition in a very specific way of the millions of different antigens that can be contacted throughout life. These receptors discriminate between the self and the nonself, deciding whether to tolerate, in the first case, or destroy it, in the second.

Lymphocytes have memory, which enables them to remember, when they recognize an antigen, if it is the first time they have met it, or if there has already been a previous interaction. Like episodic memory that is highly dependent on slow-wave sleep (Chap. 6), the immune memory is also maximal in NREM sleep.

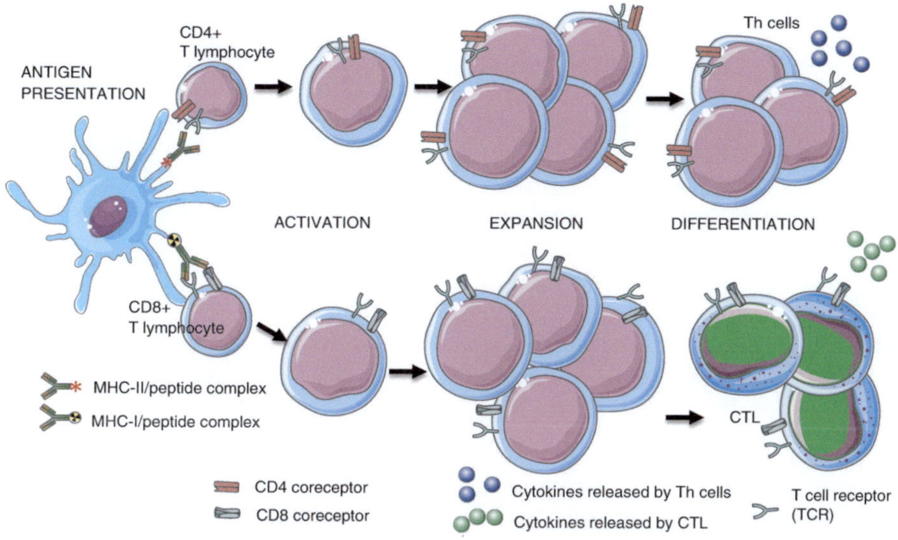

Fig. 4.23 Antigen presentation and activation, expansion, and differentiation phases of T cells. The figure was prepared in part using image vectors from Servier Medical Art (www.servier.com), licensed under the Creative Commons Attribution 3.0 Unported License (http://creativecommons.org/license/by/3.0/)

T helper (Th) lymphocytes constitute a fundamental component of acquired immunity and their phenotypes determine the type of response (Fig. 4.22). Based on the cytokine environment, expression of transcription factors and the cytokine secretion pattern, Th lymphocytes can be differentiated into four phenotypes: Th1, Th2, and Th17 (effector phenotypes; Fig. 4.24), and T regulatory (Treg), which regulate any excessive response of effector lineages.

Th1 cells play a key role in the development of the inflammatory process through the production of proinflammatory cytokines (IFN-γ, IL-1, IL-6, IL-12, IL-18).

Th17 cells are a subset of CD4+ T cells, which are highly relevant in inflammatory processes. They express the transcription factor RORγt and produce IL-17 involved in autoimmunity and in eliminating extracellular pathogens. The current concept is that the inflammatory response involves Th1/Th17 control.

Th2 cells produce cytokines of anti-inflammatory activity (IL-4, IL-5, IL-10, IL-13) and are responsible for the anti-inflammatory response.

Treg cells are a subpopulation of CD4+ (mainly CD25 +) lymphocytes expressing the transcription factor Foxp3 that controls the functions of effector cells.

The profile of the sympathetic predominance characteristic of prolonged wakefulness includes the increase in innate immunity, as revealed by an augmented number of NK cells, a reduced proliferation of T lymphocytes, increased apoptosis of T lymphocytes and memory cells and a reduced number of "naive" cells. Thymic involution and increased production of autoantibodies are characteristic of the sympathetic predominance. Th2 cytokines (IL-4, IL10, IL-13) increase.

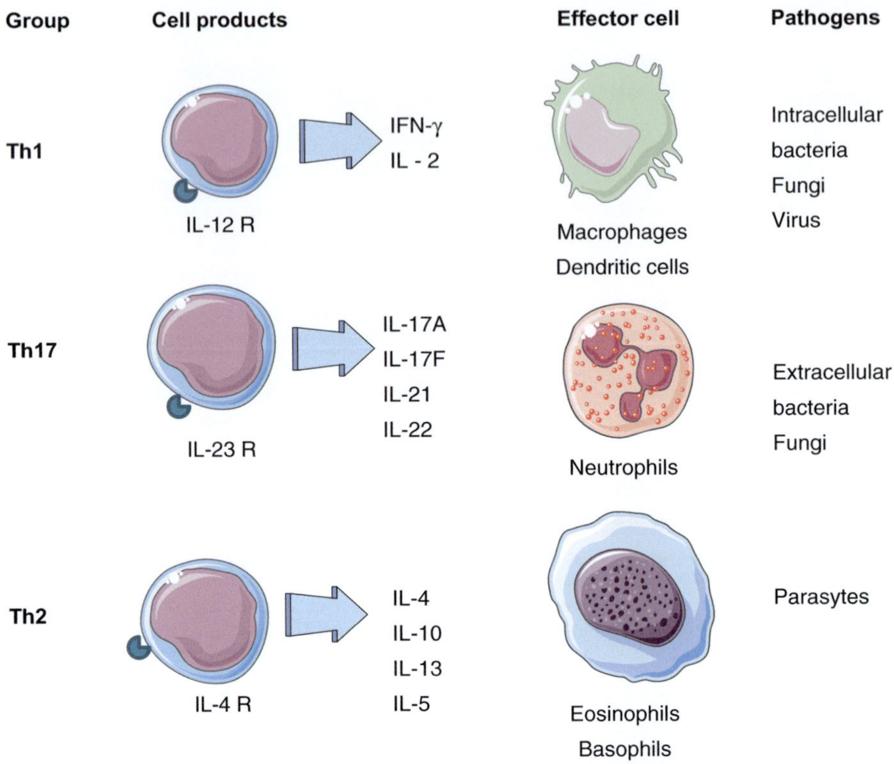

Fig. 4.24 Based on the cytokine environment, transcription factor expression, and cytokine secretion pattern, Th lymphocytes can be differentiated into three effector phenotypes: Th1, Th2, and Th17. The figure was prepared in part using image vectors from Servier Medical Art (www.servier.com), licensed under the Creative Commons Attribution 3.0 Unported License (http://creativecommons.org/license/by/3.0/)

The sympathetic innervation of lymphoid organs largely explains these effects (Fig. 4.25). A crucial role of peripheral innervation in regulating cell behavior and response to the microenvironment has recently emerged. In the hematopoietic system, the ANS regulates stem cell niche homeostasis and regeneration and fine-tunes the inflammatory response [36].

The ANS is activated in response to infection or injury and its products have major effects on immune function and inflammation [37, 38]. The parasympathetic nervous system also modulates inflammation by acting as an anti-inflammatory neural circuit. The vagus nerve senses peripheral inflammation and transmits action potentials to the brainstem, the area postrema, and the NTS.

Excessive activation of the immune system is prevented by anti-inflammatory mediators such as corticosteroids and anti-inflammatory cytokines. Moreover, the brain not only senses peripheral inflammation through vagal afferent nerve fibers, including those originating in the carotid bodies [39], but also provides an integrated response dampening the immune system through vagal efferents [37]. This

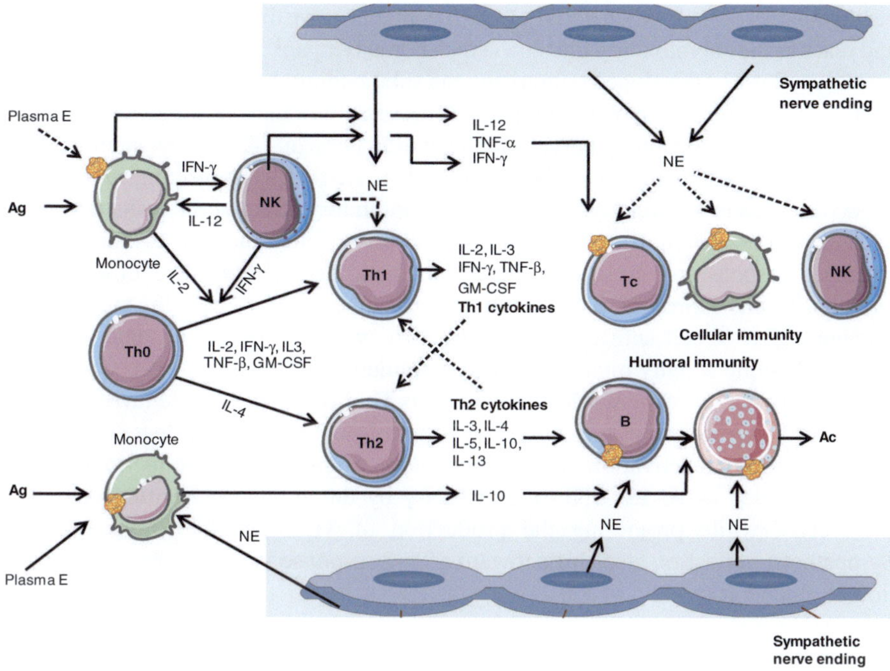

Fig. 4.25 The sympathetic nervous system promotes Th2 responses (humoral immunity) and inhibits Th1 responses (cellular immunity). *Ag* antigen, *Ac* antibodies, *R* receptor. The figure was prepared in part using image vectors from Servier Medical Art (www.servier.com), licensed under the Creative Commons Attribution 3.0 Unported License (http://creativecommons.org/license/by/3.0/)

so-called anti-inflammatory pathway has been introduced as a second system by which the immune system is modulated by ANS. In sepsis, the anti-inflammatory effect is mediated by the modulation of splenic macrophages, whereas in the gut, vagal nerve fibers synapse with enteric cholinergic neurons interacting with resident intestinal macrophages [40]. In the gastrointestinal tract, the microbiome contributes significantly to the activation and inhibition of autonomic control and has an impact on the set of the neuroinflammatory inhibitory reflex mediated by the cholinergic nervous system [41].

Valuable information on autonomic nerve regulation of immune function has derived from studies in the thymus and submaxillary lymph nodes. The thymus plays a critical role in establishing and maintaining the peripheral T-cell pool. It does so by providing a microenvironment within which T-cell precursors differentiate and undergo selection processes to create a functional population of major histocompatibility complex-restricted, self-tolerant T cells. These cells are central to adaptive immunity. Thymic T-cell development is influenced by sympathetic noradrenergic signaling [42].

As for thyroid and parathyroid glands (Chap. 3), submaxillary lymph nodes receive sympathetic innervation from the neurons located in the SCG, whereas their parasympathetic innervation derives from the lingual nerve, branch of the facial nerve, and reaches the submaxillary glands via the chorda tympani. Based on this, an experimental model was developed comprising the submaxillary lymph nodes and the ipsilateral local manipulation of the sympathetic nerves and/or the ipsilateral manipulation of regional parasympathetic nerves via the chorda tympani [43]. In the submaxillary lymph model, reactive immune homeostasis was studied by subjecting unilaterally denervated rats to different types of stress. An inhibitory effect of the ipsilateral sympathetic nerve ablation on the sympathetic-driven immunosuppression and an increased response to stress after parasympathetic denervation were observed. In addition, T cells express choline acetyltransferase, the ACh-synthesizing enzyme and immunological T cell activation enhances ACh synthesis, suggesting that lymphocytic cholinergic activity might be related to immunological activity [44].

The sympathetic system selectively inhibits Th-1 responses, while favoring Th-2 responses, whereas the parasympathetic system has the opposite effect (Fig. 4.26). NE and E, through activation of β-adrenoceptors, suppresses the production of Th-1 cytokines, such as IL-12, TNF-α, and IFN-γ by the antigen-presenting cells and Th-1 cells, while promoting the production of Th-2 cytokines (e.g., IL-10). Sympathetic activation also shifts toward a Th-2 response via the increased production of glucocorticoids.

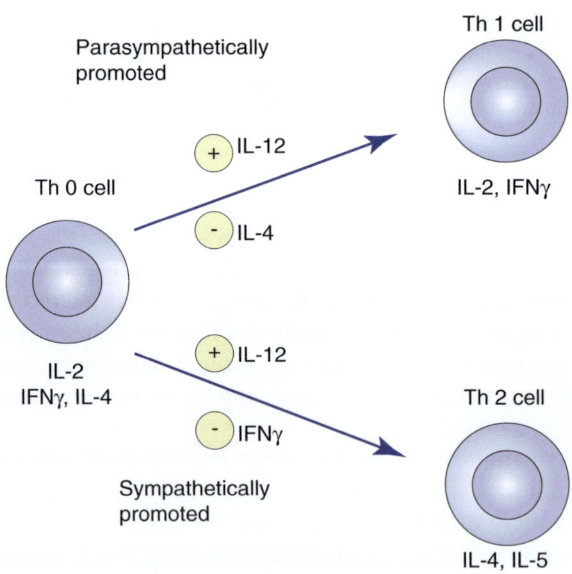

Fig. 4.26 Differentiation of T helper cells in Th1 and Th2 cells. Repeated stimulation in the presence of IL-12 produced by macrophages causes differentiation to Th1 cells that produce IL-2 and interferon (IFN) γ, both cytokines that are very effective at increasing immune responses involving macrophages and other phagocytes and cellular immunity. Stimulation in the presence of IL-4 promotes the development of Th2 cells, effective in producing cytokines acting on mast cell and eosinophil responses and on humoral type immunity. Modified with permission from Cardinali [3]

The first half of sleep represents a proinflammatory state characterized by the downregulation of the two reaction systems, the hypothalamic–pituitary–adrenal axis, and the sympathetic nervous system. Mediators that serve cell growth, differentiation, and restoration, such as growth hormone (GH), prolactin (PRL), and melatonin, show increased blood levels during early sleep (Fig. 4.27). During this period, plasma leptin, which is released by adipocytes as a signal indicative of the size of the lipid deposits, also increases.

Melatonin, GH, PRL, and leptin are synergistic and support the activation, proliferation and differentiation of immune cells, with the production of proinflammatory cytokines (IL-1, IL-12, TNF-α) and Th1 cytokines (IFN-γ). In contrast, cortisol and catecholamines generally suppress these immune functions in an anti-inflammatory manner. In rat submaxillary lymph nodes, a daily rhythm of lymph cell proliferation and response is found with maxima during the resting phase of the 24-h cycle.

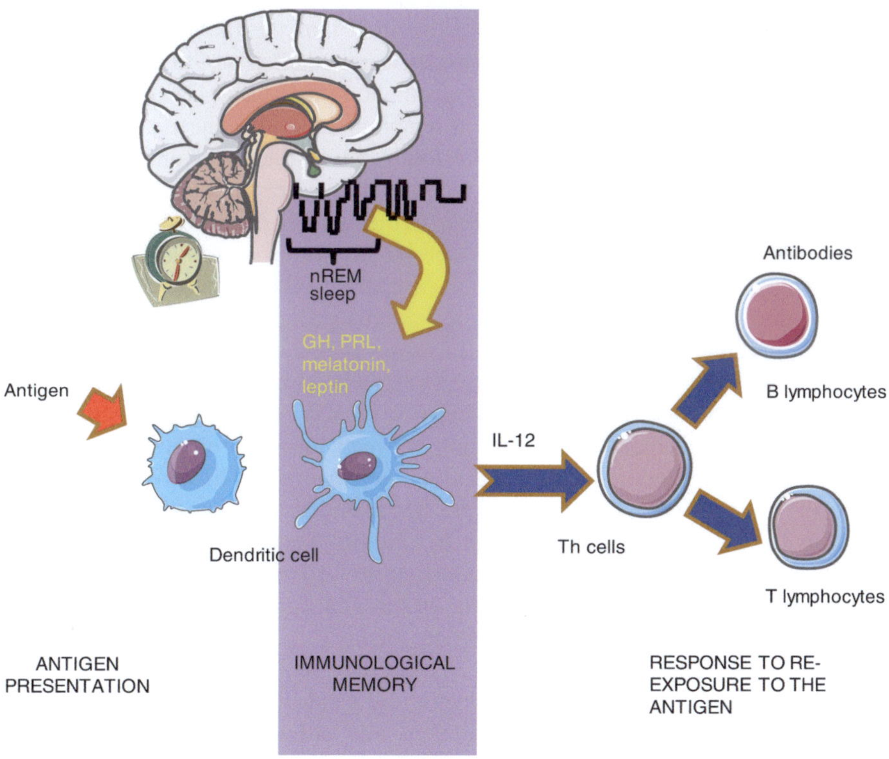

Fig. 4.27 Endocrine changes during early sleep are critical for immune memory-related phenomena. These include the interaction between antigen-presenting cells and T cells, which increases IL-12 production, the shift of the Th1/Th2 cytokine balance toward Th1 predominance, the migration of virgin T cells to lymph nodes, and the proliferation of Th cells. The figure was prepared in part using image vectors from Servier Medical Art (www.servier.com), licensed under the Creative Commons Attribution 3.0 Unported License (http://creativecommons.org/license/by/3.0/)

There is growing evidence that indicates a significant diurnal rhythm for all blood cell types in humans [45]. Circulating lymphocytes show peak levels during the night, whereas granulocytes and NK/dendritic cells peak during the late afternoon. In addition, circulating granulocyte rhythm is severely affected by prolonged wakefulness, with higher circulating levels, lower amplitude, and loss of rhythmicity.

Additional data indicates a shift in the Th1/Th2 cytokine balance toward Th1 predominance during nocturnal sleep in humans, tracked by IFN-γ/IL-4-producing CD4+ cells. There is a shift toward Th1 cytokines (cell-mediated immunity) during early sleep, when NREM sleep is predominant, favoring the cellular aspects of adaptive immune responses over anti-inflammatory activity, or Th2 cytokines (humoral immunity). The effect is reversed during the late part of sleep, when REM sleep is dominant [46].

There were two possible mechanisms by which the CNS is regulating circadian activity in immune organs. One could involve purely neuroendocrine signals as circulating melatonin or glucocorticoids. The other is neural, involving the local autonomic nerves. In several studies, it was verified that the daily changes in cell proliferation in the submaxillary lymph nodes were linked in part to a circadian signal reaching the tissue through local sympathetic innervation [47]. In addition, in pinealectomized rats, cell proliferation in lymph nodes was reduced by half, maintaining its daily maximum at midday. Melatonin administration restored the levels of cell proliferation in the submaxillary lymph nodes of both pinealectomized and SCGx rats. A significant effect of melatonin in maintaining the normal diurnal rhythmicity of neurotransmitter synthesis and release in various neuroimmune territories of the sympathetic nervous system was also reported [48]. Accumulating evidence indicates that melatonin exerts a biphasic immune effect: in basal conditions or a depressed immune response, melatonin increases immune activity; in conditions of augmented immune and inflammatory reaction, melatonin decreases it [48].

The endocrine milieu during early sleep is critical for several phenomena linked to immunological memory: the interaction between antigen-presenting cells and T cells to increase production of IL-12, shifting the balance of Th1/Th2 to Th1 dominance, the migration of naïve T cells to the lymph nodes, and the proliferation of Th cells (Figs. 4.27 and 4.28). In other words, imitating the role that slow-wave sleep plays in the neural mechanisms of episodic memory (Chap. 6), the endocrine milieu during early sleep promotes the initiation of the Th1 immune response, leading to the formation of long-lasting immunological memory.

Sleep facilitates extravasation of T cells and their redistribution to lymph nodes (Fig. 4.28). Sleep improves IL-12 production by dendritic cells, which are precursors of mature medullary antigen-presenting cells and monocytes. IL-12 is a key cytokine for the induction of Th1-type adaptive immune responses. Production of the main anti-inflammatory cytokine IL-10 by monocytes is reduced, whereas that of IL-17 (which stimulates the growth and differentiation of T cells) increases in sleep. The increase in cortisol, E, and NE in wakefulness produces a strong anti-inflammatory effect that disrupts the sleep proinflammatory response. IL-10 is produced by stimulated monocytes and reaches peak levels in the morning [46].

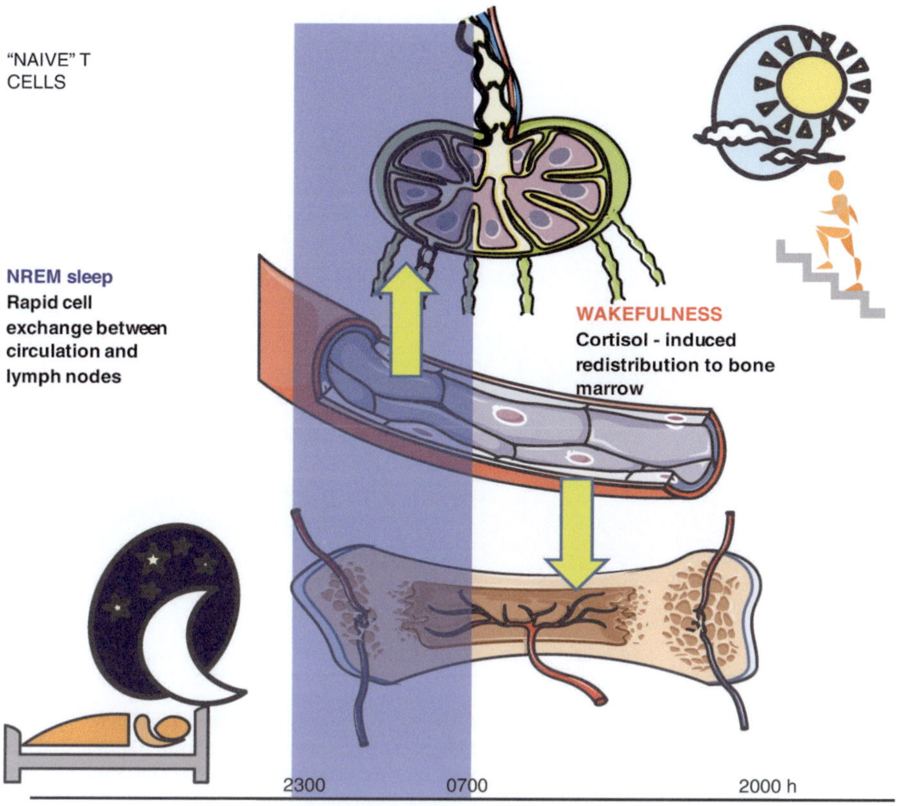

Fig. 4.28 In slow-wave sleep, the virgin lymphocytes enter the lymphoid tissue. In wakefulness, they pass to the blood by the action of cortisol and are found in bone marrow. The figure was prepared in part using image vectors from Servier Medical Art (www.servier.com), licensed under the Creative Commons Attribution 3.0 Unported License (http://creativecommons.org/license/by/3.0/)

Slow-wave sleep is the right time to initiate immune responses because of the proinflammatory neuroimmune environment described above. If the sleep-related production of proinflammatory cytokines were to occur during wakefulness, it would cause discomfort, fatigue, immobility, pain, etc., adaptatively incompatible with the requirements of the waking state. The encounter with the pathogen usually occurs during waking and is adaptatively controlled by the innate immune defense, predominant during wakefulness, which protects the peripheral tissues and spleen before the slow process of adaptive immunity in the lymph nodes. Cytotoxic effector cells are at their maximum during the waking period [49, 50].

The predominance of the sympathetic system in wakefulness mobilizes a subset of white blood cells ("stress" leukocytes) with potent cytotoxic effect (Fig. 4.29). They are a phylogenetically primitive group of immune cells with a high level of expression of β-adrenoceptor that multiply in peripheral blood during wakefulness. "Stress" leukocytes are mature cytotoxic cells with a long replicative history and a very short

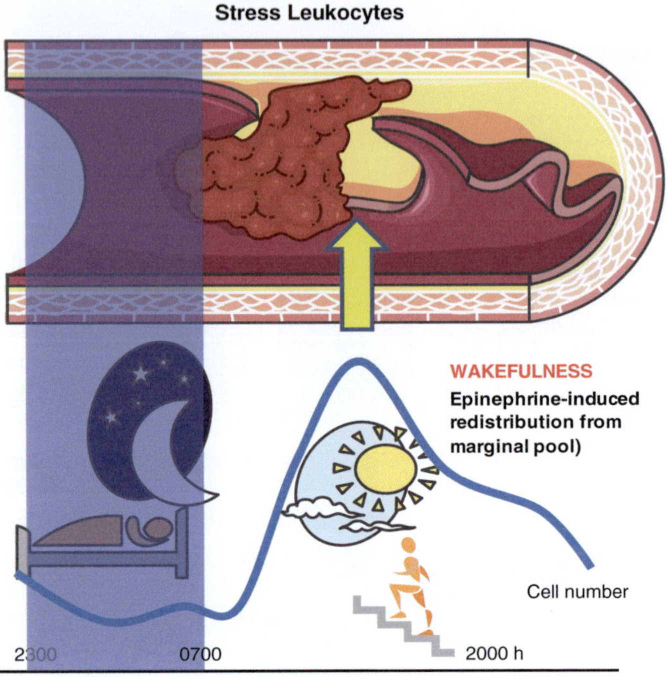

Fig. 4.29 In wakefulness, the ANS mobilizes a subset of leukocytes ("stress leucocytes "), with potent cytotoxic effect. The figure was prepared in part using image vectors from Servier Medical Art (www.servier.com), licensed under the Creative Commons Attribution 3.0 Unported License (http://creativecommons.org/license/by/3.0/)

telomere. They release toxic substances (granzyme, perforin, IFN-γ, Fas ligand apoptosis inducer) that eliminate infected or malignant cells. "Stress" leukocytes reside in the marginal pool and are attached to the endothelium of post-capillary venules. They are mobilized within seconds after stimulation of β_2-adrenoceptor by E and do not recirculate (Fig. 4.29). They have a short half-life and do not need costimulation, which explains their immediate effector potential. Proinflammatory neutrophils, monocytes, and dendritic cells are mobilized after sympathetic activation via α-adrenoceptors. Overall, these effects explain the "adrenergic leukocytosis," which is part of immunosurveillance for rapid responses during wakefulness [50].

Sleep manipulation strongly affects distinct immune parameters, including leukocyte numbers, cytokine production, and cytotoxic activity of immune cells. Whereas short-term total sleep deprivation of only a single night seems to primarily compromise adaptive immune functions, as has been revealed in several human studies relying on vaccination as an experimental model of acute infection, prolonged sleep restriction induces an immunological condition mainly characterized by small but distinct increases in inflammatory markers [51]. This is often referred to as a "low-grade inflammation," and it is of major clinical relevance because it has been associated with an increased risk for developing, for example, diabetes mellitus and cardiovascular diseases. Increases in inflammatory markers have also been observed in habitual short sleepers and in patients with primary insomnia and may represent a mechanism that also mediates increased susceptibility to clinical signs of common cold infections.

One night of total sleep deprivation strikingly decreases the number of dendritic cells producing IL-12, the main inducer of the Th1 response to approximately 40% of normal sleep levels. Regarding other cytokines, IL-2 is increased in sleeping subjects, compared with waking conditions, further supporting the notion that sleep facilitates Th1 immunity. The acute enhancing effect of sleep on lipopolysaccharide-stimulated monocytic TNF-α production adds to the notion that nocturnal sleep favors an immune defense to a microbial challenge [52].

In humans, the number of circulating T cells shows a circadian rhythm with peak counts during the night and a steep decline in the morning. Sleep per se appears to counter this rhythm by acutely reducing the number of total T cells in circulation. The T cell population, however, is rather heterogeneous, comprising various subpopulations with distinctive features and functions and different circadian rhythms. When eight different T cell subsets (naïve, central memory, effector memory, and effector CD4+ and CD8+ T cells) over a 24-h period under conditions of sustained wakefulness were compared with a regular sleep–wake cycle in 14 healthy young men, sleep reduced the number of all T cell subsets during the nighttime [46]. The changes are comparable with changes seen for example after vaccination and are therefore likely to be of physiological relevance [53]. The changes in immune response after a short period of sleep deprivation in humans are summarized in Table 4.1.

Reduced sleep duration for a longer period (e.g., 10 days) caused increased IL-6 levels, which are correlated with increased pain sensitivity. Corroborating this finding, cross-sectional analyses indicated that shorter sleep is associated with higher levels of inflammatory markers, such as C-reactive protein, and IL-6. Interestingly, it has been shown that a 2-h nap during the daytime is able to reverse the effects of one night of sleep deprivation, particularly the increased IL-6 and cortisol levels, and consequently improve alertness and performance in the sleep-deprived subjects, indicating that short naps counterbalance sleep deprivation by restoring the immune and hormonal milieu necessary for proper immune response [54].

Table 4.1 Changes in immune parameters after short sleep deprivation in humans

Sleep deprivation	
Circulating cells	
Th cells, cytotoxic cells, activated T cells, NK cells	↑
Monocytes, dendritic cell precursors	↓
B lymphocytes	↔
Cytokines	
IL-2, IL-7, IL-12, IL-18, TNF-α, IFN-γ/IL-4 ratio	↓
IL-10, IL-4, IL-6, TNF-α, IFN-γ	↑
Other	
T cell proliferation	↓
Regulatory T cell activity	↓
NK cell activity	↑
Complement system	↓
Response to vaccines	↓

Extended periods of partial or total sleep deprivation are associated with alterations in many aspects of immunity, such as increased counts of white blood cells, granulocytes, and monocytes. The analysis of cytokine activity indicates the activation of innate immunity after extended sleep deprivation. Although during normal sleep, proinflammatory cytokines such as IL-6 and IFN-α do not change, it is noteworthy that both are involved in the regulation of early innate immune response and both increase after sleep deprivation. As we have discussed in Chap. 2, various cytokines influence the quality and timing of sleep. The most frequently studied in this respect are IL-1, IL-6, and TNF-α. They are somnogenic and induce fatigue, besides being proinflammatory. It is interesting that these cytokines augment in sleep disorders that cause excessive daytime sleepiness, such as sleep apnea or narcolepsy.

Hence, a reciprocal interaction between sleep and the immune system occurs in humans. On one hand, sleep pressure (or sleep demand) and NREM are increased during an experimental immune challenge, suggesting that inflammatory mediators released during the immune response might be modulators of both physiological and pathological sleep, by acting on neurotransmission systems. To date, the effects of IL-1β and TNF-α effects (produced either in periphery or the CNS) have been extensively studied and are implicated in the modulation of sleep response in both animals and humans (Fig. 4.30). On the other hand, sleep is required for a proper

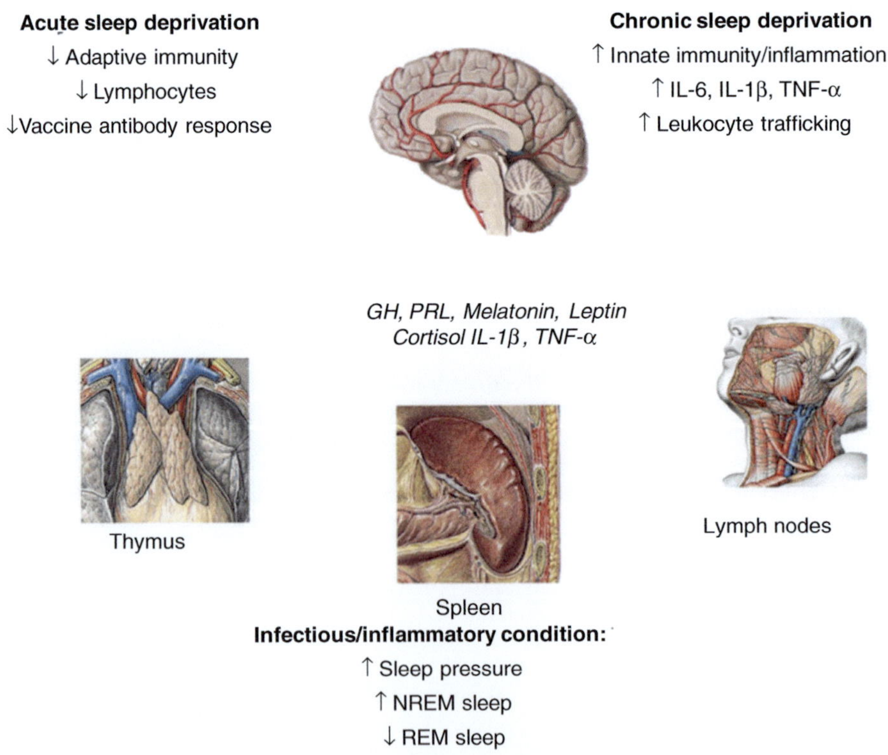

Fig. 4.30 Interactions between acute or chronic sleep deprivation and inflammatory processes

immune response, possibly because of the endocrine milieu elicited by NREM during nighttime sleep. Hormones such as GH, PRL, leptin, and melatonin peak during NREM sleep and have facilitatory effects in the adaptive immune response, particularly on Th1 immunity. Short or prolonged periods of sleep loss have a negative impact on NREM sleep and disrupt this hormonal pattern, preventing these immunosupportive actions and leading to immune suppression. During an ongoing immune response, pro-inflammatory and Th1 cytokines feed back to the brain to enhance NREM sleep [46].

Therefore, it is possible to speculate that NREM sleep acts as an adjuvant to the optimal immune response, creating a hormonal milieu that favors type 1 cytokines and Th1 immunity. In turn, the minimum sleep requirement (or sleep pressure) increased during an infection or immune response, possibly resulting in immune suppression if this minimum requirement is not reached. These outcomes create a vicious cycle between sleep loss, immune modulation, and infectious–inflammatory diseases (Fig. 4.31).

As stated, enough data support the association between NREM sleep and immune responsiveness. None is known concerning REM sleep itself, because of the methodological problems of monitoring immune responses at short intervals.

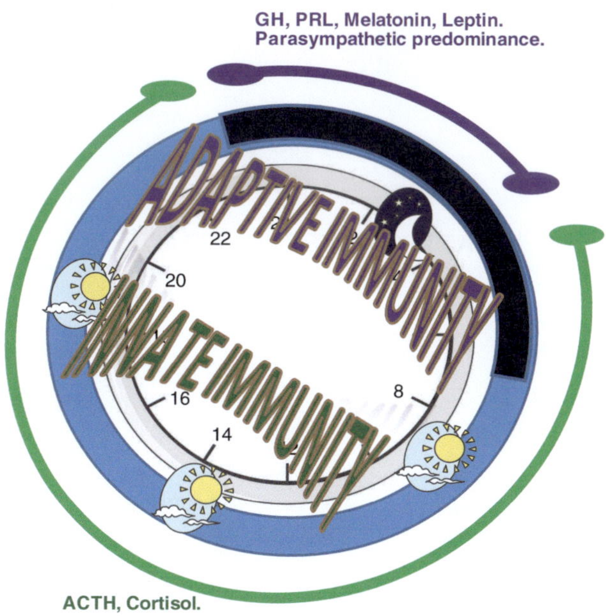

Fig. 4.31 During the first half of the night there is a proinflammatory state, with downregulation of the two reaction systems, the hypothalamic-pituitary–adrenal axis and the sympathetic nervous system. During the second half of the night, cortisol and catecholamines generally suppress immune functions and act in an anti-inflammatory manner

24-h Rhythms in Gastrointestinal Function

Symptoms of several gastrointestinal tract conditions, e.g., gastroesophageal reflux disorder, peptic ulcer disease, biliary colic, and cyclic vomiting syndrome, primarily manifest or are worse nocturnally, as are the serious events of peptic ulcer perforation and esophageal, gastric, and congestive gastropathic variceal hemorrhage [55–58]. All these emphasize the necessity to analyze the changes in enteric ANS function in the two physiological stages of sleep, NREM and REM sleep, compared with wakefulness.

The changes in gastrointestinal functioning during sleep are different depending on the organ studied and its function within the gastrointestinal system. For example, spontaneous esophageal function is markedly reduced during sleep because there is no need for this organ to function except in the event of food entering the upper esophageal area. This rarely occurs without volitional swallowing, which is diminished significantly during sleep. On the other hand, rectal motor activity persists during sleep, which appears to be a mechanism necessary to preserve continence during sleep.

All activity of the gastrointestinal tract is periodic, affecting motility, gall bladder function, the gastrointestinal blood flow, gastric pancreatic and bile secretions, the rate of nutrient absorption, and many other physiological events (Fig. 4.32). This

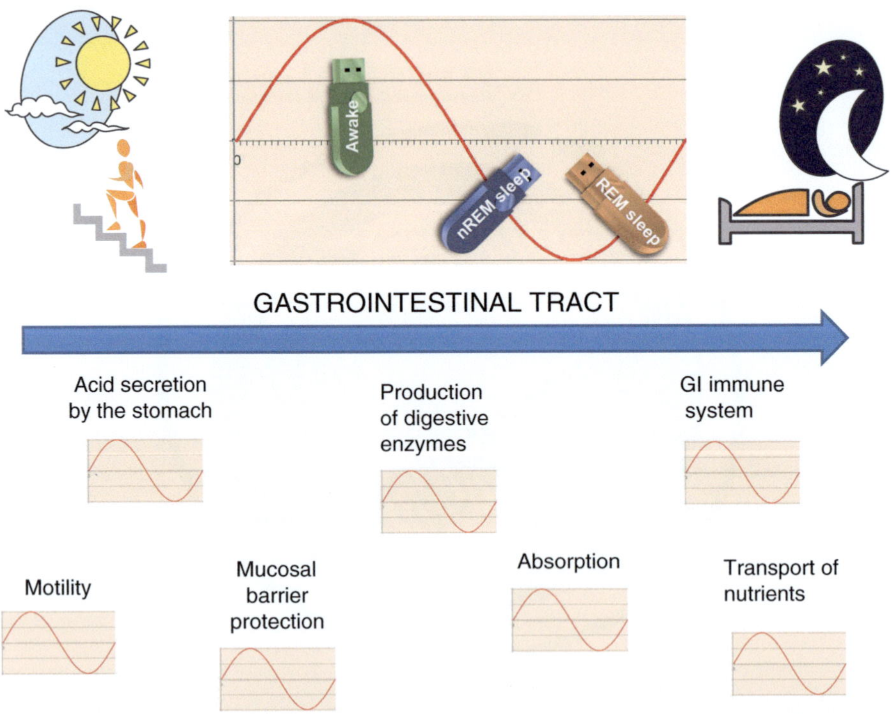

Fig. 4.32 Circadian and ultradian variations are present in most functions of the gastrointestinal tract. The figure was prepared in part using image vectors from Servier Medical Art (www.servier.com), licensed under the Creative Commons Attribution 3.0 Unported License (http://creativecommons.org/license/by/3.0/)

periodicity is the consequence of autonomous rhythms of different frequencies, both ultradian and circadian, and of masking effects of meal schedule synchronizers.

To discern the relative roles of the circadian pacemaker and the effects of sleep, it is necessary to study the changes in circadian patterns in the absence of the effects of sleep, such as during sleep deprivation or when it occurs in atypical hours. A summary of the data obtained regarding the digestive function is shown in Table 4.2.

In addition to the circadian aspects, ultradian rhythms of 90–120 min deserve special attention, and are characteristically found in most of the functions of the digestive tract. This period coincides with that of the alternating phases of NREM–REM overnight sleep phenomenon and continues during the day in cycles of activity–rest known as the basic rest–activity cycle, i.e., cycles at the alert level that occur approximately every 90–120 min [59]. As discussed in Chap. 2, this frequency value is a harmonic of the 24-h cycle.

Digestion begins in the mouth with the process of chewing, salivation, and swallowing. These initial digestive processes are related to the waking state, with little information about their circadian links. The swallowing rate decreases from 25/h during waking to 5/h during sleep. Swallows usually follow sleep awakenings [55, 56].

Swallowing must overcome the pressure of the upper esophageal sphincter for the bolus to enter the esophagus (Fig. 4.33). The upper esophageal sphincter pressure decreases during N3 sleep to 25% of wakefulness. During sleep, the airway is protected from aspiration by the cricopharyngeal muscle. This muscle maintains a pressure barrier in the proximal esophagus to inhibit food aspiration. Like the diaphragm and the muscles in middle ear, the cricopharyngeal muscle differs from the rest of the striated muscles to keep the tone throughout all stages of sleep. The upper esophageal sphincter is tonically contracted, and a pressure between 40 and 80 mmHg usually exists because of the function of this sphincter. Swallowing induces a reflex relaxation to allow the positioning of food and liquids in the upper esophagus, where the normal peristaltic mechanism transports these materials into the stomach.

During wakefulness, gastroesophageal reflux may occur in normal people. This occurrence is primarily postprandial in normal subjects and is associated with multiple episodes of reflux that are neutralized relatively rapidly (in less than 5 min; Fig. 4.34). Swallowing triggers the primary esophageal peristalsis, which decreases in the deepest stage of slow-wave sleep and REM sleep. The inferior esophageal sphincter decreases in tone during sleep, but remains above intragastric pressure.

Table 4.2 Circadian and homeostatic influence on digestive function

Digestive function	Circadian influence (C process)	Homeostatic influence (S process)
Salival output	Possible on pH	Yes
Swallowing/esophageal motility	No	Yes
Gastric acid secretion	Yes	Controversial
Gastric motility	Possible	Yes
Intestinal absorption	Yes	Unknown
Intestinal motility	Possible	Yes
Colon motility	Possible	Yes
Rectus–anal function	Possible	Yes

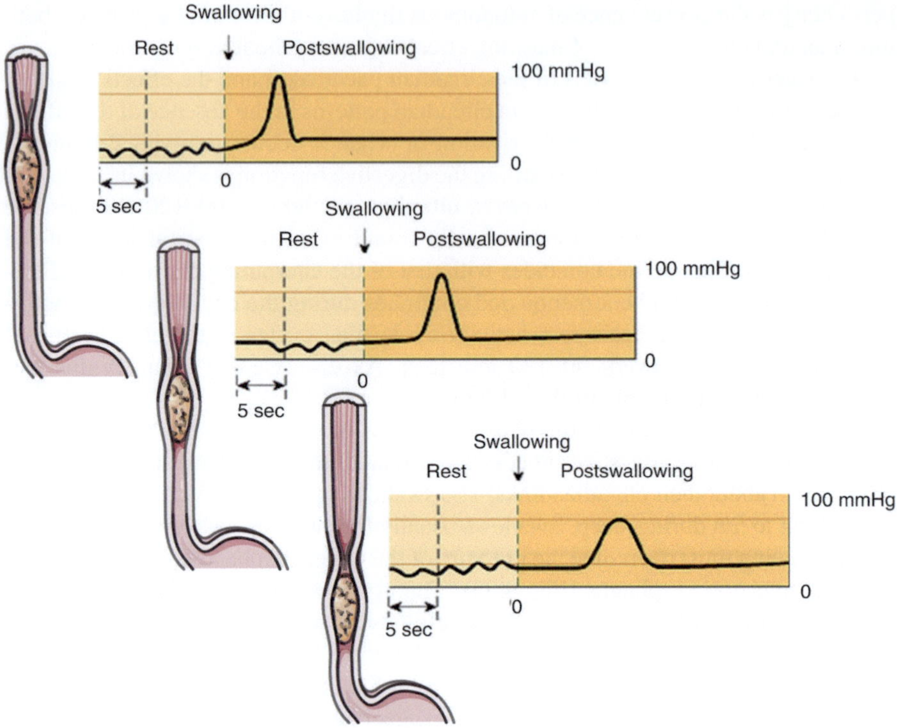

Fig. 4.33 Swallowing starts primary esophageal peristalsis. It must overcome the pressure of the upper esophageal sphincter as the bolus enters the esophagus. The figure was prepared in part using image vectors from Servier Medical Art (www.servier.com), licensed under the Creative Commons Attribution 3.0 Unported License (http://creativecommons.org/license/by/3.0/)

A drop of 5–30 s in inferior esophageal sphincter pressure with sudden increases in intra-abdominal pressure cause gastroesophageal reflux during sleep.

Gastroesophageal reflux is exacerbated during sleep [58]. The clearance of acid into the esophagus occurs in two distinct phases: (a) a first phase (volume clearance) is caused by swallowing, leading to primary esophageal contractions; (b) a second phase (acid clearance) neutralizes any acidic residues of pH <4 (the most important function of saliva). Both phases are inhibited during sleep.

During sleep, prolongation in acid clearance is due to several factors. First, and perhaps most importantly, there is a delay in the conscious response to acid in the esophagus during sleep. Studies have documented that an arousal response almost invariably precedes the initiation of swallowing. An inverse relationship has been described between the acid clearance time and the amount of time the individual spends awake during the acid clearance interval. If an individual responds with an awakening and subsequent swallowing when acid is infused in the distal esophagus during sleep, clearance is substantially faster than if the individual has a prolonged latency to the initial arousal response [58].

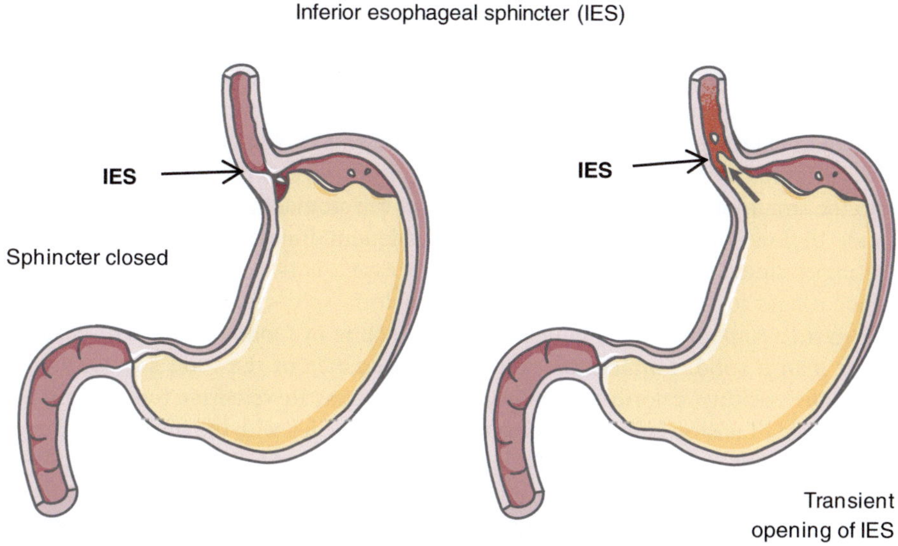

Fig. 4.34 The inferior esophageal sphincter (IES) decreases in tone during sleep, but remains above intragastric pressure. A drop in IES pressure with sudden changes in intra-abdominal pressure cause gastroesophageal reflux during sleep. The figure was prepared in part using image vectors from Servier Medical Art (www.servier.com), licensed under the Creative Commons Attribution 3.0 Unported License (http://creativecommons.org/license/by/3.0/)

Another important aspect of complete neutralization of the acid in distal esophagus is the salivary flow. Saliva is essential for the neutralization of the acidic esophagus. Salivary flow stops completely with the onset of sleep, which would therefore substantially retard the acid neutralization.

Although swallowing frequency diminishes during sleep, esophagic peristalsis is maintained. The frequency of primary contractions in the esophagus (peristaltic contractions preceded by a swallow) diminish progressively from stage N1 to stage N3 sleep. Secondary peristaltic contractions (spontaneous contractions) showed a similar decline from waking to stage N3 sleep, but showed a significant recovery during REM sleep.

The inferior esophageal sphincter prevents reflux of the stomach contents into the esophagus. In the upright position, postprandial gastric distention causes brief relaxation of the inferior esophageal sphincter with transient reflux that is quickly cleared (Fig. 4.34). Several factors assist in the rapid clearance or neutralization of stomach contents: (a) volume clearance: two or three swallows induce clearance of reflux material; (b) acid neutralization: saliva itself buffers the acidity of the refluxed material. These mechanisms quickly return the distal esophagus to pH 5.5–6.5 (normal esophageal pH) during wakefulness.

Nocturnal gastroesophageal reflux is common, with up to 10% of the population reporting symptoms of nocturnal reflux in survey studies. As noted, nocturnal gastroesophageal reflux is potentially more injurious than a diurnal one because acid clearance mechanisms are impaired during sleep. When episodes of gastroesophageal reflux occur during sleep, acid contact time is prolonged. During sleep, salivary flow virtually stops and the frequency of swallowing is very decreased. The clearance of refluxed material occurs only after arousal from sleep. Most nocturnal gastroesophageal reflux episodes occur during prolonged waking or after arousal from stage N2 sleep [58].

The stomach plays an important role in food storage and its controlled emptying into the small intestine. The main functions of the stomach are: (a) acidification of food; (b) flow control of the bolus into the duodenum. Ingested food accumulates in the upper stomach, the fundus, and once processed, is driven into the antrum and hence passes to the duodenum through the pyloric sphincter in discrete amounts. The gastric emptying rate is influenced by the nature of food: a liquid is discharged faster than a solid. Furthermore, other characteristics of the food such as liquid osmolality, acidity, caloric content, etc., can affect gastric retention time [55, 58].

The rate of gastric emptying has a rhythm and changes with the time of day, varying considerably between meals. For example, breakfast and dinner have very different gastric retention. In the morning, the gastric emptying rate is higher than at night.

By nasogastric recording, there is a basal rate of production of HCl, with a maximum around the beginning of the night, both in normal healthy individuals and in patients with duodenal ulcers. Ulcer patients produce a greater amount of acid than normal, but the basic pattern of daily secretion tends to be the same. In general, gastric acidity increases at night, between 22:00 and 02:00 h, and decreases in the morning. This rhythm persists in the absence of sleep, indicating their circadian nature. The electrical activity of the stomach also has a very strong daily rhythm [55, 58].

Patients with duodenal ulcers lose rhythms of acid secretion, with overproduction of acid throughout the day and night. Overproduction of HCl is linked to chronic *Helicobacter pylori* infection. *Helicobacter pylori* is a bacterium that lives in the mucous layer of the stomach, where the enzyme urease is active with the conversion of urea into CO_2 and ammonia, which is a buffer of luminal acid that protects the microorganism. *H. pylori* also secretes proteins that modulate the immune response and directly alter cell signaling pathways of the mucosa. More than half of the world's population is infected with *H. pylori*. In most cases, the infection, although chronic, is mild and causes no symptoms. In some individuals, however, infection leads to increased acid secretion and symptomatic inflammation causes ulceration of the stomach or duodenum. Almost all duodenal peptic ulcers and about half of gastric peptic ulcers show *H. pylori* infection [55, 58].

The fall of somatostatin induced by *H. pylori* increases the parietal cell mass and overproduction of acid without a circadian pattern. *H. pylori* infection interferes with the release of clover protein, increasing the risk of mucosal damage. Trefoil protein 1 (clover) is a protein from the digestive tract encoded by the TFF1 gene that shows circadian rhythmicity. The release of clover protein is reduced by sleep deprivation, indicating a relationship with both circadian and homeostatic processes.

Melatonin is a strong inhibitor of acid secretion and increases the release of gastrin, resulting in stimulation of the lower esophageal sphincter activity (Fig. 4.35). Both actions protect the esophagus, minimizing contact with reflux. Furthermore, melatonin reverses inflammatory lesions and reduces lipid peroxidation produced by contact with

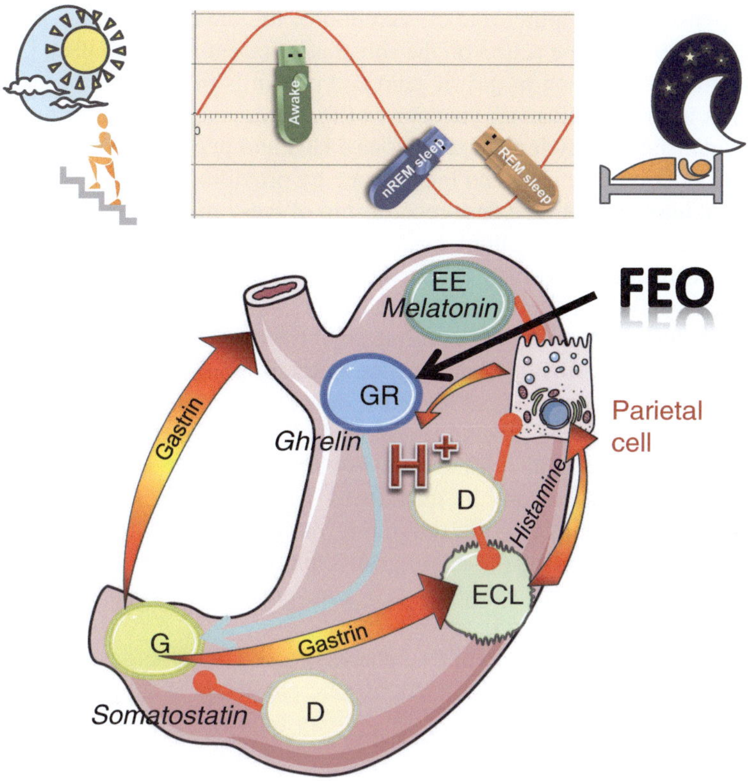

Fig. 4.35 Food-entrained oscillators (FEO) in the stomach. FEO activity precedes and promotes food and eating behavior. A location of FEO in the stomach is the oxyntic glandular cells (GR) co-expressing ghrelin and circadian clock proteins. Entero-endocrine cells (EE) release melatonin, which is a strong inhibitor of acid secretion antagonizing the effect of histamine and increasing the release of gastrin, resulting in the stimulation of lower esophageal sphincter activity. *D* somatostatin-releasing cell, *ECL* enterochromaffin-like cells. See also Fig. 3.25. The figure was prepared in part using image vectors from Servier Medical Art (www.servier.com), licensed under the Creative Commons Attribution 3.0 Unported License (http://creativecommons.org/license/by/3.0/)

gastric juice. Melatonergic enteroendocrine cells produce about 500 times more melatonin than the pineal gland. There is an entero-hepatic circulation for melatonin [60, 61].

The origin of rhythmicity in gastric secretion of HCl does not appear to be endocrine as it is not coincident with the parallel rhythms of plasma gastrin (a major hormone that stimulates gastric secretion). Nerve stimulation is more important than endogenous circadian rhythmicity as rhythmicity in gastric secretion of HCl disappears in vagotomized patients.

Pepsin, a protease that degrades food proteins in the stomach, is secreted by cells of the gastric mucosa as an inactive pepsinogen, passing to the active form, pepsin, by the action of HCl and subsequently also by the action of activated pepsin itself. Like HCl, protease pepsin activity follows a daily rate, with a maximum after dinner. As the presence of HCl and very acidic pH are essential requirements for the action of pepsin, the coupling rate of secretion of the protease with the rhythm of HCl and pH optimizes the action on the substrates present in the stomach [55, 58].

The nocturnal secretion of gastric acid and pepsin is in principle beneficial for the body, as it helps to clean up the debris of food that may occur after meals. In a healthy organism, the rhythms of secretion of HCl and pepsin activity are in phase with the production of mucus and bicarbonate by providing a natural defense. The pathological alteration of the phase relationship of these rhythms increases the vulnerability of the gastric epithelium, leading to ulcerogenesis.

Gastrointestinal tract motility is characterized by rhythmic waves of membrane depolarization and contraction of the muscle fiber, with a shape and frequency characteristics of each portion of the gastrointestinal tract. In the stomach, for example, the depolarization consists of a spike action potential followed by a plateau phase before repolarization. In the small intestine, in contrast, the action potentials are in spikes, and in the colon the spikes are prolonged (haustrations).

Like many other variables of the digestive tract, such as motility, the stomach presents ultradian fluctuations in its rate of gastric secretion. These ultradian rhythms are closely related to the migrating motor complex (MMC) during the interdigestive period. During the interdigestive phase, increases in gastric acid and in pepsin production occur every 90–120 min, preceding the appearance of the MMC in the duodenum [55, 58].

The MMC (Fig. 4.36) consists of four phases: phase 1, rest; phase 2, irregular contractions; phase 3, 9–12 regular contractions/min; phase 4, transition between phases

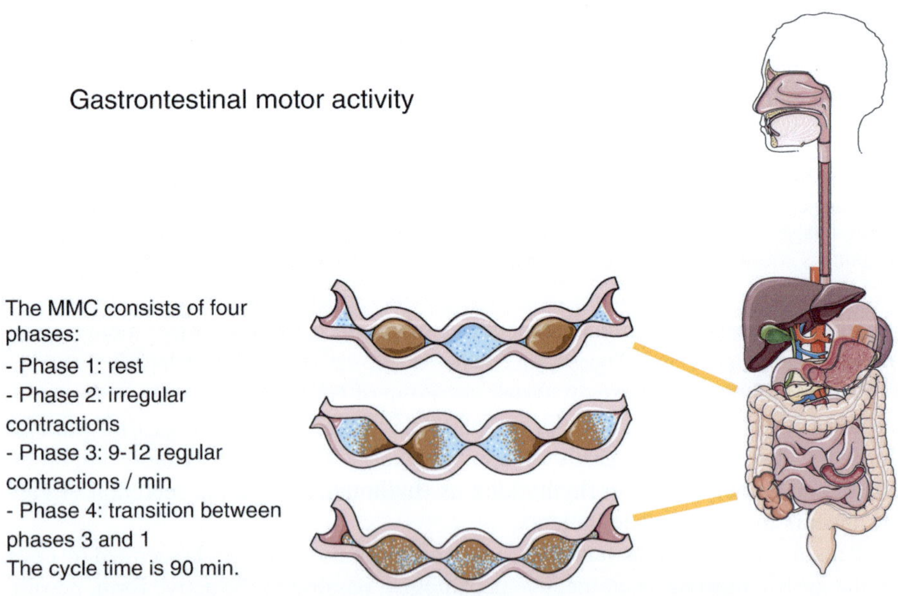

Fig. 4.36 The four phases of migratory motor complexes. The figure was prepared in part using image vectors from Servier Medical Art (www.servier.com), licensed under the Creative Commons Attribution 3.0 Unported License (http://creativecommons.org/license/by/3.0/)

3 and 1. The cycle time is about 90 min, such as NREM/REM sleep alternation. Food ingestion establishes a pattern of vigorous contraction throughout the distal stomach and small bowel. If no food enters the stomach, the MMC cycle has a period of about 90 min (Fig. 4.36).

Gastric motor function is controlled by a gastric pacemaker located in the smooth muscle of the greater curvature of the stomach. By electrical recordings it was shown that the amplitude of gastric engine cycle decreases in NREM sleep and is restored in REM sleep. In normal subjects, acid secretion, water secretion, and the fractional rate of emptying all showed significant decrements during sleep. There did not appear to be any differences between REM and NREM sleep, but these measures demonstrated a significant difference between presleep waking and REM sleep. Data obtained by the use of radionuclide emptying assessments suggest that this difference might be circadian, rather than sleep-dependent. Some studies have shown a marked delay in the gastric emptying of solids in the evening, compared with the morning.

Motility of the small intestine is characterized by a high frequency of targeting movements against peristalsis. Asynchronous contraction of the circular muscle layers and outer longitudinal internal (segmentation) help the mixing of food with digestive secretions and improve absorption by allowing a constant renewal of food contact area of intestinal epithelium. During the interdigestive periods, intestinal motility does not stop, but continues as a characteristic rhythmic pattern.

The MMCs are generated in many ultradian pacemakers distributed along the digestive tract and coordinated by the enteric nervous system. If the small intestine is cut into several pieces and subsequently subjected to reanastomosis, each fragment starts a separate cycle and the coupling of the MMC reappears within 45–60 days, coinciding with the intrinsic regeneration of nerves. Control of the rhythms of the 24-h propagation velocity of the MMC resides in the CNS, as vagotomy results in the suppression of these rhythms [55, 58].

In the mouse colon, a functional circadian clock and a subset of rhythmically expressed genes have a direct impact on colonic motility, such as the contractile response of colonic tissue to ACh, stool output, and intracolonic pressure changes. These effects are attenuated in mice with disrupted clock function [62]. Clock-controlled transcription of genes, such as choline acetyltransferase and neuronal NO synthase, leads to the rhythmic release of ACh and NO, which initiate diverse biochemical, cellular, and physiological processes within the colonic circular muscle, which may in turn, through a cascade of second-order messengers and various signaling pathways, lead to enhanced colonic motility (Fig. 4.37). We discussed in Chap. 3 how ACh and NO participate in causing peristalsis.

The propagation velocity generated by the MMC is reduced by more than 50% during sleep by the suppression of phase 2. The spread of contractions decreases in proportion to the depth of sleep, and is absent during N3 sleep. The frequency of the spread and the pressure in the colon during REM sleep is like those found in NREM N2 sleep. It can be concluded then that the MMCs are the basic rhythmic pattern of intestinal motility, of endogenous origin, with meals having an effect on them (food entrained oscillators). In addition, MMCs also appear to be associated with changes in secretory phenomena such as gastric, pancreatic, and biliary secretion.

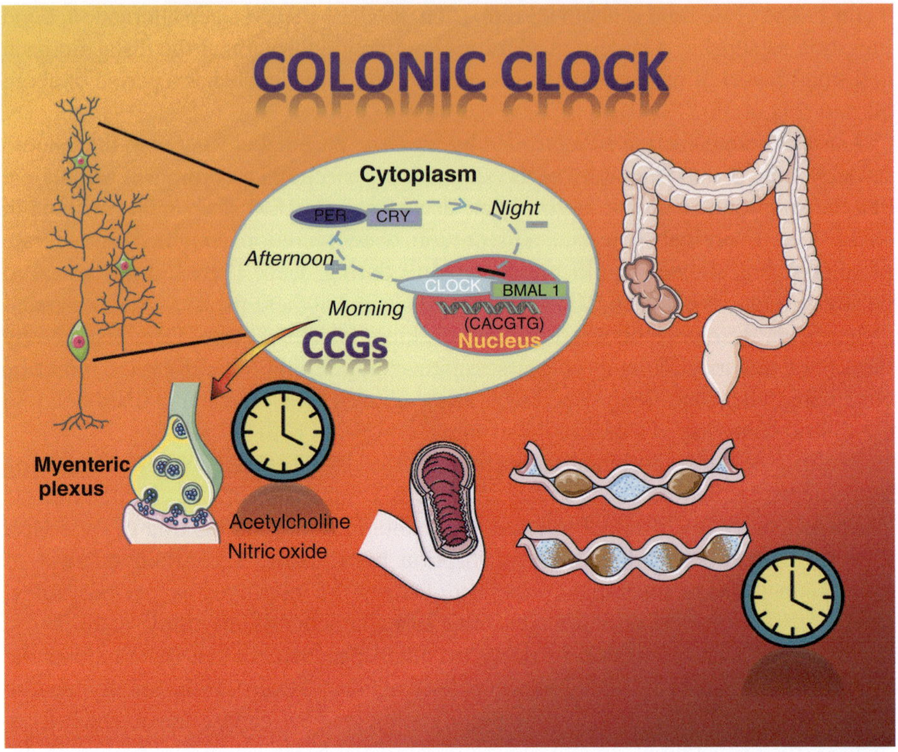

Fig. 4.37 Circadian regulation of colonic motility. The rhythmic expression of clock genes within the neurons of the myenteric plexus modulate colonic motility through clock-controlled transcription of genes such as acetylcholine (ACh) transferase and neuronal nitric oxide synthase (nNOS). Transcription of ACh and nNOS leads to the rhythmic release of ACh and nitric oxide (NO), which initiates diverse biochemical, cellular, and physiological processes within the colonic circular muscle. The figure was prepared in part using image vectors from Servier Medical Art (www.servier.com), licensed under the Creative Commons Attribution 3.0 Unported License (http://creativecommons.org/license/by/3.0/)

In humans, the MMCs are interrupted 10–20 min after food intake, staying in a kind of permanent phase III. The duration of this interruption appears to be proportional to the caloric content of the food and its chemical composition, the most prolonged interruption with fat and sugar-rich foods, and less with protein-rich foods. The mediators of this disruption are fundamentally hormonal, as hormones released during the digestive process (secretin, gastrin, CCK, pancreatic polypeptide) can inhibit the MMC cycle. On the contrary, the interruption is not affected by vagotomy, splenectomy, or celiac and SCGx or a combination of procedures. Differing from enteral feeding, parenteral nutrition does not interrupt the MMC cycle, indicating that it is the presence of nutrients directly into the digestive tract that interrupts this process. Endocrine activity of the small intestine is periodic on some regulatory peptides, and is manifested by the existence of cycles in the release of pancreatic polypeptide, motilin, somatostatin, and secretin in blood circulation [55, 58].

The MMC plays an important role in maintaining the intestinal mucosa, providing adequate evacuation of debris and ensuring a cephalocaudal flow of the intestinal content. The association of cycles in motility with increases in gastric acid and pepsin in the stomach and bicarbonate secretion, amylase, bile acids, trypsin, and lipase in the exocrine pancreas, and bile, promote antibacterial action and cleaning of the digestive tract. In fact, when the MMC is disturbed, pathological changes associated with bacterial infections, myotonic dystrophy, intestinal pseudo-obstruction, etc., usually occur.

Pancreatic juice originates primarily in the acini of the exocrine pancreas and is modified by ductus cells before being secreted at the duodenum with a rich content of bicarbonate and digestive enzymes. Pancreatic juice has an alkaline pH because of its high bicarbonate content, and neutralizes the acid gastric juices from the stomach, raising the duodenal pH to values of 6–7. In addition, pancreatic juice is rich in digestive enzymes: trypsin, chymotrypsin, elastase, carboxypeptidase, lipase, and amylase. Pancreatic secretion presents both ultradian and circadian rhythms. Ultradian rhythms are in phase with MMCs originating in the duodenum. There is a direct relationship between the rate of secretion of pancreatic juice and intestinal changes in contractility at each phase of the MMC: during phase I, resting, the basal secretion rate remains low; during phase II, pancreatic secretion increases in synchrony with muscle contractions; during phase III, an intense muscle contraction, pancreatic secretion reaches maximum values. There is also a large overlap between the periods of peak pancreatic secreting and gastric pH changes.

In contrast to what happens with the ultradian rhythms, circadian oscillations in pancreatic secretion are of small amplitude. Overall maximum pancreatic juice secretion occurs overnight, and in this period the greatest amount of amylase is also produced.

In the case of bile secretion in the duodenum, gallbladder emptying shows, like the gastric and pancreatic secretions, both ultradian and circadian rhythms. Ultradian rhythms are like those discussed above and in fasting individuals they match the duodenal MMC. Peak secretion of bile precedes maximum duodenal motility and pancreatic secretion by about 30–40 min.

The gallbladder also shows a circadian rhythm in the rate of secretion of bile, with a maximum during the early hours of the morning, between 07:00 and 09:00 h, coinciding with the maximal rate of formation of bile salts by the liver and in the activity of the key enzymes in the bile formation, e.g., cholesterol 7-hydroxylase. Thus, the increased rate of synthesis of bile salts coincides with the release of bile into the duodenum; it has been postulated that the increased activity of the 7a-hydroxylase would be initiated by the decrease in the content of bile salts in the liver.

Cholesterol concentrations in bile secretion and those of other lipids follow a circadian rhythm, with high nocturnal values. The index of cholesterol saturation (lithogenic index, as calculated from the concentrations of cholesterol, phospholipids, and bile salts) also shows daily fluctuations, with higher values in the early morning than in the afternoon.

Several experimental studies have shown that a large number of digestive enzymes present biological rhythms. An example is the intestinal disaccharidases, which have a synchronization with the rhythm of food, but not with the light–dark periods. Regardless of the changes that may occur in enzyme activity as a function of time, absorption processes can also be modulated rhythmically. The expression of glucose transporters shows circadian rhythms. In rats fed ad libitum, glucose absorption is low during the day and high overnight. This rhythmic pattern is independent of the darkness and light cycle and is synchronized by the feeding schedule. Periodicity is anticipatory and held for a few days of food deprivation and during discontinuous parenteral nutrition.

A period of REM sleep is associated with increased pressure on the colon and frequency of contractions at similar levels to N2 sleep. Arousals stimulate an increase in the spread of contractions in all segments of the colon. Intestinal and colonic activity are generally decreased during sleep. Rectal motor activity increases during sleep, but propulsion is retrograde. This and the fact that the anal sphincter tone (although reduced) remains higher than the rectal tone prevent anal leakage during sleep [55, 58].

Irritable bowel syndrome is a homeostatic and circadian dysfunction of small and large intestinal rhythms that entails a constellation of symptoms, including alternating constipation and diarrhea. It affects 8–25% of the Western population, with a predominance in women of 2–3 times more than in men. Exaggerated abdominal pain, altered intestinal function, inflammation of the mucosa, exaggerated response to stress, and an increase of proinflammatory cytokines are signs of irritable bowel syndrome. Stress plays a key role in the triggering and exacerbation of symptoms [55, 56].

Several studies indicate alterations of the autonomic tone in irritable bowel syndrome that results in abnormal central processing of pain with allodynia as a major sign. Allodynia refers to central pain sensitization following normally nonpainful, often repetitive, stimulation. Allodynia leads to the triggering of a pain response from stimuli that do not normally provoke pain. Allodynia is different from hyperalgesia: an extreme, exaggerated reaction to a stimulus that is normally painful.

Polysomnography studies have indicated that patients with irritable bowel syndrome have an increased autonomic sympathetic activity during REM sleep. Neuroplasticity changes such as the learning of an abnormal visceral response, are consolidated in REM sleep (procedural memory, Chap. 6). Hence, alterations of the autonomic tone seen during REM sleep could be consolidated as a memory of anomalous visceral pain (allodynia).

Irritable bowel syndrome is predominant in people who undertake rotating shift work. In several studies, logistic regression analysis indicated an association between this pathology and rotational shift work, even after controlling for other confounding factors, such as poor sleep quality (chronodisruption). Greater abdominal pain was observed in the groups with rotating shifts, which could be due to the interruption of the circadian factors that modulate the visceral sensitivity.

Inflammatory bowel disease comprises the chronic inflammation of the large intestine, and sometimes of the small intestine, whose main forms are Crohn's disease and ulcerative colitis. The main difference between the two is the location and nature of the inflammatory changes. Disruption of the circadian regulation of the intestinal immune system is a possible aggravating cause. Circadian clock gene disruption can affect the intestinal permeability of epithelial cells and the phenomenon initiates the inflammatory cascade observed in patients during exacerbation of the disease [55, 56].

There is evidence that sleep disorders can affect inflammatory processes in the colon. Both acute and chronic sleep deprivation aggravates colon inflammation in murine models of colitis.

Concerning sleep, its disturbances coincide with digestive discomfort. NREM sleep decreases more than REM sleep. During REM sleep, neuronal reprogramming of the enteric nervous system occurs (neuroplasticity changes) and these changes can lead to allodynia and visceral hyperalgesia.

References

1. Fernandez-Gil MA, Palacios-Bote R, Leo-Barahona M, Mora-Encinas JP. Anatomy of the brainstem: a gaze into the stem of life. Semin Ultrasound CT MR. 2010;31:196–219.
2. Tseng CJ, Cheng PW, Tung CS. Chapter 29 – Pharmacology of the nucleous tractus solitarii. In: Biaggioni I, Burnstock G, Low PA, Paton JFR, editors. Primer on the Autonomic Nervous System. 3rd ed. San Diego: Academic; 2012. p. 141–4.
3. Cardinali DP. Neurociencia Aplicada. Sus Fundamentos. Buenos Aires: Editorial Médica Panamericana; 2007.
4. Verriotis M, Chang P, Fitzgerald M, Fabrizi L. The development of the nociceptive brain. Neuroscience. 2016;338:207–19.
5. Wehrwein EA, Joyner MJ. Chapter 8 – Regulation of blood pressure by the arterial baroreflex and autonomic nervous system. In: Buijs RM, Swaab D, editors. Handbook of clinical neurology. Autonomic nervous system. New York: Elsevier; 2013. p. 89–102.
6. Macefield VG, Henderson LA. Chapter 3 – Imaging of brainstem sites involved in cardiovascular control. In: Biaggioni I, Burnstock G, Low PA, Paton JFR, editors. Primer on the autonomic nervous system. 3rd ed. San Diego: Academic; 2012. p. 13–6.
7. Mischel NA, Subramanian M, Dombrowski MD, Llewellyn-Smith IJ, Mueller PJ. (In) activity-related neuroplasticity in brainstem control of sympathetic outflow: unraveling underlying molecular, cellular, and anatomical mechanisms. Am J Physiol Heart Circ Physiol 2015;309:H235–43.
8. Gordan R, Gwathmey JK, Xie LH. Autonomic and endocrine control of cardiovascular function. World J Cardiol. 2015;7:204–14.
9. Wieling W, Groothuis JT. Chapter 39 – Physiology of upright posture. In: Biaggioni I, Burnstock G, Low PA, Paton JFR, editors. Primer on the autonomic nervous system. 3rd ed. San Diego: Academic; 2012. p. 193–5.
10. Lombard JH, Cowley J. Chapter 38 – Neural control of blood vessels. In: Biaggioni I, Burnstock G, Low PA, JFR P, editors. Primer on the autonomic nervous system. 3rd ed. San Diego: Academic; 2012. p. 187–91.
11. Halcox JPJ. Chapter 66 – Endothelial dysfunction. In: Biaggioni I, Burnstock G, Low PA, Paton JFR, editors. Primer on the autonomic nervous system. 3rd ed. San Diego: Academic; 2012. p. 319–24.
12. Smith PM, Ferguson AV. Recent advances in central cardiovascular control: sex, ROS, gas and inflammation. F1000Res. 2016;5.
13. Chatham JC, Young ME. Regulation of myocardial metabolism by the cardiomyocyte circadian clock. J Mol Cell Cardiol. 2013;55:139–46.
14. Young ME. The circadian clock within the heart: potential influence on myocardial gene expression, metabolism, and function. Am J Physiol Heart Circ Physiol. 2006;290:H1–16.

15. Smolensky MH, Hermida RC, Portaluppi F. Circadian mechanisms of 24-hour blood pressure regulation and patterning. Sleep Med Rev. 2017;33:4–16.
16. Chouchou F, Desseilles M. Heart rate variability: a tool to explore the sleeping brain? Front Neurosci. 2014;8:402.
17. Portaluppi F, Tiseo R, Smolensky MH, Hermida RC, Ayala DE, Fabbian F. Circadian rhythms and cardiovascular health. Sleep Med Rev. 2012;16:151–66.
18. Alibhai FJ, Tsimakouridze EV, Reitz CJ, Pyle WG, Martino TA. Consequences of circadian and sleep disturbances for the cardiovascular system. Can J Cardiol. 2015;31:860–72.
19. Gilbey MP. Chapter 30 – Entrainment of sympathetic rhythms. In: Biaggioni I, Burnstock G, Low PA, Paton JFR, editors. Primer on the autonomic nervous system. 3rd ed. San Diego: Academic; 2012. p. 147–9.
20. Guyenet PG. Regulation of breathing and autonomic outflows by chemoreceptors. Compr Physiol. 2014;4:1511–62.
21. Costa KM, Accorsi-Mendonca D, Moraes DJ, Machado BH. Evolution and physiology of neural oxygen sensing. Front Physiol. 2014;5:302.
22. Prabhakar NR. Chapter 68 – Oxygen sensing. In: Biaggioni I, Burnstock G, Low PA, Paton JFR, editors. Primer on the autonomic nervous system. 3rd ed. San Diego: Academic; 2012. p. 331–3.
23. Barnes PJ. Chapter 41 – Autonomic control of the lower airways. In: Biaggioni I, Burnstock G, Low PA, Paton JFR, editors. Primer on the autonomic nervous system. 3rd ed. San Diego: Academic; 2012. p. 201–4.
24. Mazzone SB, Canning BJ. Chapter 18 – Autonomic neural control of the airways. In: Buijs RM, Swaab D, editors. Handbook of clinical neurology. Autonomic nervous system. New York: Elsevier; 2013. p. 215–28.
25. Mansukhani MP, Kara T, Caples SM, Somers VK. Chemoreflexes, sleep apnea, and sympathetic dysregulation. Curr Hypertens Rep. 2014;16:476.
26. Harper RM, Bandler R, Spriggs D, Alger JR. Lateralized and widespread brain activation during transient blood pressure elevation revealed by magnetic resonance imaging. J Comp Neurol. 2000;417:195–204.
27. Alexander MP, Gillingham S, Schweizer T, Stuss DT. Cognitive impairments due to focal cerebellar injuries in adults. Cortex. 2012;48:980–90.
28. Xu F, Frazier DT. Role of the cerebellar deep nuclei in respiratory modulation. Cerebellum. 2002;1:35–40.
29. Strata P. The emotional cerebellum. Cerebellum. 2015;14:570–7.
30. Zhu JN, Yung WH, Kwok-Chong CB, Chan YS, Wang JJ. The cerebellar-hypothalamic circuits: potential pathways underlying cerebellar involvement in somatic-visceral integration. Brain Res Rev. 2006;52:93–106.
31. Li B, Zhu JN, Wang JJ. Histaminergic afferent system in the cerebellum: structure and function. Cerebellum Ataxias. 2014;1:5.
32. Li B, Zhuang QX, Gao HR, Wang JJ, Zhu JN. Medial cerebellar nucleus projects to feeding-related neurons in the ventromedial hypothalamic nucleus in rats. Brain Struct Funct. 2017;222:957–71.
33. Zhu JN, Wang JJ. The cerebellum in feeding control: possible function and mechanism. Cell Mol Neurobiol. 2008;28:469–78.
34. Ciriello J, Caverson MM. Hypothalamic orexin-A (hypocretin-1) neuronal projections to the vestibular complex and cerebellum in the rat. Brain Res. 2014, 1579:20–34.
35. Lu JH, Wang XQ, Huang Y, Qiu YH, Peng YP. GABAergic neurons in cerebellar interposed nucleus modulate cellular and humoral immunity via hypothalamic and sympathetic pathways. J Neuroimmunol. 2015;283:30–8.
36. Hanoun M, Maryanovich M, Arnal-Estape A, Frenette PS. Neural regulation of hematopoiesis, inflammation, and cancer. Neuron. 2015;86:360–73.
37. Kox M, Pickkers P. Modulation of the innate immune response through the vagus nerve. Nephron. 2015;131:79–84.
38. Sankowski R, Mader S, Valdes-Ferrer SI. Systemic inflammation and the brain: novel roles of genetic, molecular, and environmental cues as drivers of neurodegeneration. Front Cell Neurosci. 2015;9:28.

39. Fernandez R, Nardocci G, Navarro C, Reyes EP, Acuna-Castillo C, Cortes PP. Neural reflex regulation of systemic inflammation: potential new targets for sepsis therapy. Front Physiol. 2014;5:489.
40. Boeckxstaens G. Chapter 11 – The clinical importance of the anti-inflammatory vagovagal reflex. In: Buijs RM, Swaab D, editors. Handbook of clinical neurology. Autonomic nervous system. New York: Elsevier; 2013. p. 119–34.
41. Parekh PJ, Nayi VR, Johnson DA, Vinik AI. The role of gut microflora and the cholinergic anti-inflammatory neuroendocrine system in diabetes mellitus. Front Endocrinol (Lausanne). 2016;7:55.
42. Leposavic G, Pilipovic I, Perisic M. Cellular and nerve fibre catecholaminergic thymic network: steroid hormone dependent activity. Physiol Res. 2011;60(Suppl 1):S71–82.
43. Cardinali DP, Esquifino AI. Neuroimmunoendocrinology of the cervical autonomic nervous system. Biomed Rev. 1998;9:47–59.
44. Kawashima K, Fujii T, Moriwaki Y, Misawa H. Critical roles of acetylcholine and the muscarinic and nicotinic acetylcholine receptors in the regulation of immune function. Life Sci. 2012;91:1027–32.
45. Nobis CC, Labrecque N, Cermakian N. Circadian control of antigen-specific T cell responses. ChronoPhysiol Ther. 2016;6:65–74.
46. Besedovsky L, Dimitrov S, Born J, Lange T. Nocturnal sleep uniformly reduces numbers of different T-cell subsets in the blood of healthy men. Am J Physiol Regul Integr Comp Physiol. 2016;311:R637–42.
47. Cardinali DP, Esquifino AI. Sleep and the immune system. Curr Immunol Rev. 2012;8:50–62.
48. Cardinali DP. Ma Vie en Noir. Fifty years with melatonin and the stone of madness. Switzerland: Springer; 2016.
49. Geiger SS, Fagundes CT, Siegel RM. Chrono-immunology: progress and challenges in understanding links between the circadian and immune systems. Immunology. 2015;146:349–58.
50. Bollinger T, Bollinger A, Oster H, Solbach W. Sleep, immunity, and circadian clocks: a mechanistic model. Gerontology. 2010;56:574–80.
51. Irwin MR. Why sleep is important for health: a psychoneuroimmunology perspective. Annu Rev Psychol. 2015;66:143–72.
52. Dimitrov S, Besedovsky L, Born J, Lange T. Differential acute effects of sleep on spontaneous and stimulated production of tumor necrosis factor in men. Brain Behav Immun. 2015;47:201–10.
53. Born J, Lange T, Hansen K, Molle M, Fehm HL. Effects of sleep and circadian rhythm on human circulating immune cells. J Immunol. 1997;158:4454–64.
54. Faraut B, Nakib S, Drogou C, Elbaz M, Sauvet F, De Bandt JP, Leger D. Napping reverses the salivary interleukin-6 and urinary norepinephrine changes induced by sleep restriction. J Clin Endocrinol Metab. 2015;100:E416–26.
55. Kibble JD, Halsey CR. Gastrointestinal physiology. In: Medical physiology: the big picture. New York: McGraw-Hill Education; 2015.
56. Hasler WL, Owyang C. Approach to the patient with gastrointestinal disease. In: Kasper D, Fauci A, Hauser S, Longo D, Jameson JL, Loscalzo J, editors. Harrison's principles of internal medicine. 19th ed. New York: McGraw-Hill Education; 2015.
57. Mills JC, Stappenbeck TS. Gastrointestinal disease. In: Hammer GD, McPhee SJ, editors. Pathophysiology of disease: an introduction to clinical medicine. 7th ed. New York: McGraw-Hill Education; 2013.
58. Barrett KE. Gastrointestinal physiology. 2nd ed. New York: McGraw-Hill Education; 2014.
59. Kleitman N. Basic rest-activity cycle--22 years later. Sleep. 1982;5:311–7.
60. Chen CQ, Fichna J, Bashashati M, Li YY, Storr M. Distribution, function and physiological role of melatonin in the lower gut. World J Gastroenterol. 2011;17:3888–98.
61. Bubenik GA. Thirty four years since the discovery of gastrointestinal melatonin. J Physiol Pharmacol. 2008;59(Suppl 2):33–51.
62. Hoogerwerf WA. Role of clock genes in gastrointestinal motility. Am J Physiol Gastrointest Liver Physiol. 2010;299:G549–55.

Third Level: The Hypothalamus

5

Abstract

The hypothalamus is a phylogenetically ancient part of the CNS, and its structure has remained relatively constant in terrestrial vertebrates throughout evolution. The hypothalamus participates as the third level of the ANS hierarchy in the coordination of specific behaviors (defense behavior, thermoregulation, feeding, sexual and maternal behavior). This Chapter discusses the hypothalamic mechanisms that control hormone secretion and how they vary during a 24-h cycle. The nature of the defense behavior, feeding behavior and temperature regulation is described.

Keywords

Allostasis • Body temperature regulation • Brown adipose tissue • Constitutive and reactive secretion of hormones • Defense behavior • Feeding behavior • Hydroelectrolytic balance • Hypothalamic nuclei and areas • Hypophysiotropic area • Sexual and maternal behavior

Objectives

After studying this chapter, you should be able to:

- Name the areas and principal nuclei of the hypothalamus and describe how the hypothalamus participates as the third level of the ANS hierarchy in the coordination of specific behaviors
- Describe the neural mechanisms that control the constitutive and reactive secretion of hypophysiotropic, adenohypophyseal, pineal, thyroid, and parathyroid hormones, and how they vary in the three body configurations (wakefulness, NREM sleep, REM sleep) during a 24-h cycle.
- Describe the nature of the defense behavior as a paradigm of reactive homeostasis and differentiate homeostasis from allostasis.

D.P. Cardinali, *Autonomic Nervous System*, DOI 10.1007/978-3-319-57571-1_5

- Identify the principal physiological components of the feeding behavior and the location and function of the participating neurotransmitters and hormones. Describe how they vary in the three body configurations (wakefulness, NREM sleep, REM sleep) during a 24-h cycle.
- Describe the neural and hormonal mechanisms that control 24-h rhythms in plasma osmolality and the intravascular volume by modifying water and electrolyte intake behavior.
- Understand how the SON–PVN–neurohypophyseal system, the renin–angiotensin system, and the secretion of ANP vary in the three body configurations (wakefulness, NREM sleep, REM sleep) during a 24-h cycle.
- Identify the homeostatic components of body temperature regulation and the location and function of the participating neural and hormonal mechanisms of shivering and nonshivering thermogenesis in the three body configurations (wakefulness, NREM sleep, REM sleep) during a 24-h cycle.
- Name the hypothalamic basis of sexual and maternal behavior

Hypothalamic Behaviors Comprise Coordinated Mechanisms, Including Autonomic, Neuroendocrine, and Motivational Components

The hypothalamus is a phylogenetically ancient part of the CNS, and its structure has remained relatively constant in terrestrial vertebrates throughout evolution, unlike other brain regions, such as the neocortex or limbic system, which show marked interspecies differences.

The hypothalamus is the main governing center for homeostatic functions. A decerebrate animal, in which the brainstem is cut between the upper and lower colliculi, keeps the regulation of its internal environment intact, as the cut leaves the hypothalamus intact. In contrast, an animal with hypothalamic lesions requires extreme care to survive, as its homeostatic functions are greatly diminished or suppressed [1].

The hypothalamus is organized to perform autonomic, endocrine, and somatic functions. To do this, the region connects with the various components of the autonomic motor hierarchy, and with the somatosensory, motor, endocrine, and immune systems.

The principal hypothalamic nuclei are depicted in Figs. 5.1 and 5.2. In humans, the hypothalamus weighs about 5–8 g. Its anatomical limits are rather diffuse. As a ventral part of the diencephalon, the hypothalamus borders the third ventricle, below the thalamus. Caudally, it limits with the mesencephalon and rostrally, with the lamina terminalis, anterior commissure, and optic chiasm. Lateral to the hypothalamus the optic tract, the internal capsule, and the subthalamic structures are found.

From a functional point of view, the hypothalamus should be considered part of a continuous neuronal component that extends from the midbrain to the basal regions of the telencephalon (limbic cortex). Together, this portion of the brain is anatomically linked to the old olfactory, limbic system by the medial forebrain bundle.

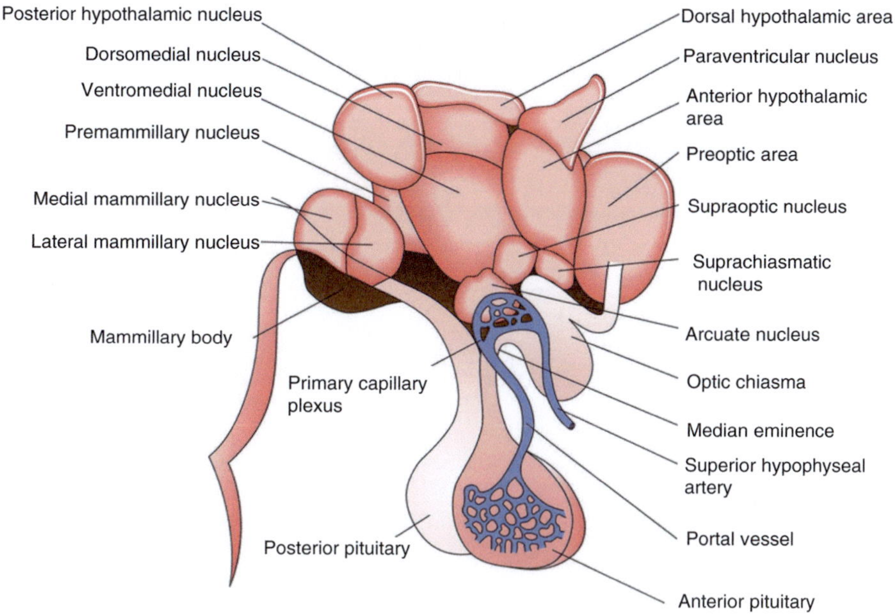

Fig. 5.1 Principal hypothalamic nuclei. Modified with permission from Cardinali [1]

The hypothalamus comprises three zones (Fig. 5.2):

- The periventricular area is a thin layer of nerve tissue adjacent to the third ventricle.
- In the medial zone, there are several nuclei. The hypophysiotropic area is the portion of the medial zone participating in adenohypophyseal regulation. In the supraoptic nucleus (SON) and paraventricular nucleus (PVN), the perikarya of the fibers forming the hypothalamic–neurohypophyseal tract and secreting arginine vasopressin (AVP) and oxytocin are located.
- The lateral hypothalamic area (LHA) contains no distinguishable nuclei, but instead Golgi type II neurons, which surround the medial forebrain bundle. This bundle is continued rostrally with the basolateral structures of the limbic system and caudally with the rostral structures of the brainstem. The medial forebrain bundle is one of the largest CNS fiber bundles, which interconnects forebrain structures with brainstem structures, forming an extensive network profile.

The LHA is an area of greatest interconnectivity of the hypothalamus, allowing the modulation of various functions, from cognitive to autonomic. The LHA, also called interstitial nucleus of the medial forebrain bundle, is one of the regions of the CNS with maximal biochemical and cellular heterogeneity (more than 35 types of neurons described) [2].

The afferent and efferent connections of the hypothalamus indicate that it is an important center of integration for autonomic, somatic, and neuroendocrine

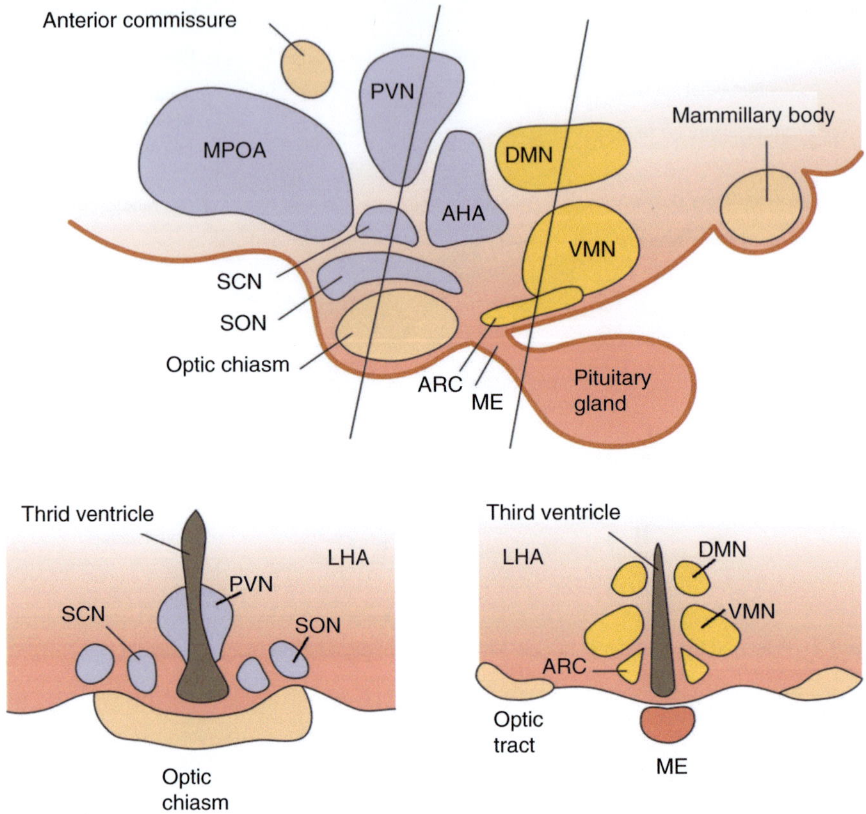

Fig. 5.2 Sagittal view of the hypothalamus and at the level of the indicated cuts (optic chiasm and median eminence, ME). *MPOA* medial preoptic area, *PVN* paraventricular nucleus, *DMN* dorsomedial nucleus, *AHA* anterior hypothalamic area, *VMN* ventromedial nucleus, *SCN* suprachiasmatic nucleus, *SON* supraoptic nucleus, *ARC* arcuate nucleus, *LHA* lateral hypothalamic area

functions. In general, LHA is reciprocally connected with the upper portion of the brainstem and upper limbic structures. It also receives somatic, inter- and exteroceptive impulses through the thalamus, cerebellum, and limbic system.

The medial hypothalamus has abundant reciprocal connections to the lateral hypothalamus, but receives few projections from other brain areas. Its function is mainly neuroendocrine. It contains receptors for humoral signals from the internal environment (glucose and other metabolites, temperature, osmolality, various hormones), its efferences being neuroendocrine (hypophysiotropic and neurohypophyseal peptides).

From a functional point of view, the hypothalamus is the level of the autonomic hierarchy that provides the complex program of the various homeostatic reactions, with their autonomic, neuroendocrine, and behavioral components. To achieve this, the hypothalamus uses the various elementary, segmental (medullary level) or system (brainstem level) programs, contained at lower levels of the autonomic hierarchy.

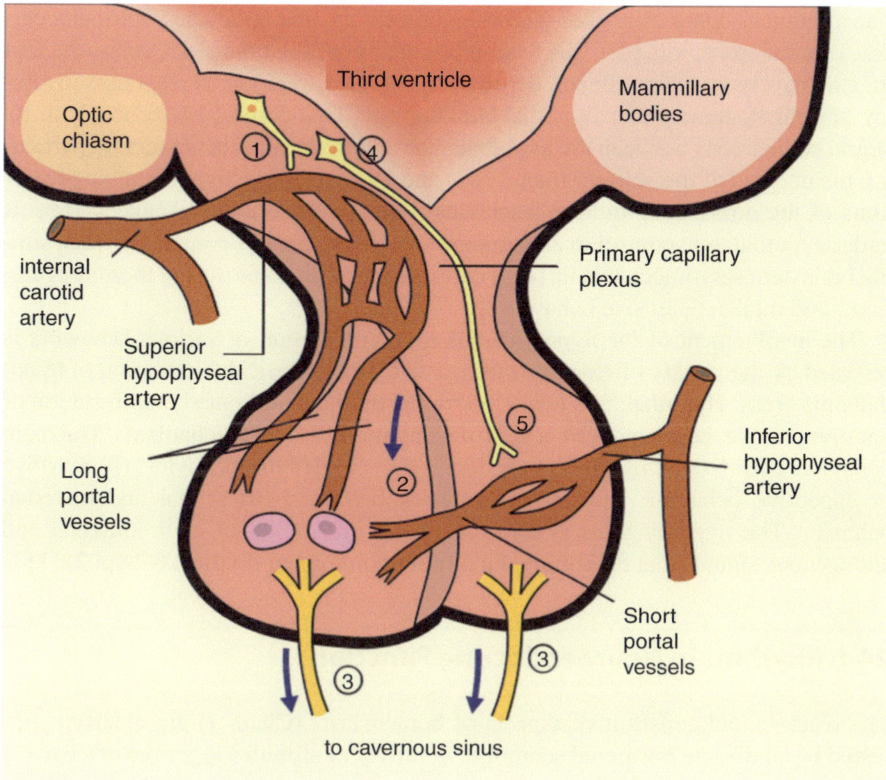

Fig. 5.3 Neuroendocrine connections of the medial hypothalamus. Hypothalamic hypophysiotropic hormones ① reach the adenohypophysis through the pituitary portal system ②③. The hypothalamic–neurohypophyseal system is neurally connected to the neurohypophysis ④⑤

There are four hypothalamic functions: (a) neuroendocrine function; (b) regulation of the ANS; (c) hypothalamic behavior regulation; (d) control of biological rhythms [1].

The neuroendocrine function is mediated by the SON/PVN neurohypophyseal systems and by the hypophysiotropic area (Fig. 5.3).

Regarding autonomic regulation, the electrical stimulation of almost all hypothalamic regions produces complex ANS responses (cardiovascular, respiratory, digestive, piloerection, etc.). These results are the basis for the hypothesis that the complex motor programs of the autonomic responses are contained in the hypothalamus, which are conveyed through the sympathetic and parasympathetic centers of the brainstem and spinal cord [3].

Let us take the cardiovascular function as an example. We discussed in Chap. 4 the cardiovascular servocontrol system (BP, cardiac output, vasomotor) that resides in the brainstem and acts through the autonomic segmental reactions in the spinal cord. The efferent components of this servocontrol system are the vagus nerve and sympathetic descending pathways to the intermediolateral columns of the spinal cord.

The afferences come from baro- and chemoreceptors, and atrial and ventricular cardiac mechanoreceptors [4]. This level of regulation of the brainstem, self-sufficient for exclusively vascular reflexes, depends on complex stereotyped responses involving several systems under hypothalamic control. This control is exerted both by neural connections between the hypothalamus and the brainstem centers (in particular, the nucleus of the solitary tract, NTS) and by descending hypothalamic projections of the dorsal longitudinal fasciculus type [5]. The cardiovascular system is under hypothalamic control in all responses involving greater complexity than simple brainstem servomechanisms (e.g., cardiovascular changes during thermoregulation, food intake, defensive behavior) [6, 7].

The involvement of the hypothalamus in the regulation of various behaviors is revealed by the variety of responses triggered by the electrical stimulation of hypothalamic areas. Hypothalamic behaviors comprise the coordinated manifestation of neurovegetative, neuroendocrine, somatic, and motivational mechanisms. The main behaviors coordinated by the hypothalamus are: (a) defense behavior; (b) nutritive or appetitive behavior; (c) thermoregulatory behavior; (d) sexual and maternal behavior. The hypothalamus is also the site of integration of environmental and endogenous signals that determine the different biological rhythms (Chap. 2) [1].

24-h Rhythms in Neuroendocrine Function

The reactive and constitutive aspects of homeostasis (Chap. 1) are clearly manifested by studies on hormonal secretion. A variety of stimuli trigger the reactive (or facultative) secretion of a hormone, whereas periodic changes are the manifestation of the constitutive secretion (predictive homeostasis) linked to biological rhythms [8].

A broad spectrum of periodic processes characterizes the endocrine system, including infradian rhythms, whose length ranges from months to years (like the menstrual cycle or seasonal rhythms in reproduction), circadian 24-h rhythms, and secretions that occur with intervals of hours (ultradian rhythms). In many cases, ultrafast or high-frequency variations, with periods of 5–15 min, are superimposed onto the hourly variations, caused by pulsatile variations or episodic secretions of pituitary hormones.

The 24-h temporal pattern of adenohypophyseal hormone release is controlled by the circadian system (C process) and the sleep homeostat (S process) as discussed in Chap. 2 (Fig. 5.4). The daily rhythmicity of the pituitary hormones occurs mainly by the circadian modulation of the amplitude of the secretory pulses, whereas the sleep/wake homeostat exerts its influence mainly on the frequency of pulse secretion.

To discern the relative roles of the circadian pacemaker and the effects of sleep on daily hormonal variation, strategies are based on the notion that the circadian rhythmicity needs several days to adapt to a sudden change in the sleep/wake cycle. Studies conducted in individuals studied under normal sleep–wake cycles cannot establish whether the rhythm is circadian or is driven by external factors such as

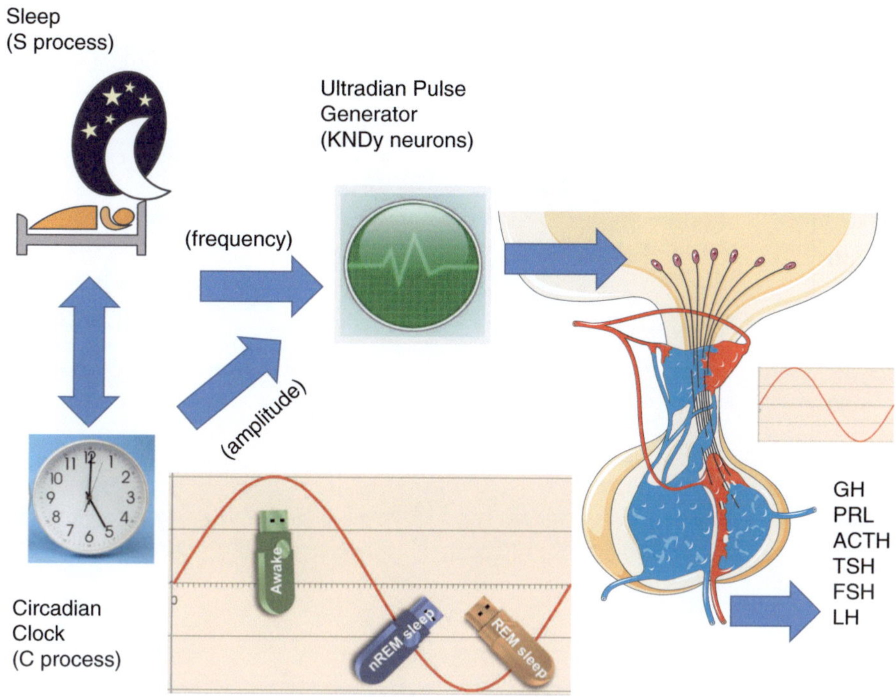

Fig. 5.4 The 24-h temporal pattern of adenohypophyseal hormone release is controlled by the circadian system (C process) and the sleep homeostat (S process). The daily rhythmicity of the pituitary hormones occurs mainly by the circadian modulation of the amplitude of the secretory pulses, whereas the sleep/wake homeostat exerts its influence mainly on the frequency of pulse secretion. *KNDy* neurons expressing kisspeptin, neurokinin B, and dynorphin. The figure was prepared in part using image vectors from Servier Medical Art (www.servier.com), licensed under the Creative Commons Attribution 3.0 Unported License (http://creativecommons.org/license/by/3.0/)

sleep–wake. A "gold standard" to disclose between the two is the constant routine protocol. Participants remain awake in bed for up to 50 h in a semirecumbent posture under dim light conditions and are fed hourly isocaloric meals.

A less precise but more practicable experimental design to discern the type of control of the constitutive secretion of a hormone is blood sampling (every 20 min) during normal sleep (day 1), night deprivation (day 2), and recovery of lost sleep from 11 am on at day 3 (Fig. 5.5) [9, 10]. By means of this strategy, it was verified that cortisol shows its maximum level at the expected time (the last part of the night and early morning) regardless of whether sleep was allowed or not (Fig. 5.5, upper panel), whereas growth hormone (GH) is secreted whenever N3 NREM sleep is detected, i.e., in the first part of the night of day 1 or during the recovery nap on day 2 (Fig. 5.5, lower panel).

For hormones controlled by the hypothalamic-pituitary axis, the pulsatile variation of their plasma levels is caused by intermittent discharges of an ultradian pulse generator (Fig. 5.4). Gonadotropin pulses were the result of gonadotropin-releasing

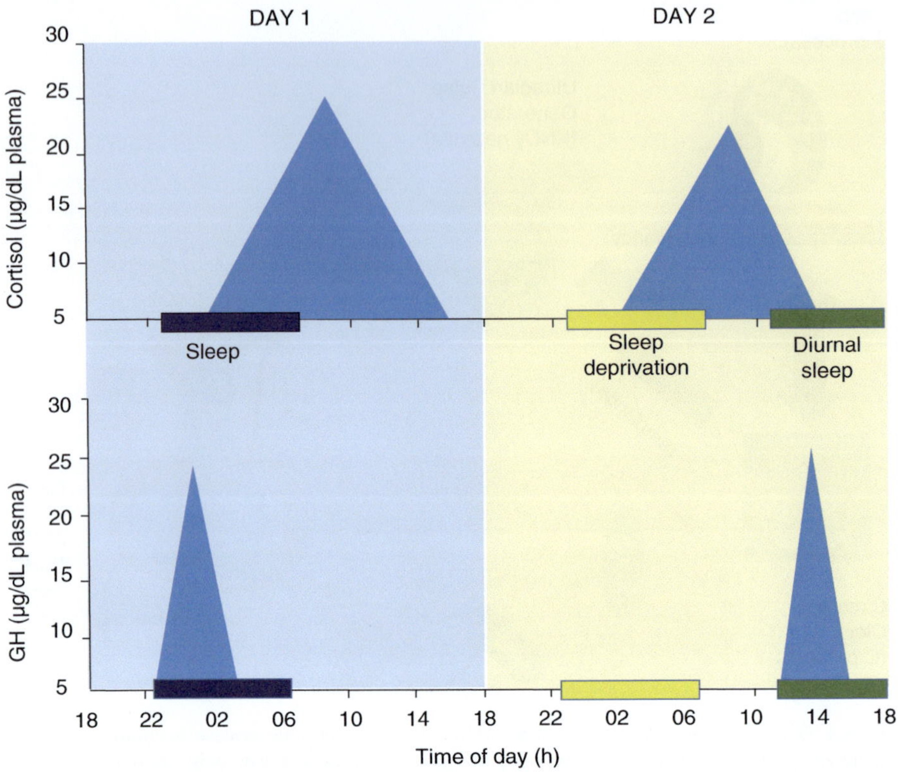

Fig. 5.5 Schematic representation of a daily hormonal rhythm controlled by the circadian clock or C process (cortisol) or by the S process (growth hormone, GH). The subject was deprived of sleep for one night and could recover the lost sleep from 11 am the next day. Cortisol is secreted in the phase established by the circadian clock preceding wakefulness, in the presence or absence of sleep. GH is secreted whenever N3 sleep occurs. Data from Van Cauter and Refetoff and Copinschi et al. [9, 10]. Ultradian variations in hormone secretion and values <5 µg/dL in both hormones are not represented

hormone (GnRH) pulses originating from synchronous discharges of GnRH-producing neurons, controlled by a pacemaker located in the anteroventral periventricular nucleus (AVPV) and the arcuate nucleus (ARC; Fig. 5.6) [11]. In primates with hypothalamic lesions, normal levels of gonadotropins are only restored by pulsatile, but not continuous, administration of GnRH. Rather, the continuous administration of GnRH inhibits gonadotropin release. These findings have been applied to the treatment of a wide variety of disorders of the pituitary–gonadal axis, using the pulsating administration of GnRH to correct deficient production of endogenous GnRH, and the continuous administration of GnRH analogs to inhibit pituitary gonadotropin synthesis [12].

The concept of the hypothalamic GnRH pulse generator emerged in the 1980s, but remained undefined for more than 20 years. The discovery of a hypogonadotropic hypogonadism associated with loss of function of GPR54, the receptor for a

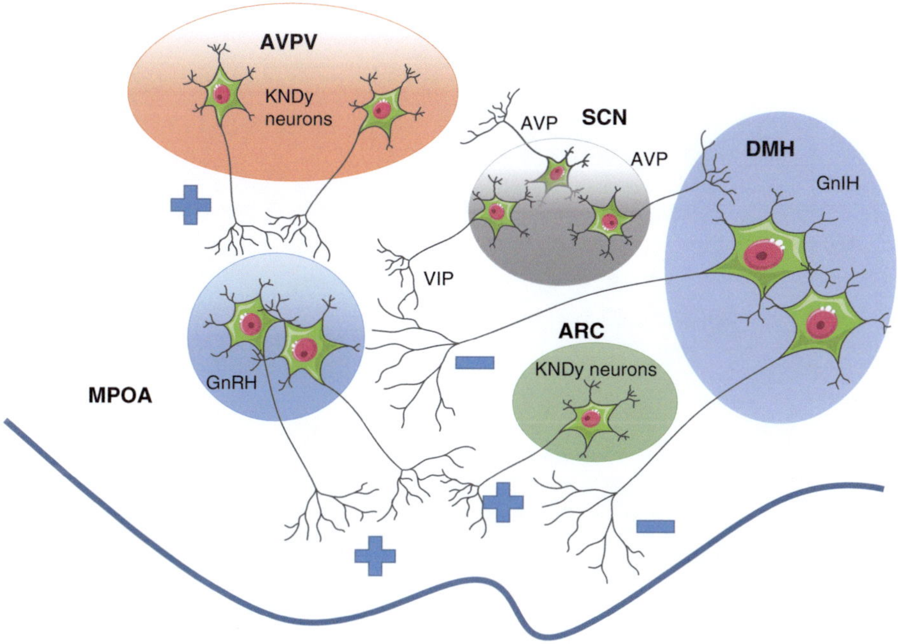

Fig. 5.6 Hypothalamic nuclei involved in the regulation of gonadotropin release. The gonadotropin-releasing hormone (GnRH) neurons of the MPOA project to the ME. Modulatory inputs upstream of the GnRH neurons include neurons that express kisspeptin (Kiss1; stimulatory of GnRH release), neurokinin B (stimulatory of GnRH release), and dynorphin (inhibitory of GnRH release; KNDy neurons). They are in the anteroventral periventricular nucleus (AVPV) and ARC. A major inhibitory signal derives from gonadotropin-inhibiting hormone (GnIH) releasing neurons of the dorsomedial hypothalamus (DMH). The SCN influences multiple sites involved in the control of gonadotropin release via monosynaptic projections from the SCN shell to the positive (KNDy and GnRH) and the negative signal of GnRH release (GnIH). The figure was prepared in part using image vectors from Servier Medical Art (www.servier.com), licensed under the Creative Commons Attribution 3.0 Unported License (http://creativecommons.org/license/by/3.0/)

54-amino acid peptide called kisspeptin (Kiss1), which is a very potent GnRH secretagogue, changed the field dramatically. Kiss1 is critical for the onset of puberty, the regulation of sex steroid-mediated feedback, and the control of adult fertility. Further studies indicated that other two peptides, neurokinin B (stimulatory of GnRH release) and dynorphin (inhibitory of GnRH release) are co-expressed in a subset of neurons in the AVPV and ARC (Fig. 5.6). The acronym, KNDy, was coined to describe these neurons. Current knowledge holds that the hypothalamic timing mechanism is initiated in the KNDy neuronal network of the nucleus by the reciprocating interplay of stimulatory kisspeptin/neurokinin B signals and inhibitory dynorphin one (Fig. 5.6) [13]. Kisspeptin neuron activity oscillates on a circadian basis, these neurons expressing clock genes that regulate their rhythmic activities [14].

The daily rhythmicity of the pituitary hormones occurs mainly by the circadian modulation of the secretory pulse amplitude, whereas the sleep/wake homeostat

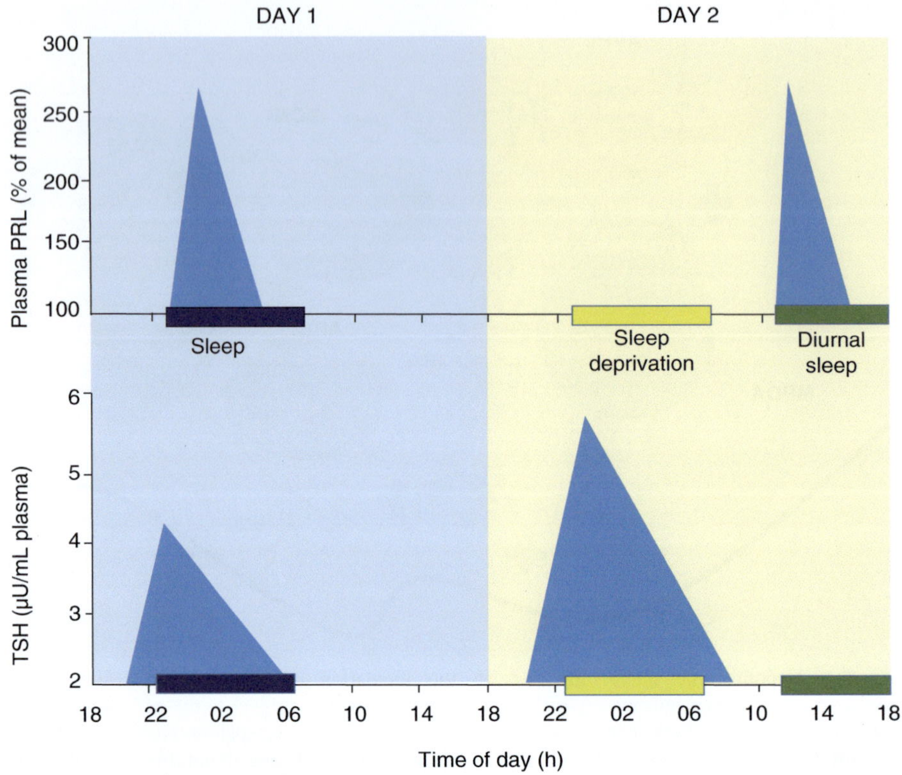

Fig. 5.7 Schematic representation of the daily rhythm of prolactin (PRL; *upper panel*) controlled by the S process, and that of thyroid-stimulating hormone (TSH), under the double control of C and S processes (*lower panel*). The volunteer was deprived of sleep for one night and could recover the lost sleep from 11 am the next day. Ultradian variations in hormone secretion or values <100% of mean plasma PRL or <2 μU of TSH/mL plasma were not represented. Data from Copinschi et al. [10]

exerts its influence mainly on the frequency of secretion of these pulses (Fig. 5.4). As stated, sleep dependence on rhythm is indicated because the effects of sleep occur independently of the time of day when the waking to sleep transition occurs. The persistence of rhythms in the absence of periods of sleep revealed the dependence of the circadian pacemaker on the generation of these hormonal rhythms.

Constitutive GH (Fig. 5.5) and prolactin (PRL; Fig. 5.7) secretion is controlled by the sleep/wake homeostat. In normal adult subjects, the daily profiles of GH in plasma are characterized by stable low levels interrupted abruptly by peaks of secretion [10]. Thus, GH release is pulsatile throughout the day, although in adults the most frequent secretory pulses (approximately 75% of pulses) occur in the early hours of sleep associated with slow-wave sleep (N3), with very low secretory activity during REM sleep.

The episodic secretion of GH depends on the specific rhythms of growth hormone-releasing hormone (GHRH) and on the hypothalamic GH inhibitory hormone, somatostatin, whose releases are 180 degrees out of phase. Variations in the secretion of somatostatin explain different sensitivities to GHRH during sleep [15]. Thus, stage N3 sleep is associated with low levels of hypothalamic somatostatin and REM sleep with high levels. Somatostatin secretion fixes the time (frequency and duration) of the release of GH, whereas the secretion of GHRH determines its magnitude. Humans are the only species that shows a close relationship between GH secretion and sleep. It has been suggested that the association between sleep and GH secretion in humans is the result of the consolidation of the sleep–wake cycle [16, 17].

The total amount of GH secreted daily is closely related to the individual's age. In the prepubertal phase, GH pulses occur predominantly between 22:00 and 04:00 h, whereas at puberty the GH pulses are larger in number and amplitude, distributing throughout the 24-h period. Prepubertal children have daily mean values of GH concentration like those post-puberty and in adult life, whereas at the end of puberty, the total amount of GH secreted daily reaches its maximum value. As age increases, nocturnal GH pulses decrease in frequency and amplitude, with no appreciable changes in plasma GH concentration during the day [10].

In normal young adult males, most of the daily GH secretion (>70%) occurs shortly after sleep onset during NREM N3 sleep. In young women, there are also pulses during the day that are more frequent and of greater amplitude than the men, related to the estrogens. From the age of 50–60 years, there is no constitutive release of GH, which is responsible for the loss of lean muscle mass (about 2 kg per decade); this coincides with the disappearance of N3 sleep. The major primary alteration that accounts for GH hyposecretion in old age is the increased secretion of hypothalamic somatostatin. There is also a decrease in the effectiveness of GHRH at releasing GH. Women generally secrete more GH than men of the same age and in women the effect of aging on GH secretion correlates with the decrease in circulating estradiol concentration [17].

Under normal conditions, the daily profile of PRL plasma levels follows a pattern with minimal morning concentrations, a gradual increase in the afternoon, and a higher nocturnal elevation that begins after the onset of sleep and culminates after around half of it [18]. This is outlined in Fig. 5.7, top panel. In addition, over 24 h, there are episodic pulses of PRL that can range from 7 to 24 throughout the day. Modulation of the amplitude of the secretory pulses gives rise to the daily pattern of hormonal secretion.

The constitutive secretion of PRL is strongly related to sleep homeostasis, as daytime sleep or naps are associated with an increase in the release of the hormone. Similarly, studies after reversal of the sleep–wake cycle have shown that sleep onset is associated with an increase in PRL secretion, whereas sleep deprivation prevents a nocturnal increase in PRL. However, in changes in the sleep–wake cycle and in sleep deprivation, it has been shown that the temporary organization of PRL release also presents some inherent circadian rhythmicity shown by a slight hormonal elevation corresponding to the moment of omitted sleep. PRL secretion plays a potential role in circadian REM sleep and N3 sleep control, which increases in patients with hyperprolactinemia and in lactating women [17].

The circadian clock controls the constitutive secretion of cortisol and melatonin. The periodicity of 24 h in the secretion of the pituitary–adrenal axis is considered a paradigm of circadian rhythmicity in humans (Fig. 5.4, upper panel). It begins at the age of 6 months and persists in adults until old age. The daily profiles of plasma cortisol are parallel and show a 2–3 h delay to those of adrenocorticotropic hormone (ACTH). This pattern shows a maximum at the beginning of the morning (04:00–08:00 h), a decrease throughout the day, followed by a low concentration during the night period (giving rise to a quiescent period from 24:00 at 02:00 h) and an abrupt rise at the end of the sleep period [19].

During the 24 h, about 15 pulses of cortisol and ACTH occur, which indicates that the daily variation of cortisol levels reflects a modulation of the amplitude of the pulses rather than the frequency. That is, the 24-h rhythm of cortisol is mainly dependent on the circadian pattern of ACTH release, which is amplified by a daily variation in the adrenal response to it. This depends on the autonomic innervation of the adrenal, a case like that described for thyroid innervation in Chap. 4. In turn, the rate of ACTH release is the result of periodic changes in the release of corticotropin-releasing hormone (CRH) [19].

The rhythm of secretion of the hypophyseal–adrenal axis persists in elderly subjects, although in general, the elevation in cortisol levels occurs earlier, with reduced latency to REM sleep. In addition, an increase in free plasma cortisol levels has been observed in older people, attributed to a decrease in the concentration of transport proteins and/or a decrease in the binding capacity of these proteins [18].

The circadian pacemaker directly controls the daily rhythm of cortisol, as demonstrated by the rapid phase shift observed after exposure to bright light at certain circadian moments. This rhythmicity of cortisol, which is of an endogenous nature, persists during sleep deprivation, although a reduction of 10–20% of the amplitude is observed. Nighttime sleep exerts an inhibitory effect on the secretion of cortisol, superimposed on its endogenous circadian rhythmicity [20].

In old age, free cortisol levels increase by 20–50% compared with those in young people, and typically, the minimum (nadir) of the rhythm of cortisol in an individual over 70 years old is higher than that of a young adult. There is also a phase advancement of the rhythm of cortisol with age [18]. Studies in animals and in human clinics have indicated important neurodegenerative effects of cortisol, especially at hippocampal levels, which are more pronounced in the nadir of the rhythm than in its crest. We will discuss them further in this Chapter. Therefore, even modest elevations of evening cortisol in the elderly facilitate the development of disturbances associated with excess glucocorticoids, such as memory deficit and insulin resistance associated with old age [18].

The plasma concentration of melatonin has peak nocturnal values ranging from 0.4 to 0.8 pmol/mL and a minimum during the day (0.02 pmol/mL) in humans. In saliva, its values are 0.6–0.8 pmol/mL at the time of the nocturnal maximum (Fig. 5.8). In man, peak melatonin secretion is not related to the sleep phase. In the same way as cortisol, the daily rhythm of melatonin secretion remains in conditions of sleep deprivation (Fig. 5.8). Owing to its marked intraindividual reproducibility, the rise in melatonin in the afternoon in dim light (dim light melatonin onset, DLMO) is the most accurate marker of τ, the circadian rhythm period (Fig. 5.8) [21].

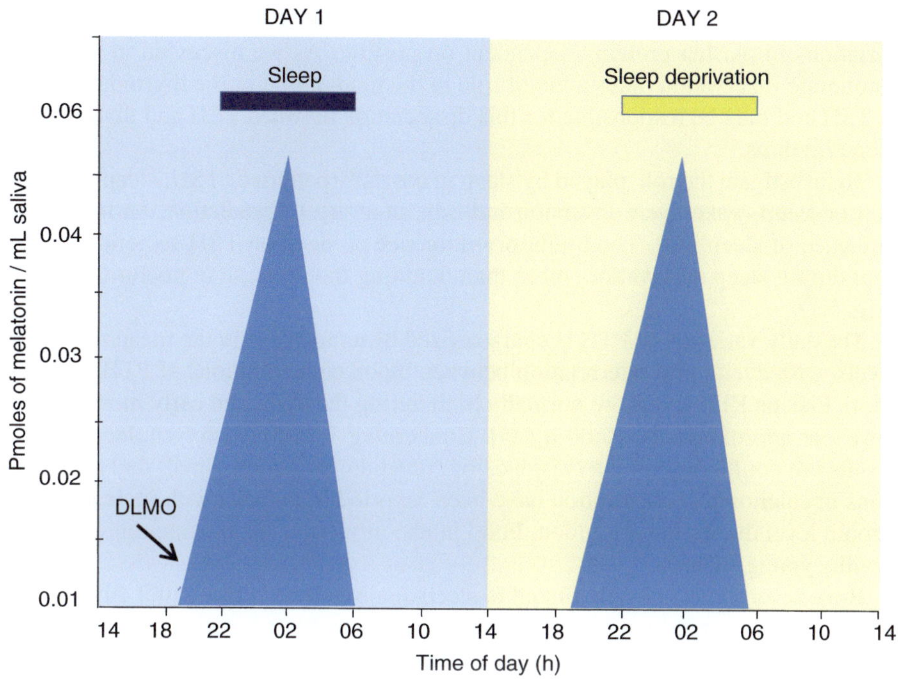

Fig. 5.8 Saliva melatonin levels in a volunteer subjected to a night of sleep deprivation. Saliva samples were taken every 20 min. Melatonin was assayed by radioimmunoassay. *DMLO* dim light melatonin onset. Values of melatonin <0.01 pmol/mL of saliva are not depicted. Cardinali et al., unpublished data

The hypothalamic–pituitary–thyroid axis presents a complicated temporal structure with rhythmic variations of multiple frequencies at all levels of the system, from the hypothalamic neurons to the cells of the peripheral effector tissues. The range of frequencies consists from rapid neural discharges to circadian and circannual rhythms, superposing on these rhythmic variations according to age. Rhythmic and nonrhythmic variations interact, modulate, and are modulated by variations of other neuroendocrine, metabolic, and immune functions. In normal subjects, thyroid-stimulating hormone (TSH) is secreted from the pituitary in a pulsatile form, demonstrating that the pituitary secretion of TSH responds better to an intermittent continuous stimulation of thyrotropin-releasing hormone (TRH). The average frequency is nine pulses in 24 h, increasing in amplitude and frequency at night, which leads to a nocturnal rise in hormone levels [22].

Under normal conditions, TSH levels are low throughout the day, experiencing a sudden increase at 20:00 h, reaching the maximum levels at night, this maximum being circadian pacemaker-controlled (Fig. 5.7, lower panel) [23]. High hormone levels remain throughout the night, corresponding to an increase of approximately 175% of daytime levels, to decrease rapidly in the morning, reaching a nadir in the early morning hours.

The existence of circadian rhythms of thyroid hormones is debatable. A 12-h hemicircadian rhythm with morning and evening minima may occur [24]. It has been suggested that these variations of low amplitude might be related to daily variations in plasma proteins dependent on posture. As we discussed in Chap. 4, autonomic innervation plays a major role in the modulation of the thyroid response to TSH and may be responsible for this dissociation between TSH and thyroid hormone rhythms.

To investigate the role played by sleep in the daily pattern of TSH, sleep deprivation or sleep–wake cycle inversion indicate, contrary to prediction, an inhibitory influence of sleep [23]. The inhibitory influence of sleep on TSH secretion is evident during sleep deprivation, more than doubling the increase in nocturnal values (Fig. 5.7).

The daily variation of PTH is characterized by an increase in the mean nocturnal levels, with a temporal interrelation between blood concentrations of PTH and calcium. Plasma PTH levels are normally high during the night and early morning and lowest at approximately 10:00 h [25]. Concerning 1,25-dihydroxycholecalciferol, results are conflicting on the existence of circadian variations. Small daily fluctuations in calcitonin concentration have been reported in humans with increased calcitonin level during the afternoon. Food intake influences serum calcitonin level in healthy young subjects.

Bone resorption is closely linked to sleep quality. Bone cells exhibit 24-h cycles with nocturnal expression of genes that regulate osteoblast function, bone mineralization and ossification, and bone resorption markers. Interruption in melatonin-mediated signaling is responsible for the increased risk of hip and wrist fracture and low bone density in shift workers (Fig. 5.9) [21].

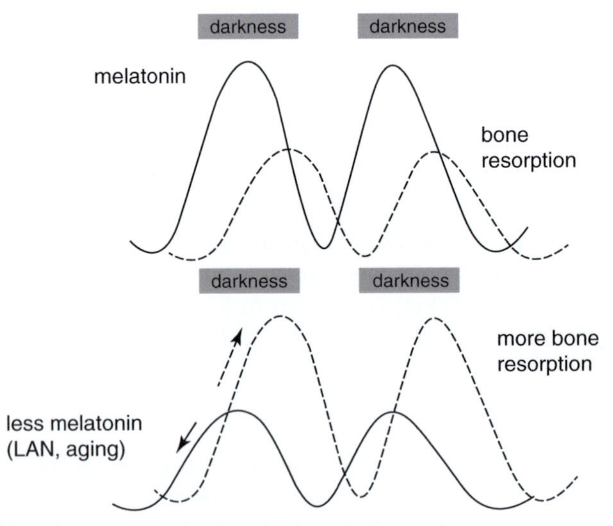

Fig. 5.9 Relationship between melatonin secretion and bone resorption in a 24-h cycle. As the image shows, both bone resorption (*dotted line*) and melatonin (*solid line*) show a daily rhythm, with peaks occurring during dark hours. Suppression of nocturnal melatonin levels, either by exposure to light at night (LAN) or in aging, increases bone resorption. The restoration of nocturnal melatonin maximum protects the bone loss

By means of studies in volunteers in the sleep laboratory to which several neuropeptides were administered in pulsatile form, the link between neuropeptides and the different stages of sleep determined by polysomnography (PSG) could be verified [26]. During the first half of the night and coincident with slow-wave sleep, the mechanisms associated with GH secretion predominate, whereas in the second half of the night there is a coincidence in the percentage increase in REM sleep with the secretion of the hypothalamic–pituitary–adrenal axis hormones. Based on this, the homology of GH secretion with the S process and activation of the ACTH/cortisol axis with the C process has been proposed (Fig. 5.10). Neuropeptide perfusion studies indicated that: (a) GHRH promotes GH release and slow-wave sleep; (b) CRH promotes the activation of the pituitary–adrenal axis and REM sleep; (c) galanin

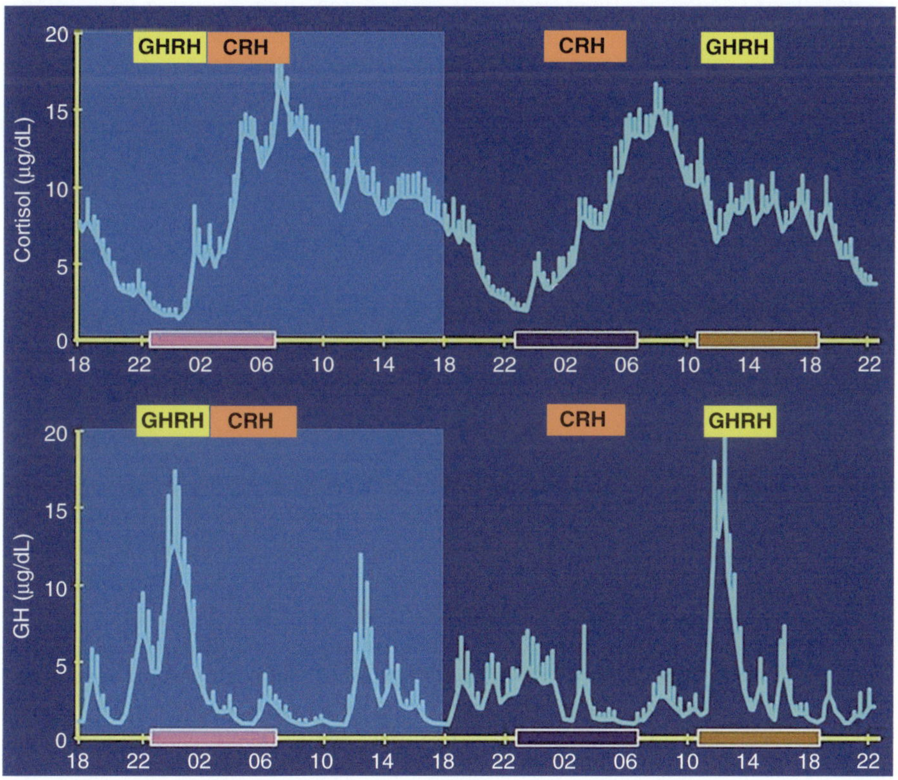

Fig. 5.10 Neuropeptide perfusion studies in healthy volunteers during a PSG show that GHRH perfusion promotes GH release and NREM sleep, whereas infusion of corticotropin-releasing hormone (CRH) promotes pituitary–adrenal axis activation and REM sleep. A reciprocal interaction of GHRH and CRH as a key to sleep regulation was proposed. GHRH predominates during the first half of the night, resulting in slow sleep, GH secretion, and minimal secretion of adrenocorticotropic hormone (ACTH) and cortisol, whereas the second half of the night is dominated by CRH (ACTH and cortisol secretion). These periods are superimposed on the results of an experiment in which a healthy volunteer was deprived of sleep for one night and could recover the lost sleep from 11 am the next day. Modified with permission from Cardinali [1]

increases slow-wave sleep in the absence of changes in GH and cortisol; (d) somatostatin inhibits GH and slow-wave sleep and promotes REM sleep; (e) NPY is an endogenous CRH antagonist and contributes to fixing the time of onset of sleep; (f) ghrelin increases GH, NREM sleep, and cortisol [26].

It should be noted that changes in EEG of sleep following administration of GHRH and CRH are not due to GH or cortisol secretion, as slow sleep decreases after GH administration, whereas slow-wave sleep and GH increase after injecting cortisol. Therefore, the results are best explained by the inhibitory feedback on GHRH and CRH respectively.

These studies have been of interest in linking normal aging with depression [26]. In both situations, there is deterioration of slow-wave sleep, shortened REM sleep and increases in REM sleep, in addition to changes in the continuity of sleep (prolonged sleep latency, frequent nocturnal awakenings, early morning awakening). Endocrine changes are similar in both situations: there is an elevation of ACTH and cortisol and suppression of GH. This is because the GHRH/CRH ratio changes in favor of CRH during an episode of depression because of CRH hypersecretion, whereas in aging the GHRH/CRH ratio changes in favor of CRH because of the reduction of GHRH activity [26].

The rhythmic aspects of the activity of the hypothalamic–pituitary–gonadal axis, the neurohypophyseal system, and the secretion of hormones associated with energy homeostasis and blood volume control are discussed below with their specific behaviors.

Defense Behavior as a Paradigm of Reactive Homeostasis

Nothing in physiology illustrates the concept of reactive homeostasis better than defense behavior. Hans Selye named it "general adaptation syndrome" to define the set of changes arising as the body's response to a wide variety of noxious stimuli (stressors) [27]. Stressors are those stimuli whose perception by the nervous system does not match the neural representation of past experiences, and to which a change in coping strategy (e.g., a particular behavior) is not successful (Fig. 5.11).

In general, the stimuli stressors can be classified into four groups: (a) physical/chemical stressors (heat, cold, intense radiation, noise, vibration, toxic substances, etc.); (b) psychological stressors (emotional and behavioral changes, such as anxiety, fear, frustration); (c) social stressors (hostile environment, disruption of relations); (d) those that alter ANS homeostasis (exercise, orthostatic, body tilt, hypoglycemia, bleeding, etc.).

An animal that is faced with a threatening situation must decide, among different strategies, what is apparently the best choice for the conservation of its life and, in a broad sense, of its species. For example, when sympathetic activation occurs, cardiac output and systemic blood flow increase and, although vasodilation occurs to increase the blood supply to active skeletal muscles for posture and intended movements, vasoconstriction in the skin can reduce blood loss in a region that is susceptible to damage [28]. A decreased time for blood clotting avoids a large

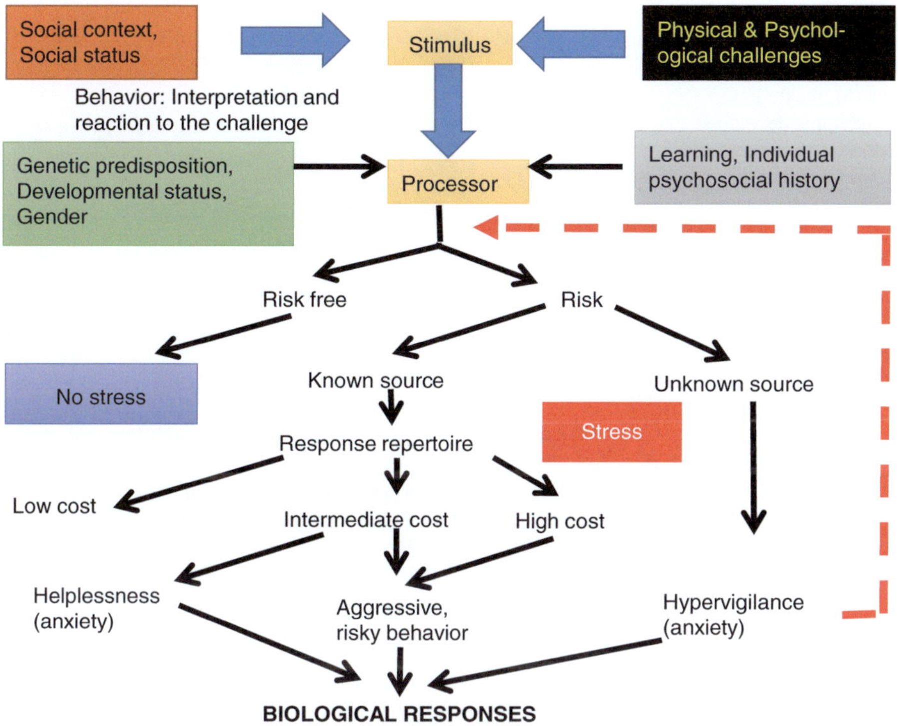

Fig. 5.11 Behavioral and biological aspects of the defense behavior. The sequence of phenomena following exposure to a novel situation is described. Strategies are of two types, high or low energy level. Their objective is adaptation. If adaptation does not occur, the stress reaction appears, with stimulation of the pituitary–adrenal axis

hemorrhage, and higher sudoresis makes the skin moistened and the animal difficult to catch. At the same time, there is enhanced breathing movements and airflow volume for gaseous exchanges and pH regulation. Liver glycogenolysis is stimulated, along with the mobilization of fatty acids from adipose tissue [29]. Muscular strength and muscle glycolysis are also improved and, for additional energy production, a higher oxygen–hemoglobin dissociation and oxygen delivery to activated cells also occur. Wave frequencies in the EEG increase, which reflects diffuse neuronal activation, and mydriasis serves to provide more visual information and to decide the best behavioral strategy to execute, such as skipping out of a dangerous place. By the action of α-adrenoceptor receptors, the secretion of insulin is inhibited (it would be useless to restore energy reserves at this moment), whereas β-adrenoceptors stimulate glucagon secretion. Cortisol, in addition to other stress-related hormones, can indirectly affect the consolidation of memories. Cells of the immunological system, lymphoid tissues, and cytokines are also involved in this integrated response (Fig. 5.12).

Thus, coping with stressors includes two strategies (Figs. 5.11 and 5.13). An active coping strategy (fight-or-flight) is evoked if the stress is predictable,

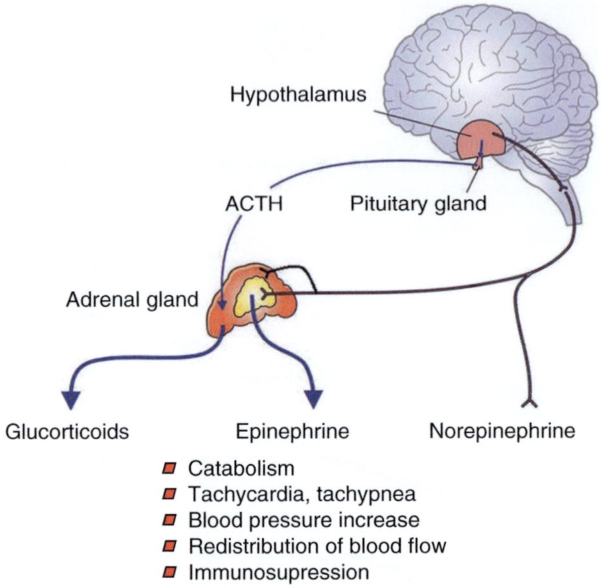

Fig. 5.12 Via ACTH release and activation of adrenal sympathetic nerves, the defense behavior is developed. Modified with permission from Cardinali [1]

controllable, or escapable. A passive coping strategy (immobility or decreased responsiveness to the environment) is evoked if the stress is inescapable [1].

The active strategy is associated with sympathoexcitation (hypertension, tachycardia, thermogenesis), whereas the passive strategy is associated with sympathoinhibition and/or parasympathetic activation (hypotension, bradycardia). The passive strategy also helps to facilitate recovery and healing.

The active strategy is called the "fight-or-flight" response from a behavioral point of view or the "defense response" from an autonomic point of view. The passive strategy is sometimes called "paradoxical fear" or "playing dead." Parts of the neural substrates that mediate active versus passive emotional coping have been identified within the brainstem [30, 31].

If strategies to new situations (high- or low-energy consumption) are not successful, the stress reaction is triggered (Figs. 5.10 and 5.12). The responses correspond to different functional configurations of the limbic system (amygdala dominance in the fight situation; predominance of septo-hippocampal components in defeat; Fig. 5.13).

During stress, the physiological processes that do not pose a short-term benefit are inhibited, such as inflammation, digestion, reproduction, and growth. When the intensity or duration of stressors exceed certain limits, pathological changes such as hypertension, gastric ulcers or neurological abnormalities ensue.

The defense capability or adaptation of an organism to stress depends on the magnitude and duration of the stressor, and on the sensitivity to the stressor. There are differences in the ways in which individuals face situations of stress. These differences are related to genetics, environmental influences, previous experience in such situations, training, social support, and physical and mental health.

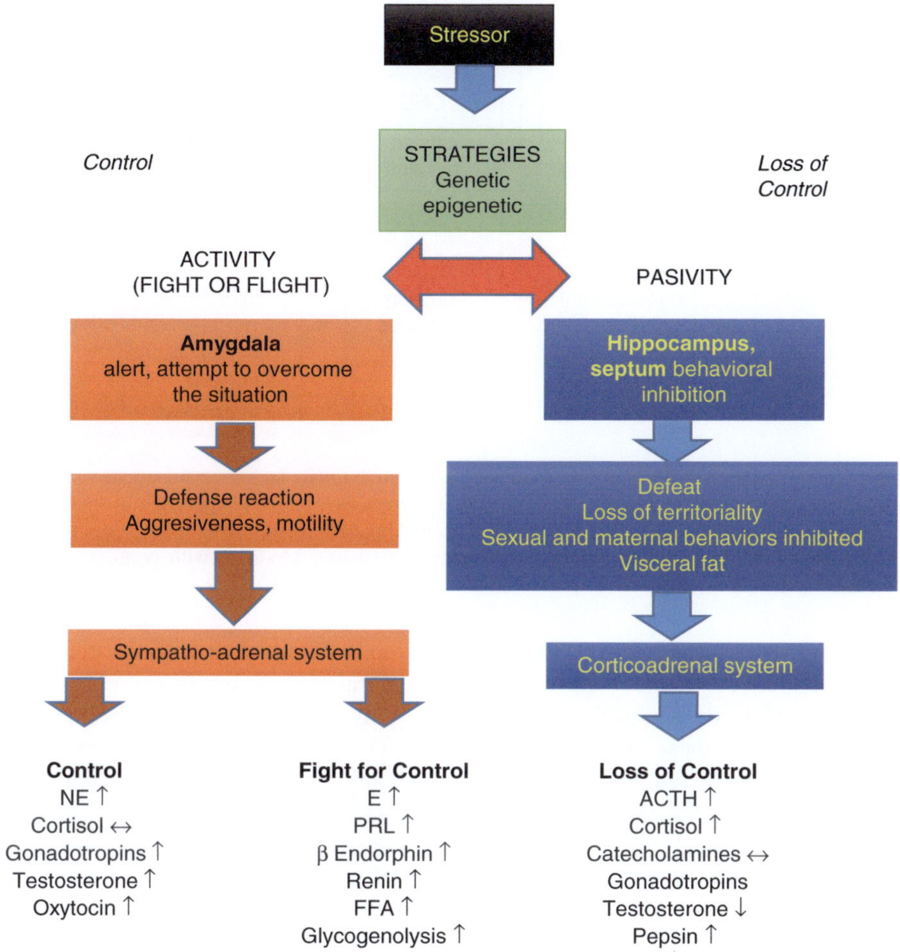

Fig. 5.13 The different neuroendocrine profiles of defense behavior

An important difference is the programming of ANS connections during early extrauterine life (Fig. 5.14) [32]. The conditions of reaction to stress have clear individual nuances, arising from fixed initial experiences in a learning phase, during the early stages of life. This explains why the same stressful situation, which is not harmful to many individuals, leads some to myocardial infarction, others to peptic ulcer or ulcerative colitis, and others to hyperthyroidism. That is, the responses of the ANS depend on the previous history and individual experience.

Stress responses involve a complex neurobehavioral cascade, which is elicited when the organism is confronted with a potentially harmful stimulus. As this stress cascade consists of a range of neural and endocrine pathways, stress can be conceptualized as a communication process on the descending branch of the brain–body axis. Therefore, interoception and stress are associated via the bi-directional

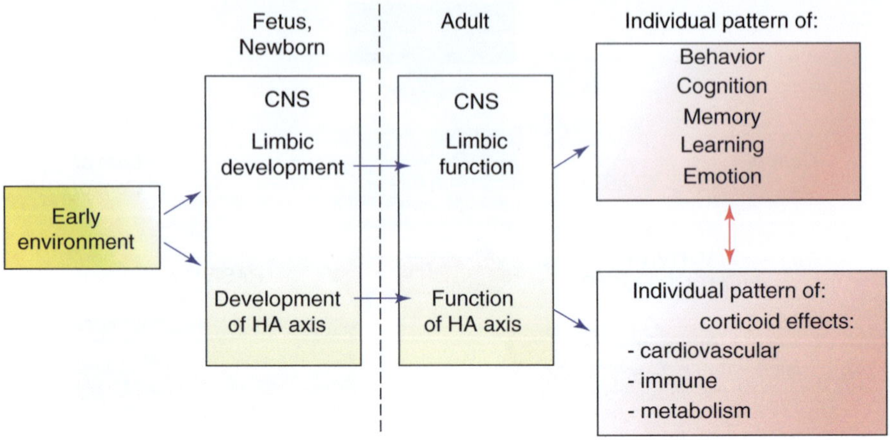

Fig. 5.14 There are differences in the ways in which individuals face stressful situations. These differences are related to genetics, environmental influences, previous experience in such situations, training, social support, and physical and mental health. An important difference is the programming of ANS connections during early extrauterine life. Modified with permission from Cardinali [1]

transmission of information on the brain–body axis. The excessive and/or enduring activation (e.g., by acute or chronic stress) of neural circuits, which are responsible for successful communication on the brain–body axis, induces malfunction and dysregulation of these information processes. Therefore, interoceptive signal processing may be altered, resulting in physical symptoms contributing to the development and/or maintenance of body-related mental disorders, which are associated with stress. A positive feedback model involving stress (early life or chronic stress, and major adverse events), the dysregulation of physiological stress axes, altered perception of bodily sensations, and the generation of physical symptoms, which may in turn facilitate stress, have been proposed [33, 34] (Fig. 5.14).

In the event of a challenge to homeostasis (stress), recovery may occur with disappearance of the stressor. For example, in a high salt diet, homeostatic mechanisms promote the elimination of excess sodium to restore equilibrium. However, even in situations when the stressor is not eliminated, it is possible to restore and maintain homeostasis (Fig. 5.15). To achieve this, animals must have to preserve the stability of their variables through changes of state related to challenging circumstances. In this sense, the concept of "allostasis" (*allo*, meaning different, and *stasis*, meaning constancy, i.e., "achieving stability through change") was introduced to take into account the regulatory systems that develop variable set points of control, showing individual differences in expression according to the capacity of the animal to cope with new situations [30]. This is associated with anticipatory behavioral and physiological responses, and is vulnerable to physiological overload and the breakdown of regulatory capacities.

The allostatic mechanisms can maintain homeostasis in the presence of a stressor (Figs. 5.15 and 5.16). As important as the mobilization of functional and

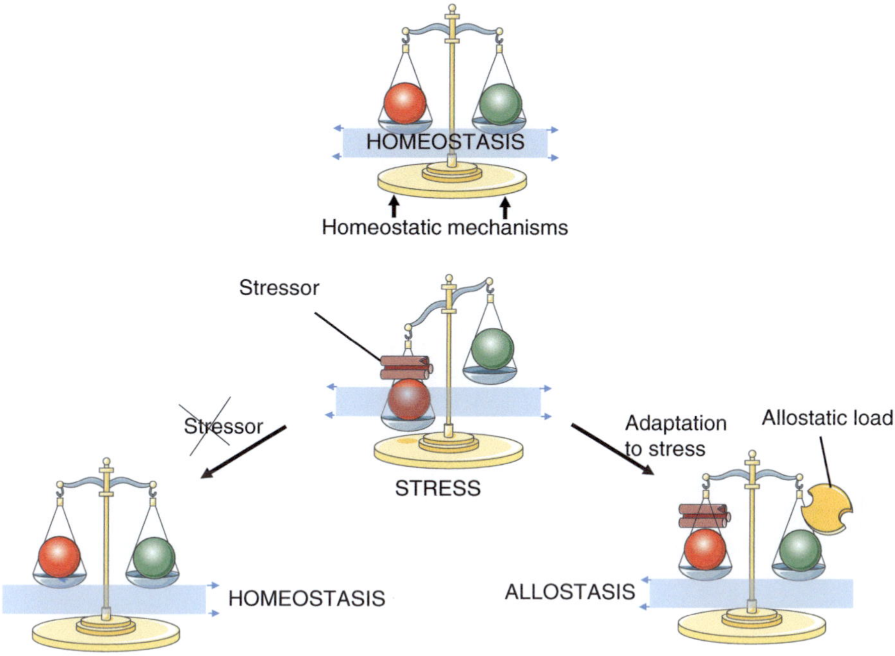

Fig. 5.15 In the event of a challenge to homeostasis (stress), recovery may occur with disappearance of the stressor (*left*). However, even in situations when the stressor is not eliminated, it is possible to restore and maintain homeostasis (*right*). The concept of "allostasis" was introduced to consider the regulatory systems that develop variable set points of control, showing individual differences in expression according to the capacity of the animal to cope with new situations. The figure was prepared in part using image vectors from Servier Medical Art (www.servier.com), licensed under the Creative Commons Attribution 3.0 Unported License (http://creativecommons.org/license/by/3.0/)

behavioral processes to maintain the homeostatic balance in response to the stressor, is the appropriate demobilization of these processes when terminating the stressor stimulus. Thus, both the mobilization and the inadequate demobilization of allostatic mechanisms (Fig. 5.16) can cause adaptation diseases, such as hypertension, obesity, diabetes, stroke, autoimmune diseases, inflammatory disorders, and gastric ulceration [30].

The coping responses during stress can be defined as cognitive and behavioral responses to managing the stress. The primary objectives of the success of coping responses are: to eliminate or mitigate the harmful environmental conditions and to enhance the prospect of recovery; to tolerate and adjust the body to negative events; to maintain a positive self-image; to maintain emotional balance; and to preserve social relations [35].

Sleep implies a recovery process from previous wakefulness. Not only the duration of the waking period affects sleep architecture and sleep EEG, the quality of wakefulness is also highly important. Studies in rats have shown that social defeat stress, in which experimental animals are attacked and defeated by a dominant

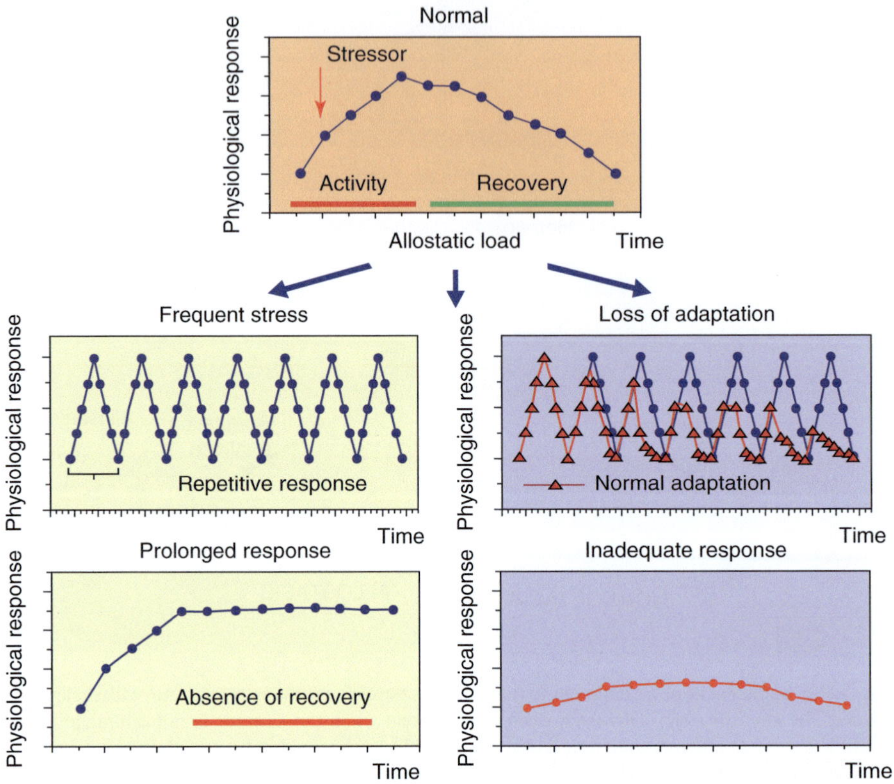

Fig. 5.16 Allostatic load. The *upper panel* illustrates the normal allostatic response, which begins with a stressor-induced response, maintained for a suitable period, and then reversed. The remaining panels illustrate four conditions that give rise to allostatic load: (1) repetitive appearance of multiple stressors (e.g., repeated elevation of BP leads to arteriosclerosis); (2) lack of adaptation (e.g., those individuals who have to speak in public and do not reduce cortisolemia with repetition); (3) Prolonged responses due to a delayed reversal (e.g., catecholamine and cortisol secretion due to stress returns to basal levels more slowly in the elderly), (4) an inadequate, low response resulting in the compensatory hyperactivity of other mediators (e.g., autoimmune disorders with inadequate secretion of corticosteroids and a high concentration of cytokines, which are usually counter-regulated by corticosteroids). Modified with permission from Cardinali [1]

conspecific, is followed by an acute increase in NREM sleep on EEG and the suppression of REM sleep. This occurred in both winners and losers, indicating that in rodents a social conflict with an unpredictable outcome has quantitatively and qualitatively similar acute effects on subsequent sleep, regardless of the results of the conflict [36].

There is evidence that acute or chronic exposure to a stressor can start or cause recurrence of psychiatric disorders such as depression, bipolar disorder, post-traumatic stress disorder, anxiety, and schizophrenia. In situations of melancholic depression, anorexia nervosa, panicky anxiety, obsessive–compulsive disorder, chronic active alcoholism, excessive exercise, abstinence from alcohol and narcotics,

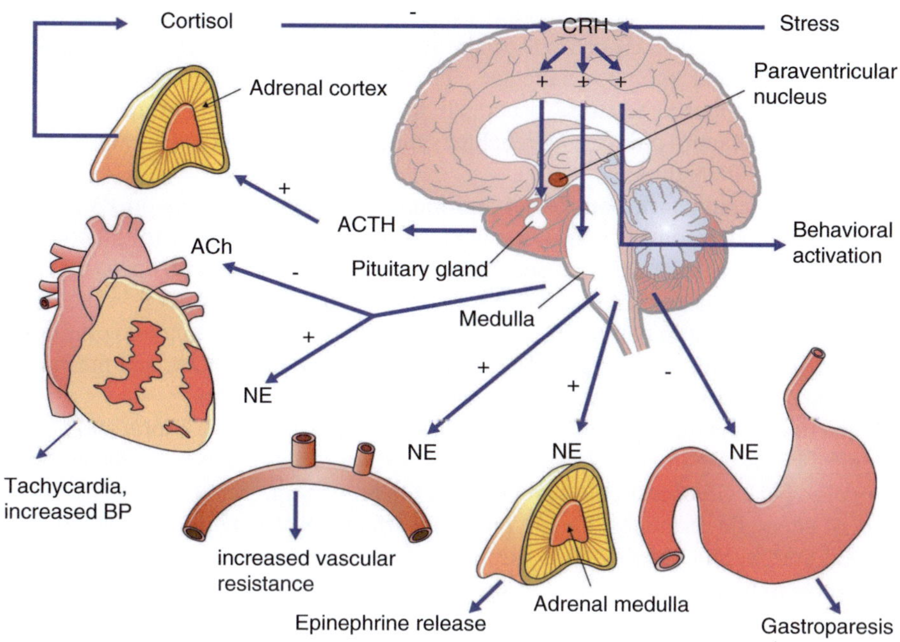

Fig. 5.17 The CRH neurons of the paraventricular nucleus of the hypothalamus command the different manifestations of defense behavior. Modified with permission from Cardinali [1]

malnutrition, sexual abuse, the chronic activation of the hypothalamic–pituitary–adrenal axis occur. On the other hand, in situations such as seasonal or atypical depression, the postpartum period, the period after smoking cessation, fibromyalgia syndrome or chronic fatigue, CRH secretion is diminished and symptoms occur such as increased appetite and weight gain, somnolence, and fatigue [37, 38].

The parvocellular neurons of the hypothalamic PVN that synthesize and release CRH and/or AVP are the final common pathway for the regulation of ACTH secretion (Fig. 5.17) [30]. The relative proportion of active CRH/AVP neurons increases significantly in stress [39]. Among other brain areas, these neurons project the median eminence, to noradrenergic neurons of the brainstem, and to the ARC. In the latter, they activate pro-opiomelanocortin -containing neurons, which, in turn, reciprocally innervate CRH/AVP PVN neurons. Dendritic release of neuropeptides has been described as a novel interpopulation signaling modality in the PVN [40].

A high density of neuronal cell bodies containing CRH is also found in the central amygdaloid nucleus and the bed nucleus of the stria terminalis. CRH and its receptors are found in extra-hypothalamic structures such as the limbic system, basal forebrain, PBN, and paragigantocellular nuclei of the medulla oblongata, LC, and other groups of noradrenergic neurons of the pons and medulla (sympathetic noradrenergic system) and spinal cord [39].

There are reciprocal neural connections between CRH neurons in the PVN and noradrenergic neurons in the brainstem. Cholinergic, serotonergic, and glutamatergic neurons have stimulatory action in those circuits. The brainstem catecholaminergic pathways to the PVN CRH neurons can be activated in bleeding, hypotension, respiratory distress, and immune challenges. The LC is one of the regions of the brain that is most responsive to stress, especially to hemorrhage. It also exerts its effects on the hypothalamic–pituitary–adrenal axis through central limbic structures and it innervates the PVN directly.

Glucocorticoids have both direct and indirect inhibitory actions on PVN neurons. The hippocampus, which shows large numbers of binding sites for glucocorticoids and mineralocorticoids, is one of the structures that inhibit PVN activity [19]. Hippocampal damage increases mRNA expression for CRH and AVP in the PVN and hippocampal stimulation decreases the activity of the hypothalamic–pituitary–adrenal axis in rats and humans. The lateral septal area, the mPOA, and prefrontal cortex are also structures that have inhibitory actions on the PVN.

The role of the medial prefrontal cortex as a coordinator of behavioral and physiological stress responses across multiple temporal and contextual domains has been proposed [41]. Glucocorticoids act as one of the primary messengers in the reallocation of energetic resources, having profound effects locally within the medial prefrontal cortex, and shaping how the brain region acts within a network of brain structures to modulate responses to stress.

Several neurotransmitters are released in the brain during stress, but CRH is the main coordinator of the psychological and behavioral changes found (Fig. 5.17) [31]. Experimental studies have shown that exogenous administration of CRH stimulates the release of ACTH by the pituitary, can change the EEG, and induces psychological and behavioral changes such as those observed in stress, for example, reduced feeding behavior. Hormones of the pituitary–adrenal axis do not block behavioral effects, but they can be reversed by the neutralization of CRH, indicating that CRH can modulate behaviors independently of the pituitary–adrenal axis (Fig. 5.18).

Therefore CRH produces activation of ANS via direct central action, rather than through the activity of the hypothalamic–pituitary–adrenal axis. CRH is part of a family of peptides that includes urocortin and urotensin. These peptides act via two receptor types: CRH-R1 and CRH-R2. The CRH-R1 receptors are most abundant in the brain and in peripheral tissues and have a higher affinity for CRH and urocortin. The CRH-R2 receptors are less abundant in the nervous system, are more significant in peripheral tissues and exhibit higher affinity for urocortin, urocortin II, and urocortin III than for CRH [31].

The predominant receptor in the activation of the hypothalamic–pituitary–adrenal axis is CRH-R1, whereas CRH-R2 plays a prominent role in the energy expenditure processes involved in homeostatic responses. Animals genetically deficient in CRH-R1 show a failed stress response, whereas those deficient in CRH-R2 are hypersensitive to stress and exhibit increased anxiety. It is possible that the acute stress response phase is linked to rCRH-1, whereas the final recovery stage includes rCRH-2 effects. Selective antagonists of rCRH-1 inhibit the anxiogenic action of CRH and have been proposed to be a new type of anxiolytics. Hypersecretion of CRH is as a phenotypic element of emotional disease vulnerability (Fig. 5.18).

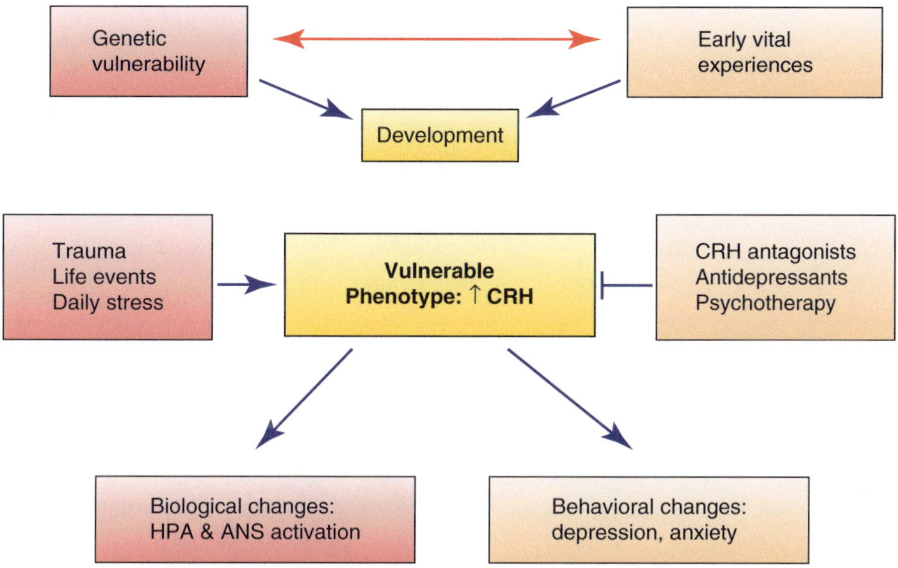

Fig. 5.18 Partly for genetic and epigenetic reasons, a vulnerable phenotype of CRH hyper-responsiveness to stress leads to emotional disorders. The use of pharmacological agents or psychotherapy counteracts the consequences of this phenotype. *HPA* pituitary-adrenal axis. Modified with permission from Cardinali [1]

Significance has also been given to the early effects of glucocorticoid hypersecretion, with lasting changes in these systems in the life of the individual [31].

The hierarchical relationship between the hypothalamus and the limbic system that controls it as the last level of autonomic motor hierarchy is clear when the defense behavior is tested. An animal bearing a hypothalamic deafferentation that disconnects the limbic system responds to the appearance in the visual field of any object with the reaction of "false rage." This indicates that the hypothalamus contains the "program" of the defense reaction, which acquires emotional meaning and purpose under the control of the limbic system. Maintaining high levels of cortisol in stress leads to verifiable chronic damage of hippocampal neurons with significant cognitive deficits (Fig. 5.19) [30, 42].

Stress interferes with the hypothalamic circuitry regulators of appetite and satiety. CRH is a potent anorexic substance whereas NPY is the most potent known orexinergic substance. NPY neurons stimulate CRH neurons, constituting a feedback loop involved in the control of food intake. Simultaneously, NPY inhibits the LC, and stimulates food intake [43]. Glucocorticoids induce gluconeogenesis and insulin resistance, by inhibiting lipolytic enzymes and stimulating enzymes that facilitate the deposition of fat, thus antagonizing the actions of GH and gonadal hormones on lipolysis and muscle and bone anabolism. Chronic stress can cause increased visceral adiposity, the suppression of osteoblastic activity, and the reduction of lean mass (muscle and bone). Monkeys subjected to chronic social stress show the classical symptoms of metabolic syndrome, such as high deposition of

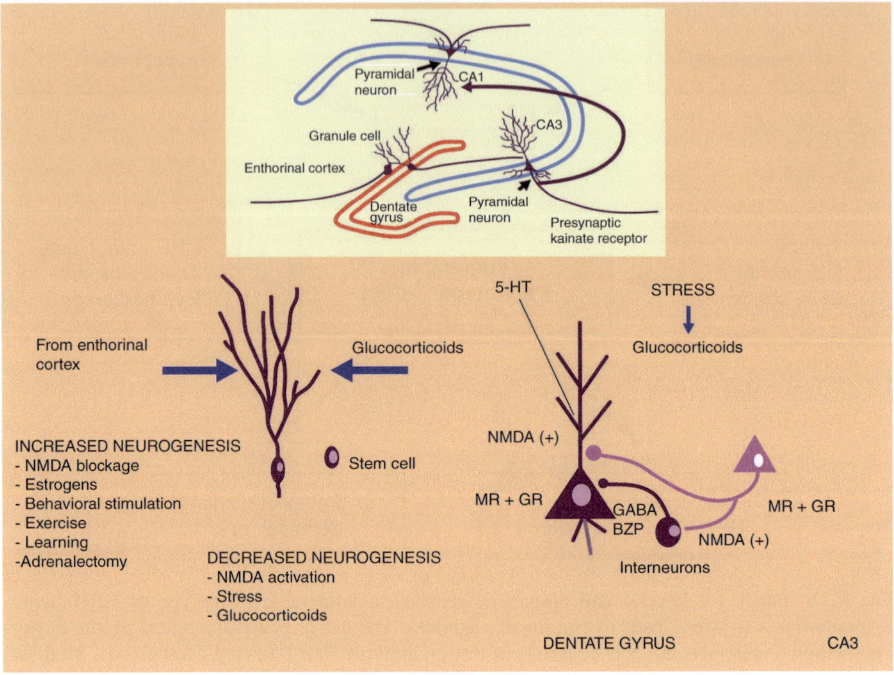

Fig. 5.19 Glucocorticoids and hippocampal neurogenesis. Several factors are listed that decrease or increase neurogenesis. *MR*, *GR* receptors for mineralo- and glucocorticoids. Modified with permission from Cardinali [1]

visceral fat, higher incidence of coronary atherosclerosis, hyperglycemia, and insulin resistance. Hypertension, adrenal hypertrophy, hypersensitivity to adrenal ACTH, hypogonadism and high cholesterol and lipid levels in the blood are also found. The correlation between chronic stress and metabolic syndrome is also observed in humans [44]. The phenotype of central or visceral obesity and a decrease in lean body mass are seen in patients with Cushing's syndrome, and in patients with melancholic depression and chronic anxiety, all situations of hypercortisolemia. Visceral adiposity establishes a vicious cycle of an increased need for insulin, hyperglycemia, and hypercholesterolemia.

Chronic stress affects therapeutic efficacy in cancer. For example, epidemiological studies revealed strong correlations between long-term survival/cancer progression and β-adrenoceptor blocker use in patients [45, 46]. The contributions of stress to immunosuppression in the tumor microenvironment and the implications of these findings for the efficacy of immunotherapies have been discussed [47]. Effects of norepinephrine (NE) on immune cells, such as those discussed in Fig. 4.25, are presumably involved.

Studies in rodent tumor models have identified adrenergic receptor expression on various cancer cells including mammary carcinogen-induced tumors, melanoma, and pituitary tumors. In humans, the Ewing sarcoma, neuroblastoma, rhabdomyosarcoma, lymphoma, melanoma, and pancreatic, lung, breast, and prostate cancer cells all displayed detectable levels of adrenoceptors. Several studies demonstrate that the activation of adrenergic receptors promotes tumor progression. For example, chronic activation of G protein-coupled receptors, such as the α1B-adrenergic receptor, can induce malignant transformation in normal cell lines, promote DNA damage, and enhance tumor formation. In addition to a role in cell survival, adrenergic receptor signaling has also been widely studied for mediating metastasis. The regulation of metastasis by adrenergic receptors occurs at multiple levels and involves not only cancer cells, but also cells in the tumor microenvironment and in the metastatic niche. [47, 48]. Sympathetic activation modulates gene expression programs that promote the metastasis of solid tumors by stimulating macrophage infiltration, inflammation, angiogenesis, epithelial–mesenchymal transition, and tumor invasion, and by inhibiting cellular immune responses and programmed cell death. Hematological cancers are modulated by ANS regulation of stem cell biology and hematopoietic differentiation programs [49].

In the author's laboratory, two breast cancer tumor lines were used to assess the reactivity of sympathetically denervated murine skin [50]. M3 tumors had a relatively high capacity for local growth and a low capacity for metastasis, whereas MM3-LN tumors grew locally at a slower rate, but metastasize very early to the lung. After local implantation in the ear, the growth of M3 and MM3-LN tumors was significantly slowed in the previously sympathetically denervated skin territory; however, their metastatic capacity remained unaffected by SCGx [50].

Stress related to inflammatory and infectious processes can alter the reproductive function by the inhibitory action of substances released by the immune system cells on the hypothalamic–pituitary–gonadal axis [51]. CRH suppresses GnRH release, whereas glucocorticoids act on the pituitary and gonads causing inhibition of the secretion of luteinizing hormone (LH), follicle-stimulating hormone (FSH), and gonadal steroids. In stressed women, a decreased frequency of LH pulses occurs together with increased cortisol secretion, which are assigned respectively to the decreased activity of the GnRH pulse generator and increased CRH activity. The central administration of cytokines reduces plasma LH levels, probably by increasing opioids and prostaglandins, which inhibit the secretion of GnRH. Additionally, cytokines can inhibit the activity of enzymes necessary for gonadal steroidogenesis.

The interaction of immune activity with the stress system is illustrated by the fact that vaccination may not be completely effective at establishing the immune defense to immunization that occurs during stress. Antigens are stressful stimuli (noncognitive), and the antigenic challenge features a stressful situation (Fig. 5.20). Infection stressor is a stimulus that induces the production of cytokines by the immune system and glial cells. Inflammatory and infectious processes stimulate visceral vagal afferents that activate the CRH stress system via the LC [52]. The hyperactivation of the hypothalamic–pituitary–adrenal axis induces the hypersecretion of glucocorticoids and immunosuppression, which are among the main consequences of stress (Fig. 5.20). The hypothalamic–pituitary–adrenal axis produces glucocorticoids, inhibiting the immune system. This, in turn, produces cytokines that are stimulatory

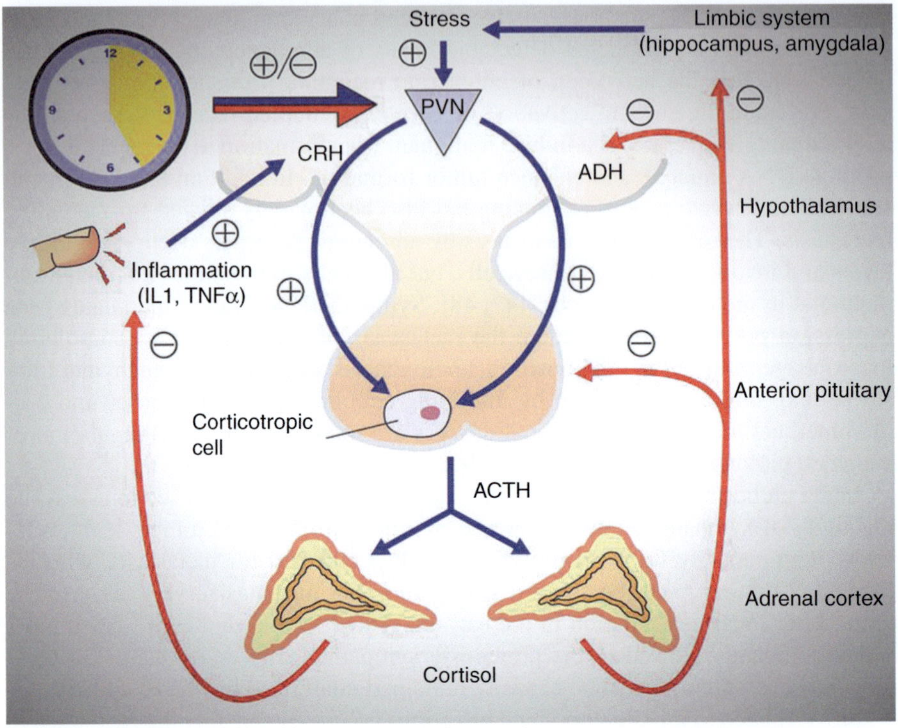

Fig. 5.20 Antigens are stressful stimuli (noncognitive), and the antigenic challenge features a stressful situation

for CRH release. Thus, this constitutes a mechanism for feedback between the neuroendocrine and immune components in stress. The prevalence of the neuroendocrine component causes immunosuppression, whereas the prevalence of the immune component causes autoimmune disease [19, 53].

The above-mentioned mechanisms also include the peripheral portion of the ANS. From studies in the SCG territory, it can be concluded that during the augmented NE release from peripheral sympathetic nerves innervating the hypothalamic–hypophyseal unit, the modulatory role of peripheral nerve terminals is differentially exerted on the release of hypophysiotropic hormones, inhibiting the release of all pituitary hormones, except for ACTH (Fig. 5.21) [54]. In animals with chronic SCGx, both ACTH and corticosterone rhythms were suppressed. The mechanism through which peripheral sympathetic neurons are capable of modifying the function of the median eminence remains undefined. As SCG efferences play an important role in regulating the cerebral and choroidal blood flow, the outcomes of SCGx on the median eminence may represent a particularly sensitive vasomotor effect of the peripheral noradrenergic terminals. Otherwise, they may involve a more complex interrelationship, such as a direct modulatory effect on hypophysiotropic hormone release [54].

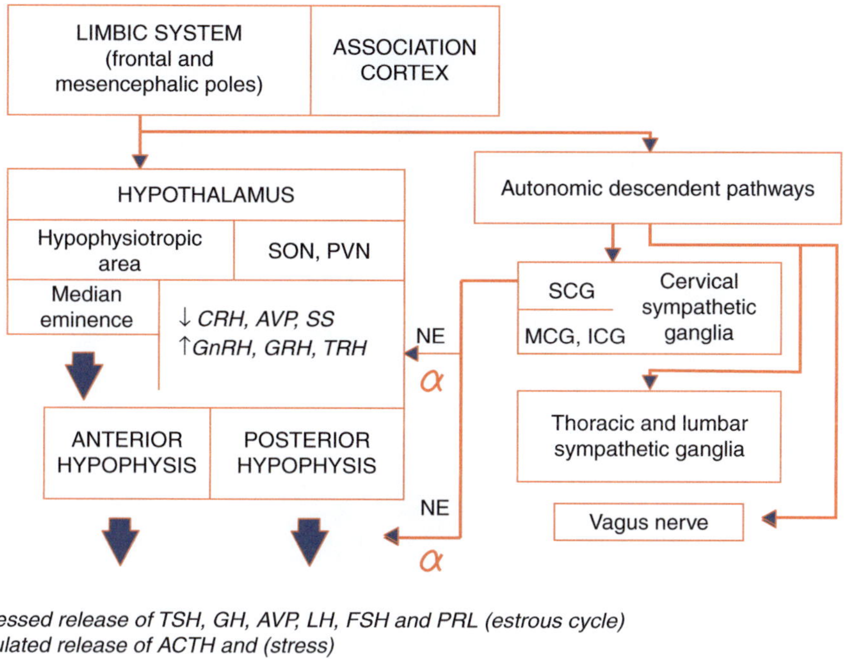

Fig. 5.21 Diagram that summarizes the changes in the hypothalamic–pituitary unit after cervical sympathetic activation. *SON* supraoptic nucleus, *PVN* paraventricular nucleus, *SCG* superior cervical ganglion, *MCG* medial cervical ganglion, *ICG* inferior cervical ganglion. Reproduced with permission from Cardinali [21]

24-h Rhythms in Food Intake, Energy Storage, and Metabolism

The first observations about daily changes related to nutrition and feeding were those of Sanctorius in the seventeenth century, who conducted the first autorhythmometry study by building a large scale in which he lived for months, recording his weight and the ingested food, and collecting his feces and urine. Among other observations, he detected a daily rhythm in body weight.

The complexity of the circadian system, described in Chap. 2, is exemplified when the rhythmicity of food and nutrition is studied (Fig. 5.22). The 24-h rhythm in feeding activity is controlled by the SCN. On the one hand, feeding time has synchronizing effects on the circadian system and food restriction can entrain certain rhythms through a food-related pacemaker. Finally, food, besides being a synchronizer and a variable controlled by the clock, food is also a masking stimulus that modifies many rhythms related to digestion, absorption, and metabolism of nutrients.

Feeding behavior is the first element to consider in the nutritional process of organisms. Most of the studies have focused on explaining how the homeostatic regulation (of the quantity and quality of food) takes place, with the hypothalamus

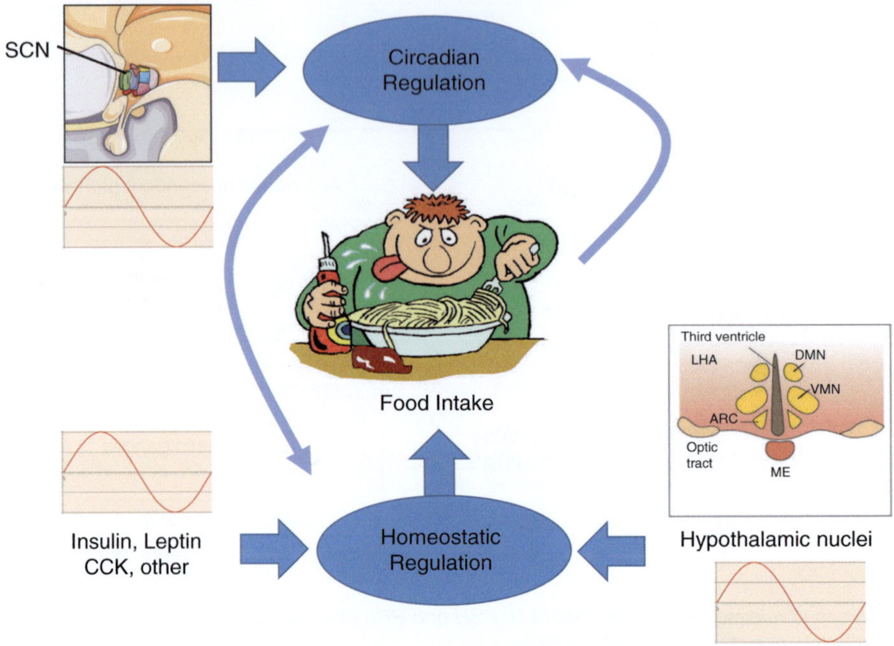

Fig. 5.22 Rhythmicity of feeding and nutrition. On the one hand, the feeding time has synchronizing effects; on the other, the food restriction can synchronize certain rhythms through a food-controlled pacemaker. Food activity is in turn controlled by the SCN. In addition to being a synchronizer and a variable controlled by the clock, food is a masking agent to directly modify numerous rhythms related to the digestion, absorption, and metabolism of nutrients. The figure was prepared in part using image vectors from Servier Medical Art (www.servier.com), licensed under the Creative Commons Attribution 3.0 Unported License (http://creativecommons.org/license/by/3.0/)

as the main neural structure involved in close coordination with the release of peripheral hormones such as leptin, CCK, and insulin [55]. Fewer studies have been devoted to the temporal aspects of such regulation.

Eating is a consummatory behavior that involves the participation of various organs, tissues, and systems. The control of eating behavior involves at least two organizational levels: (a) a homeostatic one, in which afferent information of nutritional status/energy reaches the CNS derived from adipose tissue, plasma, stomach, the small and large intestines, the pancreas, and the liver; (b) a hedonistic one, in which the CNS evaluates both the peripheral afferents and those from other brain areas, including the visual, olfactory, gustatory, oral and lingual, general sensitivity, and visceral sensitivity, pondering the nutritional value, palatability, and how pleasant the food is (Fig. 5.23) [55].

Regulating appetite and adiposity involves short-term signals originating from meals (quantity and taste) that determine the start and end of a meal; the level of energy reserve does not trigger them. Other, long-term signals are related to adiposity and intake coupled with caloric expenditure (Figs. 5.24 and 5.25). The regulation of each meal is based on the ingested volume, as the post-absorptive signals have little influence on the intake that produces them. Instead, the post-absorptive signals

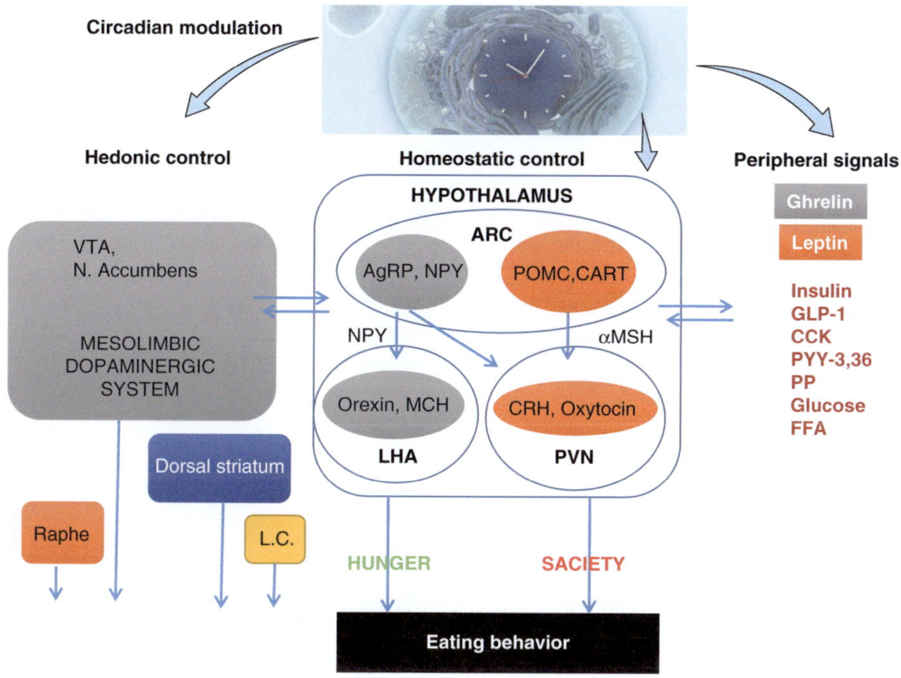

Fig. 5.23 Food behavior is regulated by homeostatic and hedonic mechanisms and by peripheral signals. *VTA* ventral tegmental area, *LC* locus coeruleus, *AgRP* peptide related to the agouti gene, *NPY* neuropeptide Y, *MCH* melanin concentrating hormone, *POMC* proopiomelanocortin, *CART* transcript regulated by cocaine-amphetamine, *GLP-1* glucagon-like peptide, *PP* pancreatic polypeptide

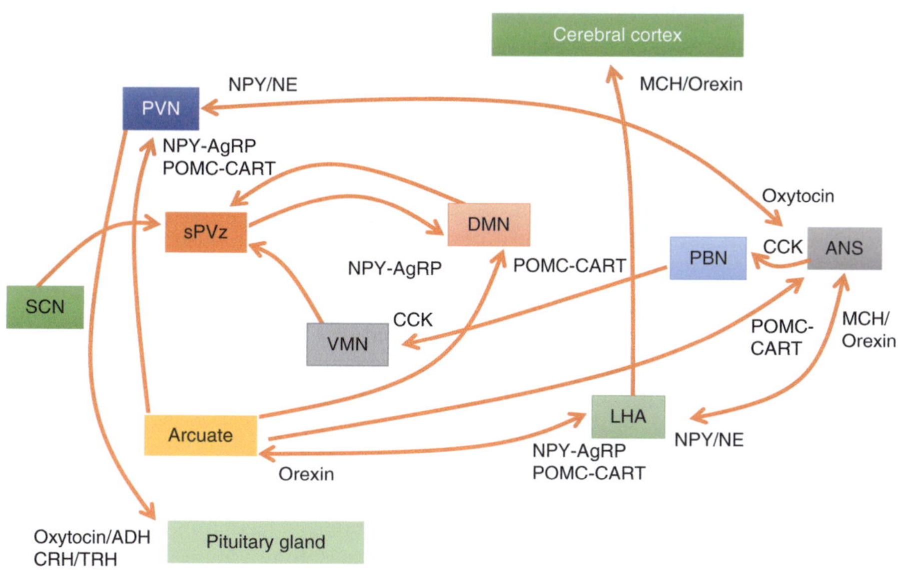

Fig. 5.24 Circuits involved in the control of food intake. The pituitary gland and the CNS regions with their respective neuropeptides and/or hormones are represented. *SPVz* subparaventricular zone. For other abbreviations see the text

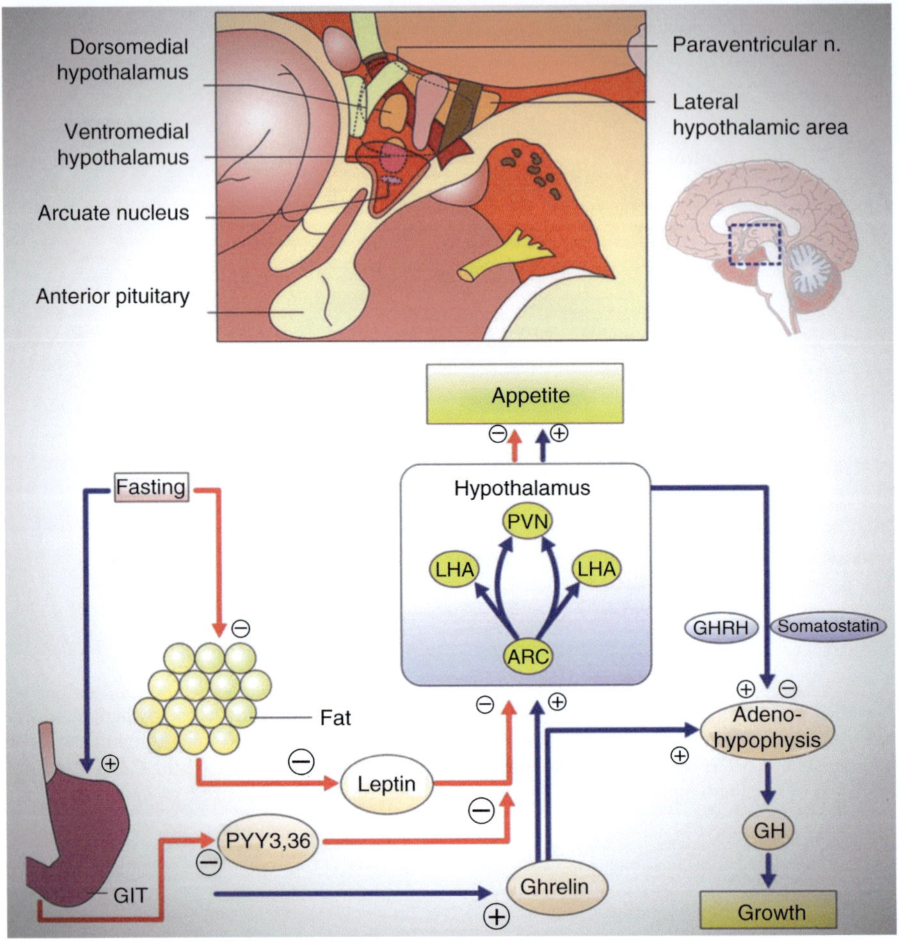

Fig. 5.25 Peripheral signals in appetite control. Leptin and PYY 3.36 inhibit it; ghrelin increases it. Modified with permission from Cardinali [1]

do influence the start of the next feeding period. Thus, satiation (or psychosensory satiety) is determined by the total volume of food ingested; it also includes specific satiation for different nutrients (carbohydrates, fats, and proteins). Satiety (or metabolic satiety) includes the suppression of the sensation of hunger and food intake and the duration of this phenomenon (interprandial period) [56].

There are three groups of afferent signals that modulate intake: (a) signals originating in the sensory systems (vision, smell, taste); (b) afferent signals originating from the gastrointestinal tract; (c) post-intake signals caused by nutrients or metabolism. Long-term intake is adjusted by metabolic indicators related to the degree of filling of fat reserves (e.g., leptin or visceral afferents originating in the adipose tissue). Homeostatic and hedonic controls are closely interrelated and often act in unison at the unconscious level to achieve biologically adaptive responses [57].

In relation to the signals involved in satiety, they are elicited by dilation and activity of the digestive tract and by various nutrients and the gastrointestinal hormones produced (Fig. 5.25). They include the anorexic leptin signal produced by adipose tissue and the peptide PYY 3-36 synthesized by the intestinal wall. Orexinergic signals such as ghrelin are also produced by the gastrointestinal tract; ghrelin stimulates appetite in addition to increasing GH and PRL release (Fig. 5.25) [58].

Classical studies have shown that experimental animals subjected to the electrical stimulation of LHA develop a typical food-seeking behavior (Fig. 5.26) [1, 59]. The reaction includes somatic motor activity, salivation, increased intestinal motility and blood flow, and decreased muscle blood flow. Conversely, stimulation of the ventromedial hypothalamus (VMH) produces satiety and catabolic-type behavior.

Stimulation of LHA produces hunger, increased parasympathetic activity, and from a metabolic point of view, glycogen synthesis, inhibition of gluconeogenesis, hypoglycemia, insulin release, and lipogenesis. VMH stimulation produces satiety, increased sympathetic activity, and from a metabolic point of view, glycogenolysis, gluconeogenesis, hyperglycemia, glucagon secretion, and lipolysis (Fig. 5.26).

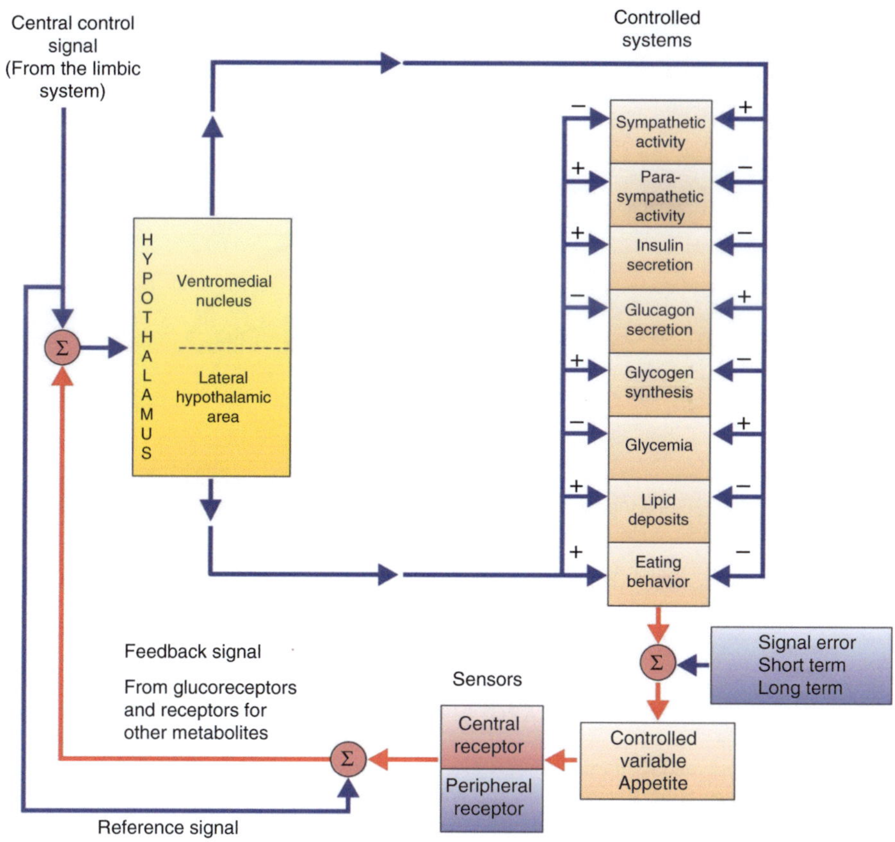

Fig. 5.26 Analysis of hypothalamic control of appetite according to the systems control theory. Modified with permission from Cardinali [1]

By contrast, lesions of each of these areas have opposite effects. VMH lesions produce hyperinsulinemia, hyperphagia, overeating, and weight gain. Reaching a certain level of obesity, the animal properly regulates intake and maintains its weight. In this model, vagotomy blocks hyperphagia, obesity, and hyperinsulinemia. LHA lesions produce opposite effects, with decreased food intake and weight. There is a close correlation of anhedonia with reduced dopaminergic neurotransmission in the nucleus accumbens (Chap. 6) [1].

The amygdala, nucleus accumbens, and orbitofrontal cortex, as parts of the limbic system, integrate the homeostatic information with motivational cognitive aspects (aversive, hedonic) of eating (Fig. 5.23). Different limbic association areas are stimulated synchronously with the LHA and VMH during food intake. Specific lesion experiments indicate that the limbic areas provide purpose to the conduct regulated by the hypothalamus.

Figures 5.23, 5.24, and 5.27 summarize the orexigenic and anorexigenic circuits found in the hypothalamus. They include [2, 56]:

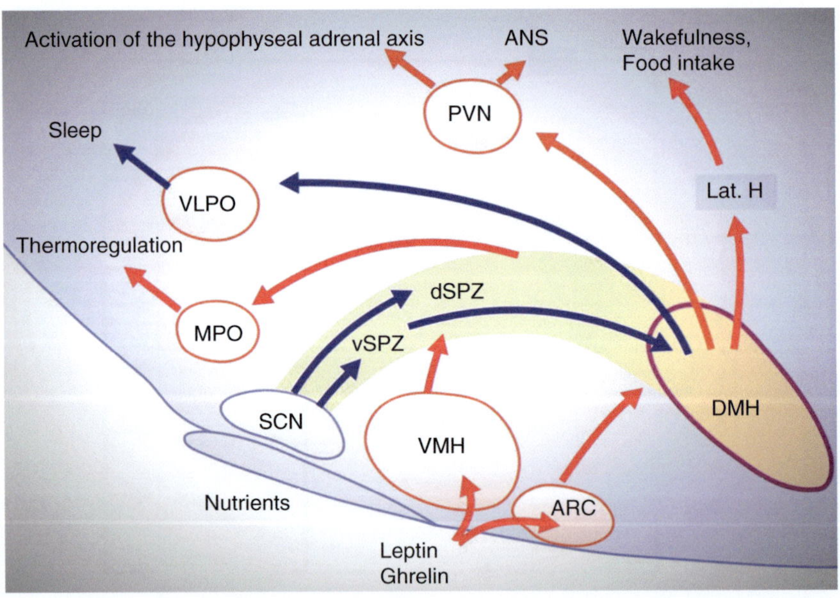

Fig. 5.27 The SCN projects on the subparaventricular zones, both the ventral (vSPZ) and the dorsal (dSPZ), and on the DMH. vSPZ neurons receive information for the organization of the daily sleep/wake rhythm and project it to the DMH, which integrates information with other sources (e.g., leptin and hhrelin in the ARC). The DMH is the source of projections that regulate circadian sleep/wake rhythms, pituitary–adrenal axis activity, and descending ANS pathways, in addition to vigilance and food intake (at the level of orexinergic neurons and MCH of the lateral hypothalamus. Integration stations allow the adaptation of circadian rhythms to environmental stimuli such as food availability, and cognitive and emotional influences of the limbic system. Modified with permission from Cardinali [1]

- A neural network of orexigenic activity that includes neurons having NPY, GABA, galanin, orexin, opioids, agouti-related peptide (AgRP) or melanin-concentrating hormone (MCH) as a transmitter and that translate and release appetite stimulus signals. The orexinergic/MCH neurons also participate in the alertness processes (Chap. 2).
- Different anorexigenic pathways, including neurons using CRH, glucagon-like peptide-1 (GLP-1), α-melanocyte stimulating hormone (α-MSH) or cocaine- and amphetamine-regulated transcript (CART), which are responsible for the extinction of appetite by interrupting the action of the orexinergic network.
- Nuclei of the medial hypothalamus (dorsomedial hypothalamus, DMH, and the VMH) that tonically restrict orexigenic signals in the interprandial interval. They are activated by leptin, a hormonal signal that decreases orexigenic activity. Ghrelin exerts opposite effects to leptin.
- A circadian regulation, coordinated by the SCN (Figs. 5.23, 5.24, and 5.27).

The integration of metabolic information requires multiple specialized areas of the CNS: (a) brainstem (NTS, dorsal vagal complex), integrating peripheral neural information with limbic structures (nucleus accumbens, amygdala, orbitofrontal cortex); (b) mesencephalon and thalamus, receiving sensory information from different levels of the gastrointestinal tract; (c) anterior brain, providing aversion and gratification generated by food [56].

Glycemia is one of the stimulatory signals important in regulating short-term appetitive behavior. The nervous system has two types of receptors for glucose. The peripheral receptors are located on the tongue, portal vein, duodenum, intestine and pancreas, and generate changes in the activity of their respective visceral afferents [60]. Central glucose receptors are located in the hypothalamus [61].

The first relay point of peripheral information on glycemia is the brainstem at the level of the NTS. The integrated information at this level follows two directions: (a) to return caudally via a reflex including neurons of the dorsal motor nucleus of the vagus to affect the viscera; (b) in a rostral direction, the information reaches the hypothalamus, where it undergoes a second process of integration, adding to the information provided by central glucose receptors, which monitor the concentration of blood and CSF glucose.

The ARC and NTS are the only brain areas with neurons expressing pro-opiomelanocortin (POMC), a precursor of several neuropeptides with signaling capacity in food intake control (Figs. 5.24, 5.28, and 5.29). The most important neuropeptide in feeding control is the POMC product α-MSH, which is associated with the reduction of food intake [62]. The NTS and dorsal nucleus of the vagus also have the highest concentrations of the α-MSH receptor, melanocortin receptor type 4 (MCR4). The injection of an agonist of MCR4 (i.e., α-MSH) at the fourth ventricle or directly into the NTS is effective in reducing food intake and body weight, whereas administration of an MC4R antagonist (e.g., AgRP) increases food intake and body weight [62].

The ARC is in a dorsal position immediately adjacent to the median eminence (Fig. 5.2). A group of neurons located in the medial portion of the ARC expresses

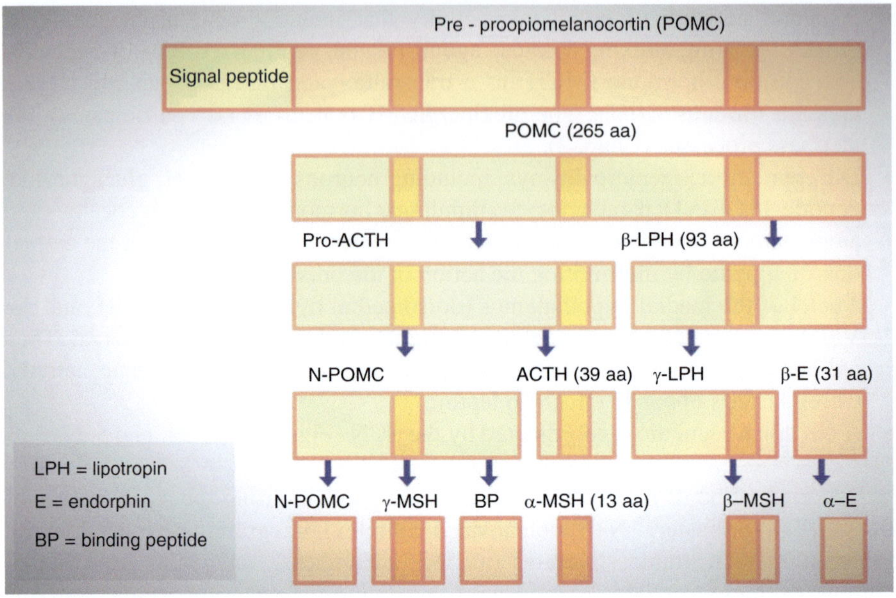

Fig. 5.28 Processing of POMC

NPY, which is a major stimulator of feeding behavior [63]. The injection of NPY in the lateral ventricles or in the area around the fornix evokes a substantial increase in food consumption. ARC neurons that express NPY project to the PVN and the LHA. Another group of ARC neurons synthesizes POMC and mainly project to the PVN, the DMH, and to brainstem neurons controlling the sympathetic activity via the intermediolateral column of the spinal cord [63].

Leptin conveys information on the nutritional status of the individual and reaches the neurons of the ARC through the bloodstream, specifically via fenestrated vessels of the median eminence. Leptin, ghrelin, and insulin modulate NPY expression in ARC neurons. There is strong evidence that many of the actions of leptin are mediated by stimulation of the melanocortin system, decreasing NPY. In the hypothalamus, NPY is synthesized by neurons of the ARC and secreted from their terminals in the PVN and lateral hypothalamus [64].

Administration of leptin to fasted rats (in which circulating leptin levels are decreased) decreases the augmented synthesis of NPY found in ARC. Ghrelin, on the other hand, acts in an opposite way to leptin, by increasing the expression of NPY mRNA and ultimately inducing an increased consumption of food. Insulin acts synergistically with leptin to decrease the augmented expression of NPY in fasted animals. Another peptide, PYY 3-36, which is released into the bloodstream by the GIT cells proportionally to the amount of food ingested, acts as another feedback signal inhibiting the release of NPY [64].

Neuropeptide Y neurons express AgRP. This agouti protein is a paracrine secretion molecule acting as an antagonist of melanocortin receptors [65]. The MC-4 receptors have been largely related to the control of feeding behavior,

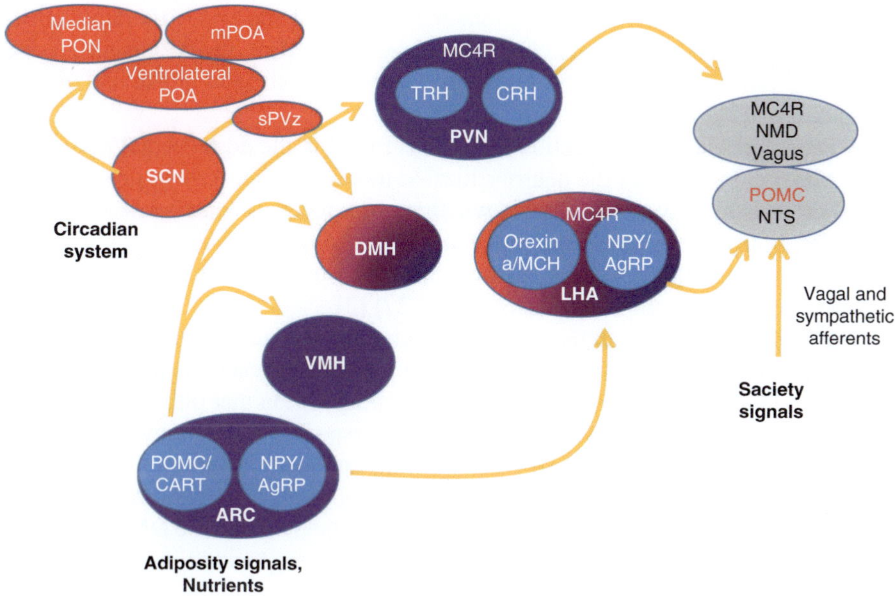

Fig. 5.29 Anatomical circuits involved in energy homeostasis. The circadian apparatus (in *red*), the centers of control of the intake (in *blue*) and links with the ANS (in *gray*) are represented schematically. Overlays are indicated in mixed *red/blue*. The master clock of the circadian network is the SCN. The key structure of the energy network that integrates adiposity (leptin and insulin) and signals related to nutrients (glucose, fatty acids) from the periphery is the ARC. Both circadian and energetic networks are indirectly connected through the DMH and the LHA, which are both sensitive to nutritional and circadian information. The nucleus of the solitary tract (NTS) receives signals of satiation from the peripheral organs via vagal and sympathetic afferents and is innervated by the PVN and the LHA; thus, it also receives signals from the circadian and energetic networks of the hypothalamus

and mutations in their genes cause obesity in rodents and humans (Figs. 5.23, 5.24, and 5.27). As already mentioned, the MC4R preferential agonist ligand is α-MSH. Intracerebroventricular injections of α-MSH decrease food intake in normal animals. In fasted animals, POMC mRNA is decreased, which can be restored by the administration of exogenous leptin or insulin.

The PVN and LHA exhibit a significant amount of MC4R, and a relatively dense innervation of fibers containing α-MSH. Fibers containing AgRP and NPY also innervate these regions. Thus, the PVN and the LHA contain two distinct populations of nerve terminals releasing POMC or AgRP, both responsive to circulating leptin, whose neurotransmitters would act on the same receptors (MC-4, MC-3) in diametrically opposed ways, acting as agonists (α-MSH) or antagonists (AgRP) on the same neural system (Figs. 5.23, 5.24, and 5.27) [62].

Another participating neuropeptide in PVN and LHA is CART, which is produced by the same neurons that express POMC. Its co-location in POMC neurons indicates a conjoint action in the same areas on the MC-4 receptors. Studies using neural mapping techniques have shown that ARC neurons that express CART and

POMC also project to the intermediolateral column of the upper thoracic spinal cord and participate in the regulation of thermogenesis. Thus, the ARC neurons can act on the maintenance of body weight by regulating both food intake and energy expenditure [63].

A large concentration of leptin receptors is found in the dorsomedial portion of the VMH, where many of the neurons located therein are sensitive to glucose. The VMH projects heavily to the sPVNz, which in turn, receives dense projection from the SCN. Many of the VMH neurons that project to the sPVNz are responsive to circulating leptin, thus providing an anatomical substrate by which leptin can control the circadian variation of feeding (Fig. 5.27) [2].

The VMN participates in the control of endocrine and autonomic systems, indirectly modulating the information starting from neurons located in sPVNz and targeting the DMH, which is essentially an integrating region that ultimately projects to the PVN, modulating endocrine and visceral responses [40]. Many studies suggest the involvement of the DMH in controlling the intake of water and food, and body weight. Stimulation of DMH results in changes in the activity of the pancreatic nerves, and DMH lesions induce hyperglycemia, indicating that the DMH regulates insulin secretion via projections to the autonomic centers. In addition, several studies suggest DMH involvement in the control of the cardiovascular system, stress and anxiety, and locomotion. Because of this complexity, it has been proposed that the DMH is one of the main components of a hypothalamic pattern generator for the visceromotor system [66].

A population of neurons from the LHA that contain MCH projects extensively to various regions of the CNS of mammals (Figs. 5.23, 5.24 and 5.27) [2]. The MCH receptor (MCH-1R) is widely distributed in the CNS, with particularly dense expression in the cerebral cortex (including the orbitofrontal, pre-limbic and sensorimotor regions, and the rhinencephalon), the nucleus accumbens, hippocampal formation, the NTS, and the LC. Several studies have demonstrated the occurrence of an increase in food intake following injection of MCH. Furthermore, the increase in MCH expression leads to obesity and insulin resistance. Another population of neurons in the LHA express orexin and project to regions of the brainstem and spinal cord, such as the LC and the dorsomedial nucleus of the vagus nerve (Fig. 2.16) [2]. When orexin is injected into the cerebral ventricles, it causes increased food intake, whereas orexin receptor antagonists decrease the intake. Mice whose orexin gene was deleted exhibit narcolepsy and hypophagia. Thus, this peptide signals to the neural systems that play a significant role in feeding behavior and the sleep–wake cycle, possibly by coordinating a set of responses to complex behaviors and complementary autonomic responses. Its link to sleep was discussed in Chap. 2.

In neuroimaging studies, hungry individuals show greater activity in the prefrontal cortex and decreased activity in the hypothalamus, thalamus, insular cortex, the cingulate gyrus, orbitofrontal cortex, the basal ganglia, temporal cortex, and cerebellum. In obese and satiated individuals, activation of the prefrontal cortex was greater than that found in normal individuals, and there is a greater reduction in the activity of the limbic and paralimbic cortex compared with normal individuals of both sexes [67, 68].

There are separate mechanisms for controlling the balance of different nutrients [55]. For the carbohydrate balance, the system operates to increase the intake of carbohydrates when their peripheral utilization decreases. GABA, NPY, and NE in the hypothalamic VMH, and glucocorticoids specifically increase carbohydrate intake, with a circadian rhythm of maximum effect at the beginning of the activity phase, coinciding with the peak plasma cortisol. 5-HT, CCK, insulin, and leptin inhibit the effect.

For fat balance, the neuropeptide galanin, opioids, and mineralocorticoid hormones are involved. These substances enhance specific fat intake by acting on VMH. NPY and leptin in particular mediate in this mechanism. The effect has a characteristic daily rhythm, with the maximum toward the end of the phase of activity.

In humans, an appetite for fat is verified rising toward the afternoon. Dopamine (DA) antagonizes the effect of galanin, opioids or aldosterone on fat intake. This is the basis of the anorexic action (appetite suppressant) of amphetamine that acts by releasing endogenous DA. Dopamine antagonists (neuroleptics) increase the appetite for fat.

For protein balance, GHRH and opioid peptides, leptin, and NPY are involved. The circadian rhythm of protein preference is similar to that of fat, with the maximum occurring late afternoon.

Estrogens play a role in stimulating the appetite for carbohydrates. In women, a greater preference for carbohydrate diets is detected. In men a greater preference for predominantly protein diets occurs. The appetite for fat increases during puberty, in both boys and girls, coinciding with increased hypothalamic galanin detected in experimental animals. High levels of estrogen in obese women further increase the synthesis of galanin in the hypothalamus, with increasing preference for fats. Thus, there is a significant correlation between the content of hypothalamic galanin and body weight in different species [69, 70].

Although there has been progress in determining the factors that modify the specific appetites for different nutrients, there is yet no simple physiological scheme that accounts for the long-term control of body weight. Clearly, lipid deposits play an important role and identification of leptin was a milestone in this regard. Adipocytes produce leptin and it is now known that leptin serves as a signal for adequate energy intake (Fig. 5.30). Circulating leptin is a signal indicating whether deposits of sufficient energy exist to initiate a process of great energy demand such as puberty. Leptin promotes inflammation, which provides a pathophysiological link between obesity and insulin resistance, atherosclerosis, and autoimmune processes. These processes are inflammatory diseases characterized by increased pro-inflammatory cytokines such as IL-6, which is an example of the link between energy homeostasis and immune function [71].

It is noteworthy that there is a viscerotopic representation of lipid deposits in central areas [72]. The existence of a permanent regional neural mechanism that links the load sensor of deposits to the activity of the higher centers is evident. For example, an early overload of these deposits (in obese children) produces definitive changes in the "set point" of the total mass contained, with the result of obesity in adulthood.

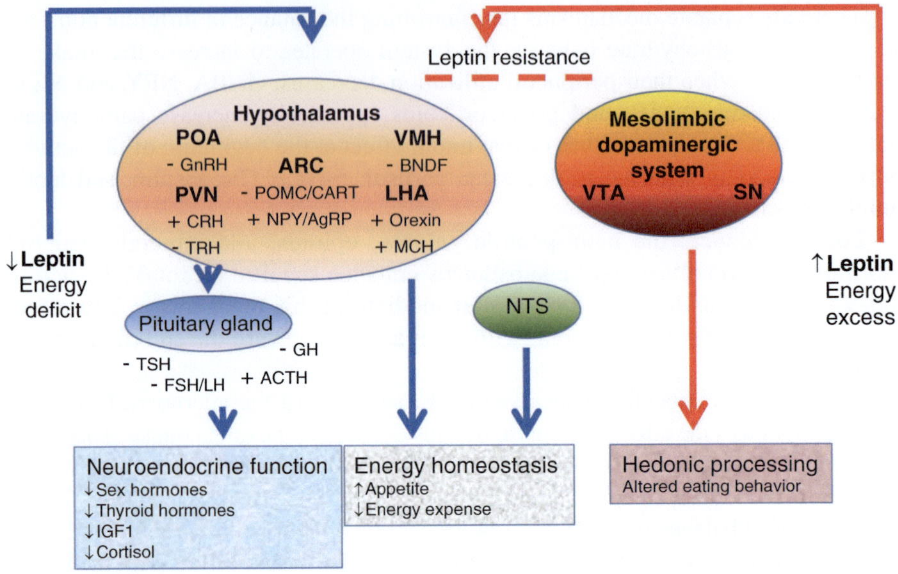

Fig. 5.30 Leptin action on the brain during states of excess energy and energy deficiency. During excess energy, access of the leptin to the hypothalamus and other areas of the brain is impaired (leptin resistance). In states of energy deficiency and therefore of leptin deficiency the neuropeptides that are normally inhibited by leptin are elevated (+) and neuropeptides stimulated by leptin are suppressed (−). Changes in the concentrations of these neuropeptides lead to alterations in neuroendocrine function and energy homeostasis. Alterations in leptin levels may also affect the hedonic aspects of eating behavior

Circadian rhythms are described for most of the factors involved in food intake [73]. Although in humans the rhythms of food intake are conditioned by cultural factors, the relative stability of dietary habits between cultures supports the existence of strong biological determinants. As mentioned, nutrients that are a source of quick energy such as carbohydrates are usually selected at the beginning of the activity period, whereas fats and proteins are preferred before the beginning of the period of rest. The rhythms of appetite, digestion, absorption, and activity of key enzymes of metabolism are largely responsible for this (Fig. 4.32).

Meal times influence factors as varied as weight gain, glucose, glucose tolerance, triglycerides, cardiovascular risk, etc. These effects are especially apparent in night-shift workers, who fail to produce a full synchronization of their metabolic rhythms at nighttime (Chap. 8).

Unlike animals used as a model for studies of feeding behavior, humans do not necessarily eat according to biological impulses, but also for cultural, religious, and hedonic reasons. However, there are situations in which these cultural conditions can be minimized. Newborns cry to demand food every 90–120 min, after 2 or 3 months they demand food only four or five times a day, and after 6 months the feeding frequency begins to resemble that of the adult. The three main meals of an adult are maintained throughout life, even in those situations in which there is temporary external isolation.

There are rhythms in neuroendocrine regulatory factors and in caloric and macronutrient intake [73]. During nocturnal sleep, levels of glucose and insulin secretion increased significantly to return to baseline in the morning. During sleep deprivation, glucose levels and insulin secretion were negligible whereas daytime sleep was associated with rises in glycemia and insulin secretion (Fig. 5.31). The diurnal variation in insulin secretion was inversely related to the cortisol rhythm; however, sleep-associated rises in glucose correlated with the amount of concomitant GH secreted [74].

Similarly, leptin is regulated by the sleep/wake homeostat and has a negative correlation with ACTH and cortisol being high during NREM sleep. The stomach secretes ghrelin, it has orexic activity, and food intake suppresses its secretion. The mean values of the day are higher than those in the evening and overnight levels are higher in the first half than those of the second half [58].

Increased leptin is one of the signals responsible for appetite suppression during sleep. Coincident with this regulation of secretion of leptin, insulin is released

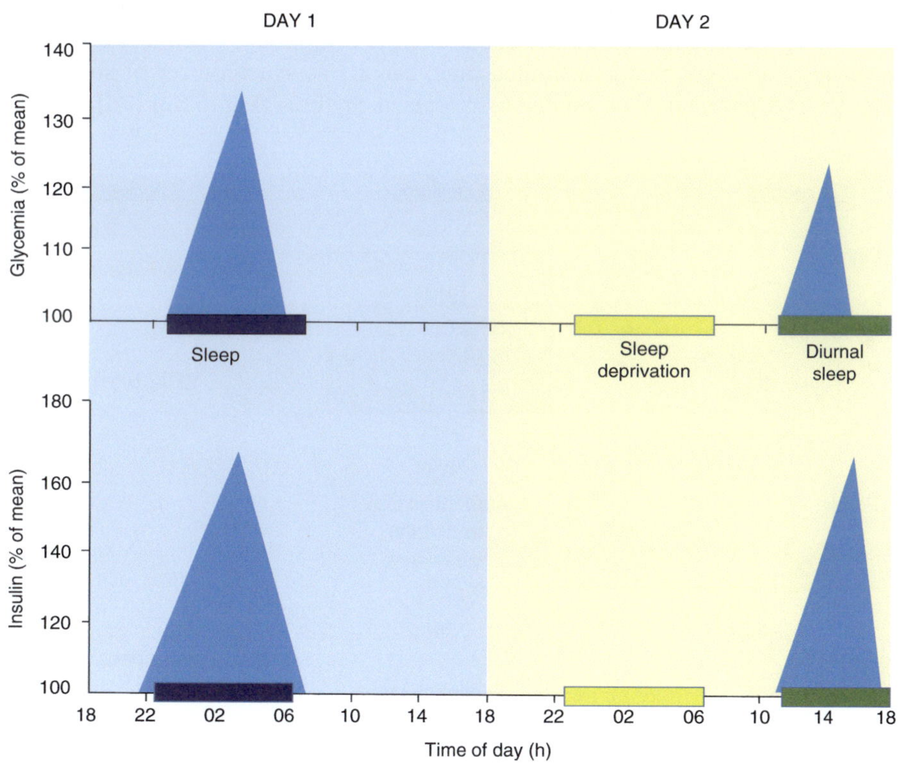

Fig. 5.31 Insulin is a hormone whose constitutive secretion is mainly controlled by the sleep–wake homeostasis. The volunteer was deprived of sleep for one night and could recover the lost sleep from 11 am the next day. Ultradian variations in hormone secretion or glycemia or values <100% of mean plasma incurred sample reanalysis or glycemia are not represented. Data from Van Cauter et al. [74]

during slow-wave sleep. Alterations in insulin and leptin secretion are present in shift workers, who tend to suffer a chronic sleep disturbance. Thus, sleep loss is associated with insulin resistance, lower levels of leptin and higher levels of ghrelin, leading to a serious risk of obesity [75].

In humans and rats, with age, a decline in melatonin levels occurs, whereas the levels of visceral fat, insulin, and leptin increase [21]. These changes are often associated with adverse metabolic consequences: glucose intolerance, insulin resistance, diabetes, dyslipidemia, and hypertension. Treatment with melatonin reverses age-associated changes in retroperitoneal and epididymal fat, and plasma concentrations of insulin and leptin to those found in young individuals, without significantly affecting food intake. These findings, together with the ability of pinealectomy to increase leptin levels, suggest that melatonin might exert an inhibitory effect on the release of this hormone [21].

Damage of the hypothalamic SCN causes immediate loss of the circadian rhythm of food intake, without alteration of the regulation of the total amount of food ingested over 24 h. The homeostatic and temporal regulation reside in different parts of the hypothalamus, but do overlap (Fig. 5.29). In addition to influencing the internal temporal structure of the body, food, when it is restricted in quantity and provided at a limited time, can act as synchronizer of structures that act as a pacemaker on numerous circadian rhythms (Fig. 5.32) [76]. In this

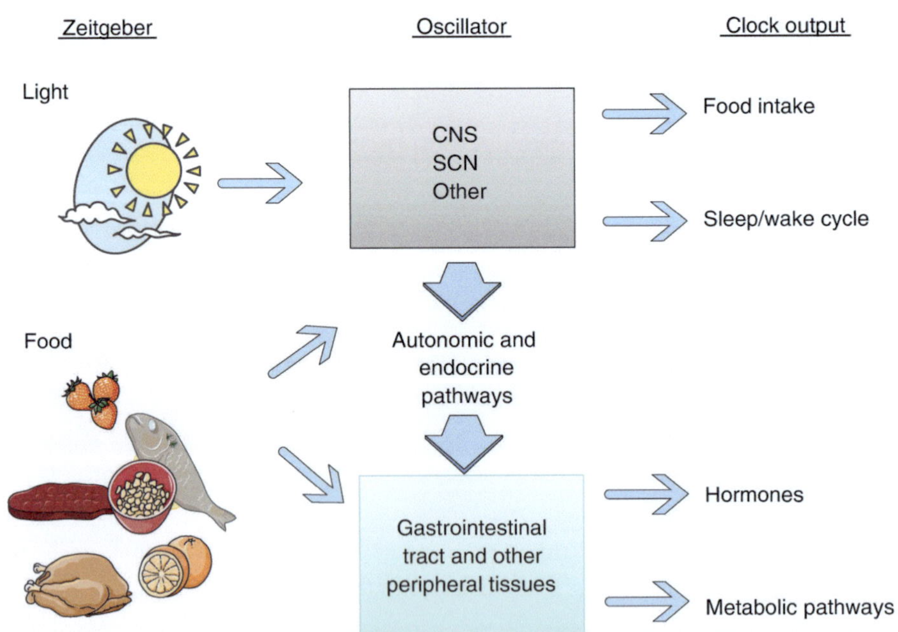

Fig. 5.32 Food as a circadian synchronizer. The figure was prepared in part using image vectors from Servier Medical Art (www.servier.com), licensed under the Creative Commons Attribution 3.0 Unported License (http://creativecommons.org/license/by/3.0/)

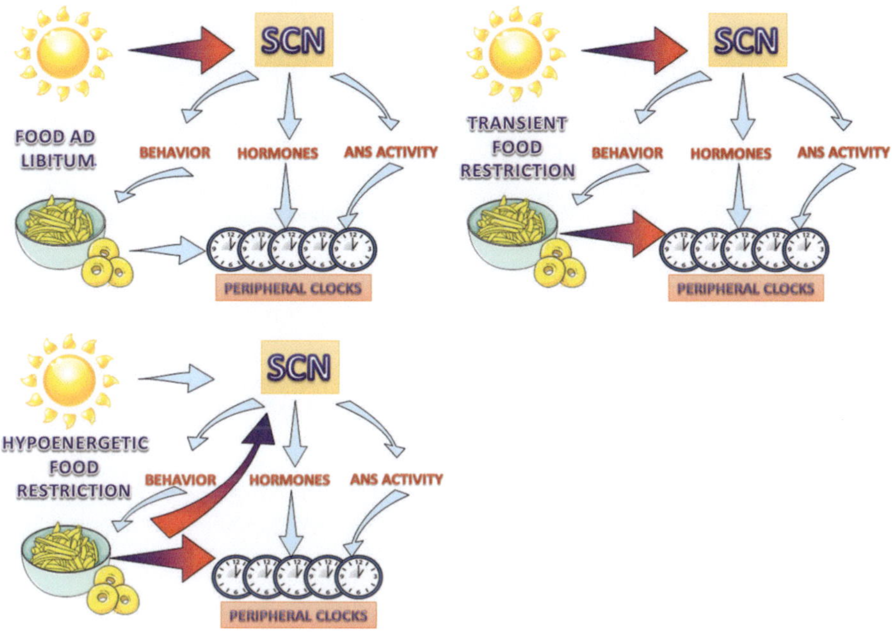

Fig. 5.33 Food-entrained oscillators. Under normal physiological conditions, synchronization of behavioral rhythms, such as feeding, is controlled by the SCN. The importance of food as zeitgeber is seen under conditions of food restriction, in particular hypoenergetic. The figure was prepared in part using image vectors from Servier Medical Art (www.servier.com), licensed under the Creative Commons Attribution 3.0 Unported License (http://creativecommons.org/license/by/3.0/)

situation, a dissociation between the rhythms controlled by light, through the SCN, and rhythms synchronized by food (food entrained oscillators, FEO) may appear (Fig. 5.33). When an animal receives a single meal with a calorie content of less than its daily needs, it progressively entrains rhythms to the feeding schedule, showing increased levels of certain variables (motor activity, gastrointestinal motility, activity of digestive enzymes, plasma cortisol) 1 or 2 h before the scheduled food availability. This anticipatory activity is controlled by an independent pacemaker to SCN, because it remains in animals with SCN lesions (Fig. 5.33) [76]. When food is abundant, the pacemaker entrained by light is directing all the circadian rhythms; however, when food is scarce and available only at specific times, a second pacemaker (FEO) takes over certain rhythms favoring the use of food, but without neglecting other rhythmic processes that continue to be controlled by the pacemaker synchronized by light (SCN). The FEO is very potent in the liver and less effective in other peripheral organs such as the lung [76].

In Chap. 2, we discussed the molecular bases of the circadian clock as a network of transcription–translation feedback loops. Rhythmic clock output is achieved through E-box elements in controlled clock genes, which affect numerous cellular processes (Fig. 5.34), including lipids and glucose metabolism (liver, muscle), adipogenesis (white and brown fat), insulin sensitivity, and lipogenesis. In turn, various

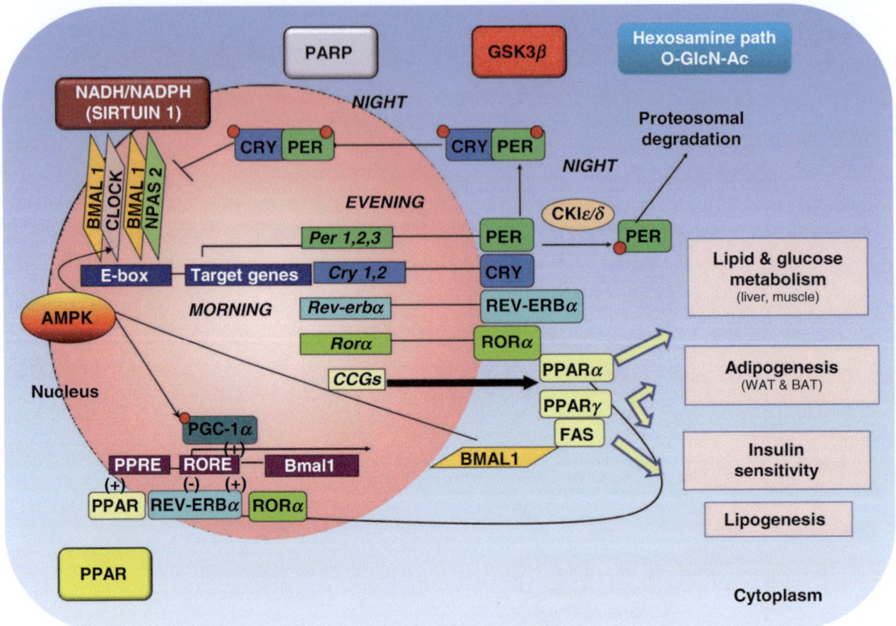

Fig. 5.34 The circadian clock is a feedback network of transcription–translation loops (Chap. 2). The rhythmic output of the clock emerges through E-box elements in clock-controlled genes (CCGs), which affect numerous cellular processes, including lipid and glucose metabolism (liver, muscle), adipogenesis (white adipose tissue, WAT, brown adipose tissue, BAT), insulin sensitivity, and lipogenesis. Several metabolic signals provide feedback to the cellular circadian system: AMPK, sirtuin 1, poly ADP-ribose polymerase (PARP), peroxisome proliferator-activated receptors (PPARs), glycogen synthase kinase 3β (GSK3β), hexosamine/O-GlcN-Ac O-β-DN-acetylglucosamine)

metabolic signals provide feedback to the cellular circadian system [73, 77]. They include:

- AMPK: AMP-activated protein kinase. metabolic indicator of cell energy charge
- NAD(P)H: NAD(P)$^+$: a metabolic indicator of a redox state (sirtuin 1)
- Poly ADP-ribose polymerase (PARP): secondary sensor cell energy charge and redox state
- Peroxisome proliferator-activated receptors (PPARs), e.g., PGC-1α (peroxisome proliferator-activated receptor γ co-activator 1-α): lipid metabolism sensors
- Metabolism of glycogen (glycogen synthase kinase 3β, GSK3)
- Via hexosamine/O-GlcN-Ac (O-β-D-N-acetylglucosamine): glucose level signaling

Sirtuin 1 is an example of a metabolic sensor. It is an NAD$^+$-dependent deacetylase (redox state sensor that measures the ratio NAD$^+$/NADH) and a regulator of clock machinery that binds to and inhibits the activity of CLOCK/BMAL1. Sirtuin 1 plays a key role in the regulation of gluconeogenesis, fat metabolism, insulin secretion, and apoptosis. Inactivation of sirtuin 1 by oxidative stress leads to the abnormal transcription of proapoptotic and proinflammatory genes, oxidative stress, inflammation, and premature cell senescence [20].

Different experiments in mice genetically modified in clock genes indicate their involvement in feeding behavior. For example, feeding *Clock*-mutant mice with a diet rich in calories doubles the accumulation of body mass and energy relative to control. Mice lacking *Per3* have a high body mass and body composition and impaired adipogenesis. Mice with knockout of *Bmal1* in the liver have altered hepatic glucose metabolism, whereas mice with knockout for pancreatic *Bmal1* show hyperglycemia, impaired glucose tolerance, and impaired insulin secretion. Also, mice bearing knockout of *Bmal1* in white adipose tissue has obesity by reducing regulatory signs to central regions of appetite [78].

As discussed in Chap. 2, the SCN can act on metabolism via both hormonal and autonomic neural pathways. In the case of the adrenal gland, glucocorticoid and blood sugar increase in the early phase of activity is the result of the activity of SCN on CRH-producing hypothalamic neurons, which stimulate the secretion of ACTH by the pituitary and this in turn activates the secretion of glucocorticoids by the adrenal cortex. In addition, the adrenal gland is subjected to a circadian rhythm of ACTH sensitivity, with maximum sensitivity during the activity phase. By neural retrograde labeling techniques, it was found that the adrenal is connected to the SCN via the ANS, through a multisynaptic path that regulates the sensitivity of the gland (Fig. 5.12) [79]. Using similar procedures, neural pathways between the SCN and the heart, pancreas, liver, thyroid, and pineal have been described (Chap. 4). Therefore, through these communication mechanisms, the SCN may activate or silence individual tissues depending on their function at different times of the day. It has been shown that both the sympathetic and the parasympathetic autonomic nervous system divisions may discriminate between different compartments of adipose tissue, such as subcutaneous and intraabdominal fat. Compartmentalization of motor neurons of the autonomic nervous system is the basis for the selective effects of the sympathetic–parasympathetic balance in the different compartments of the body. This anatomical segregation of autonomic neurons provides a physiological basis for selective changes in the sympatho-parasympathetic balance in various body compartments and at different times of the day [80].

From a circadian point of view, the body is divided into two functional autonomous compartments: (a) a thoraco-muscular compartment; (b) a visceral compartment (Fig. 5.35). In the period of wakefulness, locomotor activity requires glucose and free fatty acids. As a homeostatic reaction, the brain facilitates the release of energy from the storage organs such as the liver and adipose tissue. If this activity is repeated on a regular daily basis, the ANS is programmed to facilitate operation in the form of an anticipatory daily rhythm for energy requirements. By contrast, during the period of NREM sleep, the ANS switches into a state of anabolic control, recovery, with energy accumulation in the organs of deposit and lower peripheral glucose utilization [80].

Clinical and pre-clinical studies have established close associations between insulin resistance and sympathetic activation and thus suggest coinciding mechanisms in their development. Acute peripheral and central increases in insulin levels

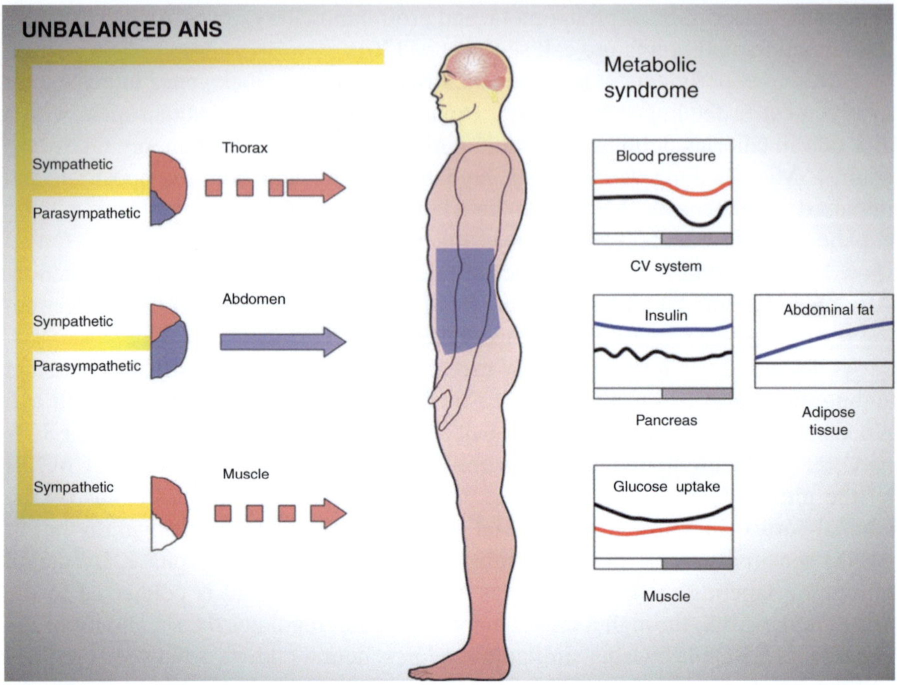

Fig. 5.35 Autonomic imbalance of the abdominal and thoraco-muscular territories in the metabolic syndrome. The result is hypertension, increased insulin resistance, and abdominal obesity. Modified with permission from Cardinali [1]

can elevate sympathetic activity through central insulin receptor action in key brain regions regulating autonomic function. Conversely, central manipulation of the sympathetic nervous system can affect insulin sensitivity peripherally through a variety of mechanisms including modulation of the renin–angiotensin–aldosterone system and pancreatic autonomic nerves. Enhanced insulin resistance and elevated sympathoexcitation likely function in a feed-forward loop in which hyperinsulinemia enhances central sympathetic output, which in turn further promotes hyperinsulinemia. In this manner, insulin and over-activation of the sympathetic nervous system interact to adversely affect insulin metabolic signaling and contribute to the development of metabolic syndrome and its associated complications [81].

What is possible to verify in the metabolic syndrome is a regional imbalance in favor of the parasympathetic in visceral territory (abdominal fat), and in favor of the sympathetic in the thoraco-muscular area (increased BP and insulin resistance; Fig. 5.35). The analysis of anatomical circuits involved in energy homeostasis illustrated in Fig. 5.29 provides the bases for the intertwined regulation of circadian and food intake mechanisms. Both networks, circadian and energy, are indirectly connected at the DMH and are sensitive to time and circadian nutritional information (Fig. 5.29). The NTS receives satiety signals from peripheral organs via vagal and sympathetic afferents and is innervated by PVN and LHA, which also receive signals from the circadian hypothalamic networks and energy [82].

Feeding time has demonstrable effects on weight gain [55]. In subjects who ate for a week one meal a day of approximately 2000 kcal. composed of 50% carbohydrates, 15% protein, and 35% fat, a weight loss of about 1 or 2 kg was observed if the food was taken in the morning. In contrast, none was observed if the food was ingested at 17:30 h. Overweight people often have increased hunger at night compared with those of normal weight; they also consume large proportions of their caloric intake in the evening. The food consumed at night induces an increase in low density lipoproteins (LDL) and decrease in high density liproteins (HDL), along with increased insulin resistance and higher levels of nocturnal glucose. In general, human late chronotypes (a) sleep 1 h less per night; (b) consumes more calories at dinner; (c) consumes more calories after 20:00 h; (d) have diets of lower quality (junk food, sugary drinks, fewer vegetables) [83].

24-h Rhythms in Plasma Osmolality and the Intravascular Volume

Both plasma osmolality and intravascular volume are the variables controlled by water and electrolyte intake behavior. The hypothalamus regulates this behavior through the supraoptic–PVN–neurohypophyseal system and AVP secretion, and via several other hormonal systems, including activation of the renin–angiotensin system and secretion of ANP [84].

The constancy of the internal medium composition is maintained primarily by controlling the intake and renal excretion of salt and water. The dehydration and the consequent need for water occur when there is a loss of water and/or an increase in effective solute, especially sodium. The role of sodium is indicated by the fact that a hypertonic solution of NaCl is more effective at raising water intake than equimolar solutions of other non-ionic solutes [85].

A large body of evidence demonstrates that the 24-h changes in blood volume and BP regulation largely depend on the interactions existing among the sympathetic nervous system, the renin–angiotensin system, and renal sodium excretion (Fig. 5.36) [86]. Twenty-four-hour rhythms in renal blood flow, glomerular filtration rate, urine volume, and urinary sodium, potassium, and chloride, with afternoon to early evening peak time in diurnally active persons, are well-known and persist independently of sleep, indicating their circadian nature [87]. Renal blood flow, vascular resistance, and glomerular filtration rate decline at night, although the decrease in urine flow, particularly in the non-elderly, is much more pronounced than expected. This indicates a 24-h periodicity of tubular reabsorption with nighttime peaks mediated by circadian rhythms of intrarenal angiotensin II and AVP [86].

Thirst is a feeling that motivates water consumption, and is triggered by cellular dehydration. A tiny increase in plasma osmolality of 1–2% can start neuroendocrine responses and the search for water. Changes in blood volume and extracellular fluid (ECF) pressure are also an important stimulus for the intake of water, but to a lesser extent, i.e., a reduction of about 10% of blood volume or BP is needed to induce the intake of water (Fig. 5.37) [88].

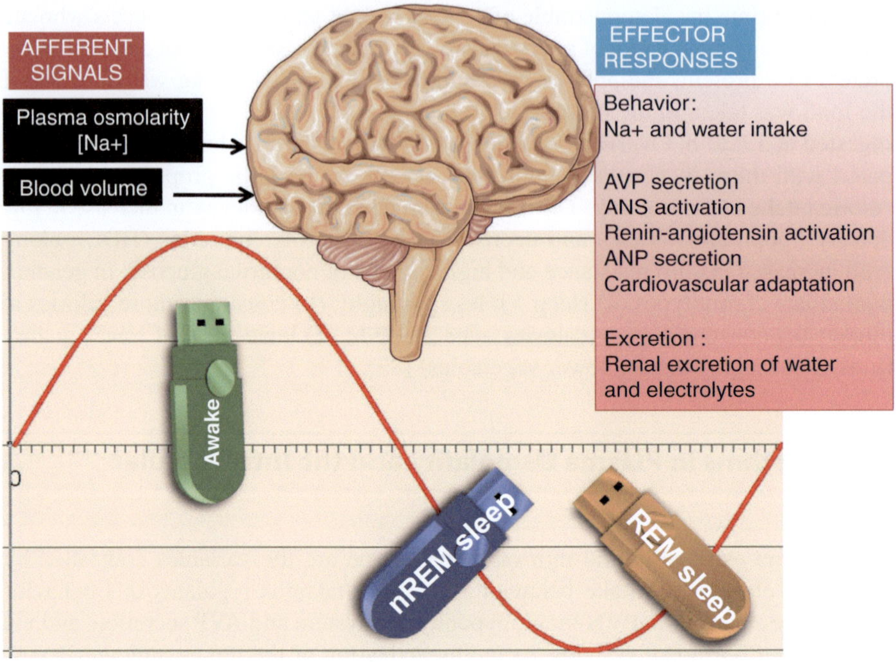

Fig. 5.36 Schematic representation of the mechanisms controlling water and salt intake. This control changes as a function of the three body configurations in a 24-h period. The figure was prepared in part using image vectors from Servier Medical Art (www.servier.com), licensed under the Creative Commons Attribution 3.0 Unported License (http://creativecommons.org/license/by/3.0/)

Osmoreceptors are highly specialized neurons, able to translate changes in external osmotic pressure into electrical signals. Current evidence supports the hypothesis that cells sensitive to changes in CSF osmolality or sodium concentration are located in the circumventricular organs in the anterior ventral region of the third ventricle (AV3V) and the area postrema (Fig. 5.37) [89]. Lesions of AV3V cause adipsia.

In addition to its location in the CNS, the sodium receptors are also present on afferent nerve terminals adjacent to the liver, kidney, and intestinal vessels. The increase in sodium concertation in the portal vein stimulates hepatic receptors leading via vagal afferents to the activation of neurons in the NTS; these, in turn, stimulate neural structures, which induce increased natriuresis and the inhibition of intestinal absorption of sodium [84].

When water is provided to an animal kept for 24–48 h with water restriction, in about 3–10 min its thirst is quenched, as if the body could predict the exact amount of water needed for the correction of osmolality simply by measuring the volume of water that passed through the mouth and the stomach (Fig. 5.37). In fact, stimuli generated in the mouth, throat or stomach are converted into afferent impulses to CNS structures involved in the corresponding integrative response to thirst [90].

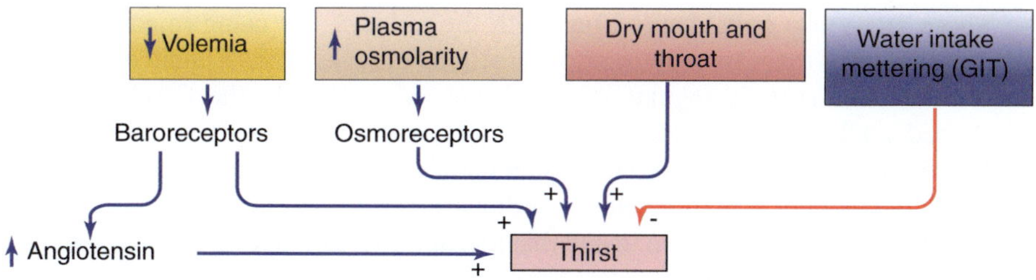

Fig. 5.37 Physiological components of the thirst response. Modified with permission from Cardinali [1]

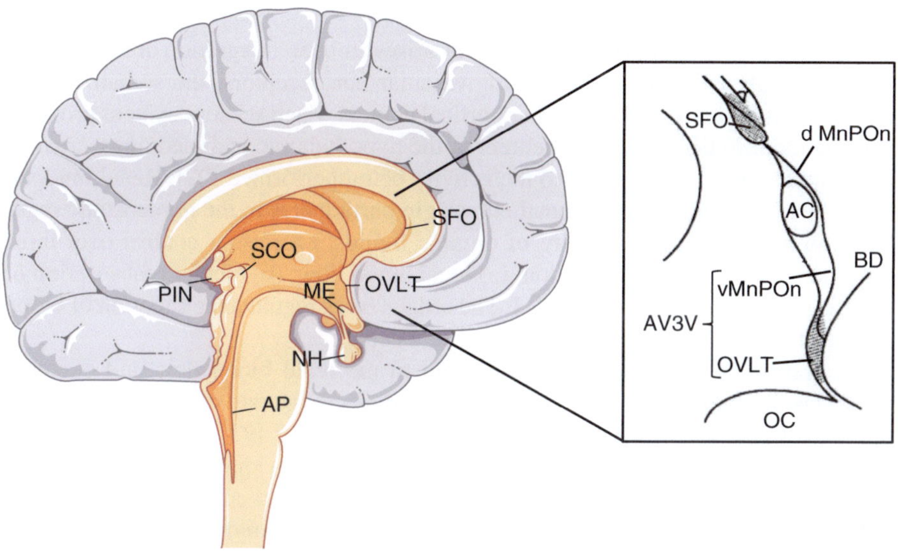

Fig. 5.38 *Left*: circumventricular organs. *AP* area postrema, *NH* neurohypophysis, *OVLT* organum vasculosus of the lamina terminalis, *SCO* subcommissural organ, *SFO* subfornical organ, *PIN* pineal gland. *Right*: the lamina terminalis is a forebrain structure that contains the SFO, the median preoptic nucleus (MnPOn), and the OVLT. The AV3V includes the ventral part of MnPOn (vMnPOn) and the OVLT. The AV3V region, the SFO, and the AP in the fourth ventricle contain neurons that are sensitive to changes in osmolality. *dMnPOn* dorsal median preoptic nucleus, *BD* diagonal band of Broca, *AC* anterior commissure, *OC* optic chiasm. The figure was prepared in part using image vectors from Servier Medical Art (www.servier.com), licensed under the Creative Commons Attribution 3.0 Unported License (http://creativecommons.org/license/by/3.0/)

The lamina terminalis is a forebrain structure containing the subfornical organ (SFO), the MnPO, and the OVLT (Fig. 5.38). The AV3V region includes the ventral part of the MnPO and the OVLT. AV3V, SFO, and the area postrema in the fourth ventricle contain neurons that are sensitive to changes in osmolality. These cells have direct projections to PVN and to the dorsal raphe and LC nuclei. These connections are important for conveying information involved in hypothalamic behavior to restore the balance of body fluids [85].

These areas also have connections with the kidneys. By using injections of a neurotropic virus that causes retrograde infections of rat kidneys, a chain of neurons involved in homeostatic regulation was identified, including the OVLT, MnPO, SFO, bed nucleus of the stria terminalis, periventricular anteroventral nucleus, SON, primary motor cortex, and the visceral area of the insular cortex [85].

The kidneys play a central role in cardiovascular homeostasis as they ensure a balance between the fluid taken in and that lost and excreted during everyday activities. This ensures the stability of extracellular fluid volume and maintenance of normal BP. Renal fluid handling is controlled via neural and humoral influences, with the former determining a rapid dynamic response to the changing intake of sodium, whereas the latter cause a slower longer-term modulation of sodium and water handling [6, 91].

Activity in the renal sympathetic nerves arises from an integration of information from the high and low pressure cardiovascular baroreceptors, the somatosensory and visceral systems, and the higher cortical centers. Each sensory system provides varying input to the autonomic centers of the hypothalamic and medullary areas of the brain at a level appropriate to the activity being performed [92]. Renal innervation considered entirely of sympathetic origin participates in the homeostatic regulation of volume and osmolality of organic liquids, exercising control over three important aspects of renal function: (a) renal blood flow; (b) tubular reabsorption of electrolytes; (c) renin secretion [91]. The renal nerves regulate the function of blood vessels, tubules, and juxtaglomerular granule cells.

Activation of renal nerves release renin to the renal blood flow and decrease urinary sodium excretion (Fig. 5.39). The basal discharge rate of renal sympathetic nerves is 0.5–2 Hz, causing a continuous release of NE [93]. Renal sympathetic nerves release renin (threshold, 0.5 Hz; β_1-adrenoceptor-mediated), increase tubular transport (threshold, 1 Hz; α_1-adrenoceptor-mediated), and constrict the renal vasculature (threshold, 2.5 Hz; α_1-adrenoceptor-mediated). Autonomic control of the kidney contributes to blood volume restoration following a positive or negative perturbation in volume status and helps balance the work load between the two kidneys. Renal efferent sympathetic nerve activity impairs sodium excretion and shifts the renal pressure–natriuresis curve to the right such that higher long-term levels of BP are required to maintain sodium excretion in balance with sodium intake [6].

In humans, the AVP involved in the water–salt balance is released from the neurohypophysis as several peaks overnight, characterized by an increase of 100–300% of the levels of the hormone in plasma. These peaks are of short duration, consistent with the short half-life of plasma AVP. No relationship was found between these episodes of AVP release and plasma sodium level, which remains constant throughout the night. Furthermore, no relationship was found between sleep stages and the levels of AVP. The activity and posture are all factors that can influence the daily AVP variations. However, in healthy subjects with constant recumbent position and fluid intake, a daily pattern of AVP remained, albeit with lower amplitude, proving the endogenous character of the rhythmicity (Fig. 5.40) [87].

One possible explanation for the nocturnal rise in AVP could be the decrease in BP pressure at these times. In this sense, there is a circadian pattern of circulating blood

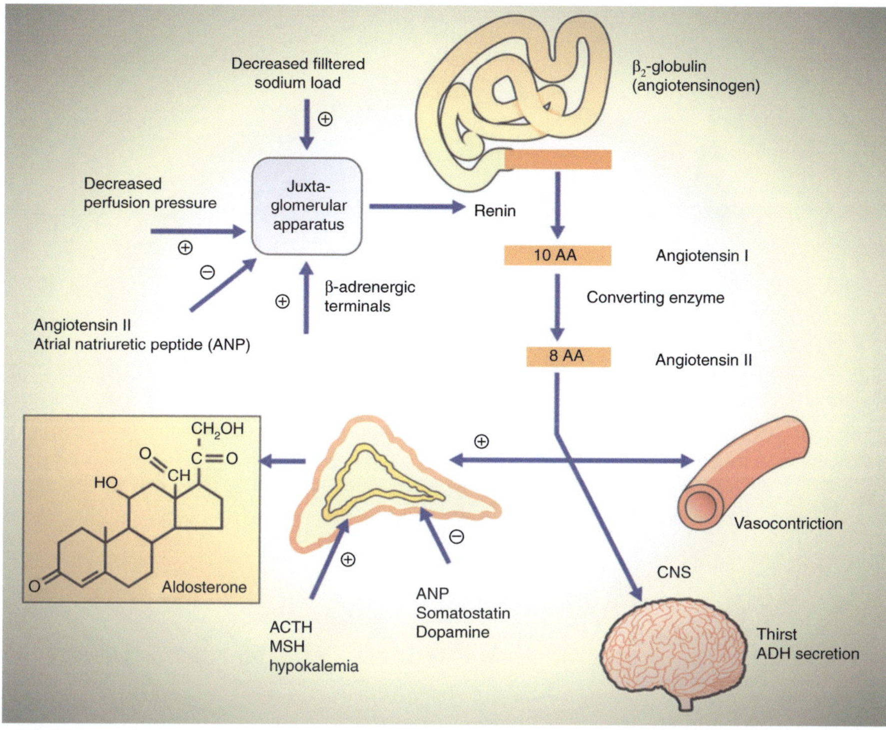

Fig. 5.39 The renin–angiotensin–aldosterone system with the different neural and endocrine interactions

volume, high during the day (12:00–18:00 h), which drops significantly (6%) at night and increases again in the morning (06:00 h). This circadian variation of blood volume could add to the BP changes to cause the increased release of AVP at night.

The amplitude of the AVP circadian rhythm declines with age. This in part explains the shift in the peak time of urine production and volume from afternoon/early evening in young persons to 00:00–0002.00 h in the elderly, commonly associated with nocturia. Another important cause of nocturia in the elderly is abnormal sleep-time BP decline, manifesting as nondipping 24-h patterning, and/or sleep-time hypertension [87].

The urine volume and electrolyte secretion is usually lower during sleep. REM sleep is associated with a decreased flow of urine and increased osmolarity. The activity of plasma renin and angiotensin levels and aldosterone are elevated during sleep (Figs. 5.40 and 5.41) [87]. ANP is elevated in untreated sleep apnea (by atrial impact of changes of negative intrathoracic pressure), resulting in nocturia and natriuresis.

These effects are mainly related to sleep rather than circadian factors. Thus, sleep deprivation inhibits the nocturnal increase in renin and aldosterone (Fig. 5.40). Renin increases during NREM sleep and decreases during REM sleep. In the transition from REM sleep to NREM sleep, renin levels rise [87].

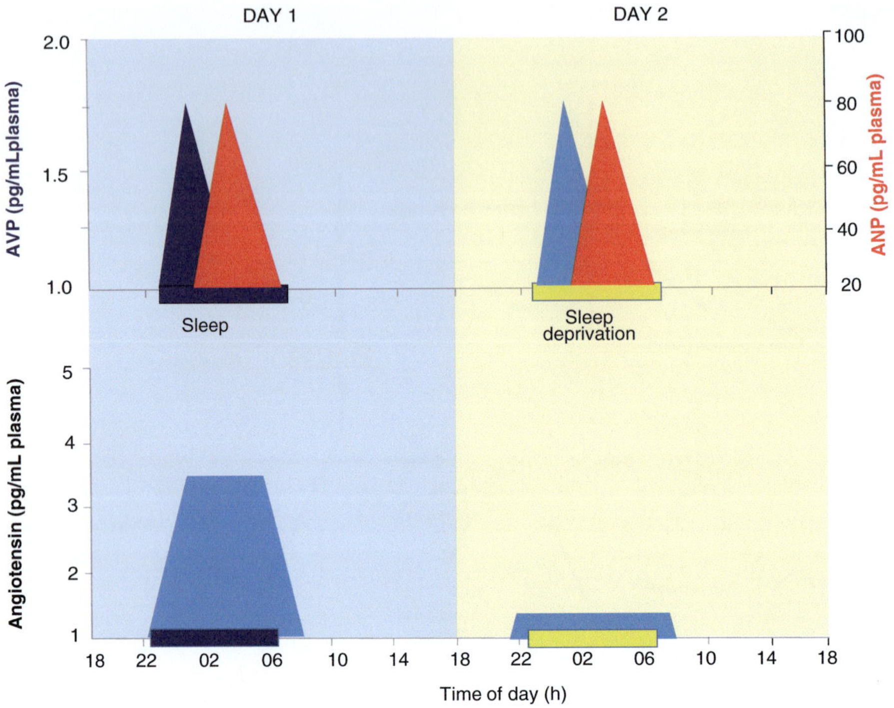

Fig. 5.40 Arginine vasopressin (AVP) and atrial natriuretic peptide (ANP) are hormones whose constitutive secretion is controlled by the C process (*upper panel*). Twenty-four hour rhythm in angiotensin production is dependent on the S process (*lower panel*). Ultradian variations in hormone secretion or values <1 pg/mL plasma (AVP, angiotensin) or <20 pg/mL plasma (ANP) are not represented. Data from Kamperis et al. [87]

The circadian rhythm of renal Na^+, K^+, and H_2O management is driven to a large extent by the circadian rhythm of aldosterone [86]. Nyctohemeral fluctuation of ANP and AVP, which both peak early during sleep, also modulates the 24-h rhythm of urinary Na excretion. The relationship between BP and natriuresis is controlled during the daytime by upright posture and activity; thus, it is mainly during the nighttime when Na^+ sensitivity, which is present to varying extents among all persons, most strongly exerts corrective effects. Thus, in acute and chronic situations when Na^+ intake is excessive or its daytime elimination compromised, the innate pressure natriuresis mechanism adjusts BP to an elevated level during nocturnal rest as a compensatory response, giving rise to abnormally elevated sleep-time BP, i.e., a nondipping 24-h pattern and/or nocturnal hypertension. This in turn promotes blood volume reduction through enhanced overnight natriuresis and diuresis [86].

Atrial natriuretic peptide belongs to a family of peptides that includes three other members, brain natriuretic peptide (BNP), C-type natriuretic peptide (CNP), and urodilatin, all encoded by different and independent genes. These peptides are present in mammals as potent protective agents against volume overload. ANP and BNP are produced mainly in the heart, but they can be found, together with their receptor

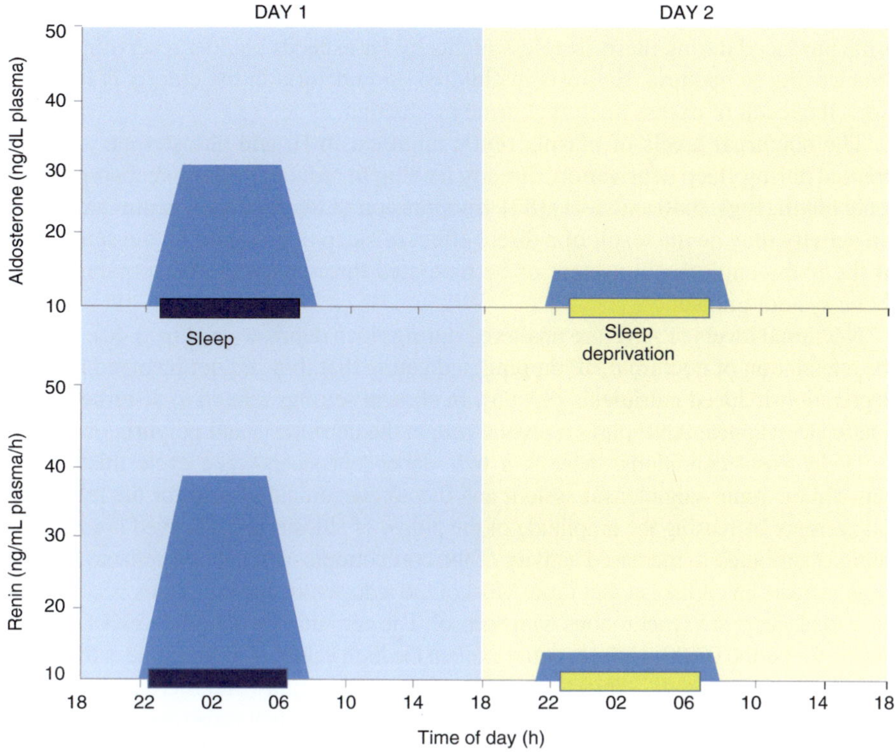

Fig. 5.41 The constitutive secretion of renin and aldosterone is dependent on the S process. Ultradian variations in hormone secretion or values <10 ng/dL plasma (aldosterone) or <10 ng/mL plasma (renin) are not represented. Data from Kamperis et al. [87]

type A, in hypothalamic and brainstem areas involved in the regulation of body fluid volume and BP [94].

Expansion of blood volume acts directly on the heart by stretching cardiomyocytes, increasing the release of ANP, and leading to an effective reduction of circulating blood volume. In addition, there is strong evidence that hypothalamic ANP is also released after the expansion of blood volume because of the increased activity of afferent pathways stimulated by baroreceptors.

ANP produced on a large scale in the heart reduces the strength of the heart contraction and the frequency of heart beats, causing relaxation in the vessels, decreased peripheral resistance, diuresis, natriuresis, and inhibition of salt and water intake [94]. Sleep deprivation markedly increased the diuresis and led to excess renal sodium excretion. Renal water handling and AVP and ANP levels remained unaltered during sleep deprivation, but the circadian rhythm of the hormones of the renin–angiotensin–aldosterone system is significantly affected (Figs. 5.40 and 5.41). Hemodynamic changes were characterized by the attenuation of nocturnal BP dipping and an increase in creatinine clearance. Acute deprivation of sleep induces natriuresis and osmotic diuresis, leading to excess urine production. The amount of urine produced during these sleepless nights by far exceeds bladder reservoir ability, thus leading to nocturia. Enuresis in children and nocturia in the elderly is in many cases the result of excess nocturnal urine production.

The nocturnal levels of plasma renin, angiotensin II, and aldosterone are suppressed during sleep deprivation, directly leading to reduced sodium reabsorption in renal tubuli (Figs. 5.40 and 5.41) [87]. Suppression of the rhythm of renin–angiotensin activity may be the result of a direct effect of sleep deprivation on the sensitivity of the renin–angiotensin system or be mediated through sympathetic–parasympathetic system imbalance.

Nocturnal levels of ANP are unaltered during sleep deprivation (Fig. 5.40), despite the attenuation of nocturnal BP dipping, indicating that they are not involved in sleep deprivation-induced natriuresis [87, 95]. In clinical settings related to disturbed sleep due to sleep apnea, ANP plays a pivotal role in the natriuresis and polyuria observed.

Under basal conditions, there is a time-dependent sleep–wake cycle relationship between the renin–angiotensin system and the adrenocortical system for the release of aldosterone. Increasing the amplitude of the pulses of aldosterone observed upon awakening is attributed to increased activity of the corticotropic–adrenal axis reflected by the large increase in cortisol at that time. Most of the aldosterone release pulses occurring at the end of sleep, are synchronous with cortisol. The permanence of high levels of cortisol during the period 07:00–15:00 h, could explain the high values of aldosterone at this time, regardless of whether the subject is asleep or awake. By contrast, aldosterone pulses during sleep periods are primarily related to major fluctuations in plasma renin [84].

DA is synthesized in the renal proximal tubule cells from filtered L-Dopa and secreted into the renal proximal tubule, where it binds to D_1 receptors inhibiting Na^+ reabsorption [96]. Renal DA acts as a paracrine substance that opposes the actions of angiotensin to increase Na^+ reabsorption via AT_1 receptors. DA affects renin release from renal juxtaglomerular cells via the D_1-like receptor family. Renal DA serves as one of several paracrine mediators in renal Na^+ excretion.

A reliable humoral biological marker of REM and NREM cycles is renin secretion. Their oscillations are strongly attached to the REM and NREM sleep cycles. NREM sleep coincides with an increase in plasma renin, whereas it decreases during REM sleep. Therefore, nights deprived of sleep are characterized by natriuresis, osmotic diuresis, and a dramatic increase in urine output. BP dipping is attenuated, and the renin–angiotensin system is clearly suppressed. It is of importance to evaluate sleep architecture and its disturbances in clinical settings with nocturnal polyuria and natriuresis, such as enuresis in children and nocturia in the elderly.

24-h Rhythms in Body Temperature Control

Humans maintain core temperature within a few tenths of a degree of 37 °C over a very wide range of environmental exposures and activity levels. During hyperthermia, heat dissipation occurs primarily via sympathetically mediated sweating and cutaneous vasodilation. During cold exposure, sympathetic cutaneous vasoconstriction helps to decrease heat dissipation; shivering increases heat generation when cooling is more extreme [97].

Centrally, thermoregulation is controlled at the preoptic/anterior hypothalamus, which acts as a thermostat by integrating central (brain) temperature with thermal information from peripheral afferents. Changes in body temperature thus elicit reflex responses for heat dissipation, heat conservation, and/or heat generation, as appropriate. These minimize changes in body temperature in classical negative feedback fashion and help to prevent the attainment of potentially dangerous body temperatures at either extreme [98]. Physiological thermoregulatory responses to heat in humans are much more efficient than responses to cold. Therefore, physiological thermoregulation is primary during hyperthermia.

Body temperature varies with the anatomical regions (e.g., axillary: 36.7 °C; scrotal: 32 °C; rectal: 37.2 °C) and has 24-h rhythm directly linked to the activity of the SCN, with a minimum in the middle of the night and a maximum toward the end of the day. The amplitude of this variation is about 0.6 °C (Fig. 5.42) [1].

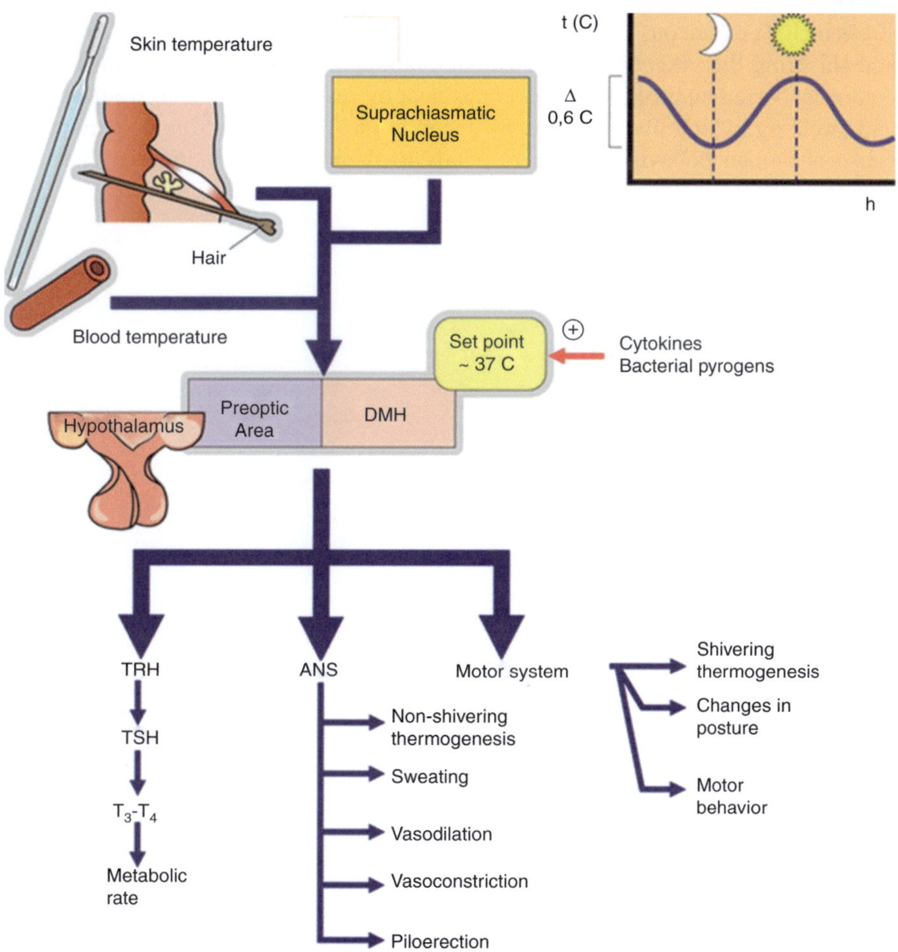

Fig. 5.42 Components of the temperature controlling mechanism

In two of the three physiological configurations in which human life elapses, i.e., wakefulness and NREM sleep, body temperature is regulated precisely. By contrast, in REM sleep, a transient disconnection of the autonomic regulatory mechanisms at the supraspinal level occurs, leading to a state of poikilothermia, that is, the central temperature tends to adapt to the ambient temperature. As an individual 75 years of age spends about 6 years in REM sleep, it can be said that we are not warm-blooded animals for an considerable part of our life. The precise mechanisms that we discuss are operational only during NREM sleep and wakefulness.

Temperature is a critical variable in health and disease. The constraint of human body core temperature within a degree or two of 37 °C, which is the optimal temperature for normal cellular function, has three causes. The first is the stable climate, which maintains temperatures across most of the surface of planet Earth within a range compatible with human life. The second cause is the ANS, which reacts robustly to thermal challenges by orchestrating a complex array of neural responses below the level of conscious awareness. The autonomic responses to cold stress include cutaneous vasoconstriction to retain bodily heat along with metabolic and shivering thermogenesis [99]. The autonomic responses to heat stress include cutaneous vasodilatation, which liberates heat by radiant and convective heat loss, and sweating, which liberates heat by evaporation. The third and least predictable cause is human behavior, which responds to thermal sensory input by seeking warmth or coolness (Figs. 5.42 and 5.43).

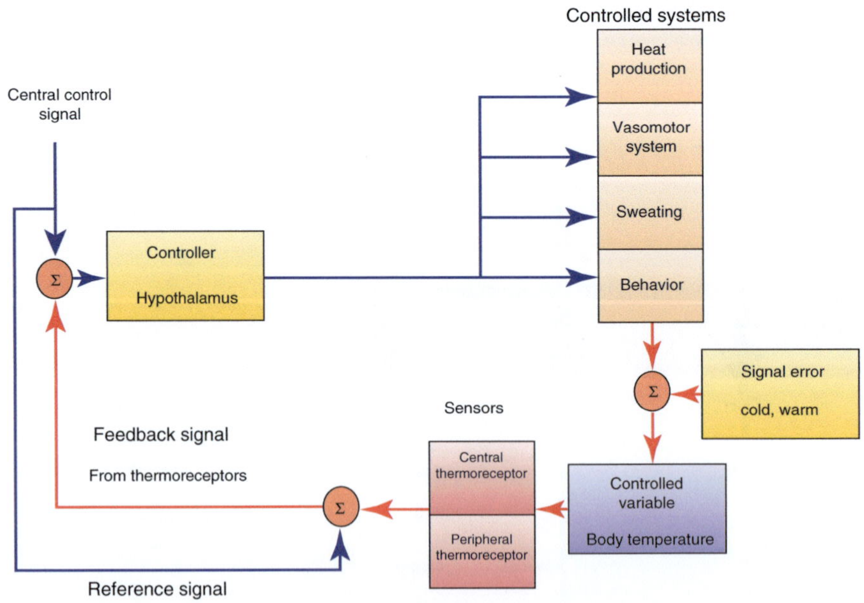

Fig. 5.43 Analysis of hypothalamic temperature control according to the systems control theory. Modified with permission from Cardinali [1]

Body temperature is controlled by central and peripheral adjustment and heat dissipation [97]. The analysis of body temperature regulation as a control theory is shown in Fig. 5.43. Maintaining core temperature is in the first instance given by vasomotor changes and only when this range is exceeded do mechanisms of heat production or dissipation start.

Heat production (thermogenesis) is under neural control. There are two types of thermogenesis, shivering and nonshivering. Shivering thermogenesis is induced by the caudal hypothalamus via projections to the brainstem nuclei of the somatic motor system. Nonshivering thermogenesis is controlled by the ANS via sympathetic β-adrenoceptor innervation of lipid deposits and involves heat production by a particular form of adipose tissue, the brown adipose tissue (BAT). The thermogenic potential of BAT is due to the presence of uncoupling protein 1, a protein uniquely found in the inner membrane of the brown adipocytes' mitochondria. The protein uncouples substrate oxidation from electron transport. In humans, BAT is found in subscapular, cervical, perispinal, mediastinal, periaortic, pericardial, and periadrenal regions [100, 101].

The sympathetic system is also responsible for vasomotor changes, which include changes in perfusion of the limbs and trunk, and sweat secretion by cholinergic terminals in the corresponding territory. Behavioral changes (looking for shelter, etc.) also contribute as an effector control system. Only in newborns are TRH release and subsequent activation of the pituitary–thyroid axis significant [99].

In addition to the cold-activated pathways (e.g., the thermoregulatory pathway), several hypothalamic nuclei have been linked to the control of BAT thermogenesis to allow diet-induced thermogenesis (e.g., the energy homeostasis regulatory pathway; Fig. 5.44). These nuclei include the ARC, preoptic area (POA), DMH, paraventricular hypothalamus, LHA, and VMH [97, 102].

Regarding temperature receptors involved in thermoregulatory behavior a dual mechanism exists, i.e., peripheral and central thermoreceptors. There are two types of peripheral thermoreceptors: (a) cold, with optimal activation from 10° to 3 °C; (b) heat, with activation above 45 °C. They are free terminations of sensory fibers such as nociceptive A δ and C fibers for cold and C fibers for heat. Temperature-sensing receptors belong to the superfamily of transient receptor potential (TRP) channels, which are located in the nerve endings of sensory cells throughout the skin [99]. The interoceptive aspects of behavioral thermoregulation have been emphasized, including the primary importance of skin temperature, the concept of thermal discomfort, and the important contribution of orbitofrontal, insular, somatosensory, and amygdala cortical regions deployed to anticipate and avoid thermal stress.

Central thermoreceptors are located at various levels of the CNS, including the spinal cord, reticular formation, and the POA itself. Most thermoreceptors in the spinal cord are heat-sensitive neurons (50–70% of neurons), whereas only 2% of them are sensitive to cold. These neurons are nonspecific, and respond to changes in osmolarity, BP, glucose, and sexual steroids (Fig. 5.44).

The signals from the peripheral receptors enter the CNS through the dorsal root of the spinal cord and send their information to the upper levels of thermal integration by the contralateral paleospinothalamic tract (Fig. 5.44). Projections of

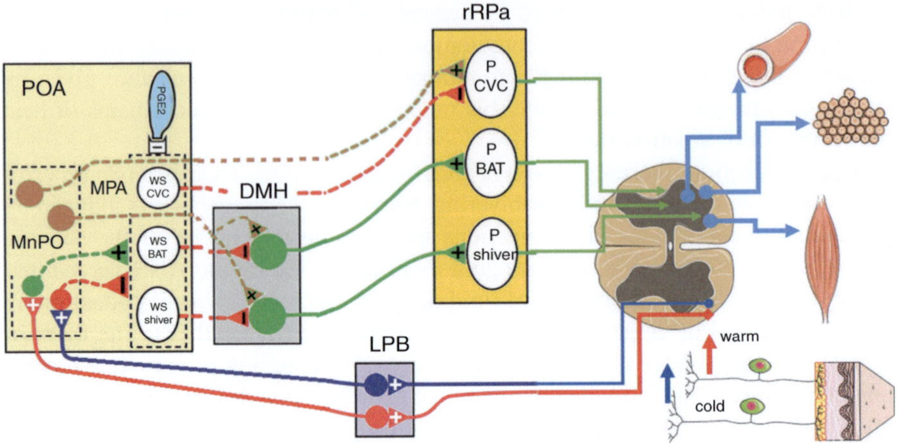

Fig. 5.44 Thermosensory signals driving thermoregulatory responses are transmitted from the lateral parabrachial nucleus (LPB) to the preoptic area (POA), which contains the microcircuitry through which cutaneous and core thermal signals are integrated to regulate the balance of POA outputs that are excitatory (*dashed green*) and inhibitory (*dashed red*) to thermogenesis-promoting neurons in the DMH. Other excitatory median preoptic area (MnPOA) neurons (*dashed brown*) either project to the cutaneous vasoconstriction (CVC) sympathetic premotor neurons in the rostral raphe pallidus (rRPa) or to DMH. Within the POA, GABAergic interneurons (*red*) in the MnPO subnucleus receive glutamatergic inputs from skin-cooling-activated neurons in the LPB and inhibit each of the distinct populations of warm-sensitive (W-S) neurons in the medial preoptic area (MPA) that control CVC, BAT, and shivering. Prostaglandin E2 binds to receptors to inhibit the activity of each of the classes of W-S neurons in the POA. Redrawn from Morrison [97]. The figure was prepared in part using image vectors from Servier Medical Art (www.servier.com), licensed under the Creative Commons Attribution 3.0 Unported License (http://creativecommons.org/license/by/3.0/)

peripheral thermoreceptors from different areas of the skin converge on the PBN at the brainstem from where it projects to the POA of the anterior hypothalamus via the medial forebrain bundle and periventricular striatum (Fig. 5.44).

There is a hierarchy of structures involved in thermoregulation, extending from the POA to the brainstem and spinal cord [97]. Tetraplegic patients who have suffered transection at the cervical level of the spinal cord can maintain their body temperature at around 37 °C, although present thermoregulatory instability when subjected to rapid changes in ambient temperature. In situations where the POA and the anterior hypothalamus are lesioned, fever in response to pyrogens is abolished.

Hypothalamic heating produces panting in anesthetized cats, and abolishes thermogenesis and produces peripheral vasodilation in dogs. All these mechanisms lead to a reduction in body temperature. The POA region is an important site for thermoregulation, being situated on top of the hierarchy for neuronal regulation of body temperature [97].

Figure 5.44 summarizes the neuronal pathways involved in thermoregulation. Retrograde transneuronal viral tracing has been of paramount importance in delineating the brain regions and the neuronal circuits connected to BAT and white adipose tissue (WAT). POA output is excitatory or inhibitory to thermogenesis-promoting neurons in the DMH and to cutaneous vasoconstriction sympathetic premotor neurons in the rostral raphe pallidus (rRPa). As already mentioned, the DMH plays a wide range of metabolic and behavioral roles, including body weight regulation [66].

The BAT and WAT send feedback information to the CNS via sensory nerves that connect adipocytes via dorsal root ganglia with the brain [103–105]. Incoming (afferent) sympathetic nerves can be distinguished from outgoing (efferent) sensory nerves with multisynaptic anterograde (*Herpes* virus) and retrograde (pseudorabies virus) viral tracers that are injected into the BAT or WAT. Many CNS sites showed both sympathetic and sensory connectivity; thus, an extensive feedback system for incoming and outgoing signals is likely.

Thermogenesis of BAT occurs as part of the basic rest–activity cycle (BRAC) and contributes to increases in body and brain temperature [105]. With ad libitum food, eating begins 15 min after the onset of BAT thermogenesis in rats. The initiation of eating is centrally programmed, and is a component of the BRAC. This increase in brain temperature that precedes eating may facilitate the cognitive processing that occurs during the search for food, when the rat engages with the external environment. Rather than being triggered by changes in levels of body fuels or other meal-associated factors, in sedentary laboratory rats with ad libitum access to food, meal initiation normally occurs as part of the centrally programmed ultradian BRAC. BRAC-associated BAT temperature increases occur in a thermoneutral environment; thus, they are not preceded by falls in body or brain temperature as a homeostatic thermoregulatory response [105].

Rats with hypothyroidism, despite having the normal diet-induced thermogenesis, are unable to survive in cold environments owing to a reduced capacity to produce heat without trembling. In addition, they also fail to increase thermogenesis in BAT in response to the infusion of NE. As discussed in Chap. 4, the SCG innervation of the thyroid gland plays a substantial role in the normal adaptive response to cold stress.

Tremor consisting of involuntary rhythmic movements of the skeletal muscles is seen in response to exposure to cold, without any change in body position. As there is no mechanical work, almost all the energy is released as heat [106]. Almost all body muscles participate in this response, except the middle ear, facial, perineal, and extraocular muscles. Premotor neurons in the rRPa play a substantial role in the efferent control of shivering thermogenesis. These, in turn, project to motoneurons in the ventral horn. The spinal cord seems to contain the basic mechanisms for these movements to occur, because cooling causes tremor in spinal animals. The ANS exerts a fine control of shivering, e.g., NE increases the sensitivity of skeletal muscle fiber to acetylcholine (ACh).

Concerning the cutaneous blood flow, the increase or reduction of skin blood flow, especially in the hands, feet, lips, ears, and nose, can facilitate or hinder the loss of heat to the environment respectively. Exposure to cold causes cutaneous vasoconstriction via NE effects on α1-adrenoceptors in arterioles and arteriovenous anastomoses, which supply the venous plexus of the skin. On the other hand, during exposure to heat, vasodilation in these regions is largely the consequence of vasoconstrictor activity removal [99].

The importance of evaporative heat loss becomes larger as the environmental heat load increases, and is the only means of heat loss when the ambient temperature rises above body temperature. Cooling by evaporation can be achieved in different ways, depending on the species: sweating in humans and cattle; panting in the dog, the sheep, and lizard; or salivation and licking in rats and kangaroo. The sweat glands respond to heat stress by sympathetic cholinergic stimulation (secretion is blocked by atropine), although they also have detectable adrenoceptors. The most potent stimulus for inducing sweat is the increase in body temperature; however, there is a modulation of the average skin temperature of the whole body and the location of the response.

Pyrogen production during bacterial infection, e.g., cytokines such as IL-1, affects the CNS in areas outside the blood–brain barrier, with increased "set-point" temperature. The absence of estrogen in perimenopausal women produces the typical hot flashes, because the thermostable zone is narrowed in the absence of estrogen (Fig. 5.45).

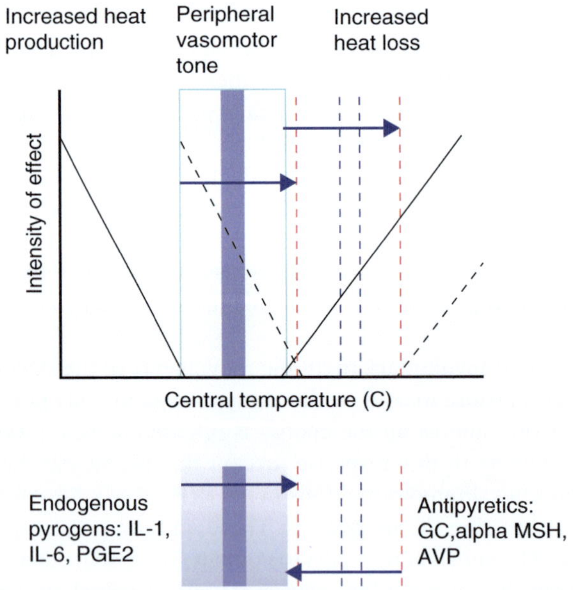

Fig. 5.45 Pyrogens move the equilibrium point of the system to the right, with increasing central temperature. Endogenous antipyretics (AVP; glucocorticoids, GC; α-MSH) or pharmacological agents re-establish the equilibrium point. The thermostable zone (*light blue*) is the central temperature range in which changes in central temperature occur without neurovegetative impact. Perimenopausal lack of estrogen reduces this thermostable zone (*dark blue*); thus, minimal changes in central temperature produce ANS symptoms ("hot flashes"). Modified with permission from Cardinali [1]

Sexual and Maternal Behavior

In the hypothalamus, there are neural command groups for sexual and maternal behaviors. For example, the same group of GnRH neurons that regulates the release of pituitary gonadotropins, and thus the central event of the sexual cycle, project to the limbic system, which regulates sexual behavior. Hence, the same group of command neurons is regulating the various components, endocrine (pituitary and gonadal hormone release), regional (erection, orgasm), and motivational (libido) of sexual behavior. A comparable case is that of oxytocin, which acts as a hormone in lactation (milk ejection reflex) and as a transmitter in the limbic pathways to induce maternal behavior.

Sexual function requires interaction of multiple levels and areas of the central and peripheral somatic and ANS. The medial preoptic area (mPOA), the ventral medullary reticular formation, the nucleus paragigantocellularis in the ventral medulla, the periaqueductal gray, the mesencephalic ventral tegmental area, neurons in the central tegmental mesencephalic region, and the medial amygdala are the main central areas involved in arousal and sexual responses [107]. Central neurotransmitters include DA (enhances sexual desire, arousal), NE (gates sensory input from the genitalia and maintains sexual arousal), ACh (mediates lubrication and vaginal engorgement), His, 5-HT (diminishes excitatory effects, interferes with arousal and orgasm), PRL (causes sexual satiety and post-orgasmic relief), and oxytocin (promotes sexual receptivity and bonding). Peripheral sympathetic and parasympathetic pathways are similar in men and women.

Hormones influence female sexual function. Progesterone furthers partner receptivity, and estrogens enhance desire, arousal, sensory thresholds, and genital arterial blood flow. Testosterone helps to initiate sexual activity [107].

Gonadotropic axis rhythms cover a wide range of frequencies, whose interaction provides a temporary program coordinated for the reproductive axis at each stage of maturation. The secretion of gonadotropins occurs in erratic pulses in the absence of hypothalamic regulation (Fig. 5.4). Thus, the pulse frequency of GnRH regulates the expression of GnRH receptors in the pituitary [12].

Neurons secreting GnRH are located within the mPOA and serve as the final output pathway regulating the LH and FSH surge [108]. GnRH neurons send axons to the median eminence of the hypothalamus where they release GnRH into the pituitary portal system, thereby triggering LH secretion and ovulation (Fig. 5.6). Females, but not males, can produce an LH surge even though there is no sex difference in the GnRH neurons themselves. The difference in ability to generate a GnRH/LH surge is believed to be upstream of the GnRH neurons and is the result of organizational processes shaped by gonadal steroid exposure during neonatal development. The circadian timing system (by acting on sexually differentiated neurons of the AVPV) regulates the dynamics of the neural circuits leading to the rhythmic generation of the GnRH/LH surge and ultimately ovulation (Fig. 5.6).

Studies in patients with defective hypothalamic GnRH secretion confirmed the close correlation between the pulsatile LH response and exogenous GnRH pulses. The synthesis and secretion of LH and FSH are differentially regulated by the frequency of the pulses of GnRH. Thus, a low frequency of GnRH pulses preferentially stimulates FSH, whereas more frequent GnRH pulses are needed for the optimal stimulation of LH [108].

The AVPV) sits upstream of GnRH neurons and is characterized by multiple sex differences (Fig. 5.6). Lesions of the AVPV prevent spontaneous and steroid-induced preovulatory surges of LH. Females have more AVPV neurons than males, and a denser projection from the AVPV) to the GnRH neurons. One sex difference that has been unequivocally implicated in the generation of the LH surge is the expression of the kisspeptin gene in AVPV neurons [14, 109]. Females have up to 25 times more AVPV kisspeptin-expressing) neurons than males. Moreover, both AVPV KiSS1 and C-FOS expression levels increase during the preovulatory LH surge. Treatment with estradiol produces a surge of LH in females but not males, and central infusion of kisspeptin receptor (Kiss1 R) antagonist into the mPOA blocks the estradiol-induced LH surge and estrous rhythmicity [14, 109].

The sex difference in AVPV kisspeptin) results from exposure to hormones at critical times during development, suggesting that hormones might have organizational effects on the AVPV in early development. Females treated postnatally with testosterone or estradiol have fewer kisspeptin neurons than untreated females and they are unable to produce the estradiol-induced LH surge in adulthood. Further, males who undergo castration shortly after birth have greater numbers of AVPV)kisspeptin neurons than intact males, and they can produce an estradiol-induced surge of LH [14].

A subset of neurons within the DMH express gonadotropin-inhibiting hormone (GnIH), a neuropeptide that has been shown to act as an inhibitor of the reproductive axis (Fig. 5.6) [110]. These neurons project directly to the GnRH neurons of the mPOA. In all mammals studied to date, GnIH rapidly suppresses LH release. The SCN sends projections to a large portion of GnIH cells, suggesting the circadian regulation of this population of neurons [14].

The effects of the circadian sleep schedule and secretion of LH and FSH are seen more clearly during puberty than in adulthood [111]. Before puberty, gonadotropins are secreted in pulses of low amplitude throughout the day and night. When children of both sexes approach puberty, the amplitude of night pulses gradually increases, and a daily hormonal rhythm becomes evident, with high levels of LH and FSH overnight [111].

Deep sleep, rather than REM or lighter sleep stages, provides a critical stimulus for LH secretion in puberty. The causal relationship between deep sleep and GnRH pulse generation is supported by neuroanatomical evidence for a direct synaptic connection between VLPO and GnRH neurons. The absence of GnRH axonal fibers in the VLPO is compatible with the hypothesis that sleep stimulates GnRH secretion, rather than the vice versa.

Going from puberty to adulthood, the amplitude of the pulses in daylight hours increases, thus eliminating the daily rhythm of LH and FSH. In girls at the age of 16 years, a characteristic LH pattern of adult women is established at the beginning of the follicular phase [8].

In studies during puberty in males, in which a change of 12 h in the sleep–wake cycle was imposed, it was shown that elevated night levels of LH, partly represent a stimulating effect of sleep itself. However, after reversing the hours of sleep, increased LH release takes place not only during the daytime sleep, but also the during the hours when sleep should take place under normal conditions, indicating the influence of circadian rhythmicity in the temporal pattern of LH release. The role of the circadian control in the nocturnal LH surge is also evident during sleep deprivation, maintaining that elevation, but on a smaller scale (Fig. 5.46) [9, 112].

In women, daily LH pulsatility changes are subject to complex modulation by the menstrual cycle. At the beginning of the follicular phase, LH pulses are large and infrequent, the night period being associated with a reduction in the frequency of the pulses; thus, the average levels of LH decrease during sleep. In the middle of the follicular phase, the pulse amplitude is decreased and the effects of sleep on modulating the frequency of the pulses is less noticeable. At the end of the follicular phase pulse amplitude increases, but sleep modulation is not present until the beginning of the luteal phase, resulting night pulsatility becoming slower [113, 114]. It is not known whether night variation in LH pulsatility in women is dependent on the sleep–wake cycle and/or the circadian pacemaker cycle.

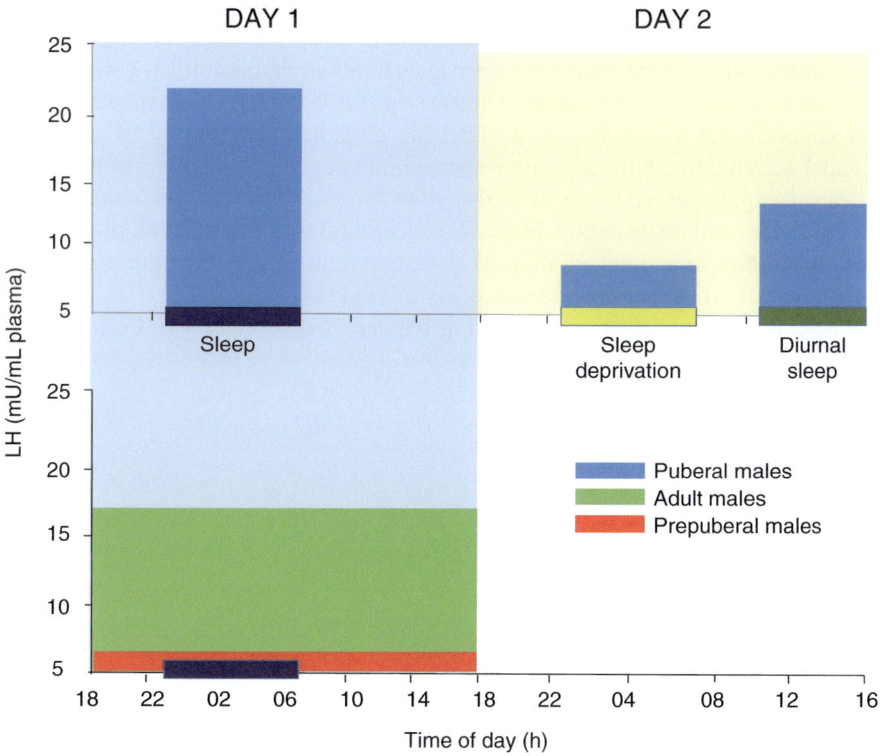

Fig. 5.46 Daily variations of plasma LH concentration in normal prepubertal, pubertal, and adult male subjects. Only puberal males showed nocturnal maxima that were partially impaired by sleep deprivation with a shift to diurnal sleep on the second day. The ultradian variations and values <5 mU/mL are not represented. Data from Van Cauter and Refetoff and Brabant et al. [9, 112]

An interaction between the menstrual cycle and circadian rhythmicity also appears to be involved in the temporal pattern of pre-ovulatory LH release, which occurs most often at the end of sleep or early in the morning. Toward the menopause, levels of gonadotropins increase their pulsatility, giving rise to high values without a consistent daily variation. In postmenopausal women, significant gonadotropin pulses, possibly reflecting pituitary secretion, are associated with large pulses of LH and FSH [115].

During puberty in children, night testosterone elevation coincides with increased serum levels of gonadotropins. In girls, puberty coincides with a circadian variation, with higher estradiol levels during the day than at night. Testosterone secretion is also pulsatile in adults, with 17–18 pulses detected per day, approximately coinciding with LH. A daily rhythm is also observed, with an amplitude of 25% of the mean value. It presents a night increase, starting shortly after midnight, reaching the maximum value at the beginning of the morning (08:00 h), and minimum values between 19:00 and 21:00 h (Fig. 5.47) [112].

As the secretory daily patterns of LH in adults are highly variable and do not show a clear nocturnal rise, in contrast to the stable testosterone circadian rhythm, it seems clear that other factors are involved. Ovarian innervation, which has a clear regulatory function, can play an important role in this phenomenon (Chap. 4).

In adults, the modulatory role of sleep and circadian rhythmicity in gonadotropic function is unclear. The nocturnal increase seems to be independent of sleep, a fact that supports the hypothesis of an intrinsic circadian rhythmicity of pituitary–gonadal activity. In adults, a positive correlation between the number of REM episodes and night average levels of testosterone occurs. With age, the frequencies of the LH pulses and testosterone decrease, canceling out most circadian rhythmicity. The morning plasma concentration of testosterone in elderly men correlated with the amount of nighttime sleep measured by PSG [116].

There is evidence for the relationship between the reproductive cycle and the secretion of oxytocin. In animal experiments, it was found that estrogens increase

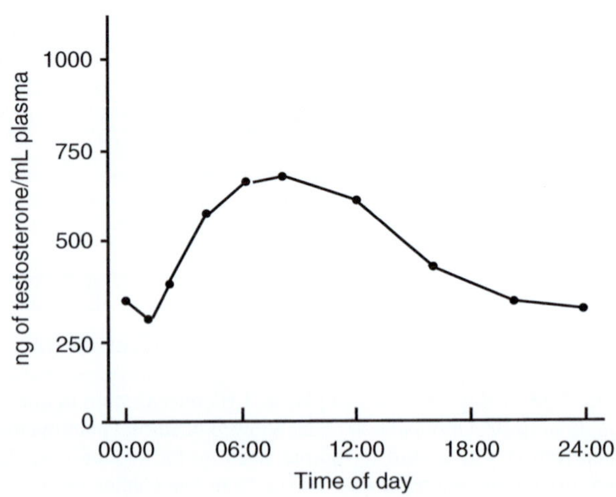

Fig. 5.47 Plasma testosterone levels in normal adult males. Redrawn from Brabant et al. [112]

the release of oxytocin, whereas progesterone inhibits it. Studies in women indicate that oxytocin levels vary throughout the menstrual cycle, characterized by a peak associated with ovulation. Although in the middle of the cycle, levels of oxytocin in the corpora lutea are far superior to those of the peripheral circulation, the data suggest that during the follicular phase, this hormone comes mainly from the pituitary gland and not from the ovaries. Moreover, in animal studies, it was found that the response to gonadal steroids creates a yearly cycle of oxytocin, which matches the corresponding seasonal periodicity of the estrous cycle [117].

In oxytocin-releasing neurons, there is evidence for the relationship between synchronous electrical activity and the pulsatile release of the hormone [118]. There is a positive correlation between the frequency of the action potential of hypothalamic neurons and the amount of oxytocin released from their axon terminals located in the posterior lobe of the pituitary. The pulsatile release of oxytocin, with rapid release and short half-life, often renders plasma levels of this hormone undetectable, even during stimulated secretion.

The transformation from a nonparental to a maternal state involves several dramatic and wide-ranging alterations, including changes in the CNS, behavior, and physiology. An interplay between the neuroendocrine system, including estradiol, progesterone, and PRL, and CNS neuromodulators, including oxytocin, DA, and AVP, helps to orchestrate multiple maternal functions [117]. Although hormonal changes occurring throughout pregnancy and at the time of parturition have been demonstrated to prime the maternal brain and trigger the onset of mother–infant interactions, extended experience with neonates can induce similar behavioral interactions [119].

Oxytocin is known to facilitate maternal behavior in many species. For example, knockout and pharmacological studies in mice suggested that oxytocin might facilitate maternal behavior, whereas a reduction in oxytocin function promoted infanticidal behavior. However, those approaches often produced global increases or decreases in oxytocin function, affecting multiple brain sites, and likely multiple oxytocin functions [118, 120].

Prolactin is known for its role in promoting maternal behavior in mammals and in promoting parental care in males and females of biparental bird species [121]. PRL regulates the onset of maternal behavior but not the maintenance, which is controlled by pup exposure. The role of PRL in mammalian paternal behavior is subtler and possibly more species-specific than for some other hormones. It has been suggested that PRL might be involved in the transition from a nonpaternal to a paternal state in male mammals [121].

References

1. Cardinali DP. Neurociencia Aplicada. Sus Fundamentos. Buenos Aires: Editorial Médica Panamericana; 2007.
2. Stuber GD, Wise RA. Lateral hypothalamic circuits for feeding and reward. Nat Neurosci. 2016;19:198–205.
3. Takahashi C, Hinson HE, Baguley IJ. Autonomic dysfunction syndromes after acute brain injury. Handb Clin Neurol. 2015;128:539–51.

4. Wehrwein EA, Joyner MJ. Chapter 8 – Regulation of blood pressure by the arterial baroreflex and autonomic nervous system. In: Buijs RM, Swaab D, editors. Handbook of Clinical Neurology. Autonomic Nervous System. New York: Elsevier; 2013. p. 89–102.
5. James C, Macefield VG, Henderson LA. Real-time imaging of cortical and subcortical control of muscle sympathetic nerve activity in awake human subjects. Neuroimage. 2013;70:59–65.
6. Zheng H, Patel KP. Integration of renal sensory afferents at the level of the paraventricular nucleus dictating sympathetic outflow. Auton Neurosci. 2016;204:57–64.
7. Shenton FC, Pyner S. Vagal afferents, sympathetic efferents and the role of the PVN in heart failure. Auton Neurosci. 2016;199:38–47.
8. Cardinali DP. In: Pandi-Perumal SR, editor. Neuroendocrine correlates of sleep/wakefulness. New York: Springer; 2006.
9. Van Cauter E, Refetoff S. Multifactorial control of the 24-hour secretory profiles of pituitary hormones. J Endocrinol Invest. 1985;8:381–91.
10. Copinschi G, Spiegel K, Leproult R, Van CE. Pathophysiology of human circadian rhythms. Novartis Found Symp. 2000;227:143–57.
11. Herbison AE. Control of puberty onset and fertility by gonadotropin-releasing hormone neurons. Nat Rev Endocrinol. 2016;12:452–66.
12. Maggi R, Cariboni AM, Marelli MM, Moretti RM, Andre V, Marzagalli M, Limonta P. GnRH and GnRH receptors in the pathophysiology of the human female reproductive system. Hum Reprod Update. 2016;22.
13. Clarke SA, Dhillo WS. Kisspeptin across the human lifespan:evidence from animal studies and beyond. J Endocrinol. 2016;229:R83–98.
14. Putteeraj M, Soga T, Ubuka T, Parhar IS. A "timed" kiss Is essential for reproduction: lessons from mammalian studies. Front Endocrinol (Lausanne). 2016;7:121.
15. Steyn FJ, Tolle V, Chen C, Epelbaum J. Neuroendocrine regulation of growth hormone secretion. Compr Physiol. 2016;6:687–735.
16. Copinschi G, Van OA, L'Hermite-Baleriaux M, Szyper M, Caufriez A, Bosson D, L'Hermite M, Robyn C, Turek FW, Van CE. Effects of the short-acting benzodiazepine triazolam, taken at bedtime, on circadian and sleep-related hormonal profiles in normal men. Sleep. 1990;13:232–44.
17. Spiegel K, Leproult R, Van CE. [Impact of sleep debt on physiological rhythms]. Rev Neurol (Paris). 2003;159:6S11–20.
18. Copinschi G, Caufriez A. Sleep and hormonal changes in aging. Endocrinol Metab Clin North Am. 2013;42:371–89.
19. Nicolaides NC, Charmandari E, Chrousos GP, Kino T. Circadian endocrine rhythms: the hypothalamic-pituitary-adrenal axis and its actions. Ann N Y Acad Sci. 2014;1318:71–80.
20. Tsang AH, Astiz M, Friedrichs M, Oster H. Endocrine regulation of circadian physiology. J Endocrinol. 2016;230:R1–R11.
21. Cardinali DP. Ma Vie en Noir. Fifty years with melatonin and the stone of madness. Switzerland: Springer; 2016.
22. Roelfsema F, Pereira AM, Keenan DM, Veldhuis JD, Romijn JA. Thyrotropin secretion by thyrotropinomas is characterized by increased pulse frequency, delayed diurnal rhythm, enhanced basal secretion, spikiness, and disorderliness. J Clin Endocrinol Metab. 2008;93:4052–7.
23. Leproult R, Copinschi G, Buxton O, Van Cauter E. Sleep loss results in an elevation of cortisol levels the next evening. Sleep. 1997;20:865–70.
24. Mazzoccoli G, Giuliani F, Sothern RB. A method to evaluate dynamics and periodicity of hormone secretion. J Biol Regul Homeost Agents. 2011;25:231–8.
25. Trivedi H, Szabo A, Zhao S, Cantor T, Raff H. Circadian variation of mineral and bone parameters in end-stage renal disease. J Nephrol. 2015;28:351–9.
26. Steiger A, Dresler M, Kluge M, Schussler P. Pathology of sleep, hormones and depression. Pharmacopsychiatry. 2013;46(Suppl 1):S30–5.
27. Tache Y, Brunnhuber S. From Hans Selye's discovery of biological stress to the identification of corticotropin-releasing factor signaling pathways: implication in stress-related functional bowel diseases. Ann N Y Acad Sci. 2008;1148:29–41.

28. Rasia-Filho AA. Is there anything "autonomous" in the nervous system? Adv Physiol Educ. 2006;30:9–12.
29. Bartness TJ, Liu Y, Shrestha YB, Ryu V. Neural innervation of white adipose tissue and the control of lipolysis. Front Neuroendocrinol. 2014;35:473–93.
30. McEwen BS, Gray JD, Nasca C. 60 years of neuroendocrinology: redefining neuroendocrinology: stress, sex and cognitive and emotional regulation. J Endocrinol. 2015;226:T67–83.
31. Contoreggi C. Corticotropin releasing hormone and imaging, rethinking the stress axis. Nucl Med Biol. 2015;42:323–39.
32. Zouikr I, Bartholomeusz MD, Hodgson DM. Early life programming of pain: focus on neuroimmune to endocrine communication. J Transl Med. 2016;14:123.
33. Schulz A, Vogele C. Interoception and stress. Front Psych. 2015;6:993.
34. Kanbara K, Fukunaga M. Links among emotional awareness, somatic awareness and autonomic homeostatic processing. Biopsychosoc Med. 2016;10:16.
35. Park CL. Meaning making in the context of disasters. J Clin Psychol. 2016;72:1234–46.
36. Kamphuis J, Lancel M, Koolhaas JM, Meerlo P. Deep sleep after social stress: NREM sleep slow-wave activity is enhanced in both winners and losers of a conflict. Brain Behav Immun. 2015;47:149–54.
37. Yamagata AS, Mansur RB, Rizzo LB, Rosenstock T, McIntyre RS, Brietzke E. Selfish brain and selfish immune system interplay: a theoretical framework for metabolic comorbidities of mood disorders. Neurosci Biobehav Rev. 2017;72:43–9.
38. Zorn JV, Schur RR, Boks MP, Kahn RS, Joels M, Vinkers CH. Cortisol stress reactivity across psychiatric disorders: a systematic review and meta-analysis. Psychoneuroendocrinology. 2016;77:25–36.
39. Herman JP, Tasker JG. Paraventricular hypothalamic mechanisms of chronic stress adaptation. Front Endocrinol (Lausanne). 2016;7:137.
40. Stern JE. Neuroendocrine-autonomic integration in the paraventricular nucleus: novel roles for dendritically released neuropeptides. J Neuroendocrinol. 2015;27:487–97.
41. Amihai I, Kozhevnikov M. The influence of Buddhist meditation traditions on the autonomic system and attention. Biomed Res Int. 2015;2015:731579.
42. Gold PW. The organization of the stress system and its dysregulation in depressive illness. Mol Psychiatry. 2015;20:32–47.
43. Maniam J, Morris MJ. The link between stress and feeding behaviour. Neuropharmacology. 2012;63:97–110.
44. Kaur J. A comprehensive review on metabolic syndrome. Cardiol Res Pract. 2014;2014:943162.
45. Fitzgerald PJ. Beta blockers, norepinephrine, and cancer: an epidemiological viewpoint. Clin Epidemiol. 2012;4:151–6.
46. Jansen L, Hoffmeister M, Arndt V, Chang-Claude J, Brenner H. Stage-specific associations between beta blocker use and prognosis after colorectal cancer. Cancer. 2014;120:1178–86.
47. Eng JW, Kokolus KM, Reed CB, Hylander BL, Ma WW, Repasky EA. A nervous tumor microenvironment: the impact of adrenergic stress on cancer cells, immunosuppression, and immunotherapeutic response. Cancer Immunol Immunother. 2014;63:1115–28.
48. Magnon C. Role of the autonomic nervous system in tumorigenesis and metastasis. Mol Cell Oncol. 2015;2:e975643.
49. Cole SW, Nagaraja AS, Lutgendorf SK, Green PA, Sood AK. Sympathetic nervous system regulation of the tumour microenvironment. Nat Rev Cancer. 2015;15:563–72.
50. Romeo HE, Colombo LL, Esquifino AI, Rosenstein RE, Chuluyan HE, Cardinali DP. Slower growth of tumours in sympathetically denervated murine skin. J Auton Nerv Syst. 1991;32:159–64.
51. Kalantaridou SN, Makrigiannakis A, Zoumakis E, Chrousos GP. Stress and the female reproductive system. J Reprod Immunol. 2004;62:61–8.
52. Boeckxstaens G. Chapter 11 – The clinical importance of the anti-inflammatory vagovagal reflex. In: Buijs RM, Swaab D, editors. Handbook of clinical neurology. Autonomic nervous system. New York: Elsevier; 2013, p. 119–34.
53. Chrousos GP. Stress and disorders of the stress system. Nat Rev Endocrinol. 2009;5:374–81.

54. Romeo HE, Cardinali DP. The autonomic nervous system of the cervical region as a channel of neuroendocrine communication. Front Neuroendocrinol 1991;12:278–97
55. Morton GJ, Meek TH, Schwartz MW. Neurobiology of food intake in health and disease. Nat Rev Neurosci. 2014;15:367–78.
56. Sohn JW, Elmquist JK, Williams KW. Neuronal circuits that regulate feeding behavior and metabolism. Trends Neurosci. 2013;36:504–12.
57. Munzberg H, Qualls-Creekmore E, Yu S, Morrison CD, Berthoud HR. Hedonics act in unison with the homeostatic system to unconsciously control body weight. Front Nutr. 2016;3:6.
58. García-García F, Juárez-Aguilar E, Santiago-García J, Cardinali DP. Ghrelin and its interactions with growth hormone, leptin and orexins: implications for the sleep wake cycle and metabolism. Sleep Med Rev. 2014;18:89–97.
59. Benarroch EE. Neural control of feeding behavior: overview and clinical correlations. Neurology. 2010;74:1643–50.
60. Mithieux G. Metabolic effects of portal vein sensing. Diabetes Obes Metab. 2014;16(Suppl 1):56–60.
61. Burdakov D, Luckman SM, Verkhratsky A. Glucose-sensing neurons of the hypothalamus. Philos Trans R Soc Lond B Biol Sci. 2005;360:2227–35.
62. Anderson EJ, Cakir I, Carrington SJ, Cone RD, Ghamari-Langroudi M, Gillyard T, Gimenez LE, Litt MJ. 60 years of POMC: regulation of feeding and energy homeostasis by alpha-MSH. J Mol Endocrinol. 2016;56:T157–74.
63. Joly-Amado A, Cansell C, Denis RG, Delbes AS, Castel J, Martinez S, Luquet S. The hypothalamic arcuate nucleus and the control of peripheral substrates. Best Pract Res Clin Endocrinol Metab. 2014;28:725–37.
64. Sobrino CC, Perianes CA, Puebla JL, Barrios V, Arilla FE. Peptides and food intake. Front Endocrinol (Lausanne). 2014;5:58.
65. Biebermann H, Kuhnen P, Kleinau G, Krude H. The neuroendocrine circuitry controlled by POMC, MSH, and AGRP. Handb Exp Pharmacol. 2012:47–75.
66. Seoane-Collazo P, Ferno J, Gonzalez F, Dieguez C, Leis R, Nogueiras R, Lopez M. Hypothalamic-autonomic control of energy homeostasis. Endocrine. 2015;50:276–91.
67. Schlogl H, Horstmann A, Villringer A, Stumvoll M. Functional neuroimaging in obesity and the potential for development of novel treatments. Lancet Diabetes Endocrinol. 2016;4:695–705.
68. Salem V, Dhillo WS. Imaging in endocrinology: the use of functional MRI to study the endocrinology of appetite. Eur J Endocrinol. 2015;173:R59–68.
69. Leibowitz SF, Akabayashi A, Wang J, Alexander JT, Dourmashkin JT, Chang GQ. Increased caloric intake on a fat-rich diet: role of ovarian steroids and galanin in the medial preoptic and paraventricular nuclei and anterior pituitary of female rats. J Neuroendocrinol. 2007;19:753–66.
70. Fang P, Yu M, Shi M, Zhang Z, Sui Y, Guo L, Bo P. Galanin peptide family as a modulating target for contribution to metabolic syndrome. Gen Comp Endocrinol. 2012;179:115–20.
71. Cermakian N, Westfall S, Kiessling S. Circadian clocks and inflammation: reciprocal regulation and shared mediators. Arch Immunol Ther Exp (Warsz). 2014;62:303–18.
72. Penicaud L. The neural feedback loop between the brain and adipose tissues. Endocr Dev. 2010;19:84–92.
73. Bailey SM, Udoh US, Young ME. Circadian regulation of metabolism. J Endocrinol. 2014;222:R75–96.
74. Van Cauter E, Blackman JD, Roland D, Spire JP, Refetoff S, Polonsky KS. Modulation of glucose regulation and insulin secretion by circadian rhythmicity and sleep. J Clin Invest. 1991;88:934–42.
75. Anothaisintawee T, Reutrakul S, Van CE, Thakkinstian A. Sleep disturbances compared to traditional risk factors for diabetes development: systematic review and meta-analysis. Sleep Med Rev. 2016;30:11–24.
76. Patton DF, Mistlberger RE. Circadian adaptations to meal timing: neuroendocrine mechanisms. Front Neurosci. 2013;7:185.
77. Cissé YM, Nelson RJ. Consequences of circadian dysregulation on metabolism. ChronoPhysiol Therapy. 2016;6:55–63.

78. Karthikeyan R, Marimuthu G, Spence DW, Pandi-Perumal SR, Bahammam AS, Brown GM, Cardinali DP. Should we listen to our clock to prevent type 2 diabetes mellitus? Diabetes Res Clin Pract. 2014;106:182–90.
79. Engeland WC, Arnhold MM. Neural circuitry in the regulation of adrenal corticosterone rhythmicity. Endocrine. 2005;28:325–32.
80. Buijs RM. Chapter 1 – The autonomic nervous system: a balancing act. In: Buijs RM, Swaab D, editors. Handbook of clinical neurology. Autonomic nervous system. New York: Elsevier; 2013, p. 1–11.
81. Johnson MS, DeMarco VG, Whaley-Connell A, Sowers JR. Chapter 64 – Insulin resistance and the autonomic nervous system. In: Biaggioni I, Burnstock G, Low PA, Paton JFR, editors. Primer on the autonomic nervous system. 3rd ed. San Diego: Academic; 2012. p. 307–12.
82. Tseng CJ, Cheng PW, Tung CS. Chapter 29 – Pharmacology of the nucleous tractus solitarii. In: Biaggioni I, Burnstock G, Low PA, Paton JFR, editors. Primer on the autonomic nervous system. 3rd ed. San Diego: Academic; 2012. p. 141–4.
83. Corbalan-Tutau MD, Gomez-Abellan P, Madrid JA, Canteras M, Ordovas JM, Garaulet M. Toward a chronobiological characterization of obesity and metabolic syndrome in clinical practice. Clin Nutr. 2015;34:477–83.
84. Mecawi AS, Ruginsk SG, Elias LL, Varanda WA, Antunes-Rodrigues J. Neuroendocrine regulation of hydromineral homeostasis. Compr Physiol. 2015;5:1465–516.
85. Antunes-Rodrigues J, Ruginsk SG, Mecawi AS, Margatho LO, Cruz JC, Vilhena-Franco T, Reis WL, Ventura RR, Reis LC, Vivas LM, Elias LL. Mapping and signaling of neural pathways involved in the regulation of hydromineral homeostasis. Braz J Med Biol Res. 2013;46:327–38.
86. Firsov D, Bonny O. Circadian regulation of renal function. Kidney Int. 2010;78:640–5.
87. Kamperis K, Hagstroem S, Radvanska E, Rittig S, Djurhuus JC. Excess diuresis and natriuresis during acute sleep deprivation in healthy adults. Am J Physiol Renal Physiol. 2010;299:F404–11.
88. Arai S, Stotts N, Puntillo K. Thirst in critically ill patients: from physiology to sensation. Am J Crit Care. 2013;22:328–35.
89. Ferguson AV. Circumventricular organs: integrators of circulating signals controlling hydration, energy balance, and immune function. Boca Raton: CRC Press; 2014.
90. Zimmerman CA, Lin YC, Leib DE, Guo L, Huey EL, Daly GE, Chen Y, Knight ZA. Thirst neurons anticipate the homeostatic consequences of eating and drinking. Nature. 2016;537:680–4.
91. Kopp UC. Role of renal sensory nerves in physiological and pathophysiological conditions. Am J Physiol Regul Integr Comp Physiol. 2015;308:R79–95.
92. Johns EJ. Chapter 17 – Autonomic regulation of kidney function. In Buijs RM, Swaab D, editors. Handbook of clinical neurology. Autonomic nervous system. New York: Elsevier; 2013, p. 203–14.
93. Jackson EK. Chapter 44 – Autonomic control of the kidney. In: Biaggioni I, Burnstock G, Low PA, Paton JFR, editors. Primer on the autonomic nervous system. 3rd ed. San Diego: Academic; 2012. p. 215–20.
94. Song W, Wang H, Wu Q. Atrial natriuretic peptide in cardiovascular biology and disease (NPPA). Gene. 2015;569:1–6.
95. Mahler B, Kamperis K, Schroeder M, Frokiaer J, Djurhuus JC, Rittig S. Sleep deprivation induces excess diuresis and natriuresis in healthy children. Am J Physiol Renal Physiol. 2012;302:F236–43.
96. Carey RM. Chapter 45 – Dopamine mechanisms in the kidney. In: Biaggioni I, Burnstock G, Low PA, Paton JFR, editors. Primer on the autonomic nervous system. 3rd ed. San Diego: Academic; 2012. p. 221–3.
97. Morrison SF. Central control of body temperature. F1000Res. 2016;5
98. Charkoudian N. Chapter 60 – Hypothermia and hyperthermia. In: Biaggioni I, Burnstock G, Low PA, Paton JFR, editors. Primer on the autonomic nervous system. 3rd ed. San Diego: Academic; 2012. p. 287–9.
99. Fealey RD. Chapter 7 – Interoception and autonomic nervous system reflexes thermoregulation. In: Buijs RM, Swaab D, editors. Handbook of clinical neurology, autonomic nervous system. New York: Elsevier; 2013. p. 79–88.

100. Cannon B, Nedergaard J. Brown adipose tissue: function and physiological significance. Physiol Rev. 2004;84:277–359.
101. Bargut TC, Aguila MB, Mandarim-de-Lacerda CA. Brown adipose tissue: updates in cellular and molecular biology. Tissue Cell. 2016;48:452–60.
102. Labbe SM, Caron A, Lanfray D, Monge-Rofarello B, Bartness TJ, Richard D. Hypothalamic control of brown adipose tissue thermogenesis. Front Syst Neurosci. 2015;9:150.
103. Roman S, Agil A, Peran M, Alvaro-Galue E, Ruiz-Ojeda FJ, Fernandez-Vazquez G, Marchal JA. Brown adipose tissue and novel therapeutic approaches to treat metabolic disorders. Transl Res. 2015;165:464–79.
104. Peirce V, Carobbio S, Vidal-Puig A. The different shades of fat. Nature. 2014;510:76–83.
105. Blessing W, Mohammed M, Ootsuka Y. Brown adipose tissue thermogenesis, the basic rest-activity cycle, meal initiation, and bodily homeostasis in rats. Physiol Behav. 2013;121:61–9.
106. Sessler DI. Perioperative thermoregulation and heat balance. Lancet. 2016;387:2655–64.
107. Hilz MJ. Chapter 48 – Physiology and pathophysiology of female sexual function. In: Biaggioni I, Burnstock G, Low PA, Paton JFR, editors. Primer on the autonomic nervous system. 3rd ed. San Diego: Academic; 2012. p. 235–8.
108. Rudolph LM, Bentley GE, Calandra RS, Paredes AH, Tesone M, Wu TJ, Micevych PE. Peripheral and central mechanisms involved in the hormonal control of male and female reproduction. J Neuroendocrinol. 2016;28.
109. Wahab F, Shahab M, Behr R. The involvement of gonadotropin inhibitory hormone and kisspeptin in the metabolic regulation of reproduction. J Endocrinol. 2015;225:R49–66.
110. Ullah R, Shen Y, Zhou YD, Huang K, Fu JF, Wahab F, Shahab M. Expression and actions of GnIH and its orthologs in vertebrates: current status and advanced knowledge. Neuropeptides. 2016;59:9–20.
111. Shaw ND, Butler JP, McKinney SM, Nelson SA, Ellenbogen JM, Hall JE. Insights into puberty: the relationship between sleep stages and pulsatile LH secretion. J Clin Endocrinol Metab. 2012;97:E2055–62.
112. Brabant G, Prank K, Schofl C. Pulsatile patterns in hormone secretion. Trends Endocrinol Metab. 1992;3:183–90.
113. Zeleznik AJ, Plant TM. Chapter 28 – Control of the menstrual cycle. In: Knobil and Neill's physiology of reproduction. 4th ed. San Diego: Academic; 2015, p. 1307–61.
114. Veldhuis JD. Neuroendocrine mechanisms mediating awakening of the human gonadotropic axis in puberty. Pediatr Nephrol. 1996;10:304–17.
115. Gore AC, Hall JE, Hayes FJ. Chapter 37 – Aging and reproduction. In: Knobil and Neill's physiology of reproduction. 4th ed. San Diego: Academic; 2015, p. 1661–93.
116. Penev PD. Association between sleep and morning testosterone levels in older men. Sleep. 2007;30:427–32.
117. Bridges RS. Neuroendocrine regulation of maternal behavior. Front Neuroendocrinol. 2015;36:178–96.
118. Miller TV, Caldwell HK. Oxytocin during development: possible organizational effects on behavior. Front Endocrinol (Lausanne). 2015;6:76.
119. Stolzenberg DS, Champagne FA. Hormonal and non-hormonal bases of maternal behavior: the role of experience and epigenetic mechanisms. Horm Behav. 2016;77:204–10.
120. Alves E, Fielder A, Ghabriel N, Sawyer M, Buisman-Pijlman FTA. Early social environment affects the endogenous oxytocin system: a review and future directions. Front Endocrinol (Lausanne). 2015;6:32.
121. Grattan DR, Le Tissier P. Chapter 12 – Hypothalamic control of prolactin secretion, and the multiple reproductive functions of prolactin. In: Knobil and Neill's physiology of reproduction. 4th ed. San Diego: Academic; 2015, p. 469–526.

Fourth Level: The Limbic System

6

Abstract

In a simplified view, brain function can be considered the product of the neocortex and the limbic system, which complement each other to generate human behavior with purpose and objective. Phylogenetically, the limbic system comprises the oldest parts of the telencephalon and the subcortical structures that derive from it. The limbic system is essential for emotionality, motivation, learning and memory. This Chapter analyzes how emotions comprise feelings and moods, and their expression in somatic and autonomic behaviors. It discusses the neurobiological mechanisms of memory and how they vary in the three body configurations (wakefulness, NREM sleep, REM sleep) during a 24-h cycle.

Keywords

Amygdala • Basal ganglia • Chronotypes • Cognitive memory • Emotionality • Hippocampus • Learning • Limbic system • Memory • Mesolimbic system • Nucleus accumbens • Papez circuit

Objectives

After studying this chapter, you should be able to:

- Understand how the limbic system is essential for emotionality, motivation, learning, and memory.
- Describe the structures, connections, and physiological significance of the limbic components of the basal ganglia.
- Underline how emotions comprise feelings and moods, and their expression in somatic and autonomic behaviors.
- Enumerate the major findings in functional neuroimaging of the ANS.
- Identify the chronotypes and linked emotional features to the three body configurations (wakefulness, NREM sleep, REM sleep) during a 24-h cycle.
- Describe the neurobiological mechanisms of memory and how they vary in the three body configurations (wakefulness, NREM sleep, REM sleep) during a 24-h cycle.

D.P. Cardinali, *Autonomic Nervous System*, DOI 10.1007/978-3-319-57571-1_6

The Limbic System Is Essential in Emotionality, Motivation, Learning, and Memory

Phylogenetically, the limbic system comprises the oldest parts of the telencephalon and the subcortical structures that derive from it. In a simplified view, brain function can be considered the product of the neocortex and the limbic system, which complement each other to generate human behavior with purpose and objective [1]. In this process of complementation, the neocortex mainly regulates precise spatiotemporal communication with the environment and executes cognitive and stereognostic functions, producing precise motor outputs (Fig. 6.1).

The limbic system has a primordial link with emotionality and motivation for action (reinforcement/reward system) and with the process of learning and memory (involving a high affective content, remembering only what we are interested in emotionally) [2]. The limbic system gives the information derived from the inner and outer world its particular emotional meaning [3]. Hence, its role as a last level in the autonomic motor hierarchy.

Another aspect to consider is the role of the limbic system as a selective inhibitor of impulses and basic needs, immediately related to survival. The selective inhibition of certain circuits of a nontopographic character, but relative to memories loaded with internal meaning can prevent the activation of (too many) lateral ways

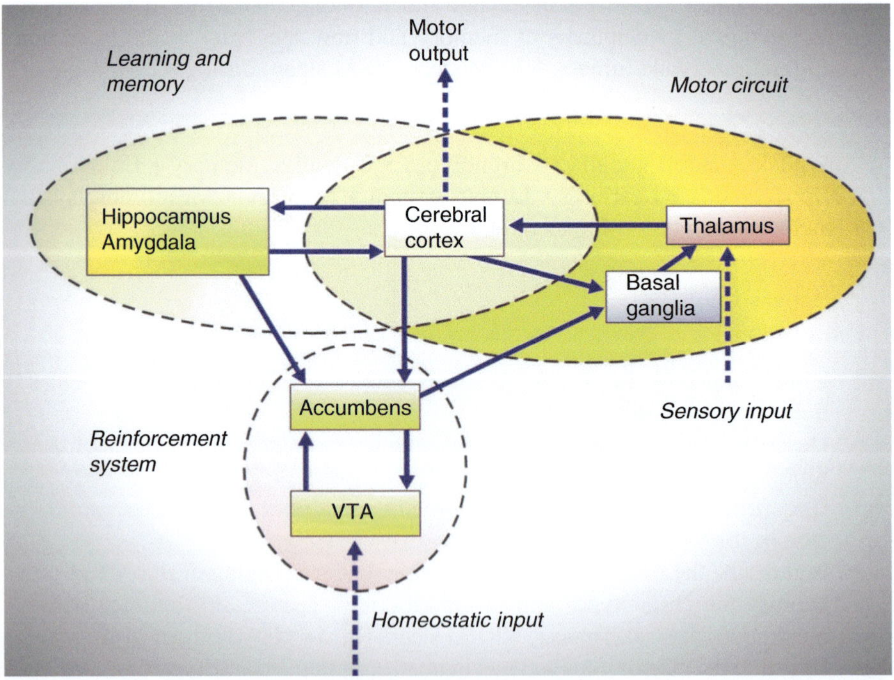

Fig. 6.1 The limbic system participates in two of the three basic functional loops of brain function (learning and memory and the reinforcement system). Modified with permission from Cardinali [1]

and thus allows the exclusive creation of relevant temporospatial associations (emotional learning). A lateral dispersion in these highly-interconnected circuits would lead to phenomena of resonance, overabundance, and/or blockade (obsessive ideas, epileptic seizures, anxiety, etc.). From a physiological perspective, the limbic system is able to carry out tasks of this type, as it repeats the basic scheme, present in many other brain structures, that different sources of information, complementary and/or opposite, are confronted in the same structure or nodal point, through intermixed circuits [4].

A cortical portion and a subcortical portion are distinguished in the limbic system:

- The cortical portion consists of the limbic gyrus, part of the ring-shaped cerebral cortex on the inner side of each hemisphere, separating the neocortex from the hypothalamus and the brainstem (Figs. 6.2 and 6.3). In 1878, Broca gave the name "limbic lobe" to this ring of cortical tissue that surrounds the hilum of each cerebral hemisphere. The limbic gyrus consists of the parahippocampal, cingulate, and subcallosal gyrus, and was called "rhinencephalon," because it was initially considered to be exclusively associated with the olfactory function. The orbitofrontal cortex is also included among the cortical areas of the limbic system.
- The subcortical portion of the limbic system consists of several nuclei (Fig. 6.4). They are: amygdala, hippocampus, nucleus accumbens, septal nuclei, epithalamus (habenula), olfactory bulb, and areas of the anterior thalamus and

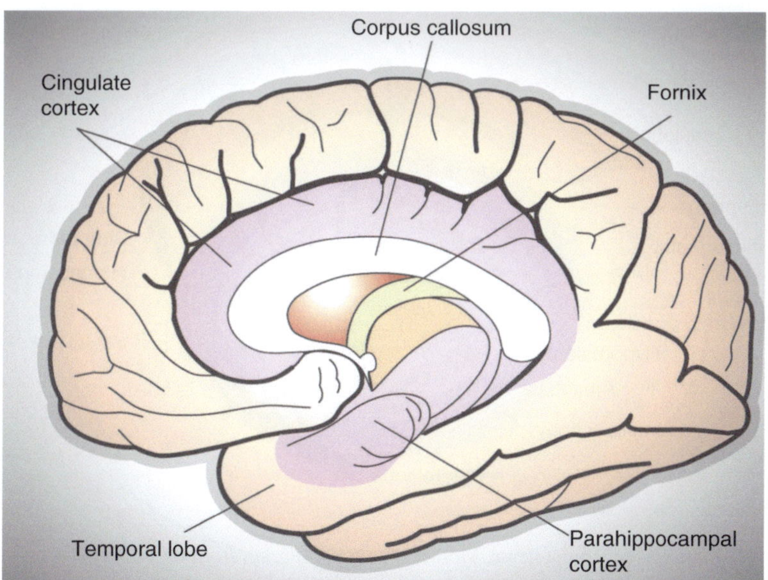

Fig. 6.2 Rhinencephalon or limbic gyrus. Modified with permission from Cardinali [1]

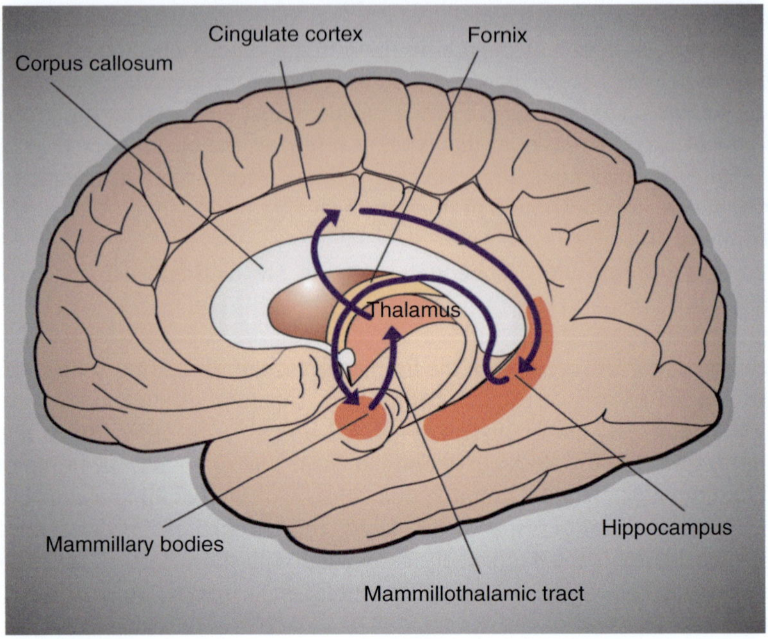

Fig. 6.3 Medial view of the structures of the limbic system. Modified with permission from Cardinali [1]

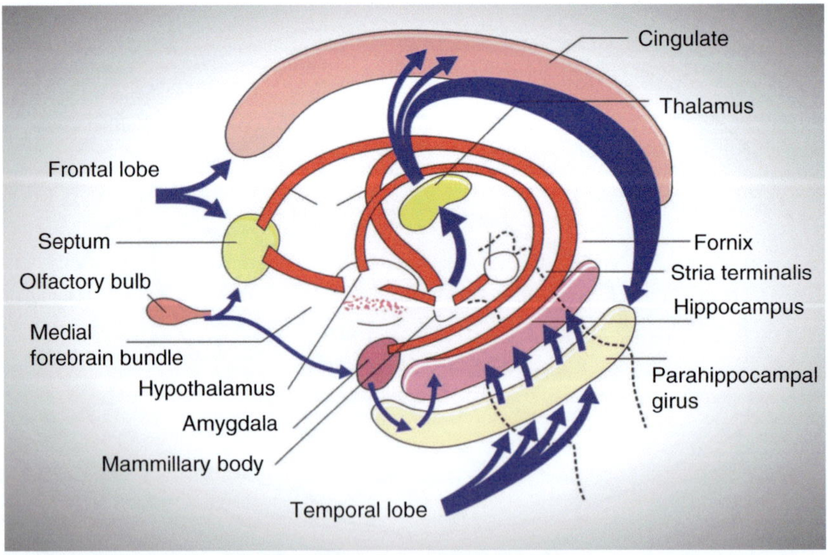

Fig. 6.4 Limbic system components. Modified with permission from Cardinali [1]

hypothalamus (preoptic area, mammillary bodies), and part of the basal ganglia (ventral pallidal region, innominate substance) [2].

James Papez's ideas about the limbic system, enunciated in the 1930s, have been confirmed by recent neuroimaging studies of brain locations [5]. For Papez, the limbic system is part of the circuit of emotional expression. As it was known that the hypothalamus was fundamental for the expression of emotional reaction programs, Papez postulated that the way in which the cerebral cortex modifies, and where these programs become conscious, is through corticohypothalamic connections via the cingulate gyrus and the hippocampus [6]. According to Papez's hypothesis, the hippocampus processes emotional information and projects to the mammillary bodies through the fornix. The hypothalamus, in turn, provides information to thalamic nuclei (through the mammillothalamic tract) and from these to the cingulate gyrus [7]. Subsequently, MacLean extended this scheme to include in the limbic system hypothalamic areas, the septal area, the nucleus accumbens, neocortical areas (orbitofrontal cortex), and the amygdala. The circuit of Papez, and the most recent modifications to it, are summarized in Fig. 6.5.

The afferent and efferent connections of the limbic system are extremely complex [8]. As we have mentioned, the most outstanding fact is a massive reciprocal connection with the hypothalamus. The hypothalamus communicates with the hippocampus and the septum through the fornix, with the amygdala through the stria terminalis and ventral amygdalofugal pathway, and with the portions of the olfactory brain through the central forebrain bundle.

Although there is no complete agreement about the anatomical composition of the limbic system, it is accepted that a set of structures located in the medial portion of the telencephalon, highly interconnected with each other, share direct projections to the hypothalamus, thus regulating the neuroendocrine, autonomic, and behavioral processes associated with this portion of the diencephalon [5].

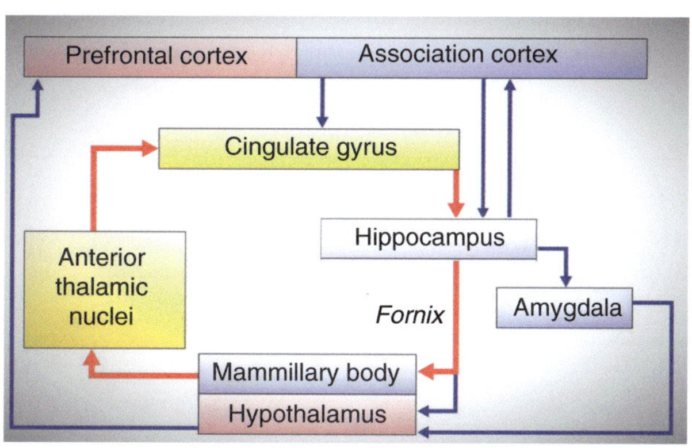

Fig. 6.5 The Papez circuit. Modified with permission from Cardinali [1]

The hippocampus is a portion of the cerebral cortex that forms a kind of horn along the curvature of the lateral ventricle (Fig. 6.6). It is subdivided into the hippocampus proper, or Ammon's horn, the dentate gyrus, and the subicular complex. The connections to hippocampal formation come from the entorhinal cortex, contralateral hippocampus, subcortical structures such as the medial septum, certain raphe nuclei, and the locus coeruleus (LC) from the brainstem. The hippocampus projects back to the subicular region and the hippocampus, in turn, extends over other cortical areas, the anterolateral thalamus, the mammillary bodies, the ventromedial and anterior nuclei of the hypothalamus, and the lateral septum. Through the fornix, the hippocampus projects over the lateral septum (Fig. 6.5) [2, 4].

Thus, the limbic system presents multiple excitation circuits, neuronal substrates of importance for both emotionality and memory. As we describe later, the fixation of memory engrams depends on the simultaneous activation of limbic system pathways [3].

At the end of the nineteenth century, Jackson proposed that a function is often represented at various levels of the nervous system. The higher levels, having appeared later, mediate the function in a more precise form than the lower levels. In addition, they are more easily excited and usually have an inhibitory action on lower levels. Therefore, a lesion of the higher levels releases the inferior levels of inhibition and allows behavioral expression [6].

With this hierarchical notion in mind, MacLean proposed in the 1950s, that the brain of modern mammals is the sum of three superimposed brains, acquired during evolution (Fig. 6.7). For MacLean, to the visceral and appetitive of the primitive reptiles, an emotional brain was added, whose functions would be

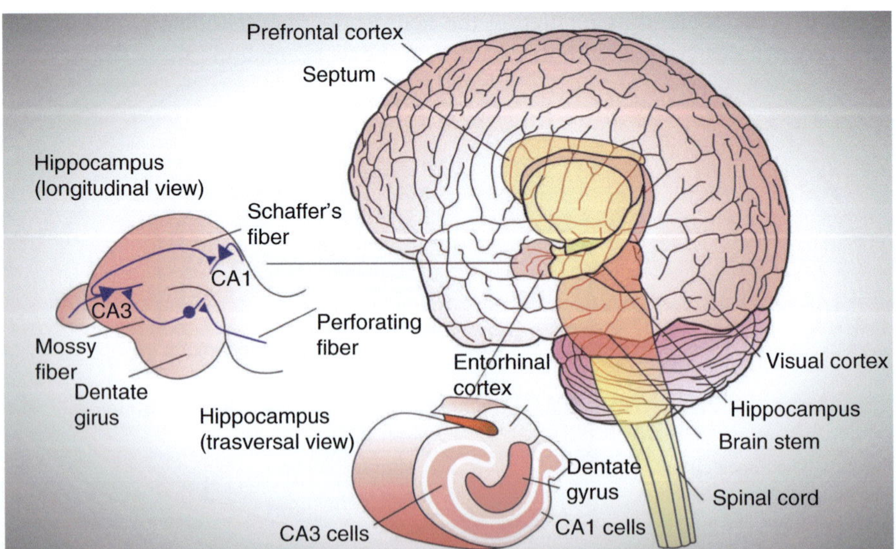

Fig. 6.6 The structure of the hippocampus. Modified with permission from Cardinali [1]

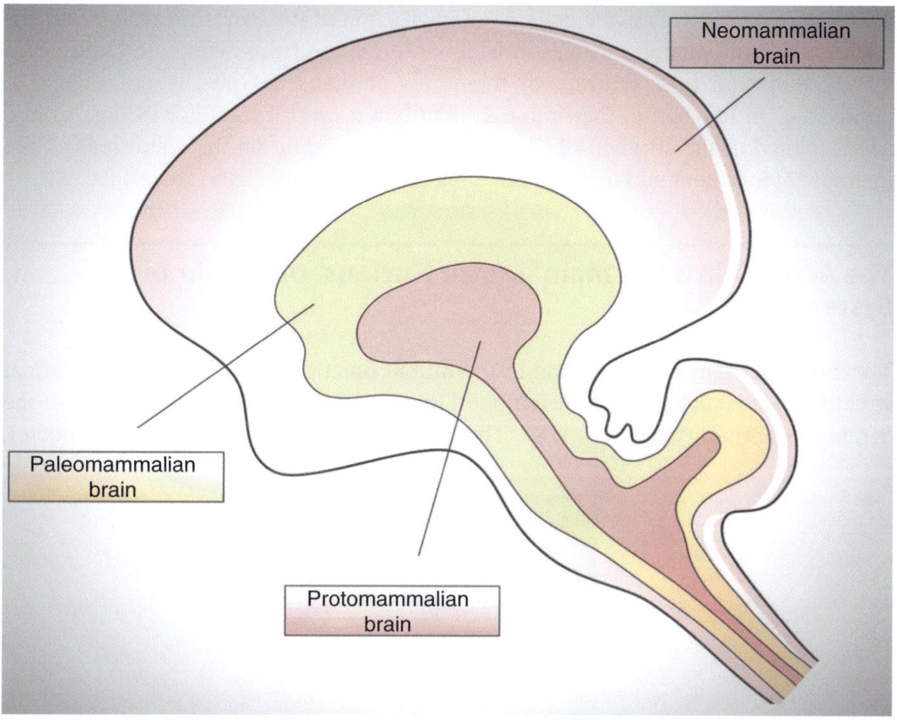

Fig. 6.7 MacLean's "trine brain" concept. Modified with permission from Cardinali [1]

assumed, finally, by the limbic system of birds and, mainly, mammals [6]. There would be:

- A first reptilian or protomammalian, vegetative or instinctive brain, formed by the upper portions of the spinal cord and part of the brainstem and the basal ganglia, with a function in the instinctive survival behaviors (mating, hunting, etc.)
- A second paleomammalian brain, emotional or limbic, and hierarchically superior to the previous one, the reason it has the possibility to block activation of primitive drives
- A third neomammalian brain, formed by neocortical structures, capable of analysis stripped of emotional elements.

However, it would be naive to take this functional division as absolute. In fact, the SNC operates with a unique behavior resulting from the function of the three hypothetical levels [1].

The limbic system determines the appearance of an inner world, a concept that is superimposed in part, but not equivalent, to that of internal environment. The internal world is not based on the presence of interoceptors or the development of homeostatic mechanisms, but on the development of internal signals of identity. For example, being able to inhibit certain desires (avoiding a food source in the presence of a predator) is the behavioral expression of the existence of internal circuits capable of generating states in which information from extero- and interoceptors is subjected to a scrutiny of memories or plans not merely contingent or immediate. In this sense, the limbic system is a powerful inhibitor of desires and needs related to the survival of the individual, depending on the conditions of the internal environment and the external environment (physical and social) [9].

The Amygdala Is the Main "Motor Nucleus" of the Limbic System

The amygdala plays a major role in the limbic function [2, 10]. It is a subcortical structure located at the tip of the temporal lobe and continuous with the uncus of the parahippocampal gyrus (Fig. 6.8). The amygdala is composed of several nuclei,

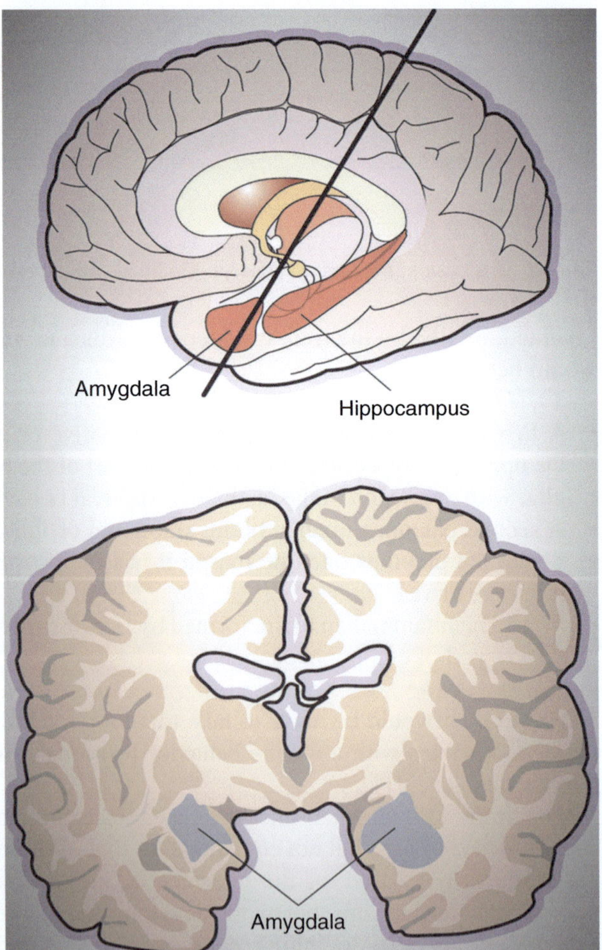

Fig. 6.8 Anatomical localization of the amygdala

reciprocally connected with the hypothalamus, hippocampus, neocortex, and thalamus (Fig. 6.9). Despite the important olfactory input it receives, the amygdala is not essential for olfactory discrimination [3].

In 1939, Klüver and Bucy described the bilateral lesion of the amygdala nuclei and part of the anterior pole of the temporal lobe as producing a behavioral syndrome in monkeys characterized by: (a) psychic blindness; (b) exaggerated oral exploratory behavior; (c) temerity or loss of fear, because, for example, they play with snakes, which they normally fear; (d) excessive indiscriminate eating behavior (hyperbulimia); (e) increased sexual behavior (self, homo-, and heterosexual); (f) hypermetamorphosis, or a tendency to react to any visual stimulus. This clinical picture suggests that, under normal conditions, the amygdala functions as an inhibitory center, preventing reckless or inappropriate behaviors in relation to feeding, sex, and exploration of the environment [11].

Given the diversity of the afferents received by the amygdala from the limbic cortex and the rest of the association cortex, in addition to its dense projection on the hypothalamus nuclei, it is easy to accept the guiding role of the amygdala for the correct structuring of most available sensory information. Thus, the pioneering study by Klüver and Bucy showed the involvement of the amygdala in numerous emotional processes, in which a complex elaboration and association of stimuli from different sensorial sources are carried out.

The electrical stimulation of the amygdala produces effects on the ANS similar to those induced by stimulation of the hypothalamus [7]. Primarily, stimulation of the central amygdala produces changes in BP and heart rate, motility, and gastrointestinal secretions, mydriasis, piloerection, etc. Stimulation of the corticomedial amygdala produces an increase in the secretion of ACTH and gonadotrophins, whereas the stimulation of the basolateral portion in some cases inhibits it. Stimulation of the amygdala also induces motor phenomena such as contralateral head spin, masticatory and swallowing movements, or clonic and rhythmic movements that may become convulsive if the stimulus is prolonged [12].

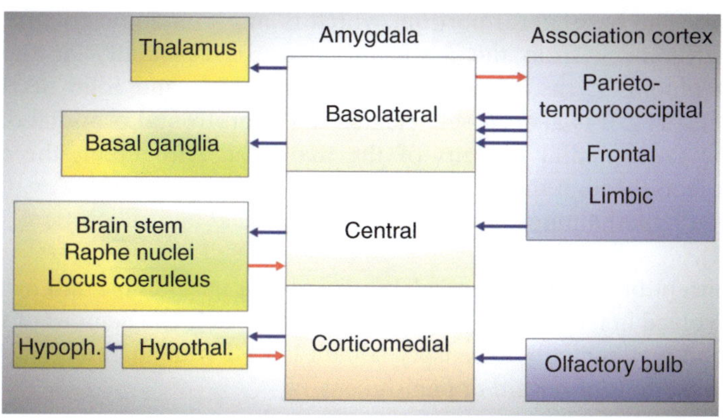

Fig. 6.9 Connections the amygdala. Modified with permission from Cardinali [1]

It is characteristic that the effects of stimulation of the amygdala depend on the functional status of the animal, its environment, and the levels of endocrine, metabolic, and autonomic variables. The same stimulus can increase ACTH levels if they are low, but decrease them if they were previously increased. This indicates the important role of context evaluation in emotional response.

The stimulation of the amygdala in humans produces auras with emotional and polysensorial content ("déjà vu," hallucinations, etc.). In animals, the selective lesion of the amygdala decreases the performance in passive avoidance tests, mainly because of the loss of fear. Animals with a lesioned amygdala show poor affective behavior, with loss of hierarchical rank [3].

Functionally, three groups of nuclei are distinguished in the amygdala (Fig. 6.9) [2]:

- Corticomedial, linked to the regulation of the hypothalamus and containing receptor sites for corticosteroids and gonadal hormones.
- Central, which projects to the brainstem nuclei, such as periaqueductal gray matter, the NTS and PBN, and the dorsal motor nucleus of the vagus nerve.
- Basolateral, with connections to the association cortexes.

Thus, the corticomedial amygdala participates in endocrine and behavioral functions related to sexual activity, the central amygdala modulates brainstem nuclei with somatic and autonomic motor responses, and the basolateral amygdala participates in processes of sensorial and behavioral association. Through the amygdala, affective behaviors that have proven to be appropriate on previous occasions are induced.

The central nucleus is closely related anatomically and functionally to the lateral hypothalamus and to various structures of the brainstem, such as NTS and PBN, which, in turn, participate in gustatory, cardiorespiratory, and visceral functions. The corticomedial portion and the periamygdala cortex receive afferents from the main and accessory olfactory bulb and project to the olfactory cortex. The basolateral portion, which is more phylogenetically modern, receives afferences from the association cortex, especially of the inferior (visual) temporal gyrus, upper temporal (acoustic), and lobe of the insula (somatosensory). It is also closely related to the prefrontal orbitomedial cortex and to the dorsomedial nucleus of the thalamus. Altogether, the amygdala nuclei project, through the stria terminalis and the ventral pathway, to various areas of the hypothalamus, apart from other cortical and subcortical structures (Fig. 6.10).

The concept of "extended amygdala" defines the mesolimbic and mesocortical circuits involved in the hedonic response (pleasure) [13]. The neurons of the amygdala respond preferentially to sensory stimuli loaded with an emotional tone, i.e., related to situations of reward or punishment.

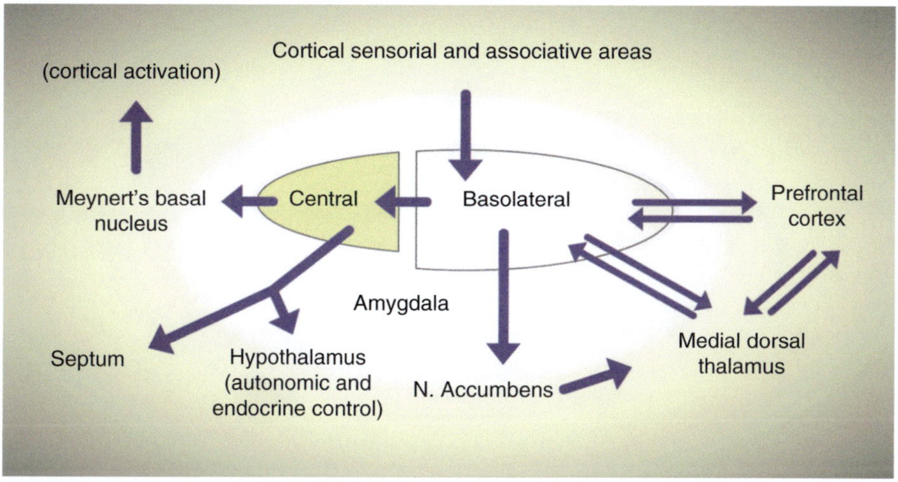

Fig. 6.10 Projections of the central and basolateral amygdaloid nuclei

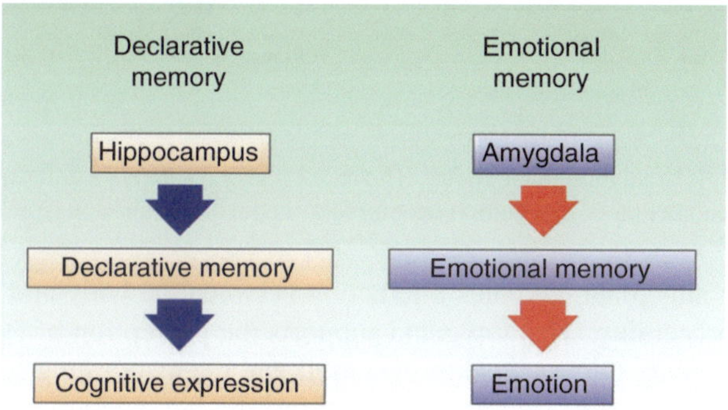

Fig. 6.11 There are cognitive circuits (based on the hippocampus) and emotional circuits (based on the amygdala) to mediate the two types of memory: declarative and emotional. Declarative memory implies what is commonly meant by "memory." The emotional memory involves those instinctive behaviors, learned or congenital, that protect life. Modified with permission from Cardinali [1]

The amygdala participates in the learning process, particularly when it comes to the association of a stimulus with an emotional response. This function is so important that we recognize today an "emotional memory," with mechanisms different from the "cognitive memory" (Figs. 6.11 and 6.12) [3]. The most conclusive contemporary evidence for the involvement of the amygdala and other limbic structures in emotional behavior has been given by PET and fMRI in individuals with affective diseases and in normal individuals in situations of anxiety [14].

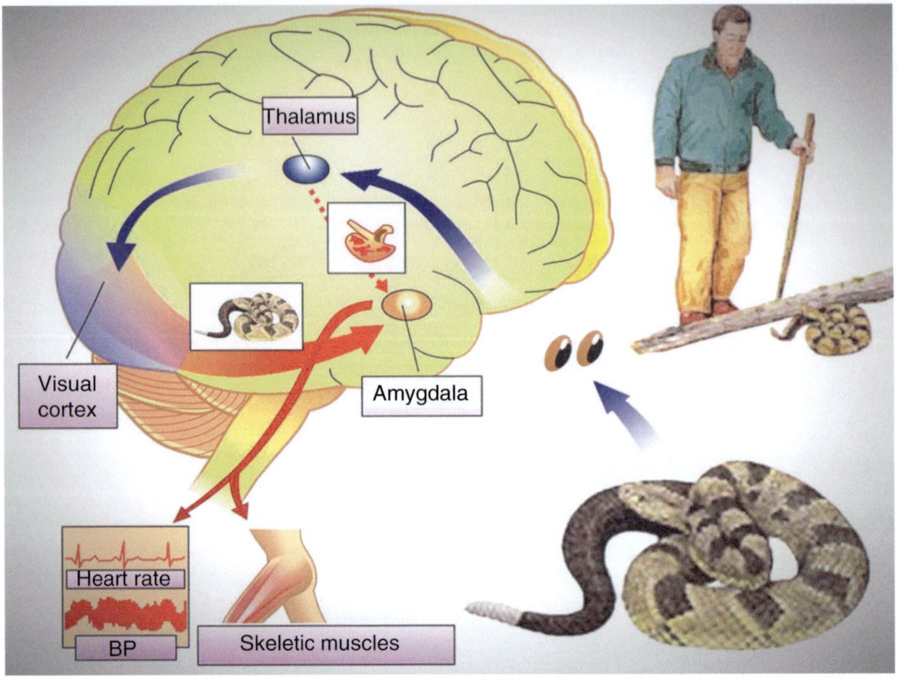

Fig. 6.12 Example of an emotional circuit. The snake's vision triggers a defense reaction before the cognitive phenomenon of recognition occurs. Modified with permission from Cardinali [1]

The human amygdala mediates interaction between the body and the brain during affective processing. The amygdala supports the perception of fear signals and threat and its activity correlates with the emotional intensity rating of affective pictures including facial expressions. Outputs from the amygdala innervate hypothalamic and brainstem autonomic circuits to trigger autonomic arousal responses to emotional challenges, particularly threats [7]. Amygdala-induced autonomic arousal is expressed as increased sympathetic activity and/or decreased heart rate variability. The amygdala is also sensitive to feedback from the periphery regarding the state of bodily arousal.

By means of various procedures (electrophysiological, autoradiographic, immunohistochemical, functional neuroimaging), the connections through which the limbic system, via the amygdala, regulate the expression of anxiety, have been schematized as follows:

- By projections to the lateral hypothalamus, the amygdala produces sympathetic activation (tachycardia, change in electrodermal response, mydriasis, increased BP, etc.).
- By projections to the dorsal nucleus of the vagus and ambiguous nucleus, the amygdala produces parasympathetic activation (gastrointestinal ulcers, urination, defecation, bradycardia).

- From projections to the nucleus parabrachialis, it produces tachypnea.
- By projections to the LC and tegmental areas, the amygdala produces activation of the noradrenergic, dopaminergic, and cholinergic activity of the reticular formation, with increased alertness.
- By projections to the motor nuclei of the ventromedial descending pathway in the reticular formation it produces hyperreflexia.
- From projections to the facial and trigeminal motor neurons, it produces changes in facial expression.
- By projections to the paraventricular nucleus, it induces CRH release, with stimulation of the adrenal pituitary axis and the descending autonomic pathways.

Because it controls emotional behavior, the limbic system controls motivation. Thus, the limbic system determines the appearance of an internal world that integrates the homeostatic functions based on the presence of interoceptors with an elaboration of internal signals of identity [5].

The avoidance of a food source in the presence of a predator is the behavioral expression of the existence of internal circuits capable of generating states in which the information coming from extero- and interoceptors is subjected to verification of the opportunity to execute it or not. It corresponds, as we discussed in Chap. 5, to an allostatic response in which physiological systems fluctuate to meet the demands of external forces. The limbic system is a powerful inhibitor of desires and needs related to the survival of the individual, depending on the conditions of the internal environment and the outside world, and therefore the main regulator of allostatic responses [2].

During emotion regulation, prefrontal control systems modulate emotion generative systems, such as the amygdala, which is responsible for the detection of affectively arousing stimuli. More specifically, these prefrontal structures include dorsal regions of the lateral prefrontal cortex that have been implicated in selective attention and working memory; ventral parts of the prefrontal cortex implicated in response inhibition; the anterior cingulate cortex, which is involved in monitoring control processes; and the dorsomedial prefrontal cortex, which is implicated in monitoring the affective state [15]. A typical pattern detected when individuals deliberately regulate affective responses (as in mindfulness meditation) is increased activation within the prefrontal cortex and decreased activation in the amygdala, suggesting that prefrontal cortex projections to the amygdala exert an inhibitory top–down influence [16].

Emotions Comprise Feelings and Moods, and Their Expression in Somatic and Autonomic Behaviors

Although it seems difficult to reach an agreement to define what is understood by "emotion," it is unanimously accepted that in situations that are tense or committed for the individual (for example, the startle that happens before the sudden presence of a predator), a nonspecific activation of the vegetative system (tachycardia, cold sweat, etc.) and the skeletal motor system (expression of terror, fight/flight) occurs, together with a greater or lesser knowledge of the cause of the shock [17].

Faced with an emotion, we can consider the internal, personal nature, which in humans also has a cognitive character. We can also consider an external, behavioral aspect, which serves as a key signal for members of the same species or related species [18].

Of course, the external expression of emotions is a consequence of the internal aspects. The conflict that arises before the unexpected evolution of a situation, such as the presence of a predator in an unexpected place, is resolved by the activation of certain autonomic and somatic motor manifestations that imply a reevaluation of available sensory data. These motor acts, mediated by both the somatic system and the ANS, are expressive of the state of the inner world [3].

Poor emotional and social adaptation to an environment in constant change characterize limbic system lesions. Without limbic connections and with an intact hypothalamus, cats or monkeys trigger complex behaviors lacking in objective or normal content, for example, the "false rabies," hyperphagia, and hypersexuality (Klüver-Bucy syndrome) mentioned above.

The limbic system acts through the programs contained in the hypothalamus, as demonstrated by electrophysiological experiments [19]. The electrical stimulation of the amygdala in the experimental animal triggers effects such as those observed after hypothalamic stimulation. Such effects include homeostatic responses and complex autonomic, endocrine, and somatic behaviors.

Bilateral ablation of the amygdala in monkeys eliminates the possibility of social functioning of the animal [20]. They cannot recognize the social meaning of the exteroceptive cues that regulate group behavior, and appear anxious and insecure. This picture is due to the interruption of the flow of information between the parietal–temporal–occipital association cortex and the hypothalamus, which occurs through the limbic system (in this case, the amygdala). The result of this alteration is the suppression of a correct evaluation of the sensorial information in the context of the affective state. Selective lesioning of the amygdaloid nuclei decreases performance in passive-type avoidance tests, probably because of the loss of fear. It should be remembered that in the basolateral amygdala there are numerous receptors for opiate and GABA, the destruction of which causes a change in the thresholds for physical pain and affective reactions. In fact, amygdala-lesioned animals present very poor affective behavior, losing their hierarchical rank in their group, and finally being rejected by it [20].

The close link among the parietal–temporal–occipital association cortex, the hypothalamus, and the limbic system is indicated by the following experiments: (a) amygdala neurons can be activated by stimulation of sensory neocortical areas; (b) temporal lobe epilepsy in humans is accompanied by various emotional, autonomic, and sensorimotor signs. Both functional neuroimaging and clinical observations in humans indicate that the connection: "parietal-temporal–occipital association cortex–amygdala" contains important neuronal substrates of motivated behaviors and emotions. That is, through this system, the sensory information is compared with the contents of memory and thus becomes significant.

Expression of emotions is primarily based on neurovegetative reactions, which are, in part, inherited and typical of the species, and partly acquired during early postnatal age. Innate emotional reactions serve as signals to the congeners and to members of other species, and are therefore of very important adaptive and evolutionary value (Fig. 6.13).

In parallel with this innate element of emotional behavior, an acquired component is identified, resulting from the first stages of contact of the newborn with his mother and the environment that surrounds him. It is through this process that the particularization of emotional responses occurs, and therefore, it influences the type of pathological condition that, if it occurs, is observed in everyone (Fig. 5.14).

The limbic cortex of a newborn child fixes engrams, depending on the type of emotional stimulation it receives in the early stages of development. Clearly, this is an active interface between neuroscience and psychology. The production of emotions is associated with the cognitive capacity of the species, and therefore with the perception and evaluation of sensorial stimuli in relation to the memory of the lived experience.

The initial works in experimental psychology carried out by Wundt at the end of nineteenth century, led to the description of the relationship between the intensity of the sensorial stimulus and the pleasure or not of the perception. Near the threshold, the stimulus is perceived as neutral, at higher intensities as pleasurable, and at even greater intensities, as unpleasant. That is, the sensorial and hedonic intensity of a given stimulus, for example, a certain taste, are not linearly related [1].

To this hedonic theory of emotion, cognitive factors were later added. According to this interpretation, the intensity of the emotion depends on the level of adaptation of the subject that perceives and its expectation before the stimulus. A relative discrepancy between these elements generates the opposite effect. Studies based on the

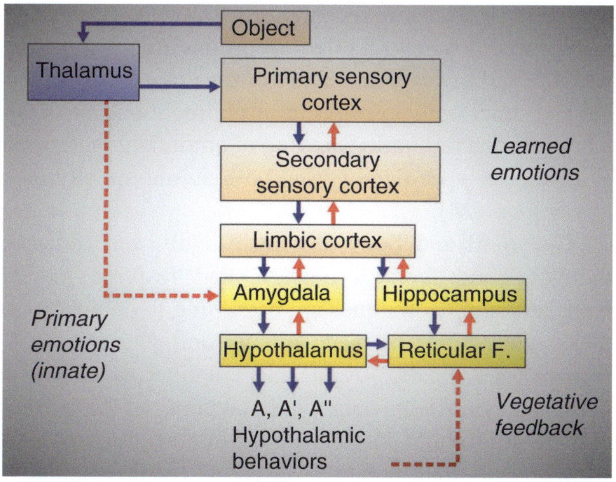

Fig. 6.13 Diagram describing the relationships between the different components involved in the congenital and acquired emotional behaviors described in the text. Modified with permission from Cardinali [1]

Fig. 6.14 Darwin's drawing to exemplify the instinctive aspects of aggressive interspecies behavior. Modified with permission from Cardinali [1]

analysis of facial expression indicated a three-dimensional aspect for emotion: pleasant–unpleasant, attention–inattention, and intensity [21].

One of the most persistent influences on the concept of emotion was that of Charles Darwin, who emphasized its genetic components. Darwin suggested that emotional expressions are evolutionary remnants of previously adaptive behaviors that persist, albeit of no use, in a moderate form (e.g., grinding teeth as a sign of aggression; Fig. 6.14).

William James was the first to propose that emotion consists of bodily changes originating in the perception of the stimulus (Fig. 6.15) [22]. This theory was called James–Lange, because of the contribution made independently by a Danish physician, Carl Lange, to its formulation. According to the James–Lange theory, emotional quality is the result of perceived changes in bodily activity triggered because of sensory perception.

The elucidation of the structure and function of the ANS marked a fundamental milestone in the study of the visceral correlates of emotion. It was Walter Cannon who first established the direct link between emotional activity and sympathetic function. In what Cannon called an "emerging theory of emotion," he described the sympathetic division of the ANS as the mediator of the reaction to stress. For Cannon, it is the CNS that triggers the emotions and not the bodily changes (Fig. 6.15) [23]. More recently, a cognitive view developed that combines both positions: there is a central effect of production of the emotional by the limbic system, which is fed by body correlates of the emotions (Fig. 6.15) [17]. Improved anatomical and functional description of bidirectional interactions between body and brain has advanced our understanding of emotional and, for some emotions, there is good evidence for specific coupling with autonomically mediated changes in peripheral physiology [24].

The mind and body are intrinsically and dynamically coupled. Perceptions, thoughts, and feelings change, and respond to, the state of the body [10]. Neuroimaging techniques are beginning to detail the neuronal substrates mediating these interactions between mental and physiological states, implicating cortical regions (specifically insular and cingulate cortices) alongside subcortical (amygdala) and brainstem (notably dorsal pons) in these mechanisms [10, 21]. For example, by combining fMRI with carotid stimulation in healthy participants, it was shown that manipulating afferent cardiovascular signals alters the central

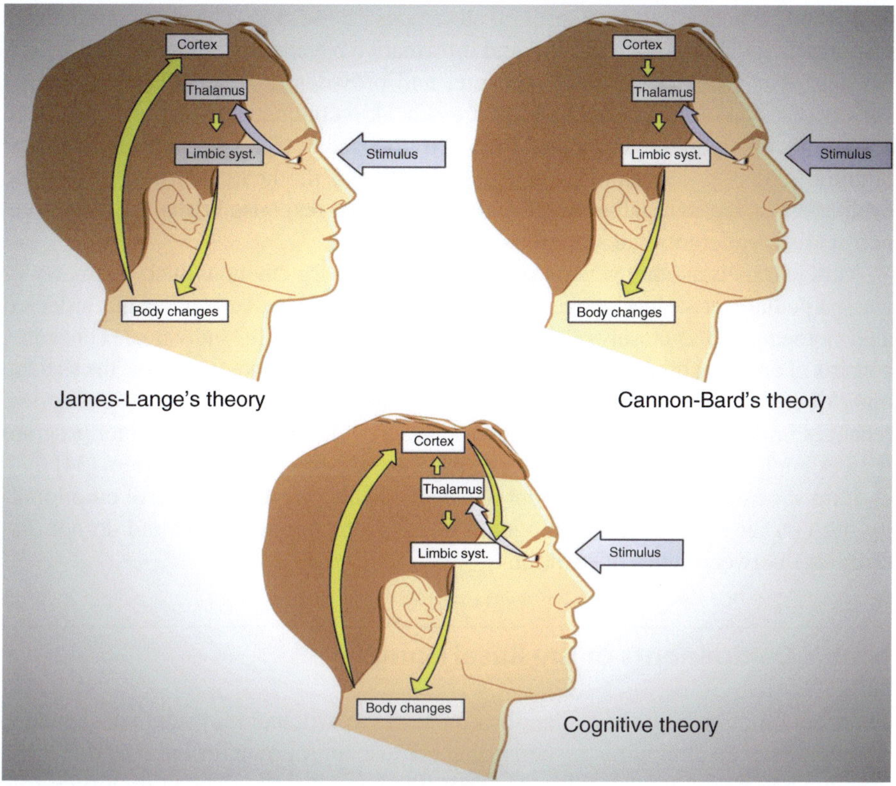

Fig. 6.15 The different theories of the production of emotions. The current (cognitive) view is eclectic between two opposing positions. Modified with permission from Cardinali [1]

processing of emotional information (fearful and neutral facial expressions) [25]. Carotid stimulation attenuated activity across cortical and brainstem regions. Modulation of emotional processing was apparent as a significant expression-by-stimulation interaction within the left amygdala, where responses during appraisal of fearful faces were selectively reduced by carotid stimulation. Moreover, activity reductions within insula, amygdala, and hippocampus correlated with the degree of stimulation-evoked change in the explicit emotional ratings of fearful faces. Across participants, individual differences in autonomic state, as assessed by heart rate variability, predicted the extent to which carotid stimulation influenced neural (amygdala) responses during appraisal and subjective rating of fearful faces [25]. Thus, cortical locations of emotional function in man have begun to be defined using PET and fMRI [26–29]. This methodology is also employed to determine the sites and effects of pharmacological treatments [30].

For the physiological detection of emotion, a sensitive, non-invasive indicator is the electrical response of the skin, also known as the psychogalvanic reflex, or electrodermal reflex. The potential or electrical resistance (or conductance) of any part of the body can be quantified through electrodes on the skin. The skin conductance response is a remarkably powerful and informative psychophysiological index [31]. Because it is relatively easy to measure, and provides reliable indices of a wide variety of psychological states and processes, skin conductance response has been one of the most popular aspects of ANS activity used to study human cognition and emotion [31, 32]. The analysis of the variability of the heart rate also allows the evaluation of the sympathetic and parasympathetic response at the thoracic level before different emotional situations (Fig. 4.13) [33].

Muscle tone is another general peripheral indicator for emotions such as anxiety or fear, particularly at the level of the face and neck muscles. The startle reflex is considered a phenomenon that is closely related to the emotional state of the individual. This reflex consists of an initial blink, with a latency close to 0.04 s. Then, contraction of the skeletal muscles ensues, with a latency of 0.1 s. Finally, after 1 s, more complex signs appear (changes in skin potential, increased BP and heart rate). This sequential motor program is an example of stereotyped reactivity of ANS and the somatic motor system [34].

It must be noted that the different physical and chemical indicators of emotionality used so far simply reflect an overall level of emotional tension and do not discriminate between types of emotions.

Limbic Components of the Basal Ganglia

The main function of the basal ganglia is to select a particular movement or sequence of thoughts or an autonomic response that is most appropriate for the situation, suppressing any possible other ones [1]. Thus, the basal ganglia play an important role in limbic function.

There are five main components of the basal ganglia (Figs. 6.16 and 6.17): (a) three subcortical nuclei: caudate, putamen, and globus pallidus; (b) a diencephalic component: the subthalamic nucleus of Luys; (c) a mesencephalic component: the substantia nigra and the ventral tegmental area [35].

The caudate and putamen have the same embryological origin, identical cellular types and are fused by their anterior part (to form the striatum). The ventral part of the striatum (ventral striatum or nucleus accumbens) has a functional identity because of its connection with the limbic system. The striatum comprises the entry nuclei to the circuit of the basal ganglia.

The globus pallidus is a diencephalic structure divided into two segments, internal and external (or medial and lateral). The substantia nigra, which is the largest nucleus of the midbrain, comprises a compact dorsal portion ("pars compacta") of pigmented dopaminergic cells, and a ventral, reticular portion ("pars reticulata") of nonpigmented GABAergic neurons.

The substantia nigra pars reticulata and the medial (or internal) globus pallidus form a functional unit as the exit sector of the basal ganglia. In these connections, and in the functional relation with the limbic system (ventral striatum), another set of dopaminergic neurons (ventral tegmental area, VTA, or A10), adjacent to the substantia nigra, participates [35].

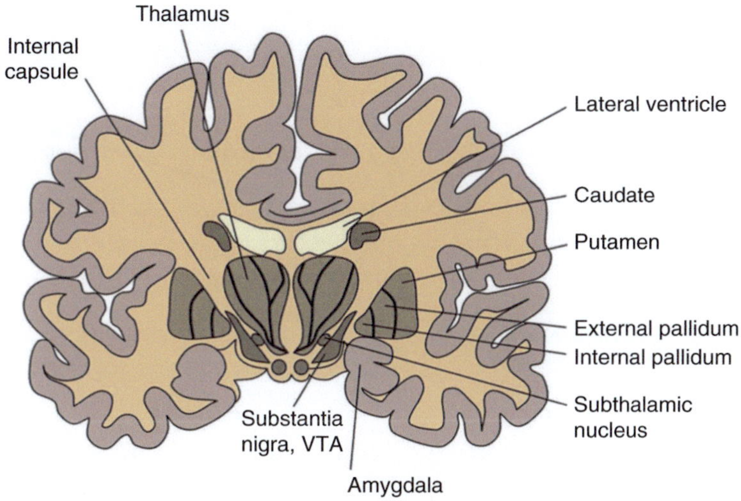

Fig. 6.16 Components of the basal ganglia. Modified with permission from Cardinali [1]

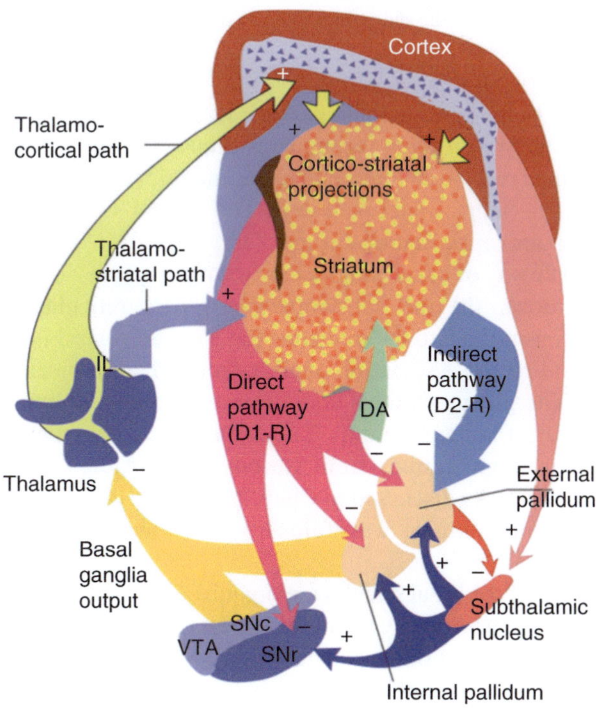

Fig. 6.17 Connections (direct and indirect) of the basal ganglia. *SNc* substantia nigra pars compacta, *SNr* substantia nigra pars reticulata, *VTA* ventral tegmental area. Modified with permission from Cardinali [1]

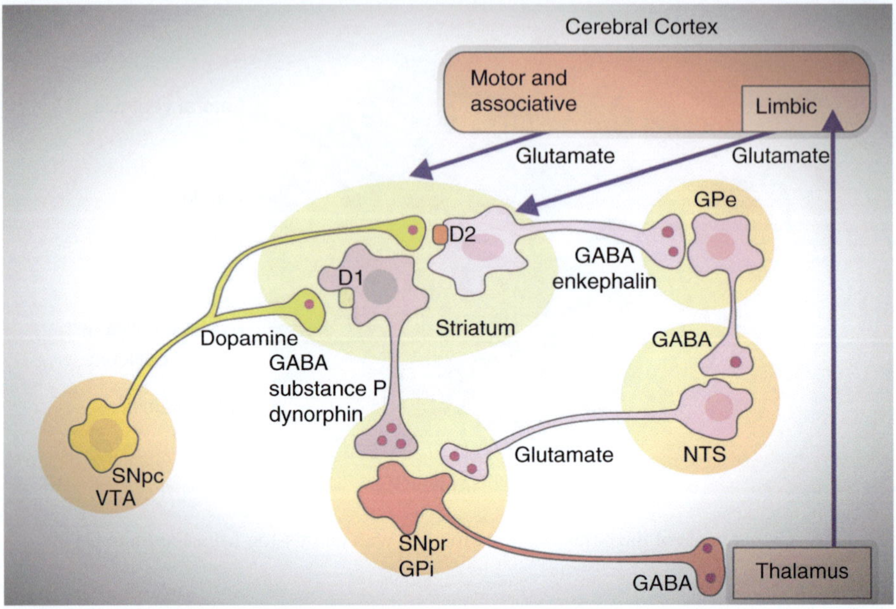

Fig. 6.18 Neurotransmitters in the direct and indirect pathways of the basal ganglia. Modified with permission from Cardinali [1]

The main entry to the basal ganglia is at the level of the striated body (caudate–putamen) and the ventral striatum (accumbens; Figs. 6.17 and 6.18). Both the caudate and the putamen receive an important dopaminergic projection of the substantia nigra pars compacta (nigrostriatal pathway). The ventral striatum receives projections from the dopaminergic neurons of the VTA.

The efferent pathway of basal ganglia has two main origins: (a) the globus pallidus (medial portion); (b) the substantia nigra pars reticulata. Both sets of GABAergic neurons project to the thalamus (on specific ventral lateral and ventral anterior nuclei and on the association nuclei), from where projections arise to the cerebral cortex. The globus pallidus also projects to thalamic intralaminar nuclei (which in turn project to the striatum). The thalamic intralaminar nuclei send and receive glutamatergic projections from various areas of the cerebral cortex and from subcortical areas. As discussed in Chap. 2, the state of this circuit determines the three body configurations: wakefulness, slow-wave sleep, and REM sleep found in a 24-h cycle.

The connections of the basal ganglia are organized in two main ways (Figs. 6.17, 6.18, and 6.19): (a) a direct pathway, involving the projection of the striatum to the globus pallidus (medial portion)/substantia nigra pars reticulata and from there to the thalamus; (b) an indirect pathway, comprising the striatum, globus pallidus (lateral portion), and the subthalamic nucleus, and from there, to the substantia nigra pars reticulata/globus pallidus (medial portion).

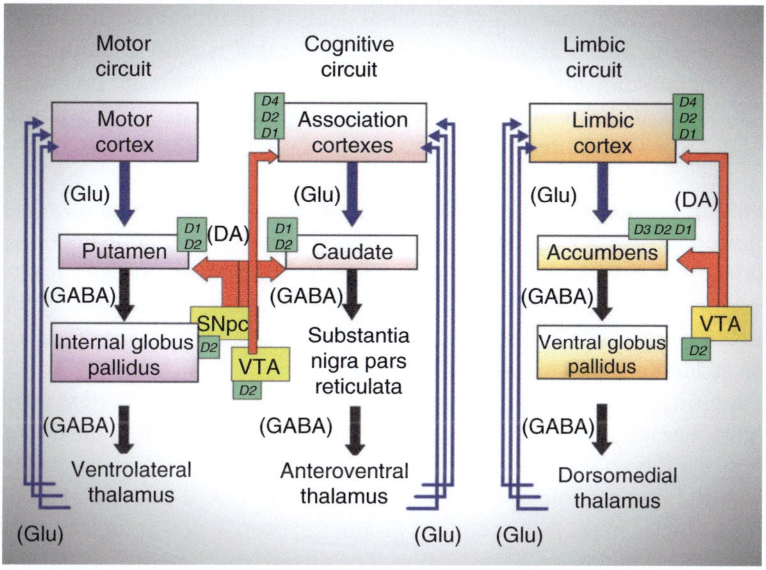

Fig. 6.19 Homologies among the motor, cognitive, and limbic circuits of the basal ganglia. Modified with permission from Cardinali [1]

In both pathways, the basal ganglia and thalamus form closed circuits that receive information from large portions of the cortical territory, and project to cortical areas of motor planning (mainly, the supplementary motor area), cognitive cortex (frontal association cortex), and limbic association cortex.

After stimulation of the cerebral cortex, the inhibitory projections of the striatum are stimulated with two consequences: (a) the direct pathway reduces the inhibitory activity of the basal ganglia nuclei (medial globus pallidus/substantia nigra pars reticulata) on the thalamus; (b) the indirect pathway, via reduction of the inhibition exerted by the lateral globus pallidus on the subthalamic nucleus, increases the excitation given by the subthalamic glutamatergic projections on the medial globus pallidus/substantia nigra pars reticulata. In this case, the inhibitory activity on the thalamus increases [36].

From the functional point of view, there is a definite anatomical segregation in the striatum: the putamen is linked to the motor functions, the caudate is primarily linked to cognitive functions (receiving the thalamic-striatal projection from the intralaminar nuclei as input) and the ventral striatum is associated with the limbic system (Fig. 6.19).

The function of the striatal–pallidal and striatal–nigral connections is to transform the excitatory input of the cortex into a balanced antagonism on the major exit neurons of the basal ganglia, i.e., the medial globus pallidus/substantia nigra pars reticulata GABAergic neurons. The dopaminergic input modulates this balance. DA acts on excitatory D1-type receptors in striatal neurons of the direct pathway, and on inhibitory D2-type receptors in striatal neurons of the indirect pathway.

As stated, the main function of the basal ganglia circuit is to select a movement or sequence of thoughts or an autonomic response that is most appropriate for the situation, suppressing any others. To achieve this, three circuits are built with functional common features and sequelae in the alterations that compromise them (Fig. 6.19) [35]. These three circuits are:

- The motor circuit originates in the regions of the motor and premotor cortex, and in the somatosensory cortex. It passes through the putamen, dorsolateral globus pallidus, and the ventrolateral nucleus of the thalamus to project back to the supplementary motor cortex. The alteration of this circuit produces hypokinetic sequelae (such as bradykinesia of Parkinson's disease) or hyperkinetic sequelae (such as Huntington's chorea).
- The cognitive circuit originates in the dorsolateral prefrontal cortex, projects to the dorsolateral portion of the caudate nucleus and from here to the dorsolateral globus pallidus and the ventral anterior and dorsomedial thalamic nuclei, to close the circuit in the dorsolateral prefrontal cortex. Lesions of this circuit (equivalent to bradykinesia in the motor circuit) produce executive deficit (apathy), with difficulty in working memory and action. The obsessive–compulsive disorder is the hyperkinetic equivalent in cognitive circuit lesions, in which stereotyped behaviors are repetitively performed (e.g., washing hands dozens of times) similar to motor tics. In Tourette's syndrome, motor tics coincide with compulsions and obsessions.
- The limbic circuit originates in the inferior and lateral portion of the frontal cortex (orbitofrontal) and projects to the ventromedial region of the caudate nucleus, the nucleus accumbens, and the dorsomedial region of the globus pallidus, to return to the cortex via the anterior and dorsomedial ventral thalamus. This circuit is especially relevant for functions of personality, socialization, restriction of impulses, empathy, etc. Abnormal hyperactivity in this circuit results in addictive behavior, irritability, impulsivity, and disinhibition. Abnormal hypoactivity is manifested by anhedonia, that is, the inability to experience pleasure.

The mesolimbic system consists of dopaminergic projections from the midbrain to the cortex (mainly the prefrontal cortex, mesocortical portion) and the ventral striatum (mesencephalic portion). The dopaminergic neurons are in the VTA (A10), adjacent to the substantia nigra, which send their axons to the cortex and the ventral striatum (nucleus accumbens, the ventral part of the caudate, and the putamen). Dopaminergic neurons of the VTA discharge in the presence of reward or of stimuli that predict reward. These projections are targets of drugs capable of generating addiction and their circuits are altered in diseases such as schizophrenia [13].

The ventral striatum receives most of its excitatory afferents from the hippocampus, basolateral amygdala, and prefrontal cortex. Another part includes the cingulate circuit that originates in the region of the anterior cingulate, and projects to the nucleus accumbens, the olfactory tuber, regions of the ventromedial caudate, and the putamen. The circuit returns to the cortex through the lateral globus pallidus and to the anterior cingulate via the dorsomedial thalamus.

Given a pattern of cortical information input, the striatum selects the most appropriate behavioral action repertoire that is triggered by the activation of the direct path. Simultaneously, and through the activation of the indirect pathway, the striatum suppresses the execution of inappropriate behavioral actions. The activity of dopaminergic neuron signals predicts environmental events of potential importance for the individual. The dopaminergic neurons of the substantia nigra or the VTA are activated only by sensorial stimuli that have a motivational meaning. By these mechanisms, the basal ganglia participate in the learning and selection of the most appropriate behavioral patterns for a determined environmental and motivational context [35].

It must be noted that routine automation reduces the computational burden of the cerebral cortex by enabling it to process other types of information "in parallel." Thus, the basal ganglia deal with the automatic (implicit) processing of information whereas the cerebral cortex deals with complex tasks of consciousness (explicit processing of information). The cortico-striatal circuits are the basis for the transformations that convert a cognitive frame of reference into an appropriate sequence of actions.

In 1954, the areas of reward and punishment were identified, mostly located in the limbic system. This was done by implanting stimulation electrodes in the central forebrain bundle and found reinforcement of induced behavior, i.e., the animal concentrated all its effort in self-stimulating without paying attention to other meaningful stimuli such as food. Other positive reinforcement points identified were the VTA, nucleus accumbens, prefrontal cortex, and lateral hypothalamus. In all these cases, the animals self-stimulated until emaciation, ceasing the effect if the dopaminergic neurons of the VTA were destroyed. The idea that the mesolimbic dopaminergic system was substantial in defining the hedonic characteristics of a stimulus was thus consolidated (Fig. 6.20).

Two circuits are distinguished in this reward system [13]: (a) a mesolimbic circuit, composed of projections of the cell bodies of the ventral tegmental area to the nucleus accumbens, amygdala, and hippocampus, which is involved in acute reinforcing effects, memory, conditioned responses, and emotional changes of the withdrawal syndrome; (b) a mesocortical circuit, which includes projections of the ventral tegmental area to the prefrontal cortex, orbitofrontal cortex, and cingulate cortex. This circuit is involved in the conscious experience of drug effects, "craving," and the compulsion to use drugs [13].

The mesolimbic and mesocortical circuits operate in parallel and are affected reciprocally and with other areas, forming what has been called the "extended amygdala." In the extended amygdala, there is interaction of the mesolimbic and mesocortical circuits by means of projections of the GABA neurons of the nucleus accumbens to the VTA and to the prefrontal cortex and glutamatergic projections from the prefrontal cortex to the nucleus accumbens and to the VTA.

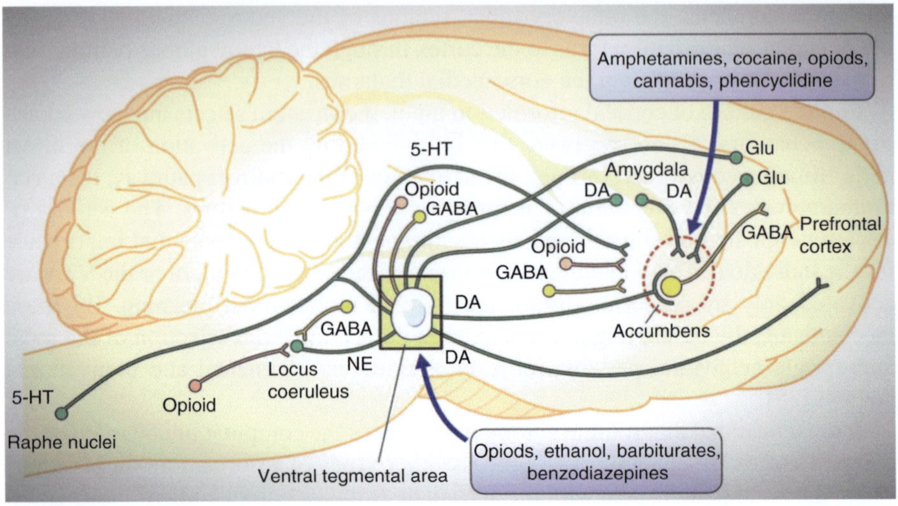

Fig. 6.20 The mesolimbic and mesocortical circuits that make up the "extended amygdala,", their neurotransmitters, and the drugs of abuse that act upon them. Modified with permission from Cardinali [1]

The dopaminergic projection modulates the flow of information in the "extended amygdala", providing a signal indicative of the presence of an important event that requires attention. This mesolimbic/mesocortical system participates not only in mechanisms of reward, but also in aversive responses, including stress.

Addictive substances induce pleasant states (euphoria in the initiation phase) and a decrease in pain. Its continued use produces adaptive changes in the CNS that lead to tolerance, physical dependence, sensitization, craving, and relapse in use. Both natural pleasures (food, drink, sex) and addictive drugs stimulate the release of DA from neurons in the VTA that project to the nucleus accumbens. This is translated into euphoria and reinforcement of behavior. In the case of natural pleasurable stimuli, there is a very rapid adaptive change (habituation). In the case of the response to addictive drugs there is no habituation (each dose of the drug stimulates the discharge of DA). During the abstinence syndrome associated with the drug abuse listed in Fig. 6.20, there is a significant decrease in DA release in the nucleus accumbens. In individuals with high limbic DA levels, there is a poor initial experience with less possibility of continued use of abuse drugs. In contrast, individuals with low levels of limbic DA, there is an increased risk of drug use because of the intense initial response [13].

Functional Neuroimaging of ANS

As discussed in previous chapters, a basic concept is that the function of the ANS is given by a highly interconnected hierarchy of neuronal structures involving brain areas from the neocortex to the brainstem, in addition to the cerebellum and the basal ganglia. The way in which those structures operate in a network was first demonstrated by neuroanatomical and neurophysiological techniques [37, 38], and more recently by the application of neuroimaging techniques, mainly fMRI [39].

In Chap. 7, we analyze several autonomic tests available to clinically assess ANS function (Table 7.2). Some can be adapted to the MRI environment, whereas others are impractical inside an MRI environment (such as orthostatic changes elicited either by a tilt table or by changing posture from sitting to standing). Procedures involving a static body position with minimal electrical equipment are most commonly used. Valsalva maneuver has been the focus of attention as a simple challenge that elicits a strong autonomic reaction (Chap. 7). The Valsalva maneuver can be performed in the supine position and can be repeated multiple times within a typical fMRI protocol [40]. A hand grip, which also involves sympathetic activity increases, is another autonomic test feasible to be used in fMRI.

Functional MRI is less suitable for identifying changes in state that last several minutes, such as the quantitative sudomotor axon reflex test. However, electrical stimulation of muscle or nerves has been performed.

By using fMRI, the pathways for sympathetic outflow in the ventral medulla were described, and the temporal patterns for such medullary activation on fMRI to foot cold pressor and Valsalva maneuver were readily apparent in healthy adolescents and adults.

Concerning the central autonomic network and limbic regions, neuroimaging has confirmed the original findings derived from recording, lesion, stroke, and physiological studies, demonstrating that cortical brain regions and other rostral brain areas participate in autonomic regulation [41], and have extended the regions we now know to be involved in autonomic regulation.

For example, the insula participates in BP challenges in a significant fashion. Forehead cold pressor, lower body negative pressure, the Valsalva, and the related forced expiratory loading all lead to insular activation. Hand grip and maximal inspiratory loading similarly recruit the anterior and posterior insula. The insula has inhibitory projections to the hypothalamus and its functional organization of the insula is asymmetrical, with the right side being preferentially active during sympathetic increases and the left side during parasympathetic action [39].

Other regions involved in autonomic regulation are the cingulate, the ventromedial prefrontal cortex, basal ganglia, and hypothalamus, along with the amygdala and hippocampus [42]. As described in Chap. 5, the hypothalamus plays a major role in regulating autonomic outflow, with substantial projections from other limbic structures and efferent projections to the brainstem. The hypothalamus shows fMRI signal responses under some conditions, but as the structure is small, differentiating local responses of the multiple subnuclei of the hypothalamus by fMRI is difficult. The ventromedial prefrontal cortex, amygdala, and hippocampus play significant roles in the sequencing of responses to BP and other ANS challenges.

By recording muscle sympathetic nerve activity at the same time as performing fMRI of the brain, the cortical structures involved in central cardiovascular control in awake human subjects can be best identified. Signal intensity and muscle sympathetic nerve activity correlated positively in the left mid-insula, bilateral dorsolateral prefrontal cortex, bilateral posterior cingulate cortex, and bilateral precuneus.

In addition, muscle sympathetic nerve activity covaried with signal intensity in the left dorsomedial hypothalamus and bilateral ventromedial hypothalamus (VMH). Construction of a functional connectivity map revealed coupling between activity in the VMH and the insula, the dorsolateral prefrontal cortex, the precuneus, and in the region of the left and right rostroventrolateral medulla [43].

In thermoneutral conditions, resting skin muscle sympathetic nerve activity is related to the level of arousal and emotional state. The identified brain regions responsible for the generation of spontaneous muscle sympathetic nerve activity include the left thalamus in the region of the ventromedial nucleus, the left posterior and right anterior insula, the right orbitofrontal cortex, the right frontal cortex, and bilaterally in the mid-cingulate cortex and precuneus [44]. Functional connectivity analysis revealed a strong positive coupling between the right orbitofrontal cortex and the right anterior insula. Signal intensity changes within the precuneus were temporally coupled with the left anterior and posterior insula, cerebellum, cingulate cortex, and thalamus. Presumably, these brain regions monitor the internal state of the body and may regulate emotional state changes [3].

One important finding in fMRI studies has been to discover the cerebellar contributions to ANS regulation (Chap. 4). The cerebellar cortex responds regionally to BP changes, including respiratory loading, lower body negative pressure, Valsalva and Mueller maneuver, cold pressor, end-expiratory breath hold, and static hand grip [39]. The data are consistent with a dampening or coordinating role for the cerebellum in the presence of significant changes in BP, which could be similar to the motor coordination role traditionally associated with the structure [45].

Autonomic functions in the brain are lateralized [42], in a manner reminiscent of other functions, including motor, sensory, and language systems. The cold pressor and hand grip challenges show multiple structures with lateralized responses to the challenges, notably in the mid and posterior insula. The amygdala, hippocampus, and ventral cerebellum show opposite responses to a Valsalva maneuver on the left and right sides. The insular cortex is of interest with respect to lateralized autonomic function, as the left-side function appears to be preferentially parasympathetic and the right-side preferentially sympathetic. Hence, resection of the left insula led to minimal autonomic changes, but resection of the right led to less sympathetic and more parasympathetic activity [46]. The lateralization of function has obvious implications for stroke or other injury, as unilateral damage would have an impact on the extent and timing of BP regulation.

The vermis participates in fear learning and memory mechanisms related to the expression of autonomic and motor responses of emotions. In humans, the cerebellar hemispheres are also involved at a higher emotional level [47].

In rodents, the reversible inactivation of the vermis during the consolidation or the reconsolidation period hampers the retention of the fear memory trace. In this region, there is a long-term potentiation of both the excitatory synapses between the parallel fibers and the Purkinje cells and of the feed-forward inhibition mediated by molecular layer interneurons (Fig. 4.19). This concomitant potentiation ensures the temporal fidelity of the system. Additional contacts between mossy fiber terminals and Golgi cells provide morphological evidence for the potentiation of another feed-forward inhibition in the granular layer. Imaging experiments show that in humans, the cerebellum is also activated during mental recall of emotional personal episodes and during learning of a conditioned or unconditioned association involving emotions [45].

Chronotypes, 24-h Rhythms and Emotion

A Gaussian distribution of the acrophase in the body temperature 24-h rhythm occurs, with a mean at 18:00 h and with two-thirds of the population within a ±1 h range. Five % of the population are out of phase 2 h before or 2 h after the mean, and these subgroups are called "larks" (morning type) and "owls" (evening type) respectively. Thus, the people with different chronotypes, the morning or larks and the evening or "owls," form the limits of the normal distribution in the human population [48].

The acrophases of the circadian rhythms of body temperature, mental performance, and sleep–wake cycles in the morning and evening groups occurs considerably before or after respectively what is considered the norm. These differences can be due to differences in sensitivity of the phase adjustment process of the circadian clock and/or to the lifestyle. Differences in lifestyle influence exposure to zeitgeber, as those individuals who go to bed and get up early are exposed much earlier to the environmental zeitgebers. However, the phase differences persist, even though the individuals studied are maintained in the same sleep–wake cycle, or in a constant routine, which would rather point to an endogenous origin, related to the synchronization mechanisms of the circadian pacemaker. The chronotypes are associated with genetic variations, lifestyle differences, mood states, cognitive function, and risks of health problems (sleep disorders, depression) [48].

The "larks" are active in the morning, reach their maximum performance during the noon hours, and enjoy little any nocturnal obligations, at which time they show tiredness and a predisposition to sleep. The "owls" rise late, are gaining energy during the day, and reach their maximal performance toward the night-time; they prefer, therefore, to prolong the vigil period.

The inclination of humans to sleep at night is probably linked to the dependence of the primate on vision, not smell, as the dominant sense. An in-day wakefulness and overnight sleep program must have been advantageous in a primitive world plagued by dangerous nocturnal predators. As a corollary, sleep must have been a positive influence on natural selection for primates, being beneficial during the hours of darkness. From this evolutionary point of view, there is a sense of population distribution in "larks" and "owls," as the existence of individuals with different sleep–wake rhythms should have allowed greater efficacy in surveillance before the possible nocturnal predators of the hominids.

Several studies in twins have shown genetic links with aspects such as circadian time and sleep/wake preferences. The "owls" suffer from a conflict between internal and external time ("social jet lag" [49]) as they have more difficulty adapting to the demand for morning activity than the "larks" or intermediate chronotypes.

Studies on fluorine-19 nuclear magnetic resonance imaging have revealed differences in the metabolic function of the brain among "owls" compared with "larks" [50]. These metabolic differences were discovered in limbic regions and may be a reason why the nocturnal chronotype has a higher risk of depression related to insomnia. The diminished whiteness of the CNS white matter can be the result of chronic social jet lag because of the stress it entails: people who are willing to stay late and sleep late are often in constant conflict with the schedule that surrounds them.

A common genetic variant has been identified that affects almost the entire human population, and is responsible for the difference of more than 1 h a day in the tendency to be an early riser or a night owl [51]. A single nucleotide in *Per1* varied between two groups that differed in their sleep–wake behavior. At this site of the genome, 60% of individuals have the nucleotide base adenine (A) and 40% have the nucleotide base guanine (G). Because one has two sets of chromosomes, in any given individual, there is a 36% chance of having two A's, a 16% chance of having two G's, and a 48% chance of having a mixture of A and G on this site. People who have the AA genotype wake up about 1 h earlier than people who have the GG genotype and the AG wake up almost exactly in the middle.

There is evidence for the relations between chronotype and cognitive ability. Nocturnity was positively related to cognitive ability, but negatively correlated with academic performance indicators. On the contrary, morningness had a negative relationship with cognitive ability and a positive relationship with academic indicators. When school performance and inductive intelligence were measured, evening subjects scored higher than morning subjects in inductive reasoning (a good estimate of general intelligence and one of the strongest predictors of academic achievement). In general, the nocturnal chronotype is more common in creative and extroverted individuals (poets, artists, inventors), whereas the morning chronotype is in individuals with logical, systematic, and deductive thinking (accountants). It is interesting that inductive reasoning is linked to innovative thinking, higher-profile occupations, and higher incomes. In general, studies have revealed that school performance of "owls" is usually lower than that of "larks" because the late chronotype is adversely affected by school schedules.

The nocturnal chronotype is also more prone to the indiscriminate use of tobacco and alcohol and to a greater caloric intake. Several studies have documented how the timing of sleeping and eating can affect body weight (Chap. 5) [52].

If our preferences for sleep and wakefulness are strongly influenced by genetics and biology, what can we do when these inclinations do not match the demands and responsibilities of our lives? Indeed, limiting exposure to artificial light at night and increasing exposure to sunlight during the day can change sleep–wake cycles, even in owls. Adequate sleep habits, not consuming alcohol before bedtime and maintaining a dark bedroom and no interfering technology can help to reinforce the sleep schedule, even if it does not perfectly align with the natural tendencies of the subject in question.

There are better times of day for the execution of tasks depending on the cognitive or physical abilities needed [53]. To know them and to try as far as possible to select the optimal schedules to develop the activities facilitate and optimize the performance. At the end of the day, the ability to concentrate and to make a material to be stored in the long-term memory are superior, whereas the ideal time to face a test occurs at noon, when the working memory and the feeling of activation and well-being are better. These daily fluctuations of behavioral parameters cannot be considered trivial, as the total change detected is about 10% and the variation in the efficiency of execution is equivalent in magnitude to the effect of sleeping only 3 h or of ingesting the legal limit of alcohol [53].

As for the measurement of biological parameters, there are numerous variables that influence them, minimizing or amplifying the presence of rhythmicity in the subjects' performance skills. These include the registration status (habitual activity or controlled in the laboratory), food intake, consumption of psychoactive substances and environmental conditions (temperature, light, noise, etc.), or social interaction. Thus, the study of circadian variations in cognitive and physical performance presents methodological difficulties added to those of the specific tasks performed [53]. If we value the response of individuals to a task, we must consider that their performance may be more influenced by the learning process and/or by their own fatigue or boredom that involves doing it repeatedly than by the very effect of the time of recording. It is impossible to carry out continuous measurements of cognitive and physical performance, selecting an interval between records that allows the recovery of the individual as per the demand, type, and duration of the tests used. A sampling interval every 2–3 h is usually adequate and, wherever possible, reliability increases by logging more than 1 day. In addition, the levels of motivation and stress of individuals, not always easy to evaluate and therefore to control, are factors that have a decisive influence in the evaluations of execution. For example, highly motivated subjects perform better and show minimal diurnal variability.

There are also considerable variations in the phase and amplitude of circadian rhythms with age, especially in childhood and in senescence. In newborn infants, circadian rhythms are poorly developed, although the existence of a circadian rhythm of low-amplitude body temperature has been described. Circadian rhythms increase in amplitude throughout the first months of life, which is believed to be due to a more robust circadian signal from the suprachiasmatic nuclei (SCN), probably because of the increase in synaptic connections among the cells of these nuclei. During this stage, the sleep pattern and the feeding pattern also mature; the subjects are more active during wakefulness and their interest in the environment around them is aroused. All these contribute to the development of circadian rhythms, as some of these variables act as zeitgebers and promote circadian clocks, all of which increase exogenous influences on circadian rhythms (masking).

In the elderly, the rhythms are also less marked [50]. There is a decrease in the clock output signals and an increase in the phase instability of the circadian rhythms in the population as a whole, alterations that persist during "constant routine" protocols. SCN neuronal degeneration itself may contribute to these modifications. With age, the ability to maintain sleep is also diminished, and the increase in the frequency of urination at night, as renal circadian rhythms deteriorate, results in a decrease in the production of urine. The phase of the circadian rhythms moves toward earlier hours of the day, and the older individuals tend to behave more like "larks" [50].

Generally, the variations in the phase and amplitude mentioned above do not cause serious problems to the individuals. Even when difficulties appear, they appear to be rather a reflection of a general pattern of development and deterioration of the body at various stages of life; thus, we can hardly call them "anomalous." However, there are examples, in both young and old, where the departure from the norm is more marked, and for which the term "anomalous" would seem more appropriate. Some children have nocturnal enuresis. It may be due to inappropriate development of the circadian rhythm of AVP secretion: very little hormone is released at night. In some elderly people, getting up at night to urinate also becomes a problem, in this case because of the decreased ability of the bladder to store large volumes of urine and an altered rhythm of diuresis.

Sleep disturbances are a symptom that causes great concern for the individual who suffers them, and in fact, some types of insomnia seem to be produced by alterations of the circadian system. One of the most frequent of these circadian alterations is the delayed sleep phase syndrome (DSPS) and related to it, although less frequent, is the advanced sleep phase syndrome (APSP) [54]. In these alterations, individuals wish to sleep at hours that are either too late (e.g., 04:00 h in the case of DSPS) or too early (e.g., at 19:00 h in the case of ASPS) for a conventional work and life schedule. In both cases, the phases of all circadian rhythms appear to be affected in the same way (temperature, melatonin, cortisol), although it is precisely the problem with sleep that generates more discomfort and therefore the one most likely to be detected in the patient. When these individuals can sleep at will, they do not seem to have problems sleeping normally (provided their rooms are isolated from the noise and outside light that normally exists during sleep during the day). The problem is derived from an anomalous phase relationship between the circadian clock and the environmental synchronizers, although the motifs are unknown. A mutation in certain clock genes may be behind these alterations [54].

In some individuals, circadian alterations occur because the circadian clock is unable to adjust to the solar day. This problem occurs occasionally in people who can see, but it is quite common in blind people, a fact that supports the idea that the normal light–dark cycle is an important synchronizer (Chap. 2). In this alteration, all circadian rhythms are maintained, including those that affect the ability to sleep and be awake, but show a period longer than 24 h. As the pace of the body rhythms shifts from that which would be appropriate for a normal lifestyle, problems with night-time sleep loss and fatigue during the day worsen, and become more marked at the time that body rhythms and lifestyle are maximally out of phase (i.e., 12 h; periodic insomnia) [54].

24-h Rhythms and Learning and Memory

The most important acquired determinant to modify human behavior is learning, and the consequence of its persistence, or memory. These processes are more persistent the earlier in life they are acquired. Hence, the relationship of the newborns with their mothers is of importance. A child with a bad or insufficient affective relationship in the early stages of his development is prone to presenting alteration in his emotional and ANS reactivity as an adult [1].

The concepts of memory and learning are closely related. Learning is a process through which new information is acquired. Memory refers to the persistence of what is learned, in a state that can be evoked later. In this sense, memory is the result of learning [55].

One commonly used classification of memory considers its persistence: (a) short-term memory; (b) long-lasting memory. The former refers to systems that retain information temporarily, whereas the latter imply permanent information (Fig. 6.21) [56].

The short-term memory is also called working memory. It is characteristic of the prefrontal association cortex and comprises a workspace in which information is maintained while it is processed. The information may come from long-term memory or from newly acquired information that is being incorporated or being used for short periods of time and then discarded.

The long-term memory is divided into two processes (Fig. 6.21): (a) declarative memory; (b) reflexive or procedural memory.

Declarative memory involves the cognitive mechanisms by which a past event is recalled, with the possibility or not of verbal expression. It constitutes the memory, which when lost, results in an individual commonly recognized as an amnesiac.

The reflexive or procedural memory refers to the learning process of the subcortical type and that does not require participation of cognition. It may include motor and ANS components. In the case of motor learning, it implies the different nuances,

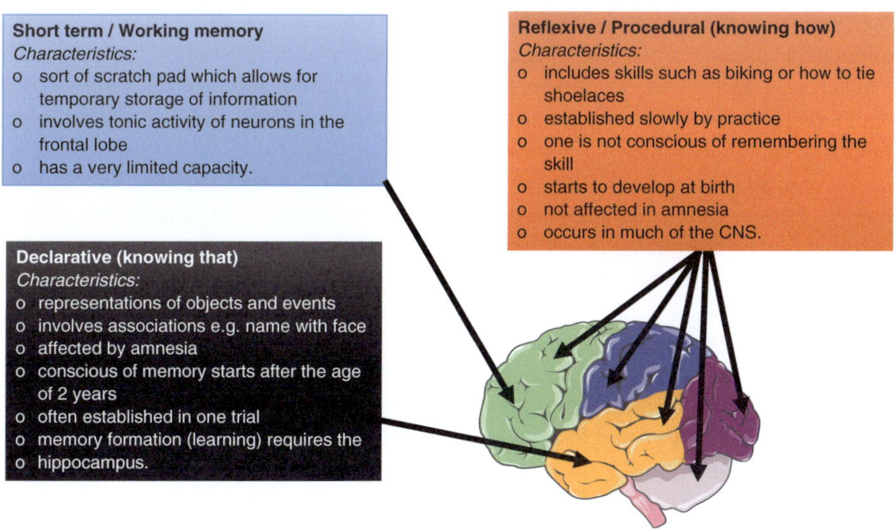

Fig. 6.21 Short-term, declarative, and reflexive memories. The figure was prepared in part using image vectors from Servier Medical Art (www.servier.com), licensed under the Creative Commons Attribution 3.0 Unported License (http://creativecommons.org/license/by/3.0/)

fixed by experience and repetition, of a motor action. Both motor and ANS learning are linked to cerebellar function.

How are these types of memory revealed? A suitable test for analyzing procedural memory is to train a person to read inverted words (i.e., reflected in a mirror). Normal individuals require on average two sessions to acquire this capacity, which is maintained for about 30 days and then extinguishes. The patient commonly recognized as an amnesiac (i.e., who suffers from declarative amnesia) performs normally on the reverse reading test, although he does not even remember that he has participated in the training sessions. That is, the mechanisms of procedural and declarative memory differ from each other and can be affected independently [57].

Classical conditioning is the basis of reflexive or procedural memory. The declarative memory, on the other hand, implies the mechanisms of fixation of the experience usually recognized as "memory" [58, 59].

Engram defines the set of neural changes that occur during the memory process. The engrams are the result of learning, and comprise biochemical and structural changes in the participating neural circuits. In general, they represent a modification of the synaptic efficacy of such circuits.

The memory lacks a cerebral location (there is no "memory center"). On the contrary, memory is the result of information processing and is a change, permanent, in the same neural circuits that process sensory information. For example, in the visual system, the inferotemporal cortex (one of the last areas in the process of form analysis) is, in addition to a secondary sensory cortex, a place for the storage of visual engrams.

One way to demonstrate the storage of engrams in high-order sensory cortical areas is by microelectrode stimulation of the cortex. Intraoperative stimulation of the primary auditory cortex (Brodmann areas 41 and 42) produces noises, that is, elemental sensory sensations. When the secondary auditory cortex, i.e., area 22 (Wernicke's area in the dominant hemisphere) is stimulated, complex sensations occur (audible words in the dominant hemisphere, melodies in the nondominant hemisphere, etc.).

The physiological reason for the fixation of memory engrams, which constitute a very small portion of the mass of information circulating in the sensory processing areas, is that the simultaneous activation of the (limbic) motivational system occurs. In fact, one remembers what has had a certain emotional, conscious or unconscious meaning [60].

When studying the correlation between the clinical picture and the underlying pathological condition in amnesiac patients, it can be verified that bilateral damage of certain brain areas makes it impossible to establish new memories (anterograde memory) and to remember (retrograde memory), although this latter type of amnesia disappears after a certain time. These brain areas are: (a) the medial temporal lobe area (hippocampus, amygdala); the medial area of the diencephalon (hypothalamic mammillary body, mediodorsal nucleus of the thalamus) [61].

The surgical lesion of the hippocampus in humans produces declarative amnesia with anterograde memory deficit. The afferents to this region come from the entorhinal cortex, the contralateral hippocampus, and from subcortical structures such as the medial septum, certain raphe nuclei, and the LC from the brainstem (Fig. 6.6). The dentate gyrus is the area where information enters the hippocampus. Both dentate gyrus and hippocampus are phylogenetically old cortical areas consisting of three layers (allocortex). Located in the dentate gyrus are the molecular layer, the grain cell layer (which are excitatory and give rise to mossy fibers), and the hilum, with a population of GABAergic interneurons. The three-layer allocortical structure of the hippocampus comprises: (a) the molecular layer, where the apical dendrites of pyramidal neurons reside; (b) the pyramidal or main layer, with its population of excitatory cells that give rise to the Schaffer collaterals; (c) a polymorphic layer containing GABAergic interneurons. There are several cornu ammonis (CA) areas: CA1, CA2, and CA3 (Fig. 6.6) [56].

It was in the hippocampus where the long-term potentiation phenomenon, which is fundamental to the learning process, was first described [56]. Long-term potentiation consists in changing the efficacy of certain synapses (as already described for synapses in the dentate gyrus or in the CA1 layer) depending on their previous activity. Potentiation is induced by brief, high-frequency stimulation of the perforating pathway (entrance to the hippocampus) both in vivo and in vitro. The process becomes manifested in a matter of seconds and its effect is maintained for weeks in the animal under physiological conditions. Long-term potentiation presents as a characteristic the possibility of associating stimuli from various sources. This property allows to the effect of a weak stimulus to be enhanced only when its presence is associated with that of a more intense stimulus. This situation is very like that of classical or Pavlovian conditioning.

Figure 6.22 outlines the basis of long-term potentiation. Normal activation of the neural pathway, which uses Glu as a neurotransmitter, and in the presence of N-methyl-D-aspartate (NMDA) and non-NMDA glutamatergic receptors in the postsynapsis (Chap. 3) produces a predominant postsynaptic response through non-NMDA receptors. This is because NMDA receptors remain inactive in the presence of Mg^{2+} at certain membrane potential values. If other depolarizing stimuli of pathways affecting the same postsynapsis are simultaneously associated, the membrane potential value is exceeded to overcome the inhibition given by Mg^{2+}, which activates the NMDA receptors (Fig. 6.22). This produces Ca^{2+} entry with a self-regenerating depolarization. This flow of Ca^{2+} through the NMDA receptor is critical for long-term potentiation to appear. Activation of protein kinase C and calmodulin kinase are the next necessary steps for this phenomenon. When the conditioned and unconditioned pathways are activated in a contingent relation, the depolarizing effect allows expression

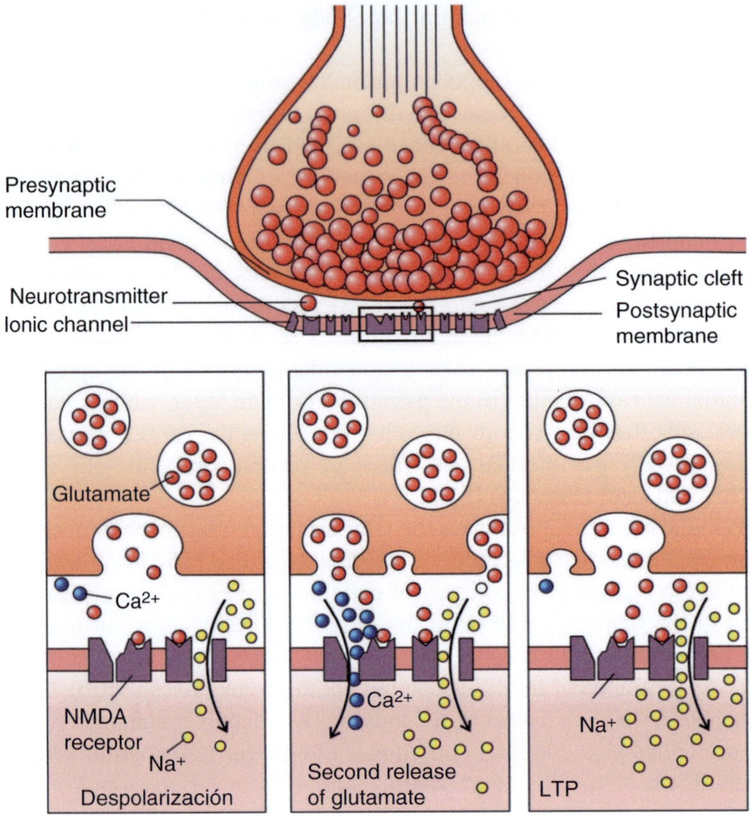

Fig. 6.22 The proposed ionic mechanism for long-term potentiation (LTP). In the second release of glutamate, calcium enters the terminal. This results in increased sodium entry during the following depolarizations. Modified with permission from Cardinali [1]

of NMDA receptor action (increase in Ca^{2+} entry) and consequently engram fixation. If two synapses fire at the same time (synchronously) they produce a larger depolarization than if they fire at different times (asynchronously). Cells that fire together wire together, whereas signals from cells that arrive at different times become weaker. This is the basis for plasticity or learning throughout the cortex [56].

The typical electroencephalography rhythm of the hippocampus is theta (4–10 Hz). This rhythm is related to projections of the medial septum and the diagonal band of Broca and coincides with low activity in the CA3 and CA1 layers of the hippocampus. When theta rhythm disappears, there are irregular waves in CA3 and CA1 concomitant with consummatory behaviors in animals, such as drinking or eating.

The electrical activity of CA3 pyramidal cells increases as a function of the novelty of the stimulus. The CA1 pyramidal cells are activated to the location of the animal in a specific and known place of its spatial environment. Through studies in animals with an injured hippocampus, it has been possible to establish the function of this region in spatial memory.

As stated above, short-term memory is intact in declarative amnesiacs. It is the long-time memory that disappears. Thus, for example, the amnesiac patient can normally hold a list of numbers for several minutes if he keeps his attention on the test, but loses it immediately if he is distracted. In lesions that produce declarative amnesia, there is no modification of the memory already acquired, but there is impairment of fixation of new engrams. Proof of this is that in patients with a bilateral lesion of the hippocampus, the stored memory does not change, nor does it alter the short-term memory, which implies mechanisms independent of the limbic system.

The capacity for the long-term storage of engrams depends both on plastic neural changes in the higher sensory processing area and on the integrity of the motivational circuits linked to the limbic system (the subject remembers only what was fixed with some emotional content). The sensory information above the cerebral cortex is fixed as a declarative engram if the simultaneous activation of the motivational system occurs. The constituent parts of the limbic system that most influence the memory process are the amygdala, the septum, the reticular formation, certain portions of the hypothalamus, and hippocampal formation [3].

The flow of information in this motivational circuit is modulated by the cholinergic projection from Meynert's basal nucleus to the neocortex and by the middle septal nucleus to the hippocampus (septo-hippocampal projections). This cholinergic projection is compromised in senile dementia or AD and these patients also show significant chronobiological and sleep disturbances.

Likewise, noradrenergic projections from LC, and beta endorphins from the hypothalamus, are involved in the process of memory fixation. Hormonal influences, such as those provided by ACTH, AVP, oxytocin or circulating catecholamines, modulate memory by action at the level of the reticular formation [60, 62]. We have already discussed how corticosteroids have a deleterious effect on hippocampal circuits (Fig. 5.19).

The homeostatic value of forgetting is essential for a normal life. It would be absurd to recall every detail of our experience, as we would not have time to live in the present! The Argentine writer Jorge Luis Borges exemplified this fact in his story "*Funes, El Memorioso*," an individual unable to forget a second of every second lived, and who remembered the last of the veins of the leaves of each tree he had seen…

Another fundamental fact is the maintenance of the engrams. We again recount Borges, who describes his memories of Adrogué, a place near Buenos Aires, where he spent vacations as a child:

> …When I think of Adrogué, I do not think of the present Adrogué, deteriorated by progress, radiotelephony and motorcycles, but in that lost and tranquil labyrinth of country houses, squares and streets that converged and diverged, full of Eucalyptus trees… Wherever I am in the world, the smell of the Eucalyptus is enough for me to return to that lost Adrogué, which now exists only in my memory…

How did the association between the aroma of Eucalyptus and Adrogué in Borges' brain did not fade? It should be noted that the neocortex is an area in continuous plastic change and that if the neural circuits are not activated, they disappear. The most logical moment for this activation is during the "hallucinosis" of REM sleep. Therefore, REM sleep is presumably responsible for the "service" of memory [1].

The role of sleep in the process of learning and memory is thus of great importance. It is noteworthy that during sleep the activation of areas used in the learning process in the previous wake period is repeated. In experiments of cortical neuroplasticity induced by monocular deprivation in the cat during the critical period of development (30 days of age) it was observed that for such plastic changes to occur REM sleep must be present. It is possible that REM sleep (by its visual imagery) provides an input analogous to the visual cortex stimulus given by the normal visual input.

There is evidence for a specific output of information from the hippocampus to the neocortex during dreaming. This phenomenon is due to cholinergic activation, which is maximal in wakefulness and REM sleep and minimal in slow-wave sleep. During wakefulness, the exit information from the neocortex is projected onto the entorhinal association cortex and hippocampus, whereas the reverse flow from the hippocampus to the cortex is attenuated (Fig. 6.23) [63].

During slow-wave sleep, information flows mainly from the hippocampus to the neocortex where, over time, memory engrams originate. During REM sleep, the exit from the hippocampus to the neocortex is blocked as in wakefulness, the flow of information from the neocortex to the hippocampus being maintained (Fig. 6.23). Thus, during slow-wave sleep the hippocampus consolidates unstable memories and transfers information to the cortex for long-term storage (these are the most favorable conditions for long-term potentiation and for synaptic plasticity).

There is evidence that the various types of memory are related differently to the type of sleep. The function of REM sleep is to reinforce the cortical plasticity involved in procedural memory and high-level cognitive processes (semantic memory), with little or no function in episodic memory [64].

In contrast, episodic memory depends on NREM sleep. The sources of memory in this period are mainly episodes experienced the previous day (episodic memory), whereas, in contrast, during REM sleep, memory sources are a mixture of episodic memories and semantics with emotional content. There are currently two hypotheses regarding the link between memory and sleep:

- For the dual process hypothesis, NREM and REM sleep facilitate different memory processes: slow-wave sleep facilitates declarative or explicit memory whereas REM sleep facilitates procedural memory, nondeclarative memory, and semantic and emotional memory. In favor of this hypothesis, NREM sleep deprivation selectively decreases performance in word association tasks or spatial memory tests whereas REM sleep deprivation damages performance in procedural tasks such as inverted writing in a mirror.

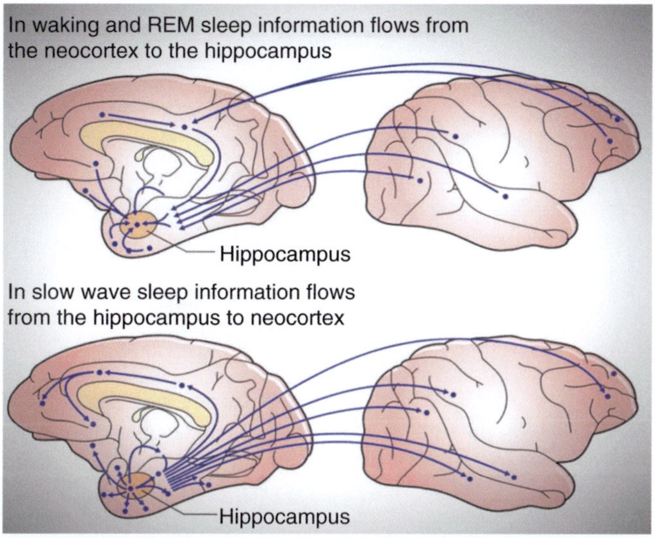

Fig. 6.23 The flow of information to and from the hippocampus is different in sleep and wakefulness. Cholinergic activation of an aminergically demodulated cortex during REM sleep, or the combined cholinergic and aminergic activation during wakefulness, block the passage of information from the hippocampus to the neocortex. In slow-wave sleep, this flow is facilitated by the aminergic effect in the presence of inhibition of the cholinergic input. Modified with permission from Cardinali [1]

- For the sequential hypothesis, the different sleep phases consolidate a memory trail in a consecutive and complementary way. There would be no functional differentiation in the role of each type of sleep in the various categories of memories.

Finally, a very important aspect regarding memory occurs during REM sleep. We suddenly remember things that happened to us a long time ago and that had "disappeared" from our daily memory. It is known that no neural network remains if unused. A patient with amputation of one of its members experiences it as a "phantom limb" for a while until the neuroplastic changes leading to the disuse of the neural circuits representing the amputated limb sensibility are completed.

How then are the distant memories kept? In part or in whole, their engrams must be activated periodically so that they do not disappear. Dreaming guarantees this "service" of memory, contributing both to its maintenance and its reworking (we remember in general only what is pleasant; thus, "all past times were better"; Fig. 6.24).

Neuroimaging methods are currently used experimentally to understand the intricacies of the memory and plasticity processes, in addition to clinically assessing the decline of several cognitive processes that result from changes in defined

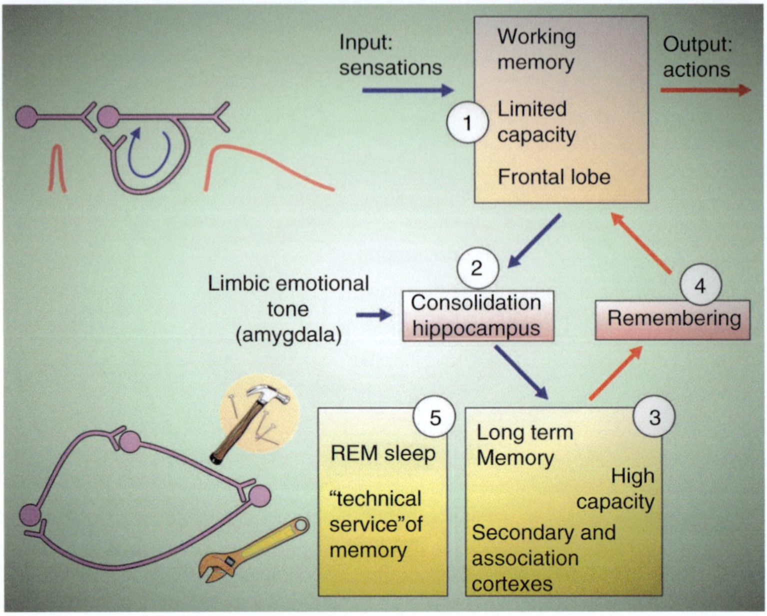

Fig. 6.24 Sensory information flowing through working memory ① is only consolidated in the hippocampus by simultaneous activation of the motivational system (limbic system) ②. Long-term memory, located in high-order or secondary sensory and associated cortexes, requires periodic maintenance by the implementation of parts or the entire engram ③. This "memory service" is most likely effected during REM sleep ⑤. Remembering ④ no longer requires the hippocampus. Modified with permission from Cardinali [1]

neural circuits in the aging brain [65–67]. Advanced magnetic resonance imaging techniques, such as functional connectivity magnetic resonance imaging, diffusion tensor imaging, and magnetic resonance spectroscopy, and molecular imaging techniques, such as ^{18}F fluoro-deoxy glucose PET, amyloid PET, and tau PET, are available for experimental and clinical use.

By using these methodologies, the functional neural architecture of working memory is described as an interaction of the frontal-parietal control network and more posterior areas in the ventral visual stream. In addition, several studies have demonstrated a link between age-related episodic memory decline and the hippocampus during active mnemonic processing, which is further supported by studies of hippocampal functional connectivity in the resting state. The hippocampus interacts with anterior and posterior neocortical regions to support episodic memory, and alterations in hippocampus–neocortex connectivity have been shown to contribute to impaired episodic memory.

American and European guidelines recommend imaging to exclude treatable causes of dementia, such as tumor, hydrocephalus, or intracranial hemorrhage, and to distinguish between different dementia subtypes, the commonest of which is Alzheimer's disease. As the hallmark feature of dementia is that of irreversible cognitive decline, usually affecting memory, and impaired activities of daily living, intervention at the preclinical stages before irreversible brain damage occurs is currently the best hope of reducing the impact of dementia [65–67].

References

1. Cardinali DP. Neurociencia Aplicada. Sus fundamentos. Editorial Médica Panamericana: Buenos Aires; 2007.
2. Waxman SG. Chapter 19. The Limbic system. In: Clinical neuroanatomy. 27th ed. New York: The McGraw-Hill Companies; 2013.
3. LeDoux JE, Damasio AR. Emotions and feelings. In: Kandel E, Schwartz J, Jessell T, Siegelbaum S, Hudspeth AJ, editors. Principles of neural science. 5th ed. New York: McGraw-Hill; 2012. p. 1079–94.
4. Ropper AH, Samuels MA, Klein JP. Chapter 25. The Limbic lobes and the neurology of emotion. In: Adams and Victor's principles of neurology. 10th ed. New York: The McGraw-Hill Companies; 2014.
5. Lovblad KO, Schaller K, Vargas MI. The fornix and limbic system. Semin Ultrasound CT MR. 2014;35:459–73.
6. Roxo MR, Franceschini PR, Zubaran C, Kleber FD, Sander JW. The limbic system conception and its historical evolution. ScientificWorldJournal. 2011;11:2428–41.
7. LeDoux JE. Emotion circuits in the brain. Annu Rev Neurosci. 2000;23:155–84.
8. Calhoon GG, Tye KM. Resolving the neural circuits of anxiety. Nat Neurosci. 2015;18:1394–404.
9. Bello-Morales R, Delgado-Garcia JM. The social neuroscience and the theory of integrative levels. Front Integr Neurosci. 2015;9:54.
10. Critchley HD, Eccles J, Garfinkel SN. Chapter 6 – Interaction between cognition, emotion, and the autonomic nervous system. In: Buijs RM, Swaab D, editors. Handbook of clinical neurology, autonomic nervous system. New York: Elsevier; 2013. p. 59–77.
11. Lippe S, Gonin-Flambois C, Jambaque I. The neuropsychology of the Kluver-Bucy syndrome in children. Handb Clin Neurol. 2013;112:1285–8.
12. Bergstrom HC. The neurocircuitry of remote cued fear memory. Neurosci Biobehav Rev. 2016;71:409–17.
13. Koob GF, Volkow ND. Neurobiology of addiction: a neurocircuitry analysis. Lancet Psychiatry. 2016;3:760–73.
14. Brooks SJ, Stein DJ. A systematic review of the neural bases of psychotherapy for anxiety and related disorders. Dialogues Clin Neurosci. 2015;17:261–79.
15. Modinos G, Ormel J, Aleman A. Individual differences in dispositional mindfulness and brain activity involved in reappraisal of emotion. Soc Cogn Affect Neurosci. 2010;5:369–77.
16. Holzel BK, Lazar SW, Gard T, Schuman-Olivier Z, Vago DR, Ott U. How does mindfulness meditation work? Proposing mechanisms of action from a conceptual and neural perspective. Perspect Psychol Sci. 2011;6:537–59.
17. Damasio AR. Descartes' error: emotion, reason and the human brain. New York: Grosset/Putnam; 1994.
18. Bader O. Attending to emotions is sharing of emotions – a multidisciplinary perspective to social attention and emotional sharing. Comment on Zahavi and Rochat (2015). Conscious Cogn. 2016;42:382–95.
19. Reznikov R, Binko M, Nobrega JN, Hamani C. Deep brain stimulation in animal models of fear, anxiety, and posttraumatic stress disorder. Neuropsychopharmacology. 2016;41:2810–7.
20. Murray EA, Gaffan D, Mishkin M. Neural substrates of visual stimulus-stimulus association in rhesus monkeys. J Neurosci. 1993;13:4549–61.

21. Dricu M, Fruhholz S. Perceiving emotional expressions in others: activation likelihood estimation meta-analyses of explicit evaluation, passive perception and incidental perception of emotions. Neurosci Biobehav Rev. 2016;71:810–28.
22. James W. Physical basis of emotion (reprinted in 1994. Psychol rev 101: 205–210). Psychol Rev. 1894;1:516–29.
23. Cannon WB. The James-Lange theory of emotions: a critical examination and an alternative theory. By Walter B. Cannon, 1927. Am J Psychol. 1987;100:567–86.
24. Harrison NA, Singer T, Rotshtein P, Dolan RJ, Critchley HD. Pupillary contagion: central mechanisms engaged in sadness processing. Soc Cogn Affect Neurosci. 2006;1:5–17.
25. Makovac E, Garfinkel SN, Bassi A, Basile B, Macaluso E, Cercignani M, Calcagnini G, Mattei E, Agalliu D, Cortelli P, Caltagirone C, Bozzali M, Critchley H. Effect of parasympathetic stimulation on brain activity during appraisal of fearful expressions. Neuropsychopharmacology. 2015;40:1649–58.
26. Lavina B. Brain vascular imaging techniques. Int J Mol Sci. 2016;18:70.
27. Im JJ, Namgung E, Choi Y, Kim JY, Rhie SJ, Yoon S. Molecular neuroimaging in posttraumatic stress disorder. Exp Neurobiol. 2016;25:277–95.
28. Biotteau M, Chaix Y, Blais M, Tallet J, Peran P, Albaret JM. Neural signature of DCD: a critical review of MRI neuroimaging studies. Front Neurol. 2016;7:227.
29. Herting MM, Sowell ER. Puberty and structural brain development in humans. Front Neuroendocrinol. 2017;44:122–37.
30. Wandschneider B, Koepp MJ. Pharmaco fMRI: determining the functional anatomy of the effects of medication. Neuroimage Clin. 2016;12:691–7.
31. Vinik AI, Nevoret ML, Casellini C. The new age of sudomotor function testing: a sensitive and specific biomarker for diagnosis, estimation of severity, monitoring progression, and regression in response to intervention. Front Endocrinol (Lausanne). 2015;6:94.
32. Guinjoan SM, Bonanni Rey RA, Cardinali DP. Correlation between skin potential response and psychopathology in patients with affective disorders. Neuropsychobiology. 1995;31:24–30.
33. Pérez Lloret S, Diez JJ, Domé MN, Alvarez Delvenne A, Braidot N, Cardinali DP, Vigo DE. Effects of different "relaxing" music styles on the autonomic nervous system. Noise Health. 2014;16:279–84.
34. Nonnekes J, Carpenter MG, Inglis JT, Duysens J, Weerdesteyn V. What startles tell us about control of posture and gait. Neurosci Biobehav Rev. 2015;53:131–8.
35. Wichmann T, DeLong MR. The basal ganglia. In: Kandel E, Schwartz J, Jessell T, Siegelbaum S, Hudspeth AJ, editors. Principles of neural science. 5th ed. New York: McGraw-Hill; 2012.
36. Waxman SG. Control of movement. In: Clinical neuroanatomy. 28th ed. New York: McGraw-Hill Education; 2017.
37. Saper CB. The central autonomic nervous system: conscious visceral perception and autonomic pattern generation. Annu Rev Neurosci. 2002;25:433–69.
38. Horn JP, Swanson LW. The autonomic motor system and the hypothalamus. In: Kandel E, Schwartz J, Jessell T, Siegelbaum S, Hudspeth AJ, editors. Principles of neural science. 5th ed. New York: McGraw-Hill; 2012. p. 1056–78.
39. Macey PM, Ogren JA, Kumar R, Harper RM. Functional imaging of autonomic regulation: methods and key findings. Front Neurosci. 2015;9:513.
40. Wu P, Bandettini PA, Harper RM, Handwerker DA. Effects of thoracic pressure changes on MRI signals in the brain. J Cereb Blood Flow Metab. 2015;35:1024–32.
41. Benarroch EE. The central autonomic network: functional organization, dysfunction, and perspective. Mayo Clin Proc. 1993;68:988–1001.
42. Harper RM, Bandler R, Spriggs D, Alger JR. Lateralized and widespread brain activation during transient blood pressure elevation revealed by magnetic resonance imaging. J Comp Neurol. 2000;417:195–204.
43. James C, Macefield VG, Henderson LA. Real-time imaging of cortical and subcortical control of muscle sympathetic nerve activity in awake human subjects. Neuroimage. 2013;70:59–65.
44. James C, Henderson L, Macefield VG. Real-time imaging of brain areas involved in the generation of spontaneous skin sympathetic nerve activity at rest. Neuroimage. 2013;74:188–94.

45. Lisberger SG, Thacht WT. The cerebellum. In: Kandel E, Schwartz J, Jessell T, Siegelbaum S, Hudspeth AJ, editors. Principles of neural science. 5th ed. New York: McGraw-Hill; 2012. p. 960–81.
46. De Morree HM, Rutten GJ, Szabo BM, Sitskoorn MM, Kop WJ. Effects of insula resection on autonomic nervous system activity. J Neurosurg Anesthesiol. 2016;28:153–8.
47. Strata P. The emotional cerebellum. Cerebellum. 2015;14:570–7.
48. Roenneberg T, Keller LK, Fischer D, Matera JL, Vetter C, Winnebeck EC. Human activity and rest in situ. Methods Enzymol. 2015;552:257–83.
49. Wittmann M, Dinich J, Merrow M, Roenneberg T. Social jetlag: misalignment of biological and social time. Chronobiol Int. 2006;23:497–509.
50. Gaggioni G, Maquet P, Schmidt C, Dijk DJ, Vandewalle G. Neuroimaging, cognition, light and circadian rhythms. Front Syst Neurosci. 2014;8:126.
51. Lim AS, Chang AM, Shulman JM, Raj T, Chibnik LB, Cain SW, Rothamel K, Benoist C, Myers AJ, Czeisler CA, Buchman AS, Bennett DA, Duffy JF, Saper CB, De Jager PL. A common polymorphism near PER1 and the timing of human behavioral rhythms. Ann Neurol. 2012;72:324–34.
52. Roenneberg T, Allebrandt KV, Merrow M, Vetter C. Social jetlag and obesity. Curr Biol. 2012;22:939–43.
53. Thun E, Bjorvatn B, Flo E, Harris A, Pallesen S. Sleep, circadian rhythms, and athletic performance. Sleep Med Rev. 2015;23:1–9.
54. Baron KG, Reid KJ. Circadian misalignment and health. Int Rev Psychiatry. 2014;26:139–54.
55. Mesulam MM. Aphasia, memory loss, and other focal cerebral disorders. In: Kasper D, Fauci A, Hauser S, Longo D, Jameson JL, Loscalzo J, editors. Harrison's principles of internal medicine. 19th ed. New York: McGraw-Hill Education; 2015.
56. Barrett KE, Barman SM, Boitano S, Brooks HL. Learning, memory, language. In: Ganong's review of medical physiology. 25th ed. New York: McGraw-Hill Education; 2016.
57. Waxman SG. Chapter 21. Higher cortical functions. In: Clinical neuroanatomy. 27th ed. New York: The McGraw-Hill Companies; 2013.
58. Sweatt JD. Neural plasticity and behavior – sixty years of conceptual advances. J Neurochem. 2016;139(Suppl 2):179–99.
59. Connor DA, Gould TJ. The role of working memory and declarative memory in trace conditioning. Neurobiol Learn Mem. 2016;134 Pt B:193–209.
60. Mammarella N, Di DA, Palumbo R, Fairfield B. Noradrenergic modulation of emotional memory in aging. Ageing Res Rev. 2016;27:61–6.
61. Ropper AH, Samuels MA, Klein JP. Chapter 21. Dementia, the amnesic syndrome, and the neurology of intelligence and memory. In: Adams and Victor's principles of neurology. 10th ed. New York: The McGraw-Hill Companies; 2014.
62. Kirsch P. Oxytocin in the socioemotional brain: implications for psychiatric disorders. Dialogues Clin Neurosci. 2015;17:463–76.
63. Walker MP. The role of slow wave sleep in memory processing. J Clin Sleep Med. 2009;5:S20–6.
64. Carskadon MA, Dement WC. Chapter 2 – Normal human sleep: an overview. In: Principles and practice of sleep medicine. 5th ed. Philadelphia: WB Saunders; 2011. p. 16–26.
65. Narayanan L, Murray AD. What can imaging tell us about cognitive impairment and dementia? World J Radiol. 2016;8:240–54.
66. Febo M, Foster TC. Preclinical magnetic resonance imaging and spectroscopy studies of memory, aging, and cognitive decline. Front Aging Neurosci. 2016;8:158.
67. Rana M, Varan AQ, Davoudi A, Cohen RA, Sitaram R, Ebner NC. Real-time fMRI in neuroscience research and its use in studying the aging brain. Front Aging Neurosci. 2016;8:239.

Clinical Implications of the Enlarged Autonomic Nervous System

7

Abstract

Since the ANS is extensively involved in the function of almost every organ system, the clinical manifestations of autonomic dysfunction are diverse. Indeed, the ANS is involved in most diseases. Any structural pathological process affecting the brain (whether infectious, inherited, neoplastic, or degenerative in nature) can result in an autonomic syndrome. This Chapter describes the ANS semiology and the different classifications of the ANS disorders, emphasizing that derived from the hierarchical organization of the ANS.

Keywords

Amyloidotic autonomic failure • Autoimmune autonomic ganglionopathy • Autonomic disturbances in spinal cord injuries • Autonomic dysfunction • Autonomic dysfunction associated with aging • Autonomic dysfunction in primary sleep disorders • Autonomic tests • Fibromyalgia and chronic fatigue • Guillain–Barré syndrome • Hereditary autonomic neuropathies • Paraneoplastic autonomic dysfunction • Peripheral neuropathies • α-Synucleinopathies

Objectives

After studying this chapter, you should be able to:

- Describe the ANS semiology, including bedside evaluation and the most important autonomic tests to perform.
- Describe the different classifications of the ANS disorders, emphasizing that derived from the hierarchical organization of the ANS discussed in this book.
- Give examples of autonomic entities, describing the functions of the ANS affected.

D.P. Cardinali, *Autonomic Nervous System*, DOI 10.1007/978-3-319-57571-1_7

Semiological Aspects of ANS Disorders

As discussed in previous chapters, the ANS is extensive and is involved in the function of almost every organ system. Therefore, the clinical manifestations of autonomic dysfunction are diverse. Indeed, the ANS is involved in most diseases. Any structural pathological process affecting the brain (whether infectious, inherited, neoplastic, or degenerative in nature) can result in an autonomic syndrome.

Although autonomic disorders are a well-defined group of conditions affecting the central and/or peripheral autonomic pathways, autonomic symptoms and signs are often seen in many other medical conditions or may be isolated manifestations of a limited autonomic instability (Table 7.1) [1–3]. To evaluate these conditions, it is necessary to cover each autonomic sector (cardiovascular, gastrointestinal, genitourinary, secretomotor, sudomotor, neuroimmunoendocrine), define the temporal profile, identify associated symptoms, recognize atypical symptoms as expression of autonomic dysfunction, and exclude other conditions that could mimic its manifestations. To properly study these disorders, multiple tests are needed, in addition to supportive data that often include neuroimaging, sleep studies, specific blood and urine testing, and tissue diagnosis [4].

Table 7.1 Clinical signs of autonomic dysfunction

Cardiovascular	Tachycardia at rest Orthostatic hypotension Arterial hypertension Arrhythmias Syncope
Gastrointestinal tract	Dysphagia, regurgitation Gastroparesis, vomiting Constipation Episodes of diarrhea
Sudomotor	Hypohidrosis, anhidrosis Hyperhidrosis Gustatory sweating
Eye	Anisocoria Nyctalopia Close blurred vision Tunnel vision Double vision
Genitourinary tract	Bladder dysfunction Urinary retention Incontinence Impotence
CNS	Anxiety Insomnia Chronic fatigue Brain fog Vertigo Dizziness Weakness

A preliminary evaluation can be done at the bedside, at least for the most disabling symptoms of autonomic dysfunction. Significant decrement in BP without compensatory tachycardia is much worse prognostically than marked tachycardia without significant BP changes (Chap. 4). Secretomotor and sudomotor functions can be assessed by observation of the mucosae and by appreciating the presence of moisture on the skin by palpation. Hyperhidrosis is easily appreciated as sweat droplets over the skin or visibly wet garments. Occasionally, simultaneous ECG monitoring can identify an ictal bradycardia or asystole, or provide confirmation that convulsive manifestations commonly seen in syncope are secondary to the hemodynamic changes.

Symptoms suggestive of autonomic dysfunction include, but are not limited to, postural hypotension, digestive discomfort, altered intestinal, bladder or sexual function, decreased or increased sweating, dry mucous membranes, and cooling or discoloration of the extremities. Insomnia, anxiety, brain fog and chronic fatigue may be present (Table 7.1). In clinical practice, the symptoms of autonomic dysfunction are often underestimated, because they are subjective, frequently transient in healthy subjects, of slow onset and evolution, of little disability in the patient (at least in the initial stages), and difficult to treat.

Table 7.2 summarizes the most important autonomic tests that allow the various autonomic functions to be explored [1–3, 5]. Autonomic tests usually consist of physical stimuli or actions that elicit changes in sympathetic and parasympathetic activity, often reflected as BP and heart rate alterations. The Valsalva maneuver is one autonomic challenge with multiple response phases that influences, to varying degrees, both sympathetic and parasympathetic outflow. Despite consisting of a relatively simple somatomotor task, namely exhaling against a resistance to a predetermined pressure (30–40 mmHg) for a defined period (15–20 s), the cardiovascular response is separated into four distinct patterns occurring over periods of time, not including preliminary inhalation:

1. Initial BP rise: on application of expiratory force, pressure rises inside the chest forcing blood out of the pulmonary circulation into the left atrium. This causes a mild rise in stroke volume during the first few seconds of the maneuver.
2. Reduced venous return and compensation: return of systemic blood to the heart is impeded by the pressure inside the chest. The output of the heart is reduced and systolic volume falls. The fall in systolic volume reflexively causes blood vessels to constrict with some rise in BP (15–20 s). This compensation can be quite marked, with BP returning to near or even above normal, but the cardiac output and blood flow to the body remain low. During this time, a compensatory tachycardia occurs.
3. Pressure release: the pressure on the chest is released, allowing the pulmonary vessels and the aorta to re-expand, causing a further initial slight fall in systolic volume (20–23 s) because of the decreased left atrial return and increased aortic volume respectively. Venous blood can once more enter the chest and the heart and cardiac output begins to increase.
4. Return of cardiac output: blood return to the heart is enhanced by the effect of the entry of blood which had been dammed back, causing a rapid increase in cardiac output (From 24 s on).

Table 7.2 Some tests of ANS function

Test	Parameter	Main part of the reflex arc tested
Cardiovascular system		
RR interval during respiration	Heart rate	Vagal afferent and efferent limbs
Heart rate variability	Heart rate	Vagal afferent and efferent limbs
Valsalva maneuver	Heart rate, BP	Afferent and efferent limbs
Mueller maneuver	Heart rate, BP	Afferent and efferent limbs
BP response to standing or vertical tilt	BP	Afferent and sympathetic efferent limbs
Heart rate response to standing	Heart rate	Vagal afferent and efferent limbs
Handgrip	Heart rate, BP	Sympathetic efferent limb
Cold pressor test	Heart rate, BP	Sympathetic efferent limb
Radiant heating of the trunk	Hand blood flow	Sympathetic efferent limb
Immersion of the hand in hot water	Blood flow of the opposite hand	Sympathetic efferent limb
Emotional stress	Heart rate, BP	Sympathetic efferent limb
Baroreflex sensitivity	Heart rate, BP	Vagal afferent and efferent limbs
Doppler	Blood flow	Sympathetic efferent limb
Plasma norepinephrine levels	Rises on tilting from horizontal to vertical	Sympathetic efferent limb
Plasma arginine vasopressin levels	Rise with induced hypotension	Afferent limb
Sudomotor system		
Sweat tests	Sweat	Sympathetic efferent limb
Quantitative sudomotor axon reflex test	Evaporation rate	Sympathetic efferent limb
Sympathetic skin response	Potentials	Sympathetic efferent limb
Pupil		
Pharmacological tests of pupillary innervation	Pupil diameter	Afferent and efferent limbs
Pupil cycle time	Pupil diameter	Afferent and efferent limbs
Pupillometry	Pupil diameter and latency	Afferent and efferent limbs
Other		
Microneurography	Potentials	Sympathetic efferent limb
Sympathetic neuroimaging	^{123}I-Metaiodobenzylguanidine uptake by NE vesicles in cardiac sympathetic nerves	Sympathetic efferent limb
Skin biopsy	Peripheral adrenergic and cholinergic fibers innervating sweat glands and arrector pili muscles	Sympathetic efferent limb

A hand grip, which also involves sympathetic activity increases, is another common autonomic test. Baroreceptor unloading tasks include lower body negative pressure, which requires a specialized suit, or the simpler to implement Müller's maneuver, which is the reverse of the Valsalva manoeuver, consisting in a negative pressure in the chest and lungs after a forced expiration following by an inspiration with closed mouth and nose [5].

The cold pressor is a passive autonomic test that has the advantage of being a consistent stimulus across subjects, although the pain component may be a confounder to BP manipulation. Other tests identify changes in state that last several minutes, such as the quantitative sudomotor axon reflex test.

Head-up tilt table testing is a provocative test designed to simulate orthostatic stress and downward gravitational fluid shifts. It is classically used to diagnose neurally mediated (reflex) syncope. Tilt testing can also be used to aid in the diagnosis of other disorders of orthostatic intolerance, including postural tachycardia syndrome and orthostatic hypotension [6].

Changes in cardiac autonomic function can be tracked by several techniques. The simplest measure of cardiac autonomic status is the resting heart rate. Greater autonomic dysfunction is associated with increasing resting heart rates over time. A more robust measure of autonomic function is heart rate variability (HRV), measured using continuous heart rate monitoring. HRV is a set of parameters that reflects interval fluctuations between sequential beats of the heart [7–9]. Measures derived from interval differences between successive beats reflect parasympathetically modulated changes in heart rate. Other HRV measures reflect the combined signaling of the two arms of the autonomic nervous system and reflect both intrinsic (e.g., baroreflex, renin–angiotensin, sleep cycles, circadian) and extrinsic (activity, rest) rhythms. In general, decreased or decreasing HRV would be a signal for worse cardiac autonomic dysfunction. However, a higher, but more disorganized, HRV pattern, detectable by certain "nonlinear" HRV measures also reflects greater cardiac autonomic dysfunction [10]. Ideally, HRV is measured using 24-h ambulatory monitoring which can capture both daytime heart rate patterns and heart rate patterns during sleep, providing insights into circadian rhythm, sleep quality, and possible sleep-disordered breathing or periodic limb movements, all of which affect cardiac autonomic functioning. However, significant clinical information can be obtained from shorter recordings performed, perhaps, at the time of clinical visits and in association with standard bedside autonomic tests.

As mentioned in Chap. 6, changes in skin conductance occur with eccrine sweating and constitute a relatively pure assay of sympathetic activity (sympathetic skin response) [11]. Alterations in sweating are mediated by cholinergic nerves and are not affected by β-adrenoceptor blockers, allowing evaluation of the sympathetic system. It consists of a potential generated by sweating in the skin in response to different stimuli, including those that produce emotional reactions [12]. An alteration in the resistance of the skin is caused and a potential is obtained.

Microneurography is a unique method of recording postganglionic sympathetic neural traffic directly from human peripheral nerves. For sympathetic microneurography, tungsten microelectrodes are inserted through the skin onto an underlying peripheral nerve, innervating either skin or skeletal muscle. Sympathetic fibers are spontaneously active and many fibers discharge in synchronized bursts of impulses. Usually, action potentials are recorded simultaneously from several fibers (multifiber activity) and presented in a mean voltage (integrated) neurogram [13].

Sympathetic neuroimaging provides an important supplement to physiological, neurochemical, and neuropharmacological approaches in the evaluation of patients with clinical autonomic disorders. Sympathetic neuroimaging to date has involved visualization of noradrenergic innervation in the left ventricular myocardium [14]. Sympathetic imaging depends on the radiolabeling of vesicles in sympathetic nerves. The most commonly used imaging agent to assess cardiac sympathetic innervation is ^{123}I-metaiodobenzylguanidine. Cardiac sympathetic neuroimaging and postmortem neuropathological findings have linked α-synucleinopathy with noradrenergic denervation in Lewy body disease. Patients with familial Parkinson's disease from abnormalities of the gene encoding α-synuclein have cardiac sympathetic denervation.

Cutaneous punch biopsies are widely used to evaluate nociceptive C fibers in patients with suspected small-fiber neuropathy [15]. Peripheral adrenergic and cholinergic fibers innervate several cutaneous structures, such as sweat glands and arrector pili muscles, and can easily be seen with punch skin biopsies. Skin biopsies allow for both regional sampling, in diseases with patchy distribution, and the opportunity for repeated sampling in progressive disorders.

Classification of ANS Disorders

There are several ways to categorize autonomic dysfunctions, depending on the points of view from which they are considered [1, 2]. Clinical disorders of ANS can be conceptualized as focal (e.g., Horner's syndrome, Chap. 3) or generalized, affecting several autonomic segments (such as progressive autonomic failure). Another possible pathophysiological classification of autonomic dysfunctions is based on their primary or secondary nature, as summarized in Table 7.3. In addition, it is

Table 7.3 Pathophysiological classification of autonomic dysfunctions on the bases of their primary or secondary nature

Primary
Pure autonomic failure
Multisystem atrophy (Shy–Drager syndrome)
Parkinson's disease
Pan-dysautonomic neuropathy
Paraneoplastic autonomic neuropathy
Secondary
General diseases: diabetes, alcoholism, renal failure, amyloidosis, neoplasms, dysautonomia of aging
Autoimmune diseases: acute and chronic inflammatory polyneuropathy, Lambert–Eaton syndrome, rheumatoid arthritis, lupus erythematosus
Metabolic diseases: porphyria, Tangier disease, Fabry disease, pernicious anemia
Hereditary disorders: familial dysautonomia, hereditary motor and sensory neuropathies, sensorial and autonomic congenital neuropathy, Friedreich's ataxia
Infections: Chagas disease, AIDS, botulism, leprosy, syphilis
Diseases of the central nervous system: medullary lesions, strokes, tumors, multiple sclerosis, amyotrophic lateral sclerosis, Adie syndrome
Intoxications: by acrylamide, heavy metals, organic solvents
Pharmacological: by antineoplastics, antidepressants, sedatives, hypotensives, adrenergic blockers, cholinergic blockers

necessary to consider whether the manifestations are predominantly due to a sympathetic, parasympathetic, or mixed dysfunction (pandysautonomia).

Some characteristic patterns, based on temporal evolution and the constellation of semiology constellation, are also important in the differential diagnosis of autonomic neuropathies. Finally, there are some neurological disorders that affect the ANS, but in most cases, they are associated with somatic nervous system involvement. ANS dysfunctions may be due to increases or decreases in autonomic control activity and may occur because of brain, spinal or peripheral nerve injuries.

The enlarged view of ANS discussed in this book leads to a classification of autonomic disorders compatible with that found in popular clinical textbooks, e.g., Low and Engstrom (Table 7.4) [3]:

- Level 1 disorders (spinal) are systematized in those where there is involvement of the spinal cord (traumatic quadriplegia, syringomyelia, multiple sclerosis, amyotrophic lateral sclerosis, tumors spinal cord) and in various autoimmune autonomic neuropathies, paraneoplastic, Guillain–Barré syndrome, botulism, and porphyria. Autonomic neuropathy by amyloidosis, diabetes, or nutritional deficiency, and dysautonomia of aging are also included in this group. Frequently, they co-exist with orthostatic intolerance disorders (syncope, postural orthostatic tachycardia syndrome, etc.).
- Level 2 disorders include alterations of the brainstem and cerebellum, such as vertebrobasilar and Wallenberg syndromes, syringobulbia, and Arnold–Chiari malformation. Disorders in BP control (hypertension, hypotension) and of heart rate and central sleep apneas are included in this group.
- Level 3 disorders consist of the alteration of specific hypothalamic behaviors with their autonomic, neuroendocrine, and behavioral consequences. They include Wernicke–Korsakoff syndrome, malignant neuroleptic syndrome, fatal familial insomnia, alterations of AVP release and of temperature regulation (hyperthermia, hypothermia), disrupted sexual function (Klüver–Bucy syndrome, Chap. 6) and appetite disturbances.
- Level 4 disorders involve the abnormal function of limbic and paralimbic circuits such as autonomic seizures or limbic encephalitis, and various disorders with involvement of the cerebral cortex (complex partial seizures, cerebral infarction of the insula).
- Multilevel ANS disorders, such as multiple system atrophy, Parkinson's disease, and diffuse Lewy body disease.

Some Clinical Autonomic Entities

Peripheral Neuropathies with Dysautonomia

Peripheral nerves are susceptible to toxic damage, metabolic disorders, trauma, or neoplasms (neuropathies). In some cases, the axon is the primary focus of injury. The myelin sheath, or both the sheath and the axon, may be involved. In some cases, the

Table 7.4 Clinical classification of autonomic disorders by organizational level (modified from Low and Engstrom [3])

Level 1 disorders (spinal cord)
Autonomic disorders with spinal cord involvement
a. Traumatic quadriplegia
b. Syringomyelia
c. Multiple sclerosis and neuromyelitis optica
d. Amyotrophic lateral sclerosis
e. Tetanus
f. Spinal cord tumors
Autonomic neuropathies
A. Acute/subacute autonomic neuropathies
1. Subacute autoimmune autonomic ganglionopathy
a. Subacute paraneoplastic autonomic neuropathy
b. Guillain–Barré syndrome
c. Lambert–Eaton syndrome
d. Botulism
e. Porphyria
f. Drug-induced autonomic neuropathies
g. Toxin-induced autonomic neuropathies
B. Chronic peripheral autonomic neuropathies
1. Distal small fiber neuropathy
2. Combined sympathetic and parasympathetic failure
a. Amyloid
b. Diabetic autonomic neuropathy
c. Autoimmune autonomic ganglionopathy (paraneoplastic and idiopathic)
d. Sensory neuronopathy with autonomic failure
e. Familial dysautonomia (Riley–Day syndrome)
f. Uremic or nutritional deficiency
g. Dysautonomia of old age
3. Disorders of reduced orthostatic intolerance: idiopathic orthostatic hypotension, reflex syncope, postural orthostatic tachycardia syndrome, associated with prolonged bedrest, associated with space flight, chronic fatigue
Level 2 disorders (brainstem and cerebellum)
a. Vertebrobasilar and lateral medullary (Wallenberg) syndromes
b. Posterior fossa tumors
c. Syringobulbia and Arnold–Chiari malformation
d. Horner's syndrome
e. Disorders of blood pressure control (hypertension, hypotension)
f. Cardiac arrhythmias
g. Baroreflex failure
h. Central sleep apnea
i. Brainstem encephalitis
Level 3 disorders (hypothalamus)
a. Thiamine deficiency (Wernicke–Korsakoff syndrome)
b. Diencephalic syndrome
c. Neuroleptic malignant syndrome
d. Serotonin syndrome
e. Fatal familial insomnia
f. Arginine vasopressin syndromes (diabetes insipidus, inappropriate arginine vasopressin secretion)

Table 7.4 (continued)

g. Disturbances of temperature regulation (hyperthermia, hypothermia)
h. Disturbances of sexual function
i. Disturbances of appetite
Level 4 disorders (focal central nervous system disorders)
a. Frontal cortex lesions causing urinary/bowel incontinence
b. Focal seizures (temporal lobe or anterior cingulate)
c. Cerebral infarction of the insula
d. Shapiro syndrome (agenesis of the corpus callosum, hyperhidrosis, hypothermia)
e. Autonomic seizures
f. Limbic encephalitis
Multilevel autonomic nervous system disorders
Multisystem degeneration: autonomic failure clinically prominent
a. Multiple system atrophy (Shy–Drager syndrome)
b. Parkinson's disease with autonomic failure
c. Diffuse Lewy body disease (some cases)
Multisystem degeneration: autonomic failure clinically not usually prominent
a. Parkinson's disease
b. Other extrapyramidal disorders (inherited spinocerebellar atrophies, progressive supranuclear palsy, corticobasal degeneration, Machado–Joseph disease, fragile X syndrome)

proximal portion of the nerve is affected, whereas in others, it is the distal portion. Neuropathies of a single nerve are called mononeuropathies, in contrast to polyneuropathies, referred to diffuse neural damage throughout the body. The most common forms of neuropathies are diabetic, renal failure, alcoholic or nutritional cirrhosis, autoimmune and traumatic diseases [16].

Many neuropathies are characterized by Wallerian degeneration of the segment distal to the lesion. Those of toxic origin produce a degeneration of the neural distal segment, especially in the limbs. This explains why the first signs of neuropathy are detected in the fingers or toes.

The demyelination processes can be primary or secondary, the abnormal demyelinated segments being localized or scattered along the axon. Indicating the importance of the myelin sheath for the transmission of the nerve impulse, the neural conduction is slowed down, or in some cases halted. These conduction abnormalities can be identified electrophysiologically by determining the neural conduction time.

On clinical neurology standpoints, it is common to consider negative or positive signs or symptoms, that is, those resulting from the inhibition or disappearance of a function, or from its increase or externalization of an abnormal function. The most common signs of peripheral neuropathies are negative: loss of strength and sensitivity. In axonal neuropathies, the signs begin in the feet and progress centrally, correlating with a symptomatology of predominance in the distal portions of the peripheral nerves. In demyelinating neuropathies, there is also a distal predominance, with a greater tendency to present foci of demyelination in the long nerves.

In addition to muscle weakness, muscle atrophy and fasciculations, both signs of denervation, are usually found. Under conditions that affect small nerve fibers, pain and heat sensitivity are preferentially lost. When neuropathy involves large myelin fibers, there is loss of proprioceptive and vibratory sensitivity. Another early negative sign is the loss of muscle reflexes [5].

The compromise of the unmyelinated fibers leads to sympathetic alterations, such as loss of sweating, cardiac arrhythmias, or alteration of cardiovascular control mechanisms. These alterations are seen only in neuropathies that attack small myelin fibers (Guillain–Barré syndrome) or unmyelinated fibers (diabetic neuropathy). Another negative symptom of peripheral neuropathies is incoordination (ataxia), of sensory origin [16].

The most prominent positive symptoms of neuropathies are sensory, or paresthesias, i.e., tingling or other sensations (burns, punctures, vibrations, etc.) originating without apparent stimulus. Spontaneous painful sensations are more frequent in small diameter fiber neuropathy.

Peripheral neuropathies may develop acutely or chronically. It may take years for the appearance of symptoms in the case of metabolic, degenerative, or genetic neuropathies. This is the case of neuropathies seen in diabetes mellitus to lead to renal failure or chronic exposure. In contrast, neuropathies that are rapidly evolving are less frequent, such as Guillain–Barré syndrome, which is an autoimmune degeneration of peripheral nerves that evolves in 1–2 weeks.

One of the challenges of managing patients with peripheral neuropathy is that there are over 100 etiological causes, and not every patient will have an identifiable etiology, even after extensive investigation. Pain is always an important symptom to elucidate, as severe burning neuropathic pain is often indicative of C-fiber involvement and may be the only complaint in patients with small fiber neuropathy [17]. Patients with painful idiopathic small-fiber neuropathy may have coexistent autonomic symptoms, but typically the autonomic involvement is subclinical or minimal at presentation. Physical examination and neurophysiological testing are helpful in determining whether large fiber or small fiber (C) involvement is predominant. Formal autonomic testing is helpful in establishing the diagnosis of peripheral neuropathy in small-fiber predominant neuropathies where nerve conduction testing is normal.

α-Synucleinopathies

Several components of the ANS network are affected in neurodegenerative disorders characterized by the presence of intracellular inclusions containing α-synuclein, a protein that plays a role in maintaining synaptic vesicles in presynaptic terminals [18]. α-Synucleinopathies include multiple system atrophy, characterized by the accumulation of glial cytoplasmic inclusions, and Lewy body disorders (i.e., Parkinson's disease, dementia with Lewy bodies, and pure autonomic failure).

Orthostatic hypotension is a major autonomic sign in α-synucleinopathies. Patients with orthostatic hypotension include: (a) those with pure autonomic failure, characterized by isolated peripheral autonomic dysfunction and decreased NE synthesis; (b) multiple system atrophy with symptoms of a central Parkinson-like syndrome and normal resting plasma NE; (c) Parkinson's disease, with lesions in postganglionic noradrenergic neurons and signs of autonomic dysfunction [19].

Pure autonomic failure is a sporadic, adult-onset, slowly progressive neurodegenerative disorder. It is characterized pathologically by the abnormal accumulation of α-synuclein in peripheral autonomic neurons and clinically by symptomatic orthostatic hypotension, variable bladder dysfunction, and sexual dysfunction, with no somatic neurological deficit. Pure autonomic failure patients usually have very low plasma NE levels when recumbent, whereas plasma NE levels when recumbent are normal in multiple system atrophy and variable in Parkinson's disease. Pure autonomic failure selectively involves efferent autonomic neurons, with the postganglionic neurons mainly affected. Afferent pathways are spared. It occurs sporadically, and progresses slowly with a relatively good prognosis. However, some cases of pure autonomic failure may develop a central neurodegenerative disorder [19].

Multiple system atrophy is a sporadic and fatal α-synuclein-linked oligodendrogliopathy manifesting with progressive autonomic failure, poorly L-Dopa-responsive parkinsonism, and cerebellar ataxia, in any combination [20]. This combined parkinsonian and autonomic disorder is also referred to as the Shy–Drager syndrome. In multiple system atrophy, involvement of the rostral ventrolateral medulla is primarily responsible for orthostatic hypotension and involvement in the pontine micturition area; the sacral preganglionic nucleus is responsible for neurogenic bladder, and involvement of the pre-Bötzinger complex and medullary raphe contributes to sleep-related respiratory abnormalities (Chap. 4). In contrast, early involvement of the enteric nervous system and cardiac sympathetic ganglia characterize Lewy body disorders. The dorsal motor nucleus of the vagus is affected both in multiple system atrophy and in the early stages of Parkinson's disease [18].

Parkinson's disease is a neurodegenerative disorder defined by its motor features: asymmetric resting tremor, rigidity, bradykinesia, and postural instability. Autonomic symptoms are present to varying degrees in patients with Parkinson's disease, but very severe dysautonomia in patients with mild motor findings suggests a diagnosis of multiple system atrophy.

There are several differing criteria in use to diagnose Parkinson's disease and parkinsonian syndromes. In Parkinson's disease, symptoms and signs of autonomic failure occur commonly, especially in cardiovascular, gastrointestinal, and genitourinary domains. Most patients with Parkinson's disease have neuroimaging evidence for cardiac sympathetic denervation. In Parkinson's disease, orthostatic hypotension can be an early finding and is associated with extracardiac noradrenergic denervation, reduced cardiovagal baroreflex, and abnormal sympathoneural responses [21]. Histopathology of postmortem tissue is typically required for definitive diagnosis. Commonly, separating Parkinson's disease from parkinsonian syndromes rests largely upon the response of the physical examination to L-Dopa replacement therapy.

Dementia with Lewy bodies is a fatal α-synucleinopathy characterized by severely affected cognition, visual hallucinations, and parkinsonism. It represents the second most common cause of neurodegenerative dementia in the elderly after

AD. Progressive cognitive decline with deficits of visuospatial ability and frontal executive function is accompanied by only mildly to moderately severe parkinsonism, which is often bilateral akinetic-rigid without the classic rest tremor. Management of patients with dementia with Lewy bodies is based on a multidimensional approach considering the cognitive decline and dementia that form the core clinical syndrome. Patients with dementia with Lewy bodies have a pronounced cholinergic deficit. Orthostatic hypotension may be a disabling feature that if present frequently exacerbates the disability arising from progressive motor disturbance [22].

Autonomic nerve system dysfunction in the form of REM sleep behavior disorder (lack of motor atonia during REM sleep) is an early manifestation of disease in α-synucleinopathies, commonly occurring in the prodromal period. This is likely due to the proximity of the cholinergic REM sleep nuclei and the autonomic nuclei in the brainstem. In the Braak model of neurodegeneration, these nuclei become impaired before the motor nuclei are affected, as the deposition of α-synuclein progresses in a rostral–caudal fashion from the lower brain stem to the cortex. REM sleep behavior disorder is a common prodromal manifestation of Parkinson's disease occurring about 12–15 years in advance.

Diabetes Mellitus Autonomic Dysfunction

In diabetes mellitus patients, perturbations in autonomic function underlie most pathophysiology, from abnormalities in pupillary function to gastroparesis, intestinal dysmotility, diabetic diarrhea, genitourinary dysfunction, among others [23]. Sympathetic nerves, which dilate the iris, show earlier and more extensive impairment than the parasympathetic nerves, which constrict the iris. Signs and symptoms of gastric emptying abnormalities in diabetes include early satiety, nausea, vomiting, large fluctuations in blood glucose, and weight loss. The prevalence of gastroparesis is 30–50% of patients with long-standing type I and type II diabetes mellitus (T2DM). Diarrhea and the development of gallstones are symptoms of gallbladder atony. Gallstones are much more likely to develop in patients with hypercholesterolemia, and are often seen in patients with diabetes. The most common gastrointestinal symptom of autonomic neuropathy is constipation. Both afferent and efferent nerves to the bladder can be affected in T2DM patients. Afferent neuropathy results in the inability to feel the need to void. Therefore, there is a decrease in the frequency of urination, which may be misinterpreted as an improvement in glucose control.

Cardiac autonomic neuropathy in diabetes mellitus has been linked to resting tachycardia, postural hypotension, orthostatic bradycardia and orthostatic tachycardia, exercise intolerance, decreased hypoxia-induced respiratory drive, loss of baroreceptor sensitivity, enhanced intraoperative or perioperative cardiovascular lability, increased incidence of asymptomatic ischemia, myocardial infarction, and decreased rate of survival after myocardial infarction and congestive heart failure [24]. Autonomic dysfunction can affect the daily activities of individuals with diabetes mellitus and may provoke potentially life-threatening outcomes. For example, intensification of glycemic control in the presence of autonomic dysfunction (more so if combined with peripheral neuropathy) increases the likelihood of sudden death and is a caveat for aggressive glycemic control [23, 25].

Autoimmune Autonomic Ganglionopathy

Autoimmune autonomic ganglionopathy consists of a panautonomic failure caused by antibodies to ganglionic acetylcholine (ACh) receptors. The clinical syndrome is characterized by significant postural hypotension, diffuse cholinergic and adrenergic impairment, gastrointestinal dysmotility, urinary retention, and pupillary dysfunction. Sicca complex (dryness of the mucous membranes, as of the eyes and mouth, in the absence of a connective tissue disease) and hypohidrosis also occur. Serological testing for ganglionic ACh receptor antibodies helps to confirm the diagnosis. These antibodies cause a similar phenotype of autonomic failure in animal models, indicating that an antibody-mediated functional impairment of ganglionic transmission is the underlying etiology [26]. The usual course of autoimmune autonomic ganglionopathy is monophasic worsening followed by incomplete recovery. Some patients experience a chronic progressive course or stable dysautonomia without recovery. The diagnosis of idiopathic autoimmune autonomic ganglionopathy is suspected in cases of acquired autonomic failure without somatic neuropathy, when toxic or paraneoplastic causes have been excluded.

Paraneoplastic Autonomic Dysfunction

Patients with paraneoplastic autonomic neuropathy may be clinically indistinguishable from those with idiopathic autoimmune autonomic ganglionopathy until a cancer, usually small-cell carcinoma of the lung, is detected [27, 28]. Immune-mediated paraneoplastic autonomic dysfunction comprises a spectrum of neurological dysfunction that may range from isolated autonomic involvement to dysautonomia with recognizable syndromes including limbic encephalitis, subacute sensory neuronopathy, neuromyotonia, and Lambert–Eaton myasthenia syndrome. The neurological dysfunction may be multifocal and widespread owing to evolving immune response to multiple onconeural antigens in a single tumor. Somatic neurological findings are of variable severity. Urgent evaluation is necessary to prevent progressive neuronal loss, especially in CNS syndromes [27, 28].

Amyloidotic Autonomic Failure

Autonomic failure is an important feature of immunoglobulin amyloidosis and hereditary systemic amyloidosis. Mutant forms of transthyretin cause the most common type of autosomal-dominant hereditary systemic amyloidosis–familial amyloidotic polyneuropathy [29]. It must be considered in all patients with familial or paraproteinemic neuropathies and autonomic dysfunction. The amyloid fibrils in immunoglobulin amyloidosis amyloidopathy are composed of immunoglobulin light chain proteins or their degradation products. Signs of autonomic failure include orthostatic hypotension with inappropriate heart rate response, impotence, dry mouth, gastrointestinal autonomic disturbances, sluggish pupillary action, and impairment of sweating. Autonomic dysfunction, rather than deposition of amyloid in the mucosa, seems to be a more frequent cause of gastrointestinal symptoms in these patients.

Autonomic Dysfunction in Primary Sleep Disorders

The ANS is integrally related to sleep initiation, maintenance, and disruption. When such disruptions become frequent or chronic, autonomic impairment may follow dysfunction [30]. In the short term, such autonomic impairment may lead to increased sympathetic drive and a sensation of hyperarousal, further perpetuating the sleep disturbance. If sustained, this impairment may result in significant morbidity and even mortality. The most prevalent sleep disorders are insomnia, sleep disordered breathing, and restless legs syndrome.

Insomnia

Insomnia is among the most prevalent health concerns in the general population and it is present at particularly high rates among people with comorbid health problems. Approximately one-third of adults have occasional difficulty with insomnia and about 10–15% have persistent problems achieving sufficient sleep. Although the word insomnia may be used colloquially to represent poor sleep, specific diagnostic criteria for insomnia disorders are detailed in the *Diagnostic and Statistical Manual, 5th Edition* (DSM-5) [31] and the *International Classification of Sleep Disorders, 3rd Edition* (ICSD-3) [32]. Essentially, insomnia disorder is a persistent difficulty in falling asleep or remaining asleep, or awakening earlier than desired in the context of an adequate opportunity for sleep, and is associated with daytime consequences (for example, fatigue, poor concentration).

Like patients with obstructive sleep apnea (OSA), many patients with insomnia have high BP and are nondippers. i.e., they do not show the normal decrease in BP at night. In addition, frequent arousals, either spontaneous or secondary to an underlying sleep disorder, can result in increased sympathetic tone. There is a typical cardiac response observed during an arousal: an initial tachycardia, which often precedes the electrocortical arousal by several seconds, followed by bradycardia. If the arousals are frequent enough, the elevation in sympathetic tone can persist long after the patient has returned to sleep (Chap. 4). This association of high BP and insomnia remained after adjusting for age, race, gender, smoking, obesity, diabetes, alcohol consumption, depression, and other sleep disorders such as OSA and periodic limb movement disorder [30].

Brain imaging studies support the increased nocturnal cardiac sympathetic drive in insomnia patients. Increased activation and hypermetabolism in the arousal networks of the hypothalamus and brainstem, in addition to their efferent projections in the medial prefrontal cortex and amygdala, have been demonstrated in insomniac patients by PET. In addition, electroencephalography studies indicate a beta (14–35 Hz) and gamma (35–45 Hz) activity, frequencies typically associated with the cortical activity of the waking state. Sleep-deprived patients without insomnia (e.g., shift workers) also exhibit signs of autonomic dysfunction and are at a greater risk of developing cardiovascular disorders, even if young and otherwise healthy [33].

Sleep-Related Breathing Disorders

This defines a wide spectrum of abnormalities of respiration during sleep, including abnormal respiratory pattern (e.g., apneas, hypopneas, or respiratory effort-related arousals) or abnormal reduction in gas exchange (e.g., hypoventilation) during sleep. The ICSD-3 has defined four major categories of sleep-related breathing disorders: (1) OSA disorders, (2) central sleep apnea disorders, (3) sleep-related hypoventilation disorders, and (4) sleep-related hypoxemia disorder [32]. For this summary, we only consider OSA. which is the most common sleep-related breathing disorder, affecting an estimated 5–10% of the general population.

In OSA patients, repeated apneas or hypopneas can have an impact on the ANS and lead to significant consequences. When a patient with OSA experiences an obstructive respiratory event during sleep, pulmonary autonomic afferents are largely inhibited because of the prolonged increase in negative intrathoracic pressure, the result of inspiring against a closed or partially closed glottis (Chap. 4). Thus, hyperventilation is prevented, baroreceptors are stimulated, and sympathetic vasomotor tone increases, leading to peripheral vasoconstriction. Another frequent concomitant phenomenon is hypoxemia, which activates chemoreceptors in the carotid bodies, and exacerbates sympathetic vasomotor tone increases. Moreover, the diving reflex is triggered by hypoxemia, with an increase in cardiac vagal tone helping to preserve blood flow to the heart and brain while limiting cardiac oxygen demand [30].

If a sleep-related hypoventilation syndrome occurs, the augmented PCO_2 stimulates central chemoreceptors and further increases sympathetic tone via a similar process. BP increases because the venous return increases during apnea and peripheral vasoconstriction persists for some time after the patient initiates a recovery breath and resumes normal breathing, resulting in large BP surges. In susceptible individuals tachyarrhythmias may occur.

Concerning BP, patients with OSA tend to be "nondipping" or can also exhibit "reverse dipping," whereby BP increases during sleep, indicative of increased sympathetic tone. Moreover, this increase, together with a diminished baroreceptor reflex sensitivity are also found in wakefulness and may contribute to the increased incidence of cardiovascular events in OSA patients. There is evidence that OSA is an independent risk factor for arterial hypertension, cardiovascular disease, and ischemic stroke. Most cardiovascular and cerebrovascular events occur in the early hours of the morning, either out of sleep or shortly after awakening. This may be related to the increased sympathetic tone at this time, as the frequency and duration of phasic REM periods increase [30].

Periodic Limb Movement Disorder and Restless Legs Syndrome

Periodic limb movements are mainly characterized by relatively simple, usually stereotyped, movements that disturb sleep or its onset. Periodic leg movements in sleep are a frequent finding in PSG; its prevalence is estimated to be 4–11% in adults. Periodic limb movement disorder is defined by the PSG demonstration of periodic limb movements of >5/h in children and >15/h in adults that cause significant sleep disturbance or impairment of functioning.

Some patients with otherwise unexplained insomnia, fatigue, or hypersomnia have PSG results that reveal an elevated number of periodic leg movements. In sleep studies, periodic leg movements are most frequently found in restless legs syndrome. They also often occur in narcolepsy, sleep apnea syndrome, and REM sleep behavior disorder.

The autonomic arousal response discussed earlier, that of a rapid rise in heart rate and arterial BP followed by rapid bradycardia and a return of BP to baseline values, has been demonstrated before the onset of periodic limb movements during sleep [30]. In fact, even without an arousal, periodic limb movements have been associated with autonomic cardiovascular response, although the magnitude of the response is greater when an arousal is present. It seems feasible that periodic limb movements result from the loss of subcortical inhibition to pacemaker cells in the spinal cord or brainstem that have phasic control of autonomic, motor, and arousal networks. Like patients with OSA and insomnia, patients with periodic limb movement disorder are at an increased risk of cardiovascular disease.

Restless legs syndrome is a common disorder affecting an estimated 5–10% of the population. Unlike periodic limb movement disorder, restless legs syndrome is a clinical syndrome that includes an urge to move the legs, symptoms that worsen with rest or inactivity, are relieved with movement, and occur predominantly in the evening hours. The most commonly ANS symptoms were sialorrhea, constipation, early satiety, heat intolerance, and orthostatic intolerance. Periodic limb movements are present in approximately 80–90% of patients with restless legs syndrome; however, not all patients with periodic limb movements have symptoms of restless legs syndrome. Although the pathophysiology of restless legs syndrome is yet to be elucidated, and is likely multifactorial, one theory involves a reduction in dopaminergic outflow to the preganglionic sympathetic neurons in the dorsal horn of the spinal cord. DA inhibits preganglionic sympathetic neurons; therefore, a reduction in DA may in turn increase sympathetic outflow [30].

REM Sleep Behavior Disorder

Rapid eye movement sleep behavior disorder is characterized by partial arousal and abnormal behaviors emerging during REM sleep that may cause injury. It is a condition whereby a patient loses the protective muscle atonia that normally occurs during REM sleep. These patients are thereby free to act out their dreams, many times with injurious consequences. REM behavior disorder is more common in individuals with neurodegenerative disorders: 50–80% of patients presenting idiopathic REM behavior disorder go on to develop a synucleinopathy over 10–15 years. Clinical manifestations in idiopathic REM behavior disorder are identical to those seen in cases secondary to Parkinson's disease, multiple system atrophy and Lewy body disease and has been associated with mitochondrial disorders, brain tumors, and many other diseases, and clinical conditions that may damage brainstem mechanisms involved in generating REM sleep atonia. REM behavior disorder is probably the strongest nonmotor predictor of future disease, with an estimated median latency of motor symptoms of 12–14 years.

Dysfunction of the ANS is an early manifestation of disease in these patients, commonly occurring in the prodromal period [30]. REM sleep behavior disorder patients suffer from common symptoms of autonomic impairment, with the greatest deficits in gastrointestinal, urinary, and cardiovascular function. Patients have decreased HRV during REM sleep and the Valsalva ratio, an indicator of cardiovagal function, was significantly lower compared with healthy controls. A postganglionic cardiac sympathetic denervation, as measured by ^{123}I-metaiodobenzylguanidine scintigraphy, further support the concept of prodromal autonomic impairment in patients with REM sleep behavior disorder [30].

Autonomic Dysfunction in Cardiovascular Disorders

The sympathetic nervous system participates in the development and progression of the essential hypertensive state, as shown by increased circulating plasma levels of NE, elevated NE spillover rate, and augmented sympathetic nerve traffic discharge detected in the high BP state. In addition, the sympathetic overdrive participates in the development of the metabolic disarray and in target organ damage [34]. The above-mentioned sympathetic abnormalities explain why adrenergic overdrive represents an important therapeutic target in the treatment of hypertension.

Hypertension is characterized by autonomic abnormalities that include sympathetic activation and parasympathetic impairment [34]. These abnormalities occur in the earlier phases of the disease, participating in high BP development and progression. In addition, these autonomic abnormalities participate in the pathogenesis of hypertension-related end-organ damage and in the development of cardiac hypertrophy, of large, medium, and small artery structural changes, and of renal dysfunction. The autonomic abnormalities characterizing high BP are potentiated in the presence of metabolic alterations, such as insulin resistance, obesity, and metabolic syndrome.

Time and frequency domain estimates of tonic and reflex vagal HRV have demonstrated consistently the loss of parasympathetic tone at the onset of heart failure [35]. As heart failure advances, the loss of sino-atrial responsiveness to neutrally released and circulating catecholamines blunt further heart rate variation, which becomes a marker of reduced life expectancy. With regard to the ventricles, vagus nerve activity is generally antiarrhythmic, as it inhibits the profibrillatory effects of sympathetic nerve activation, whereas atrial arrhythmias generally derive from heightened levels of both vagus and sympathetic nerve activity [36].

Because heart failure was conceptualized initially as a primarily hemodynamic disorder, attention was directed first at potential baroreceptor reflex-mediated contributions to sympathetic activation and vagal withdrawal [37]. Augmented peripheral chemosensitivity to hypoxia, present in about half of patients with advanced congestive failure, is associated with higher plasma NE concentrations, loss of tonic and reflex heart rate modulation, oscillatory patterns in breathing, and ventricular arrhythmias. Each of these abnormalities is exacerbated by exercise and anticipates premature death. The central sympathoexcitation induced by sleep apnea may be both causal and contributory to heart failure and its progression [30].

The absence of proper regulation of mean arterial pressure can have significant pathophysiological consequences. Low mean arterial pressure can cause inadequate blood flow to organs, syncope, and shock. On the other hand, elevated mean arterial pressure contributes to increased oxygen demand by the heart, ventricular remodeling, vascular injury, end organ damage, and stroke.

The arterial baroreflex system is a key controller of mean arterial pressure and is a complex system (Fig. 4.12). It can be considered in its entirety as an integrative physiological system or in terms of its regulated component parts. Baroreflex sensitivity for the control of heart rate is consistently decreased in numerous pathological states including chronic hypertension, coronary artery disease, postmyocardial infarction, heart failure, diabetes mellitus, obesity, and aging [38]. The prominent feature of baroreflex failure is volatile arterial hypertension. Hypertensive episodes can be explained by sympathetic activation that is unopposed by the baroreflex. Episodes can be triggered by factors such as psychological stress, physical exercise, and pain [39]. A few patients with baroreflex failure present with episodes of hypotension and bradycardia that can be observed when patients are resting and cortical input is diminished.

Postural tachycardia syndrome is a syndrome of orthostatic tachycardia associated with symptoms of cerebral hypoperfusion and/or autonomic activation [40]. Patients are usually women aged 20–50 years and some have limited autonomic neuropathy. Evidence of peripheral denervation of sudomotor fibers includes sweat loss from the legs on the thermoregulatory sweat test or the quantitative sudomotor axon reflex test (Table 7.2). Peripheral adrenergic denervation may be present, resulting in impairment of reflex vasoconstriction with baroreflex unloading.

In hyperadrenergic postural tachycardia syndrome, sympathetic tone is increased, and manifests as orthostatic hyperadrenergic response and sometimes as a spontaneous episode of excessive sympathetic activity. Patients are characterized by supine vasoconstriction, increased peripheral venous pressure, reduced stroke volume and cardiac output, blunted orthostatic vasoconstrictive responses, often supine tachycardia, and reduced blood volume of uncertain origin. There is increased supine cardiac output compared with healthy volunteers. Relative lower extremity vasodilation persists during orthostatic stress, causing venous pooling in the legs [40].

The postural tachycardia syndrome is not a single disease. It is best viewed as a "disorder" or a syndrome in which excessive orthostatic tachycardia can be a final common pathway of many underlying pathophysiological processes. These include hyperadrenergic postural tachycardia syndrome, NE transporter deficiency, mast cell activation disorder, and neuropathic postural tachycardia syndrome. These pathophysiological mechanisms are not mutually exclusive, but can coexist in an individual patient with postural tachycardia syndrome [41]. Some postural tachycardia syndrome patients have a form of dysautonomia, with preferential denervation of sympathetic nerves innervating the lower limbs [40].

Patients with delayed orthostatic hypotension show a fall in systolic BP of ≥20 mmHg or diastolic BP of ≥10 mmHg lasting longer than 3 min of standing or upright tilt table testing [19]. Delayed orthostatic hypotension can be caused by a variety of pathogenetic mechanisms, including extra-adrenal pheochromocytoma, hyperadrenergic orthostatic hypotension, hyperbradykininism, primary hyperepinephrinemia, and hypoaldosteronism with baroreflex impairment. There were greater abnormalities of both phase II and IV of the Valsalva maneuver in patients with earlier and greater decreases in BP compared with those with later and milder reductions in BP. Abnormalities were also more likely detected in HRV in patients with earlier and more severe decreases in BP. Symptoms of orthostatic intolerance in patients with delayed orthostatic hypotension are similar to those with orthostatic hypotension that occurs within the first 3 min of upright posture [19].

In pathophysiological states, such as hypertension, heart failure, and chronic renal disease, there may be an inappropriate sympathoexcitation causing sodium retention, which exacerbates the disease process. The contribution of the renal sympathetic nerves to these cardiovascular diseases is demonstrated by the long-term normalization of BP found in resistant hypertensive patients subjected to renal denervation (Chap. 5) [42].

Pregnancy increases sympathetic nerve firing and decreases both basal parasympathetic activity and baroreflex function [43]; these changes are exaggerated in women with preeclampsia. Preeclampsia is a potentially fatal hypertensive disorder of pregnancy that is initiated by reduced placental perfusion. Increased sympathetic tone contributes to the hypertension, as basal muscle sympathetic nerve activity is clearly increased above the levels observed in normal pregnant women. The changes in basal autonomic tone may counteract to some degree the profound vasodilation that is a hallmark of normal pregnancy [44].

Autonomic Dysfunction Associated with Aging

Healthy human aging is associated with several abnormalities in ANS function that can impair an older person's adaptation to the stresses of everyday life [45]. Medications that acutely lower BP may also contribute to hypotension in elderly patients, particularly benzodiazepines, diuretics, antihypertensives, α-blockers used for prostatic obstruction, L-Dopa, tricyclic antidepressants, and neuroleptics. Orthostatic hypotension is an important symptom of autonomic failure, commonly associated with diabetes, malignancy, amyloidosis, Parkinson's disease, multiple system atrophy, Lewy body dementia, pure autonomic failure, and other syndromes in elderly patients [45]. The increase in plasma NE is primarily due to an increase in NE spillover at sympathetic nerve endings and secondarily due to a decrease in clearance. When plotted continuously over time, the beat-to-beat heart rate or BP signal is highly irregular because of the interactions of multiple autonomic control systems operating over different time scales.

Guillain–Barré Syndrome

Guillain–Barré syndrome is an acute immune-mediated demyelinating neuropathy that may cause profound weakness and respiratory failure over a period of a few weeks. Although the disease affects motor fibers preferentially, paresthesias and pain are common sensory manifestations. Autonomic neuropathy of some degree, particularly involving cardiovascular and gastrointestinal function, is found in two-thirds of patients and may be a life-threatening complication of the disease [46]. Autonomic dysfunction in Guillain–Barré syndrome can affect the sympathetic and the parasympathetic nervous system. Autonomic manifestations comprise a combination of autonomic failure and autonomic over-reactivity, the latter most commonly being manifested as sinus tachycardia and systemic hypertension. Marked arrhythmias and wide fluctuations in BP may also occur. Gastrointestinal dysmotility is common, but rarely progresses to adynamic ileus. Typically, autonomic neuropathy improves in concert with motor and sensory nerve function.

Hereditary Autonomic Neuropathies

Inherited neuropathies with autonomic involvement include Fabry disease (angiokeratoma corporis diffusum), an X-linked, inherited, slowly progressive metabolic disorder with nonspecific clinical manifestations [47]. Unlike many other lysosomal storage diseases, most patients remain clinically silent during the first few years of life. Cerebrovascular disease secondary to multifocal abnormalities of large and small vessels, including transient ischemic attacks and stroke, may also occur.

Acute hepatic porphyrias (acute intermittent porphyria, variegate porphyria, and hereditary coproporphyria) are a group of autosomal dominant, inherited metabolic disorders that manifest as acute or subacute, severe, life-threatening motor neuropathy, abdominal pain, autonomic dysfunction, and neuropsychiatric symptoms [48]. The neurological manifestations of all forms of acute porphyrias are identical. Symptoms of an acute attack include severe abdominal pain, nausea, vomiting, constipation, diarrhea, urinary frequency and hesitancy, urine discoloration, labile hypertension, tachycardia, excessive sweating, pain in the limbs and back, and convulsions.

Familial dysautonomia (Riley–Day syndrome) or hereditary sensory and autonomic neuropathy type III, is an autosomal recessive disease caused by mutations in the gene that encodes for I-κ-B kinase complex-associated protein [49]. Affected patients have a complex neurological phenotype. First, because of a congenital abnormality in the afferent baroreflex pathways, BP is extremely labile, with severe episodic hypertension and orthostatic hypotension. In addition to the autonomic cardiovascular abnormalities, patients with familial dysautonomia also have decreased pain and temperature perception, an impaired sense of taste, abnormal swallowing, gait ataxia, decreased/absent myotatic reflexes and decreased ventilatory responses to hypoxia and hypercapnia [49].

Autonomic Disturbances in Spinal Cord Injuries

In cervical and high thoracic transection, the entire or a large part of the sympathetic outflow, together with the sacral parasympathetic outflow, is separated from cerebral control. Autonomic malfunction may affect the cardiovascular, thermoregulatory, sudomotor, gastrointestinal, urinary, and reproductive systems [5]. Basal heart rate is usually below normal. In patients with high cervical lesions, who need artificial ventilation because of diaphragmatic paralysis, severe bradycardia and cardiac arrest may occur during tracheal stimulation. Plasma NE levels are low and do not rise with head-up postural change, unlike normal subjects. There is a marked rise in levels of plasma renin, aldosterone, and AVP, which may contribute to the recovery of BP and account for other symptoms, such as reduced urine output. The reverse, severe hypertension, may occur during autonomic dysreflexia following stimulation below the level of the lesion, predominantly, but not always, through noxious stimuli [5].

Drug-Induced Autonomic Dysfunction

Medications frequently alter BP regulation, which may interfere with daily activities or contribute to cardiovascular complications. Hypotension is frequently accompanied by symptoms, but hypertension may go unrecognized until serious complications develop, prompting medical attention. Medications taken for anti-inflammatory, decongestant, or anorexic effects are frequently associated with an increase in BP. Although patients may take these medications to promote a healthy lifestyle, the frequent use of over-the-counter, off-label, and herbal medications may contribute to drug-induced dysregulation of BP control. A thorough medication history is needed, or an opportunity to prevent complications could be missed [3].

Autonomic Disorder in Fibromyalgia, Chronic Fatigue Syndrome and Chronic Pain

Fibromyalgia is a disabling disease affecting 2–5% of the population, mostly young and middle-aged women (7 to 1 vs men). Its major clinical signs are: generalized musculoskeletal pain and allodynia, poor sleep quality, significant levels of fatigue, and cognitive alterations, particularly problems with concentration and memory. The etiology of fibromyalgia remains unknown; however, there are often characteristic alterations in sleep patterns and changes in neuroendocrine transmitters that suggest that the pathophysiology of the syndrome might be associated with autonomic and neuroendocrine regulation. Central sensitization, dampening of inhibitory pain pathways, and changes in neurotransmitters can lead to abnormal processing of sensory signals in the CNS, as such lowering the pain threshold and amplifying sensory signals causing constant pain (allodynia) (Chap. 4) [50].

Frequent comorbidity of fibromyalgia and mood disorders suggests that stress response and neuroendocrine abnormalities might play roles in the disease process. In fibromyalgia, the stress–adaptation response is disrupted, leading to stress-induced symptoms. Patients with fibromyalgia also often suffer from comorbid anxiety disorders. Neuroimmunoendocrine alterations in fibromyalgia include: (a) decrease in cortisol secretion, which has been associated with the presence of chronic or intense physical or psychological stress; (b) decrease in GH secretion; (c) chronic activation of various cytokines by stress (higher levels of IL-8 and IL-10, which has been linked to sleep problems); (d) decreased melatonin levels. fMRI studies support the hypothesis of increased central pain, showing that patients with fibromyalgia have a decreased volume of gray matter in the prefrontal cortex, amygdala, and anterior cingulate cortex. Duration of pain or functional disability due to pain do not correlate with gray matter volumes. One possible finding is that reductions in gray matter volume may be a precondition for central sensitization in fibromyalgia [51].

Chronic fatigue syndrome is a disabling disorder characterized by persistent or relapsing unexplained fatigue, accompanied by characteristic physical, constitutional, and neuropsychological symptoms lasting at least 6 months [52]. The fatigue is not relieved by rest, nor explained by medical or psychiatric conditions, and it is accompanied by a range of cognitive and somatic symptoms. The pathophysiology of chronic fatigue syndrome remains unclear. A hyperserotonergic state and hypoactivity of the hypothalamic–pituitary–adrenal axis have also been indicated, but it remains uncertain whether these are a cause of or a consequence of chronic fatigue syndrome. Female gender genetic disposition, certain personality traits and physical and emotional stressors have been identified as risk factors. Moreover, exposure to childhood trauma has been found to increase the risk of chronic fatigue syndrome three- to eightfold. Prevalence of chronic fatigue syndrome in primary care setting ranges from 3 to 20%.

Many chronic fatigue patients experience clinical features consistent with autonomic dysfunction, such as orthostatic symptoms and tachycardia, increased sweating, pallor, sluggish pupillary responses, gastrointestinal symptoms, and frequency micturition. A particularly strong association has been observed between chronic fatigue syndrome and orthostatic intolerance, including neurally mediated hypotension and postural tachycardia syndrome.

Both fibromyalgia and chronic fatigue syndrome are chronic clinical conditions characterized by intractable fatigue and nonrestorative sleep [52]. In fibromyalgia, the prominent diffuse musculoskeletal discomfort is the distinguishing symptom. However, in general clinical terms the two conditions cannot be clearly distinguished.

Fibromyalgia and chronic fatigue syndrome, together with irritable bowel syndrome and migraine are clinical entities showing an abnormal central processing of pain [50]. Usually, nonpainful stimuli are amplified and experienced as painful (allodynia) and sleep abnormalities are present because patients are unable to achieve deep NREM sleep. Participation of the ANS in the pathogenesis of chronic pain is indicated by the decreased high-frequency HRV, i.e., the decreased parasympathetic activity, found in chronic pain. ANS dysfunction depends on pain conditions. For example, fibromyalgia, unlike irritable bowel syndrome, is characterized by greatly decreased parasympathetic activity. Fibromyalgia dysautonomia is also characterized by basal hyperactive sympathetic activity with hyporeactivity to stress. This hyporeactivity to stress is less marked in localized chronic muscle pain [53].

Hyperthermia, Hypothermia

Hyperthermia is a result of a failure of the regulatory mechanisms to prevent the body temperature from rising above euthermia values, because of high environmental or metabolic heat (Chap. 5). The hyperthermia syndrome consists of several disorders, ranging from mild to fatal and temporary. The severity of hyperthermia depends on a variety of endogenous factors that may affect the magnitude and rate of increase in body temperature to interact with the thermoregulatory system or other control systems.

Several symptoms may accompany an excessive heat load, such as swelling in the lower limbs, orthostatic dizziness or syncope, exhaustion with headache, vomiting and diarrhea, and, in more severe cases, heat stroke. Heat stroke is the breakdown of thermoregulatory mechanisms resulting in a sudden increase in body temperature to values above 39–40 °C. Among the various factors leading to heat stroke, two classes can be distinguished: physiological unsuitability and malfunction of the effector organs. During exposure to a moderately hot environment, blood flow of the body surface increased without compromising the blood supply to other tissues, a fact that is due to a rise in and/or redistribution of cardiac output [54]. Flow redistribution occurs by reducing the blood supply to tissues considered nonvital organs to increase that to organs of the splanchnic area, kidney, muscle, and adipose tissue. However, when the heat stress becomes severe, the thermoregulatory mechanisms may conflict with the mechanisms regulating blood pressure, as competing demands arise. It is believed that such a conflict in homeostatic systems is the main causative factor in heat stroke.

Hypothermia usually occurs when the body temperature drops below 35°C and can be mild (34–35°C), moderate (30–34°C) or severe (below 30°C). There is a failure of the thermoregulatory mechanisms to maintain euthermia, but they cannot prevent the decrease in body temperature either. Hypothermia can result from exposure to very cold air, immersion in cold water, metabolic disorders (hypothyroidism, hypoglycemia, malnutrition) or disorders of the central nervous system (hypothalamus), drug effects, alcohol abuse, or even attenuated behavioral thermoregulation [55]. Hypothermia produces progressive depression of mental functions, apathy starting with and ending with lethargy, and coma when the body temperature reaches 28–30°C. Except for the adrenergic activity in peripheral vessels (to induce vasoconstriction) and nerve activity involved in the tremor, hypothermia leads to a gradual reduction in the conduction velocity of nerve impulses. In severe hypothermia, carbon dioxide retention induces an acidotic state. It is important to emphasize that the reduction in temperature has several benefits in conditions of oxygen shortage. In the clinic, forced hypothermia is being increasingly used as a therapeutic strategy for some surgical procedures or brain ischemia situations.

References

1. Biaggioni I, Burnstock G, Low PA, Paton JFR, editors. Primer on the autonomic nervous system. 3rd ed. San Diego: Academic; 2012.
2. Buijs RM, Swaab D, editors. Handbook of clinical neurology. Autonomic nervous system, vol. 117. New York: Elsevier; 2013.
3. Low PA, Engstrom JW. Disorders of the autonomic nervous system. In: Kasper D, Fauci A, Hauser S, Longo D, Jameson JL, Loscalzo J, editors. Harrison's principles of internal medicine. 19th ed. New York: McGraw-Hill Education; 2015.
4. Sandroni P. Chapter 78 – Clinical evaluation of autonomic disorders. In: Biaggioni I, Burnstock G, Low PA, Paton JFR, editors. Primer on the autonomic nervous system. Third ed. San Diego: Academic; 2012. p. 377–82.
5. Ropper AH, Samuels MA, Klein JP. Chapter 26. Disorders of the autonomic nervous system, respiration, and swallowing. In: Adams and Victor's principles of neurology. 10th ed. New York: The McGraw-Hill Companies; 2014.
6. Raj SR. Chapter 79 – Tilt table studies. In: Biaggioni I, Burnstock G, Low PA, Paton JFR, editors. Primer on the autonomic nervous system. Third ed. San Diego: Academic; 2012. p. 383–7.
7. Vigo DE, Nicola-Siri L, Ladrón de Guevara MS, Martinez-Martinez JA, Fahrer RD, Cardinali DP, Massoli O, Guinjoan SM. Relation of depression to heart rate nonlinear dynamics in patients >60 years of age with recent unstable angina pectoris or acute myocardial infarction. Am J Cardiol. 2004;93:756–60.
8. Vigo DE, Guinjoan SM, Scaramal M, Nicola-Siri L, Cardinali DP. Wavelet transform shows age-related changes of heart rate variability within independent frequency. Auton Neurosci. 2005;123:94–100.
9. Vigo DE, Domínguez J, Scaramal M, Ruffa E, Solernó J, Nicola-Siri L, Cardinali DP. Nonlinear analysis of heart rate variability within independent frequency components during thesleep-wake cycle. Auton Neurosci. 2010;154:84–8.
10. Vigo DE, Castro MN, Dörpinghaus A, Weidema H, Cardinali DP, Nicola-Siri L, Rovira B, Fahrer RD, Nogués M, Leiguarda RC, Guinjoan SM. Nonlinear analysis of heart rate variability in patients with eating disorders. World J Biol Psychiatry. 2007;20089(03):183–9.
11. Liu JC, Verhulst S, Massar SA, Chee MW. Sleep deprived and sweating it out: the effects of total sleep deprivation on skin conductance reactivity to psychosocial stress. Sleep. 2015;38:155–9.
12. Guinjoan SM, Bonanni Rey RA, Cardinali DP. Correlation between skin potential response and psychopathology in patients with affective disorders. Neuropsychobiology. 1995;31:24–30.
13. Wallin BG. Chapter 80 – Sympathetic microneurography. In: Biaggioni I, Burnstock G, Low PA, Paton JFR, editors. Primer on the autonomic nervous system. Third ed. San Diego: Academic; 2012. p. 389–92.
14. Goldstein DS. Chapter 29 – Sympathetic neuroimaging. In: Buijs RM, Swaab D, editors. Handbook of clinical neurology. Autonomic nervous system. New York: Elsevier; 2013. p. 365–70.
15. Wang N, Gibbons CH. Chapter 30 – Skin biopsies in the assessment of the autonomic nervous system. In: Buijs RM, Swaab D, editors. Handbook of clinical neurology. Autonomic nervous system. New York: Elsevier; 2013. p. 371–8.

16. Kuritzky L, Espay AJ, Gelblum J, Payne R, Dietrich E. Diagnosing and treating neurogenic orthostatic hypotension in primary care. Postgrad Med. 2015;127:702–15.
17. Chiang MC, Tseng MT, Pan CL, Chao CC, Hsieh ST. Progress in the treatment of small fiber peripheral neuropathy. Expert Rev Neurother. 2015;15:305–13.
18. Cersosimo MG, Benarroch EE. Chapter 5 – Central control of autonomic function and involvement in neurodegenerative disorders. In: Buijs RM, Swaab D, editors. Handbook of clinical neurology. Autonomic nervous system. New York: Elsevier; 2013. p. 45–57.
19. Garland EM, Hooper WB, Robertson D. Chapter 20 – Pure autonomic failure. In: Buijs RM, Swaab D, editors. Handbook of clinical neurology. Autonomic nervous system. New York: Elsevier; 2013. p. 243–57.
20. Wenning GK, Krismer F. Chapter 19 – Multiple system atrophy. In: Buijs RM, Swaab D, editors. Handbook of clinical neurology. Autonomic nervous system. New York: Elsevier; 2013. p. 229–41.
21. Kaufmann H, Goldstein DS. Chapter 21 – Autonomic dysfunction in Parkinson disease. In: Buijs RM, Swaab D, editors. Handbook of clinical neurology. Autonomic nervous system. New York: Elsevier; 2013. p. 259–78.
22. Ropper AH, Samuels MA, Klein JP. Chapter 21. Dementia, the amnesic syndrome, and the neurology of intelligence and memory. In: Adams and Victor's principles of neurology. 10th ed. New York: The McGraw-Hill Companies; 2014.
23. Vinik AI, Erbas T. Chapter 22 – Diabetic autonomic neuropathy. In: Buijs RM, Swaab D, editors. Handbook of clinical neurology. Autonomic nervous system. New York: Elsevier; 2013. p. 279–94.
24. Rodriguez-Colon SM, Li X, Shaffer ML, He F, Bixler EO, Vgontzas AN, Cai J, Liao D. Insulin resistance and circadian rhythm of cardiac autonomic modulation. Cardiovasc Diabetol. 2010;9:85.
25. Cryer PE. Chapter 23 – Hypoglycemia-associated autonomic failure in diabetes. In: Buijs RM, Swaab D, editors. Handbook of clinical neurology. Autonomic nervous system. New York: Elsevier; 2013. p. 295–307.
26. Muppidi S, Vernino S. Chapter 25 – Autoimmune autonomic failure. In: Buijs RM, Swaab D, editors. Handbook of clinical neurology. Autonomic nervous system. New York: Elsevier; 2013. p. 321–7.
27. Chefdeville A, Honnorat J, Hampe CS, Desestret V. Neuronal central nervous system syndromes probably mediated by autoantibodies. Eur J Neurosci. 2016;43:1535–52.
28. Paraschiv B, Diaconu CC, Toma CL, Bogdan MA. Paraneoplastic syndromes: the way to an early diagnosis of lung cancer. Pneumologia. 2015;64:14–9.
29. Ueda M, Ando Y. Recent advances in transthyretin amyloidosis therapy. Transl Neurodegener. 2014;3:19.
30. Miglis MG. Autonomic dysfunction in primary sleep disorders. Sleep Med. 2016;19:40–9.
31. American Psychiatric Association. Diagnostic and statistical manual of mental disorders. 5th ed. 2015.
32. American Academy of Sleep Medicine. International classification of sleep disorders. 3rd ed. Darien: American Academy of Sleep Medicine; 2014.
33. Lavigne G. Chapter 17 – Relevance of sleep physiology for sleep medicine clinicians A2 – Kryger, Meir H. In: Roth T, Dement WC, editors. Principles and practice of sleep medicine. 5th ed. Philadelphia: W.B. Saunders; 2011. p. 199–200.
34. Mancia G, Grassi G. Chapter 26 – The central sympathetic nervous system in hypertension. In: Buijs RM, Swaab D, editors. Handbook of clinical neurology. Autonomic nervous system. New York: Elsevier; 2013. p. 329–35.
35. D'Negri CE, Marelich L, Vigo D, Acunzo RS, Girotti LA, Cardinali DP, Siri LN. Circadian periodicity of heart rate variability in hospitalized angor patients. Clin Auton Res. 2005;15:223–32.
36. Tan AY, Verrier RL. Chapter 12 – The role of the autonomic nervous system in cardiac arrhythmias. In: Buijs M, Swaab D, editors. Handbook of clinical neurology. Autonomic nervous system. New York: Elsevier; 2013. p. 135–45.

37. Wehrwein EA, Joyner MJ. Chapter 8 – Regulation of blood pressure by the arterial baroreflex and autonomic nervous system. In: Buijs RM, Swaab D, editors. Handbook of clinical neurology. Autonomic nervous system. Elsevier; 2013. p. 89–102.
38. Chapleau MW. Chapter 33 – Baroreceptor reflexes. In: Biaggioni I, Burnstock G, Low PA, Paton JFR, editors. Primer on the autonomic nervous system. Third ed. San Diego: Academic; 2012. p. 161–5.
39. Jordan J. Chapter 72 – Baroreflex failure. In: Biaggioni I, Burnstock G, Low PA, Paton JFR, editors. Primer on the autonomic nervous system. Third ed. San Diego: Academic; 2012. p. 349–53.
40. Jones PK, Shaw BH, Raj SR. Clinical challenges in the diagnosis and management of postural tachycardia syndrome. Pract Neurol. 2016;16:431–8.
41. Wieling W, Groothuis JT. Chapter 39 – Physiology of upright posture. In: Biaggioni I, Burnstock G, Low PA, Paton JFR, editors. Primer on the autonomic nervous system. Third ed. San Diego: Academic; 2012. p. 193–5.
42. Johns EJ. Chapter 17 – Autonomic regulation of kidney function. In: Buijs RM, Supowit SC, editors. Handbook of clinical neurology. Autonomic nervous system. New York: Elsevier; 2013. p. 203–14.
43. Hart EC, Charkoudian N, Joyner MJ. Chapter 54 – Sex differences in autonomic function. In: Biaggioni I, Burnstock G, Low PA, Paton JFR, editors. Primer on the autonomic nervous system. Third ed. San Diego: Academic; 2012. p. 261–4.
44. Brooks VL, McCully BH, Cassaglia PA. Chapter 55 – Autonomic control during pregnancy. In: Biaggioni I, Burnstock G, Low PA, Paton JFR, editors. Primer on the autonomic nervous system. Third ed. San Diego: Academic; 2012. p. 265–8.
45. Lipsitz LA, Novak V. Chapter 56 – Aging and the autonomic nervous system. In: Biaggioni I, Burnstock G, Low PA, Paton JFR, editors. Primer on the autonomic nervous system. Third ed. San Diego: Academic; 2012. p. 271–3.
46. Anandan C, Khuder SA, Koffman BM. Prevalence of autonomic dysfunction in hospitalized patients with Guillain-Barre syndrome. Muscle Nerve. 2016 Dec 31. doi: 10.1002/mus.25551. [Epub ahead of print]
47. El-Abassi R, Singhal D, England JD. Fabry's disease. J Neurol Sci. 2014;344(15–9):5–19.
48. Arora S, Young S, Kodali S, Singal AK. Hepatic porphyria: a narrative review. Indian J Gastroenterol. 2016;35:405–18.
49. Chelimsky G, Chelimsky T. Unusual structural autonomic disorders presenting in pediatrics: disorders associated with hypoventilation and autonomic neuropathies. Pediatr Clin N Am. 2017;64:173–83.
50. Sluka KA, Clauw DJ. Neurobiology of fibromyalgia and chronic widespread pain. Neuroscience. 2016;338:114–29.
51. Morton DL, Sandhu JS, Jones AK. Brain imaging of pain: state of the art. J Pain Res. 2016;9:613–24.
52. Penfold S, St DE, Mazhar MN. The association between borderline personality disorder, fibromyalgia and chronic fatigue syndrome: systematic review. BJPsych Open. 2016;2:275–9.
53. Martinez-Martinez LA, Mora T, Vargas A, Fuentes-Iniestra M, Martinez-Lavin M. Sympathetic nervous system dysfunction in fibromyalgia, chronic fatigue syndrome, irritable bowel syndrome, and interstitial cystitis: a review of case-control studies. J Clin Rheumatol. 2014;20:146–50.
54. Periard JD, Travers GJ, Racinais S, Sawka MN. Cardiovascular adaptations supporting human exercise-heat acclimation. Auton Neurosci. 2016;196:52–62.
55. Perlman R, Callum J, Laflamme C, Tien H, Nascimento B, Beckett A, Alam A. A recommended early goal-directed management guideline for the prevention of hypothermia-related transfusion, morbidity, and mortality in severely injured trauma patients. Crit Care. 2016;20:107.

Clinical Implications of the Timed Autonomic Nervous System

8

Abstract

There are many reasons for the lack of sleep in our society that operates 24 h a day, 7 days a week, constantly. The main disturbing factor has been the technological advance of being able to light our evenings artificially. This Chapter analyzes the impact of the lack of sleep in the "24/7 Society", principally the disruption of the three ANS physiological programs. Typical examples of a desynchronized ANS are jet lag, shift work and chronodisruption, the metabolic syndrome and mental illnesses being also examples of a desynchronized ANS. The chronobiological aspects of normal and pathological brain aging and cancer are discussed.

Keywords

24/7 Society • Adjuvant chronobiological treatment • Chronodisruption • Chronodisruption in cancer • Chronodisruption in mental illnesses • Chronodisruption in metabolic syndrome • Chronodisruption of brain aging • Jet lag • Light at night • Melatonin • Shift work

Objectives

After studying this chapter, you should be able to:

- Describe the mechanisms underlying the inadequate experience of day and night in the "24/7 Society."
- Describe the disruption of the three ANS physiological programs in a 24-h cycle due to the inadequate experience of day and night.
- Describe the physiological changes observed in jet lag, shift work. and chronodisruption as examples of a desynchronized ANS.
- Understand why the chronobiological treatment of a desynchronized ANS is needed for full recovery in most cases.
- Describe the metabolic syndrome and mental illnesses as examples of a desynchronized ANS.
- Understand the chronobiological aspects of normal and pathological brain aging.
- Understand the chronobiological aspects of cancer.

D.P. Cardinali, *Autonomic Nervous System*, DOI 10.1007/978-3-319-57571-1_8

Due to the "24/7 Society," the ANS Has Lost Adequate Experience of Day and Night

There are many reasons for lack of sleep in our society, which operates 24 h a day, 7 days a week, constantly. The early hours of school or work, TV programs increasingly displacing the "prime time" ones late at night, the daily stress, or the widespread use of foods and beverages rich in caffeine are among the precipitating factors. However, the main factor has been the technological advance being able to artificially light our evening.

Our hominid ancestor, *Homo erectus*, used caves as shelter and must have begun to use fire about a half million years ago. *Homo sapiens* built artificial shelters protected from the sun's rays and manufactured lamps that allowed him to extend the daily lighting period about 70,000 years ago. The first lamp invented was made of a shell, hollowed-out rock, or other similar nonflammable objects and was filled with a combustible material (probably dried grass or wood), sprinkled with animal fat (the original lighter fluid) and ignited.

In the last 200 years, we have shifted our routines from rural environments to the cities and from outdoor life to confinement in our homes. Furthermore, with the advent of electric light we have become progressively isolated from the natural cycles of light and darkness that shaped our biological rhythms for millions of years. This is an environmental mutation, with an increasing impact on the quantity and quality of our sleep.

The internet explosion has added a further complication. Increasingly, individuals spend part of their nights in front of lit monitors (LCD, tablets, smartphones), screens that produce at least two phenomena of concern to the sleep–wake cycle: (a) plundering the natural period of sleep by reducing it to dangerous levels; (b) adding a disruptor factor, the monitor light during the circadian period causes phase delays of the biological clock, which produces a later sleep on subsequent nights, tending to perpetuate the situation of nocturnal sleep deprivation [1].

The artificial light striking the retina from dusk until dawn exerts a strong inhibitory activity of hypothalamic neurons that induces sleep and a strong excitatory activity of brain mechanisms that maintains wakefulness. As we have already discussed in Chap. 2, suppression of the nocturnal release of melatonin occurs, which is responsible for synchronizing our circadian rhythms and for the "opening of the sleep gate," (Fig. 2.12) [2]. If someone tries to sleep at 18:00 or 19:00 h, he or she would probably take a 1- to 2-h nap. If instead the time to sleep onset shifts to 21:00 or 22:00 h, a 6- to 8-h consolidated sleep tends to occur. This sharp increase in sleep propensity is due to inhibition of wakefulness promoting action of the SCN given by melatonin. Ambient light by inhibiting the secretion of melatonin, reduces sleepiness, promotes alertness, and interferes with sleep.

In daylight conditions, the SCN exerts maximum wake-up activity toward the end of the period of wakefulness, a "second wind" that keeps us awake despite the sleep debt already accumulated in wakefulness. Before the widespread use of electric light, people experienced this "second wind" in the afternoon (17:00 to 18:00 h) and possibly kept alert until nightfall. However, exposure to artificial

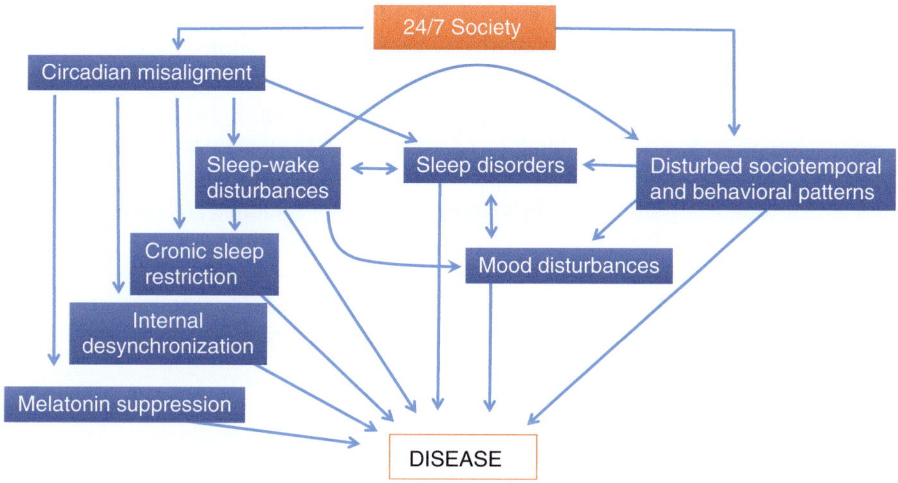

Fig. 8.1 The impact of 24/7 society. Reproduced with permission from Cardinali [75]

light after dark has incorporated a signal indicating to the SCN a time of day that is not real, deferring the "second wind" and delaying the secretion of melatonin (Fig. 8.1).

Indeed, technology has disconnected us from the natural 24-h day in which our species has evolved. A comparison of the biological effects of reading an electronic book on a light-emitting device with reading a printed book in the hours before bedtime indicated that those reading an electronic book took longer to fall asleep and had reduced evening sleepiness, reduced melatonin secretion, later timing of their circadian clock, and reduced next-morning alertness than those reading a printed book [1].

Because about 20% of electricity consumption worldwide is devoted to the production of light, many governments are eliminating traditional incandescent lamps (emitting in the red) for more efficient LEDs in search of savings. However, this white light solid state is typically rich in blue light, which is the portion of the spectrum that most inhibits photoreceptor retinal ganglion cells and hence the secretion of melatonin, which further amplifies the disruptive effect on sleep–wake rhythm [3]. Therefore, sleep topics are not only a medical problem, but substantially influence the social and organizational frame of society.

Both longitudinal statistics in northern hemisphere countries and regional data in Latin America indicate that in just 40–50 years, we have reduced our sleeping time by 25% [4]. Longitudinal statistics from the National Sleep Foundation, USA, indicate that the number of daily hours of sleep has fallen since 1960 to date from 8.2 to 6 h daily. In our study, we verified that 65% of the population, regardless of age, reported having sleep disorders in the last 12 months; 40% of these disorders were described as moderate to severe, and there was a sleep deficit of about 2 h a day. All respondents recognized the negative consequences of poor sleep for health and quality of life [5].

In the USA, 30% of employed adults and 44% of night workers sleep less than 6 h per night, versus less than 3% of the adult US population 50 years ago [6]. Globally, children sleep about 1.2 h less on weekdays before school activity than a century ago. It is noteworthy that children tend to become hyperactive instead of sleepy when they do not get enough rest and have difficulty in concentrating, focusing attention; thus, deficiency of sleep can be confused with hyperactivity/attention deficit disorder, a condition "over-diagnosed" in many societies [7].

The Disruption of the Three ANS Physiological Programs ("Body Configurations") Is a Major Consequence for the "24/7 Society"

We discussed in Chap. 2 how the three different "bodies" (wakefulness, slow-wave sleep, and REM sleep) necessarily follow each another harmoniously to ensure health. A 76-year-old man (the current life expectancy in our society) sleeping 8 h daily will live 50 years in the physiological state of wakefulness, 20 years in slow-wave sleep, and 6 years in REM sleep. However, our society has reduced by 25% the amount of time spent sleeping over the last 40 years. Therefore, the above calculation now changes to a distribution of 55 years of wakefulness, 15 years of slow-wave sleep and 6 years of REM sleep (Fig. 8.2).

A timed view of the ANS from clinical standpoints allows interpretation of symptom intensity and mortality of human diseases, conditions, and syndromes that exhibit a 24-h pattern and are thus related to the three body configurations. Twenty-four-hour patterns are characteristic of more than 100 acute and chronic common and rare human diseases (Fig. 8.3) [8]. The worst symptoms of many of these tend to be expressed either late evening/overnight or early morning, thus significantly compromising nocturnal sleep, daytime productivity, and overall quality of life, and are therefore of great relevance clinically to patient management, for example [8]:

- Cardiac—atrial premature beats and tachycardia, paroxysmal atrial fibrillation, atrial–ventricular block, paroxysmal supraventricular tachycardia, ventricular premature beats, angina pectoris, acute (nonfatal and fatal) incidents of myocardial infarction, sudden cardiac arrest, acute cardiogenic pulmonary edema, heart failure
- Vascular and circulatory system—hypertension, acute hypotension/syncope, intermittent claudication, venous insufficiency, standing occupation leg edema, arterial and venous branch occlusion of the eye, menopausal hot flash, sickle cell syndrome, abdominal, aortic, and pulmonary thromboembolism, deep venous thrombosis, cerebrovascular transient ischemic attack, stroke
- Respiratory—viral and allergic rhinorrhea, reversible (asthma) and nonreversible (bronchitis and emphysema) chronic obstructive pulmonary disease, cystic fibrosis, high-altitude pulmonary edema
- Gastrointestinal tract—esophageal reflux, peptic ulcer, cyclic vomiting syndrome, biliary colic, hepatic variceal hemorrhage
- Renal—colic, nocturnal enuresis, polyuria

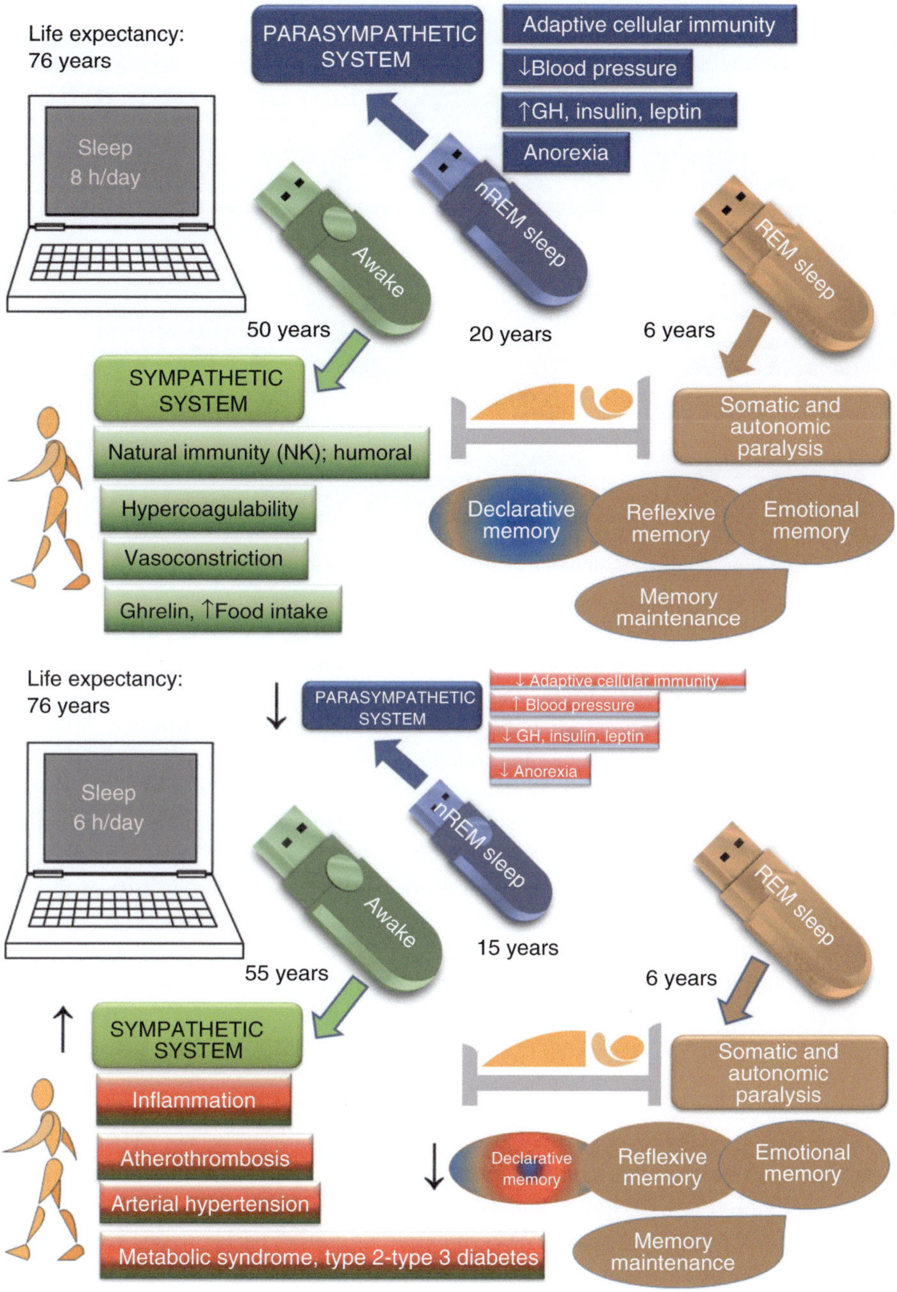

Fig. 8.2 The three different "bodies," wakefulness, slow-wave sleep (NREM sleep) and REM sleep, must necessarily follow each another harmoniously to ensure health. *Upper panel*: a 76-year-old man sleeping 8 h daily will live 50 years in the physiological state of wakefulness, 20 years in slow-wave sleep and 6 years in REM sleep. *Lower panel*: reduction of 25% of sleep over the last 40 years leads to predominance of the wakefulness state and reduction of the slow-wave sleep, associated with cardiovascular disease, metabolic syndrome, obesity, and type II and type III diabetes

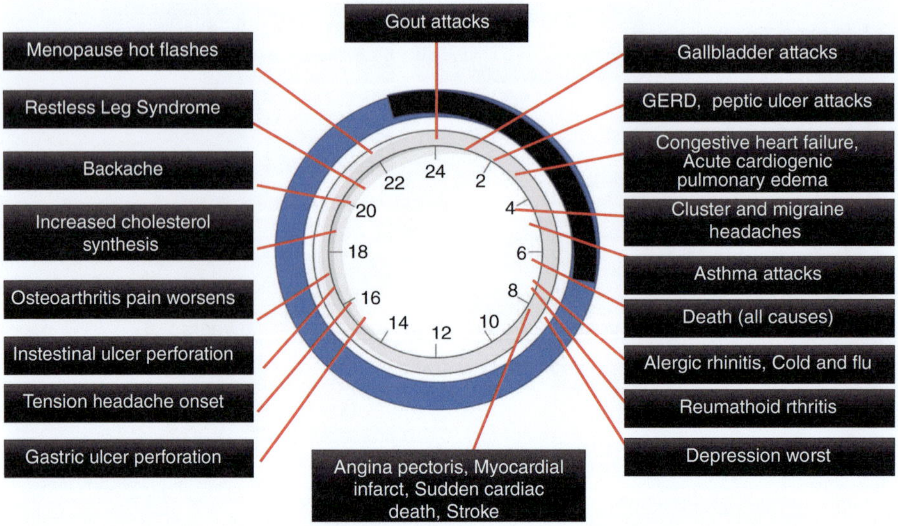

Fig. 8.3 Chronopathology. Acrophases for several clinical entities across a 24-h cycle

- Neural—frontal, parietal, temporal, and occipital lobe seizures, Parkinson's and Alzheimer's disease, hereditary progressive dystonia
- Psychiatric/behavioral—major and seasonal affective depressive disorders, bipolar disorder, dementia-associated agitation, addictive alcohol, tobacco, and heroin cravings, and withdrawal phenomena
- Pain—cancer, post-surgical, diabetic neuropathic and foot ulcer, tooth caries, fibromyalgia, sciatica, multiple sclerosis muscle spasm, and migraine (tension, cluster, hypnic, paroxysmal hemicranial headache)
- Autoimmune and musculoskeletal—rheumatoid arthritis, osteoarthritis, spondylarthrosis, gout, Sjögren's syndrome, systemic lupus erythematosus
- Infection—susceptibility, fever, mortality
- Skin—atopic dermatitis, urticaria, psoriasis, palmar hyperhidrosis
- Ocular—bulbar conjunctival redness, keratoconjunctivitis sicca, intraocular pressure, anterior ischemic optic neuropathy, recurrent corneal erosion syndrome

The imbalance at the expense of slow-wave sleep imposed by the "24/7 Society" (Fig. 8.2) can be costly in countless aspects of our life and health. The impact of sleep deprivation is widespread and affects not only the physical, but also the psychological and social wellbeing. With impaired cognitive performance in the areas of attention, memory, and executive functions, the added emotional and behavioral consequences of sleep deprivation may explain the exasperated social behavior found in our present life [9].

The predominance of the sympathetic configuration of wakefulness is a strong predisposing factor for chronic, low-degree inflammation. The term "inflammaging" has been coined to denominate the contribution of inflammatory processes to the progression of aging [10]. However, inflammation is not only a matter of normal senescence, but is also observed in several diseases that can be linked to the predominant sympathetic configuration given by prolonged wakefulness.

Chronic, mild inflammation is multiply intertwined with other potentially deteriorating processes, among which mitochondrial dysfunction is of premier importance [10]. An abnormally prolonged sympathetic predominance, as in sleep deprivation, implies several immune remodeling processes that include tendencies toward enhanced proinflammatory signaling (Chap. 4).

Numerous epidemiological studies indicate the association of sleep deprivation with cardiovascular disease, metabolic syndrome, obesity, and T2DM [4, 11]. Moreover, today the association of these symptoms is emphasized, with dementia, particularly Alzheimer's disease, which is often called "type 3 diabetes". There is also epidemiological evidence for the link between poor sleep when working shifts and cancer, especially breast cancer [12].

Jet Lag, Shift Work, and Chronodisruption as Examples of a Desynchronized ANS

Changes in the environmental timing cues (zeitgebers) are associated with alterations in the body's 24-h rhythms (Chap. 2). Under most circumstances, this adjustment to the environmental light/dark cycle is normal and consistent. However, mismatches can occur when changes in environmental demands are either sudden or severe. In shift work, there is a phase shift of the activity/rest cycle with regard to the light/dark cycle, whereas jet-lagged time zone travelers encounter a pattern of light and darkness, activity, and social schedules shifting together in time [13]. The endogenous circadian system is slow to adapt to new time cues, and until the correct phase relationship between biological rhythms and external zeitgebers is re-established, a host of physiological and behavioral problems can manifest. Similar problems are encountered by shift workers operating under new work schedules out of phase with the normal light/dark cycle, other competing zeitgebers, and their endogenous body clock.

Chronodisruption comprises the changes in amplitude and phase of circadian rhythms that are found as a comorbidity of most acute and chronic diseases. The observed chronodisruption may be due to a failure in one or more components of the clock itself, in the output signal to the different systems, in the presence of synchronizers, or in the transmission of information from the zeitgebers to the circadian clock (Fig. 8.4). In the clinic, the exact cause of an alteration is unlikely to be known, and in any case, most are multifactorial (Table 8.1) [14–17].

As discussed in Chap. 2, almost every physiological function has a circadian "phase map" consisting of an ordered sequence of peaks and valleys (Fig. 2.1).

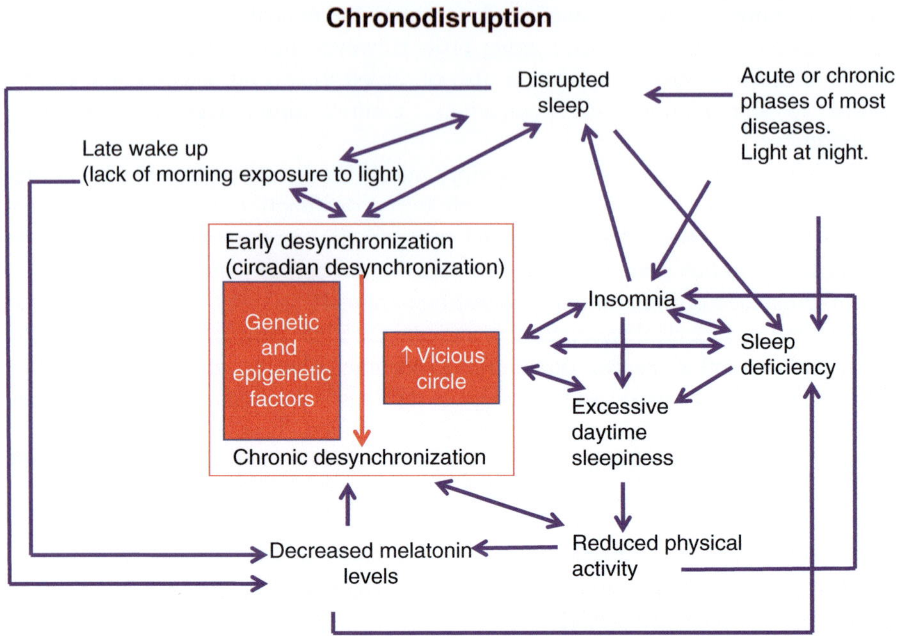

Fig. 8.4 Chronodisruption found as a comorbidity of most acute and chronic diseases. The observed chronodisruption may be due to a failure in one or more components of the clock itself, in the output signal to the different systems, in the presence of synchronizers, or in the transmission of information from the zeitgebers to the circadian clock

Table 8.1 Etiology, symptoms, and treatment of chronodisruption

Etiology	Acute or chronic phases of most diseases Sleep deprivation, light at night (LAN)	Alteration of the central circadian pacemaker indicated by the alteration in melatonin secretion
Symptoms *(phase and amplitude changes of circadian rhythms)*	ANS dysfunction	Insomnia, somnolence, metabolic syndrome, thoracic-muscular sympathetic predominance, parasympathetic abdominal predominance, gastrointestinal disorders
	Somatic alterations	Tiredness, fatigue, stiff neck/back, low back pain, headache, drowsiness
	Cognitive functions	Disorientation, loss of sociability and motivation, poor attention, performance, memory, or concentration
	Behavior	Aggressiveness, impulsivity, hyperactivity, irritability
	Psychiatric disorders	Depressive symptoms, personality disorders, anxiety
Therapeutic approach	Chronobiological treatment (exposure to morning light, melatonin supplementation at bedtime) *In the case of acute or chronic diseases, the chronobiological treatment must accompany the specific treatment for "ad integrum" recovery*	

While the period of the phase map for different physiological functions is similar, the peaks and valleys of the maps generally do not coincide. Phase maps are also sensitive to environmental changes and can be transiently affected by temporal disruptions such as jet-lag disorder or shift work, or by the disruption imposed by an acute or chronic illness, regardless of its severity.

Jet Lag

Flight dysrhythmia, more commonly known as jet lag, comprises a constellation of symptoms consisting of daytime fatigue, impaired alertness, nighttime insomnia, loss of appetite, depressed mood, poor psychomotor coordination, and reduced cognitive skills, among other (Table 8.2) [18]. These symptoms are caused by the

Table 8.2 General symptoms of jet lag

Anorexia or loss of appetite
Apathy
Bowel irregularities (constipation or frequent defecation)
Clumsiness
Daytime somnolence
Decreased vigilance and attention domains
Depression
Diminished mental abilities (i.e., cognitive performance, concentration, judgment, decision making, memory lapses)
Diminished physical performance
Disorientation
Fuzziness
Gastrointestinal symptoms (e.g., bloating and upset stomach)
General feeling of malaise
Generalized fatigue and lethargy
Glucose metabolism dysregulation
Headache
Impaired alertness
Impaired task performance (increased accidents and errors)
Inappropriate timing of defecation and urination
Irritability
Menstrual irregularity
Mood disturbances
Muscular pain
Sleep loss
Sleeping difficulties (inappropriate sleep at local time)
Slowed reflexes
Stress
Tiredness (traveler's fatigue)
Traveler's thrombosis (deep vein thrombosis)
Trouble initiating and maintaining sleep
Tumor progression is noted in chronic animal model

Please note that given the inter-individual variability and susceptibility, the core symptoms vary and not necessarily every individual experiences or exhibits the entire spectrum of symptoms temporary misalignment between the circadian clock and external time, which occurs because of rapid travel across time zones. The number of time zones crossed and the direction of travel influence the severity of jet lag symptoms. Eastward travel tends to cause difficulty in falling asleep whereas westward travel usually interferes with sleep maintenance.

In jet lag, the recovery time for restoring the normal rhythm profile (re-entrainment) can differ significantly from one physiological function to another. Although during the period of re-entrainment individual circadian rhythms generally move in a direction that corresponds with that of the environmental time shift, in some cases the circadian system moves in a direction that is opposite to that of the environmental change, giving rise to a phenomenon called "splitting." Phase map "partitioning" is then more complex, involving a partial re-entrainment by some phase maps in a direction that is opposite to that of other phase maps [13].

Jet lag affects the health status of frequent air travelers. The disruptive effect of jet lag has been documented in experimental animals at the molecular level of clock genes in the SCN and in the clock genes present in peripheral tissues [19]. Eastbound travel causes a phase advance in all the body's circadian rhythms, whereas westward flight has the opposite effect, i.e., it produces a phase delay. Consequently, travelers tend to synchronize their bodily rhythms at a speed of 1.5 h a day after westward and 1 h a day after eastward flight, irrespective of whether they travel during the day or at night. This difference in adjustment time is usually attributed to the greater ability of the internal body clock to adapt to a longer rather than to a shorter day (τ longer than 24 h, Chap. 2).

Jet-lag disorder falls under the category of circadian rhythm sleep disorders. The symptomatology of jet-lag disorder includes both physiological and psychological disturbances (Table 8.2). Several studies suggest that chronic or repeated jet-lag exposure can lead to cognitive decline and temporal lobe atrophy in humans if there is a short recovery time between flights. Model jet-lag disorder is also associated with tumor progression in rodents and elevated mortality rate in aged mice [13].

Jet-lag disorder symptoms show considerable inter- and intra-individual variability. Age is one important factor. In simulated jet-lag disorder, middle-aged male subjects had more symptoms than younger men. Moreover, those over 60 years are reported to have greater difficulty in adapting to jet-lag disorder [20]. Another important variable is the individual's "chronotype" (Chap. 2). Individuals who are "morning chronotypes" generally have less difficulty in phase-advancing their body rhythms (i.e., adjusting after a flight from west to east) than "evening chronotypes," and vice versa in the case of a phase delay. Generally, those who had "rigid" sleep habits had more severe symptoms after a transmeridian flight [21].

Several studies have examined the effects of simulated and real jet lag on physiological and psychological variables in different populations including aircrew members, and have confirmed that these frequent flyers suffer marked sleep–wake problems because of jet lag. For example, in a 2-year collaborative field study of

Spanish pilots flying the routes from Madrid, Spain, to Mexico City, Mexico (−7 time zones) or from Madrid to Tokyo, Japan (+8 time zones) we used telemetry to record pilots' activity, temperature, and heart rate [22, 23]. Subjective time estimation and other psychological variables such as anxiety, tiredness, and performance were recorded. Urinary 6-sulphatoxymelatonin and cortisol excretion (determined in 6-h intervals) were also measured. Activity/rest and heart rate rhythms, linked to a "weak" or exogenous oscillator, became rapidly synchronized, whereas temperature or 6-sulphatoxymelatonin excretion rhythms, which are closely regulated by the biological clock (Chap. 5) showed a more rigid response after the phase shift of the light/dark cycle [22]. In both young (<50 years old) and old (>50 years old) pilots arriving in Mexico or Tokyo, the activity/rest rhythm rapidly adjusted to the new schedule, whereas the acrophase of the temperature rhythm tended to fluctuate near the original temporal zone. This desynchronization was evident until the return flight (day 5) and persisted after arrival in Madrid [22]. The sequelae of desynchronization were less tolerable in older than younger pilots. Skin temperature rhythm did not become entrained neither on reaching Tokyo nor after the return flight to Madrid in the group of older pilots [22]. The changes in urinary 6-sulphatoxymelatonin and cortisol excretion were consistent with these conclusions.

Systematic and incorrectly planned work schedules of airline pilots produce a chronic disruptive condition, a higher incidence of stress-related emotional changes, and a diminished life expectancy. The optimal work strategy for this population is a compromise between two extreme possibilities: a long rest period at stopovers until full re-entrainment is achieved, or a short stop accompanied with relative isolation, maintaining the original "home" local habits to prevent re-entrainment. With the first strategy, aircrew would be systematically exposed to a re-entrainment process, with a permanent disruption to the circadian system, whereas the second approach, although less disturbing to the body clock, would probably not allow pilots to have the necessary rest and alertness for the return flight [24].

Sleep deprivation produces an allostatic overload that can have deleterious consequences [25]. Restriction of sleep to 4 h per night is associated with increases in BP pressure, decreases in parasympathetic tone, increases in evening cortisol and insulin levels, and increases in appetite through the elevation of ghrelin, a pro-appetitive hormone, and decreases in the levels of leptin, which has anorexic activity (Chap. 5). Proinflammatory cytokine levels are also increased, along with decreases in performance in tests of psychomotor vigilance after a modest sleep restriction to 6 h per night. Allostatic overload in animal models causes atrophy of neurons in the hippocampus and prefrontal cortex, the brain regions involved in memory, selective attention, and executive function. It also causes hypertrophy of neurons in the amygdala, the brain region involved in fear and anxiety, and aggression (Chap. 6). Thus, the ability to learn and remember and to make decisions may be compromised and may be accompanied by increased levels of anxiety and aggression.

Both long-term and short-term exposure to transmeridian flights have an impact on cognitive functioning. For example, in a group of individuals who were on a transmeridian flight and who underwent functional magnetic resonance imaging study, participants from the jet-lag group presented decreased activation in the bilateral medial prefrontal and the anterior cingulate cortex [26]. The results are suggestive of a negative impact of jet lag on important cognitive functions such as emotional regulation and decision-making during the first few days after individuals arrive at their destination.

Shift-Work Disorder

A wide range of work schedules is referred as "shift work." They include occasional on-call overnight duty, rotating schedules, and steady, permanent night work (Fig. 8.5) [6]. Owing to the overlap of these categories, it is difficult to generalize about shift work disorder. Over 10% of night workers and of rotating workers met the minimal criteria for shift work disorder.

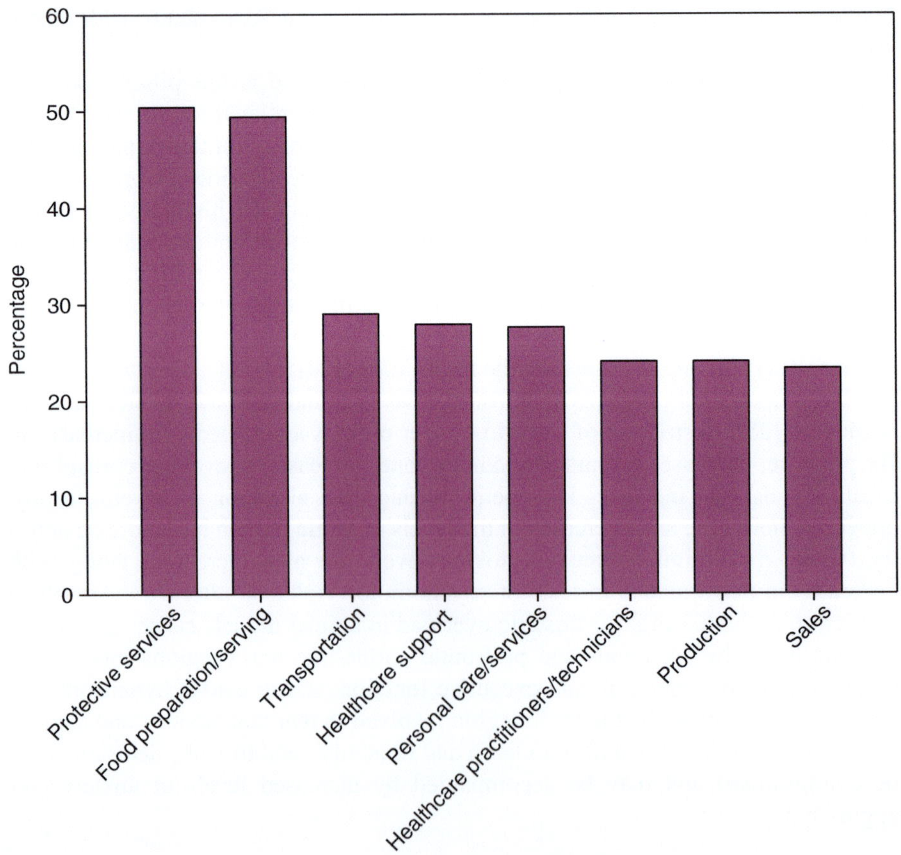

Fig. 8.5 Prevalence of shift work. Reproduced with permission from Cardinali [75]

Several studies have now confirmed that there is a relationship between cardiovascular disease and shift work (Fig. 8.6). Shift workers are at a 40% higher risk of developing ischemic heart disease. The association between shift work and metabolic syndrome, a major risk factor for cardiovascular disease, also occurs. Alternating shift work has been reported to be a significant independent risk factor for high BP, an effect that was more pronounced than that of age or body mass index [6].

In 2007, the International Agency for Research on Cancer classified shift work as a probable human carcinogen (2A). Women who work on rotating night shifts are reported to be at a moderately increased risk of breast cancer after extended periods of working night shifts [27], as are female cabin crew [28]. Because melatonin has oncostatic effects, including effects on estrogen and fat metabolism, it may play a role in both breast and endometrial cancer. Light exposure at night reduces melatonin levels, but the role of melatonin in both of these cancers remains to be defined.

Misalignment between the circadian pacemaker and the timing of sleep, wake, and work occurs in shift workers, and shift work disorder, with insomnia, reduced

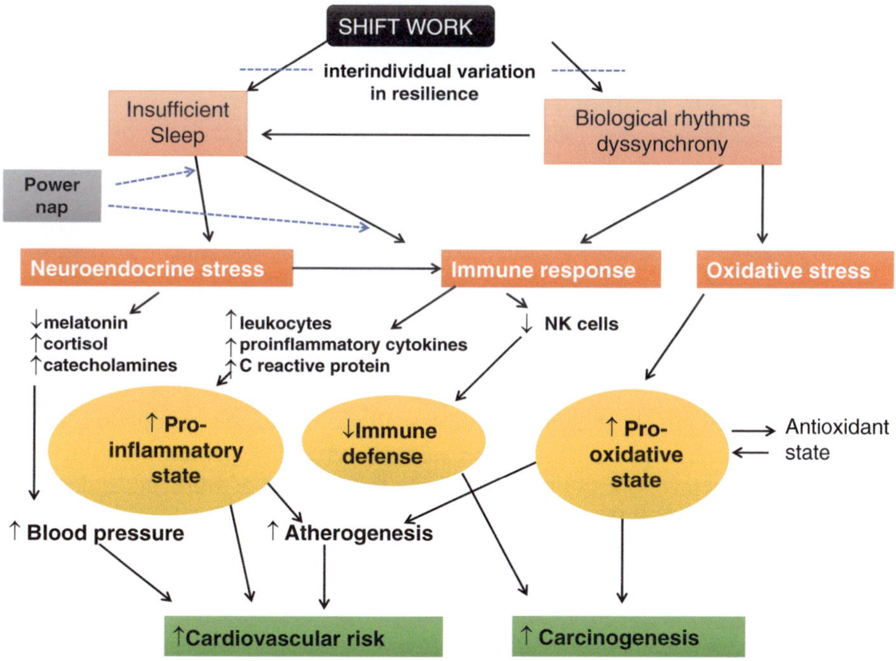

Fig. 8.6 Potential physiopathological pathways by which shift work may lead to cardiovascular disease and cancer. Experimental circadian misalignment and sleep restriction protocols disrupt and enhance the activity of neuroendocrine stress systems, reduce immune defense (NK cells), and cause inflammation and oxidative damage. Inter-individual vulnerability to the adverse effects of sleep restriction and circadian misalignment contribute to a heterogeneous tolerance to shift work. Prophylactic naps could blunt the stress response, possibly correct stress-dependent immune changes, and improve the recovery of immune homeostasis. Reproduced with permission from Cardinali [75]

sleep, and excessive sleepiness, is common. All these impair cognitive function, alertness, and mood, and increase the risk of accidents. For years, the public, media, and regulatory authorities have blamed the effects of excessive speed and alcohol as the main causes of road accidents. However, it is important to note that the lack of sleep produces the same effects on the ability to drive a vehicle as drinking alcohol. In psychometric studies, to be awake for 17–18 h disturbs the ability to drive a vehicle in a similar manner to the effect of an alcohol concentration in the blood of 0.05 g/dL [29]; moreover, both situations may add up to decreased attention. Between 20 and 25% of road accidents are caused by fatigue and sleepiness of drivers, being more frequent between 0200 and 0800 h. This trend is particularly evident on motorways and monotonous routes.

In several studies on this subject, high number of drivers refer to often being drowsy at the wheel. Clearly, there are more sleepy drivers than drunk ones on roads. For example, in representative samples of public transportation drivers in the Metropolitan Area of Buenos Aires and long-distance drivers covering various geographical corridors of the country, we conducted surveys on health and working conditions and applied objective measures of physiological variables, including evaluation of the sleep/wake rhythm using actigraphy, circadian rhythmicity by the peripheral rhythm of body temperature, alertness by determining psychomotor response to a stimulus, autonomic activity by heart rate variability, and endocrine response to stress by measuring cortisol in the saliva [30, 31].

In short-distance drivers, a high prevalence of work-related stress, overweight, obesity, physical inactivity, and hypertension was observed. The quantity and quality of sleep on weekdays was poor, with partial recovery at the weekend, a high frequency of daytime sleepiness, and high risk of apnea. The neurohormonal weekday pattern was consistent with stress and a significant drop in psychomotor performance was observed during working hours, especially the morning shift [30, 31].

In long-distance drivers, a high prevalence of cardiovascular risk factors, such as overweight, physical inactivity, and smoking, was also found. Sleep patterns of poor quality, with little sleep at home, while traveling, and at the destination, and a decrease in amplitude circadian rhythms, were observed. The pattern was consistent with high cortisol levels, with little recovery in the days out of work, and a decrease in alertness at the end of the return trips [30, 31].

In addition to the known effects on sleep, eating patterns, and alterations in social life, gastrointestinal disorders are very common in shift workers. The association between rotational work shifts and gastrointestinal disorders has several causes. Irregular ingestion habits of workers affect the synchronization of numerous circadian rhythms, in particular, those related to digestive functions and metabolism [32]. Gastrointestinal disorders may be due to ingestion of food at the "wrong" times, which induces anomalous patterns of motility and digestive secretions. In addition, the absence of hot food, which occurs frequently during the night (with a predominance of snacks), a high carbohydrate intake, caffeine and alcohol, and high consumption of tobacco have all been proposed as causes of gastrointestinal disorders in shift workers.

Chronodisruption

This term defines the changes in amplitude and phase of circadian rhythms that are found in acute or chronic illness, even in mild situations such as common flu ("poor wakefulness, poor sleep") [14–17]. The clock itself, the information pathways from synchronizing agents to the clock or the efferent pathways of the clock can be affected alone or conjointly (Fig. 8.4). The deterioration of circadian rhythms with age is an example of a change with many components, which implies a decrease in the effectiveness of many aspects of the circadian system, from a reduced influence of the synchronizers, to a deterioration of the clock itself, to a decrease in the ability to obey the clock output signal [33].

The observation that the circadian system is not functioning normally does not necessarily imply that it is the primary cause of the alteration. Rather, the primary defect may have originated elsewhere, and its effects on the circadian system be one of the many changes it causes. In this case, and although treating the circadian system could improve the individual's nocturnal sleep and diurnal activity, we would not be addressing the real cause of the problem. A combined (specific and chronobiological) treatment is needed for full recovery (Figs. 8.7 and 8.8) [14–17].

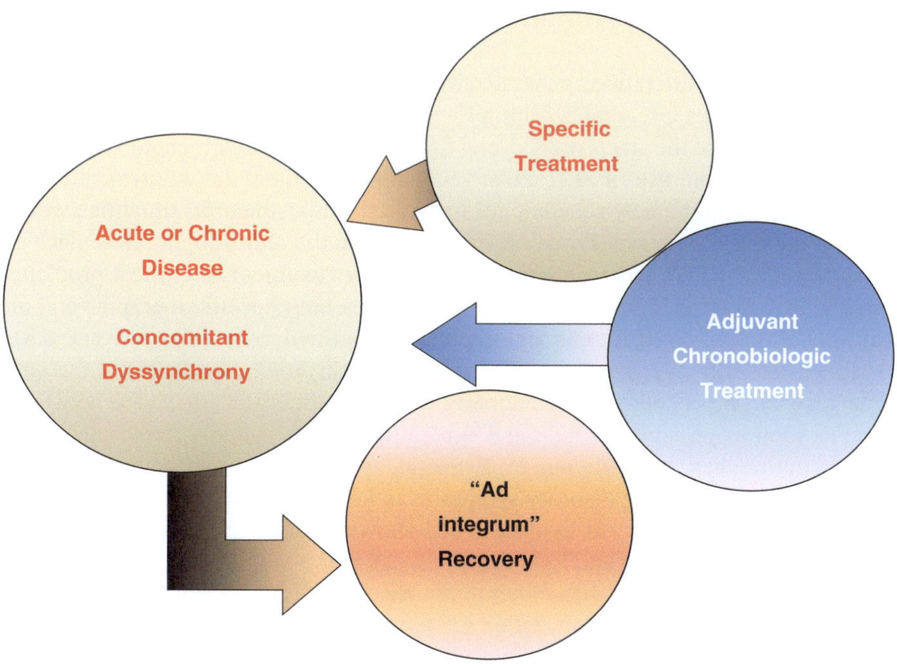

Fig. 8.7 The concomitant chronodisruption occurring in most acute or chronic diseases must be adequately treated to obtain full recovery of health. Reproduced with permission from Cardinali [75]

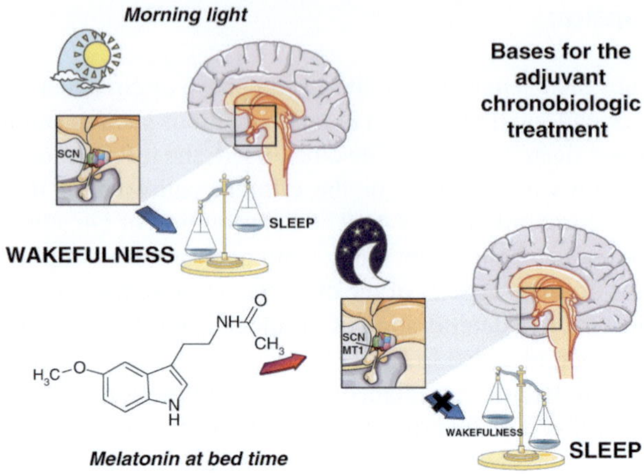

Fig. 8.8 Exposure to light in the morning and the administration of melatonin in the evening provide a strong synchronization signal and an increase in the amplitude of sleep/wake cycle. From clinical standpoints, the changes in amplitude ("poor sleep together with poor vigilance") is a paramount sign of the disease and its correction increases substantially the patient's quality of life. The morning light and melatonin in evening hours are the natural resources to restore proper rhythmicity of sleep/wake rhythm. Reproduced with permission from Cardinali [75]

Chronobiological Treatment of a Desynchronized ANS

The jet lag and the alterations generated by shift work are two situations that reflect the functioning of a circadian system of a normal subject that has not adjusted to a schedule change of the sleep/wake cycle. Consequently, the treatment of these problems consists in the use of alternative zeitgebers that allow the adjustment of the circadian clock to the new schedule, or to avoid the adjustment in situations where it is unnecessary, as in rapid rotation shifts, or when the return flight takes place as soon as within 1 or 2 days. On the other hand, the treatment of clinical problems associated with circadian rhythm alterations needs both specific therapies and an adjuvant chronobiological treatment to obtain an optimal result (Figs. 8.7 and 8.8).

To successfully overcome the effects of jet-lag disorder, adjustment to the new time zone can be encouraged by adopting the social timing of life in the new time zone as soon as possible. In field and simulation studies suitably timed melatonin administration has been shown to accelerate phase shifts and to significantly improve self-rated jet-lag disorder symptoms in large numbers of time zone travelers. In addition, exposure to light has been shown to accelerate phase shifts. The combination of melatonin and light exposure, one in the evening and the other in the morning, is more effective than either treatment alone (Fig. 8.7) [20].

The effect of light or melatonin on the circadian system can be measured by a phase response curve in which the minimum in core body temperature is used as an estimate for the crossover point of the curve (Fig. 8.9). Light pulses administered before this point delay the circadian clock, whereas light pulses after it phase

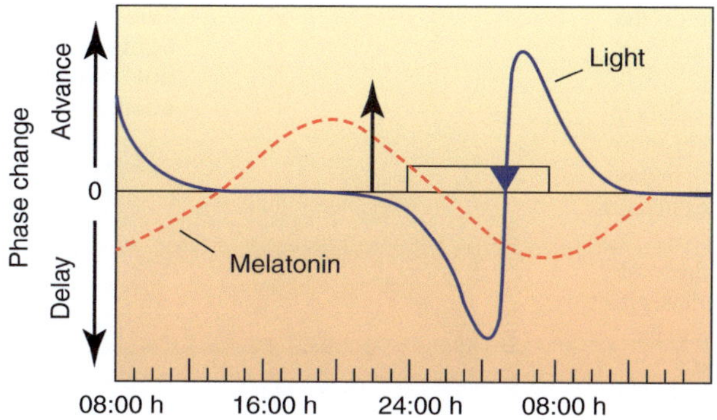

Fig. 8.9 The phase response curves to light and melatonin are opposite, but not symmetrical. Reproduced with permission from Cardinali [75]

advance the clock. Light exposure close to the minimum core body temperature produces the greatest phase shifts [34]. Phase delays of approximately 2.5–3 h per day and phase advances of 1.5–2 h per day have been observed following carefully timed exposure to bright light [35]. The phase response curve for melatonin is the opposite to that of the light, although not symmetrical (Fig. 8.9).

Melatonin is the prototype of a chronobiotic drug [36]. It is produced in most organisms from algae to mammals, but its role varies considerably across the phylogenetic spectra [37]. In humans, melatonin plays a major function in the coordination of circadian rhythmicity, remarkably the sleep–wake cycle (Chap. 2). Melatonin secretion is an "arm" of the biological clock in the sense that it responds to signals from the SCN. More particularly, the timing of the melatonin rhythm indicates the status of the clock, both in terms of phase (i.e., internal clock time relative to external clock time) and amplitude. From another point of view, melatonin is a chemical code of the night: the longer the night, the longer the duration of its secretion. In many species, this pattern of secretion serves as a time cue for seasonal rhythms (Chap. 2).

The usefulness of melatonin for ameliorating the symptoms of jet lag has been compellingly demonstrated in numerous investigations. A meta-analysis (Cochrane database) concluded that melatonin taken at bedtime in the place of destination (2200 h to midnight) was effective for decreasing the jet lag symptoms in air travelers who crossed five or more time zones [38].

There is considerable evidence that a light stimulus of sufficient intensity applied at a critical circadian phase can essentially stop the human circadian clock by resetting the circadian oscillator close to a phaseless position at which the amplitude of circadian oscillation is zero, i.e., type 0 resynchronization (Fig. 8.10) [39]. Indeed, exposure of humans to cycles of bright light, centered on the time at which the human circadian pacemaker is most sensitive to light-induced phase shifts, can markedly attenuate or reduce endogenous circadian amplitude.

Timed light and melatonin administration allowed an almost immediate resynchronization of circadian rhythms in a group of jet air travelers who had made a

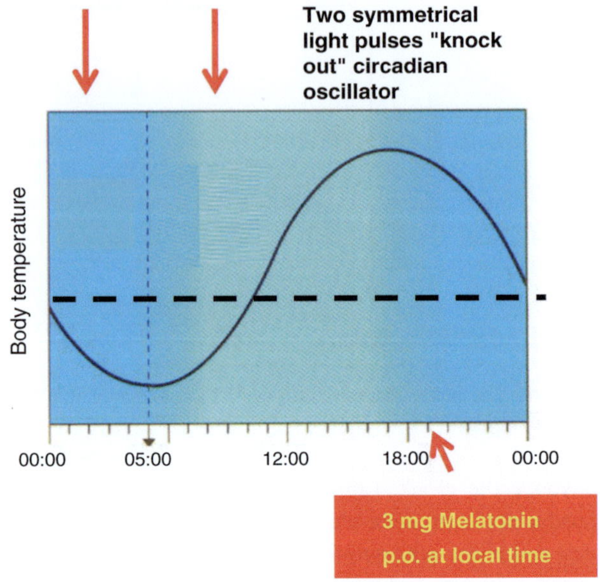

Fig. 8.10 To suppress the circadian clock oscillation, the application of symmetrical light pulses in the first and second part of the night is needed. In this "knock out" of the circadian clock, melatonin administration at the local time has an immediate effect of synchronization to the new time schedule. Reproduced with permission from Cardinali [75]

transmeridian flight over 12 time zones. Under the conditions of transfer of 12 time zones over a period of hours, a fully inverted (180°) relationship between the subjective day and the geophysical day occurs. Thus, a patterned exposure to natural light covering portions that symmetrically delay and phase advance the circadian rhythms resulted in suppression of the circadian pacemaker function. This allowed the use of melatonin at local night to resynchronize the circadian oscillator to the Tokyo time. Additionally, we administered a nonphotic stimulus (exercise) in a schedule to coincide with exposure to natural light to mask the circadian oscillator. The observed rate of resynchronization was about 2 days, significantly different from a minimum resynchronization of up to 8–10 days expected after a flight through 12 time zones [40, 41].

A positive correlation was found between the pre-flight melatonin production rates, evaluated by measuring urinary 6-sulphatoxymelatonin excretion, and sleep quality and morning alertness after a flight [40]. It is known that individuals who possess a weak circadian time structure, as revealed by the low amplitude of body temperature rhythm, are more prone to developing biological intolerance to shift work [42] and presumably to jet lag. In Table 8.3 the tentative recommendations for flights through >8 time zones are summarized.

Among the guidelines for the effective management of shift-work disorder, organizational level changes are important. Three types of intervention have been recommended: (a) switching from slow to fast rotation; (b) changing from backward to forward rotation; (c) self-scheduling of shifts [6]. There is evidence that a rapidly rotating schedule is less detrimental as it minimizes the time spent in a desynchronized state. Clockwise rotation, rather than counterclockwise rotation, was reported to be preferred by workers, probably because the body clock period is somewhat

Table 8.3 Estimated time of day for exposure to natural or artificial bright light (>1000 lux, 30 min) to blunt the circadian system of an air traveler in long (>8-h) time shifts

Number of time zones crossed	Eastbound	Westbound
8	0600–0900 h and 1100–1400 h	1400–1700 h and 1900–2200 h
9	0700–1000 h and 1200–1500 h	1300–1600 h and 1800–2100 h
10	0800–1100 h and 1300–1600 h	1200–1500 h and 1700–2000 h
11	0900–1200 h and 1400–1700 h	1100–1400 h and 1600–1900 h
12–13	1000–1300 h and 1500–1800 h	1000–1300 h and 1500–1800 h

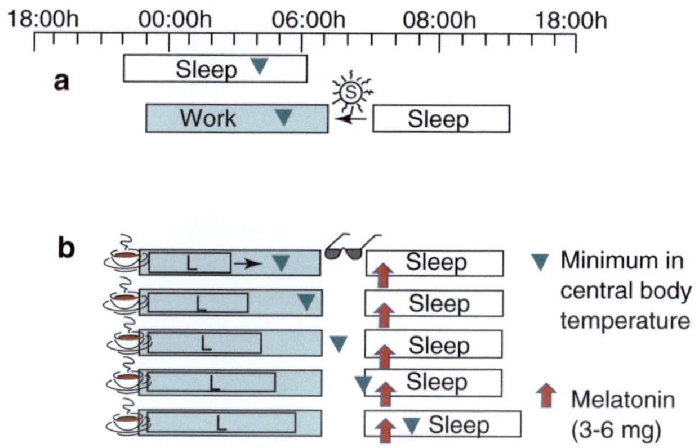

Fig. 8.11 The common situation of a night worker working in a 5-day night shift is depicted. The minimum central body temperature must be shifted toward daytime sleep to ensure a recovery sleep at home. (**a**) Light and activity at night cause a phase delay that is counteracted by sunlight on return home. Although light and physical activity during the night work shift cause some phase delay, light on the commute home opposes this process (causing phase advance). (**b**) Periods of bright light progressively longer in the first part of the night shift give a phase delay that is sustained by using dark glasses on the return home and by taking melatonin before sleep at home (melatonin treatment before daytime sleep is useful for causing sleepiness and inducing phase shifting). Planned 30-min napping just before or on the job combined with caffeinated drinks reduce sleepiness and improve alertness while working [43]

longer than 24 h. Longer duty shifts allow more time off work. It is possible that a flexible approach is best because of major individual differences in workers. The goal is to achieve at least 7 h of sleep per 24 h. Melatonin treatment before daytime sleep is used to promote sleepiness and to induce phase shifting.

Figure 8.11 summarizes the common situation of a night worker working a 5-day night shift [43]. The minimal central body temperature must be shifted toward daytime sleep by using bright light during the first part of the work period and wearing dark goggles on the commute home. Melatonin (3–6 mg) before daytime sleep causes sleepiness and phase shifting. Planned 30-min napping just before or on the job combined with caffeinated drinks reduce sleepiness and improve alertness while working. The wakefulness-promoting agents armodafinil and modafinil have been approved by the Food and Drugs Administration (FDA) for the treatment of excessive sleepiness in patients with shift-work disorder.

In conclusion, both jet-lag disorder and shift-work disorder share a similar cause and management has major similarities. In both cases, three factors are important: (a) sleep scheduling; (b) resetting the body clock with light and/or chronobiotics; (c) use of drugs to promote wakefulness if needed.

For the treatment of clinical chronodisruption it is important to note that the primary alteration may have originated elsewhere, and its effects on the circadian system are one of the many changes it causes. In other words, a chronobiological treatment could be effective because it fights some of the symptoms of the disorder rather than the disorder itself. These warnings are of little importance to those who are limited to treating patients, but they pose a problem of interpretation for those who wish to understand the substance of the problem and try to develop a more rational treatment [14–17]. The combination of the specific and chronobiological treatment is needed for full recovery (Figs. 8.7 and 8.8).

Subjects suffering from insomnia of various causes usually take melatonin, administered 1 or 2 h before the time when it is desired to sleep. In addition to insomnia associated with jet lag and shift work, melatonin is also effective in insomnia of the (otherwise healthy) elderly, in patients with senile dementia, in blind subjects, in subjects who can see but present free-course rhythms, and in patients with delayed-phase sleep syndrome. Its efficacy is demonstrated in both objective and subjective sleep time estimates and in the objective measurement of actimetry. In a meta-analysis including 19 studies and involving 1683 subjects, melatonin showed significant efficacy in reducing sleep latency and increased total sleep time [44]. Trials of longer duration and the use of higher doses of melatonin demonstrated greater effects. Several consensus statements encourage the use of melatonin to treat insomnia [45].

On average, humans cannot do well without sleep for more than a few days (about 2 or 3 days). With only 24–48 h of sleep deprivation, failure of short-term memory appears, there is an increase in the feeling of fatigue, sleepiness, and aggression, and a depressed mood. After 72–98 h without sleep, fatigue is severe and episodes of mental confusion and distortion may occur; in certain individuals, sleep deprivation may produce persecutory delusions.

To solve sleep deprivation, there is no choice but to sleep. In general, after sleep deprivation, we recover one third of the total lost sleep time, 100% of slow-wave sleep and 30–50% of REM sleep [46]. Therefore, the estimated 10 h of sleep deficit accumulated during the week if we sleep 2 h less than needed daily can be recovered by sleeping 1.5 h more on Saturday and Sunday. Hence, the extraordinary importance of not using strict schedules for sleep at the weekend and let it flow without using an alarm clock.

Some Clinical Autonomic Entities Associated with a Desynchronized ANS

Metabolic Syndrome

In the present world, food has become abundant and simultaneously, the need for physical effort has been greatly reduced. From an evolutionary perspective, this is another "environmental mutation," that, together which the advent of artificial lighting, has greatly contributed to the loss of the experience of a differentiated day and night. Chronodisruption, with a disturbed balance among the three different configurations of organ and system regulation in a 24-h cycle, is a direct consequence of this (Fig. 8.2).

Obesity has a profound impact on health, such as T2DM and hypercholesterolemia, mainly through its influence on secretion and insulin sensitivity. The metabolic syndrome comprises a group of metabolic abnormalities (hyperinsulinemia, insulin resistance, hypertension, obesity, hyperlipoproteinemia, hypertriglyceridemia) that increase the risk of cardiovascular disease and T2DM. The metabolic syndrome is also associated with an increased risk of nonalcoholic fatty liver disease and renal dysfunction. Similarly, there is evidence for the correlation of metabolic syndrome with dementia ("type 3 diabetes mellitus") and with cancers, mainly of the breast, pancreas, and bladder [47].

There is impressive information indicating that the obesity is associated with low-grade inflammation of the white adipose tissue, which can subsequently lead to

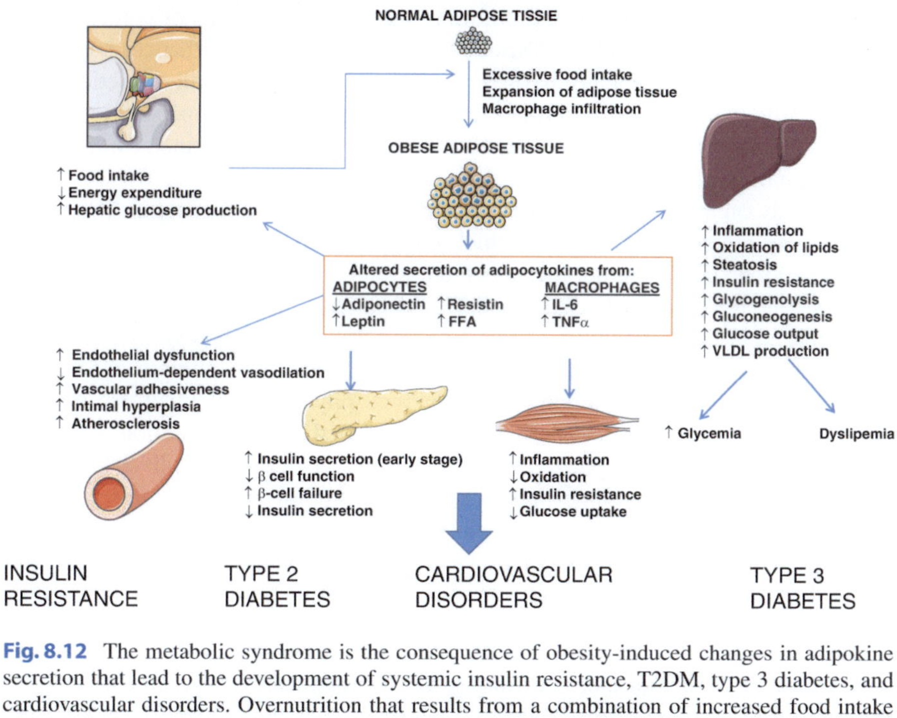

Fig. 8.12 The metabolic syndrome is the consequence of obesity-induced changes in adipokine secretion that lead to the development of systemic insulin resistance, T2DM, type 3 diabetes, and cardiovascular disorders. Overnutrition that results from a combination of increased food intake and reduced energy expenditure leads to adipose tissue expansion, increased adipocyte size and number, and increased macrophage infiltration that, together, lead to increased free fatty acid release, dysregulated secretion from adipocytes of a variety of adipocytokines, including adiponectin, leptin, and resistin, and increased release from resident macrophages of the inflammatory cytokines (TNF-α, IL-6). Dysregulated secretion of these adipokines elicits a variety of adverse effects on numerous tissues and leads to the development of systemic insulin resistance that increases the risk for development of the metabolic syndrome, a variety of cardiovascular disorders, and T2DM and type 3 diabetes (when combined with dementia). Reproduced with permission from Cardinali [75]

insulin resistance, impaired glucose tolerance, T2DM, and type 3 diabetes [48, 49]. Adipocytes actively secrete proinflammatory cytokines such as TNF-α, IL-1β, and IL-6 and trigger a vicious circle that leads to additional weight gain, largely as fat (Fig. 8.12). Increased circulating levels of C-reactive protein and other inflammatory biomarkers also support the occurrence of inflammation in obesity.

The altered production of proinflammatory cytokines modulate adipocyte size and number through paracrine mechanisms that exert an important role in the regulation of fat mass (Fig. 8.12). The amounts of proinflammatory molecules derived from adipose tissue in obese patients diminishes after weight loss [50]. Therefore, the fat cells are both a source and a target for TNF-α, IL-1β, and IL-6 (Fig. 8.12).

A number of factors play an unequivocal role in increasing the risk of developing T2DM [51]. These factors include dysfunction of pancreatic β cells, abnormal adipogenesis and absence of adequate insulin responsiveness in the liver, genetic susceptibility, physical inactivity, excessive food consumption and/or high calorie food intake, and a sedentary life-style.

In addition to these predisposing factors, there is now an increasing amount of evidence that disruption of circadian timing mechanisms is a major contributor to the development of T2DM [49]. Various types of circadian disturbances have been correlated with T2DM, including disruption of the timing of bodily functions that are normally synchronized, improper timing of food intake, dampened clock gene expression and polymorphisms, sleep loss/disturbance and impairments of the melatonin signaling pathway [52].

In a study on 593 patients with a recent diagnosis of T2DM, sleep debt resulted in long-term metabolic disruption, which may promote the progression of the disease [53]. Sleep quality rather than sleep duration played an important role in insulin resistance in these newly diagnosed T2DM patients [54]. The increased obesity rate recorded in 100,000 women of the Breakthrough Generations Study, associated with increased levels of light at night exposure, supports this assumption [55].

Glucose metabolism is among the numerous physiological functions that are governed by the circadian apparatus [56]. It is known that the distribution of glucose to all parts of the body is organized by the molecular clock present in liver. Circadian regulatory mechanisms utilize both neural and humoral communication to exert close control over insulin, leptin, and plasma glucose levels (Chap. 2).

Direct evidence for the association between circadian clock disruptions and T2DM has been provided by studies in mice in which linkages between clock gene mutations and diabetes states were examined [57]. *Per2* mutant mice show several abnormal profiles, including the absence of rhythmicity in plasma glucocorticoid levels, obesity, and low levels of neuropeptides involved in appetite regulation. Clock gene mutations promote delays in pancreatic gene expression, which in turn affects the regulation of the growth, development, and survival of insulin cells and thus of the glucose signaling pathway [58]. *Clock*-mutant mice showed a lack of rhythmicity in the action of insulin, a state that was reversible once the CLOCK protein had been reintroduced [59].

Additional confirmatory evidence for circadian control over metabolic activity was provided by a study of mice in which the *Bmal1* gene was specifically knocked in the pancreas. This produced an animal model that replicated T2DM in that blood glucose levels were found to be elevated throughout the 24-h cycle [58].

In diabetes-prone, genetically engineered rats that were maintained in continuous light and jet-lag-like conditions, circadian disruption ensues [60]. Similar findings were obtained in a study of humans subjected to a forced desynchronization protocol for a period of 28 h. The treatment produced circadian misalignment in all subjects, with 30% of individuals exhibiting a disturbed glucose metabolism that resembled diabetes [61]. Also, all subjects showed a decreased concentration of leptin, a reversal of cortisol rhythm, and high amounts of glucose (postprandial) even in the presence of increased insulin. Additionally, the increased levels of cortisol late in the day, i.e., during the end of wakefulness, were a potentiating factor for insulin resistance and hyperglycemia [59]. These findings suggest that the misalignment of clock functions accelerates the development of T2DM (Fig. 8.13).

Light exposure at night, even at low levels, has been reported to alter food timing and body mass accumulation, thus suggesting that artificial lighting is an important contributing factor to the increased prevalence of metabolic disorders [56]. As discussed in Chap. 5, circadian rhythms could be entrained by manipulating the timing

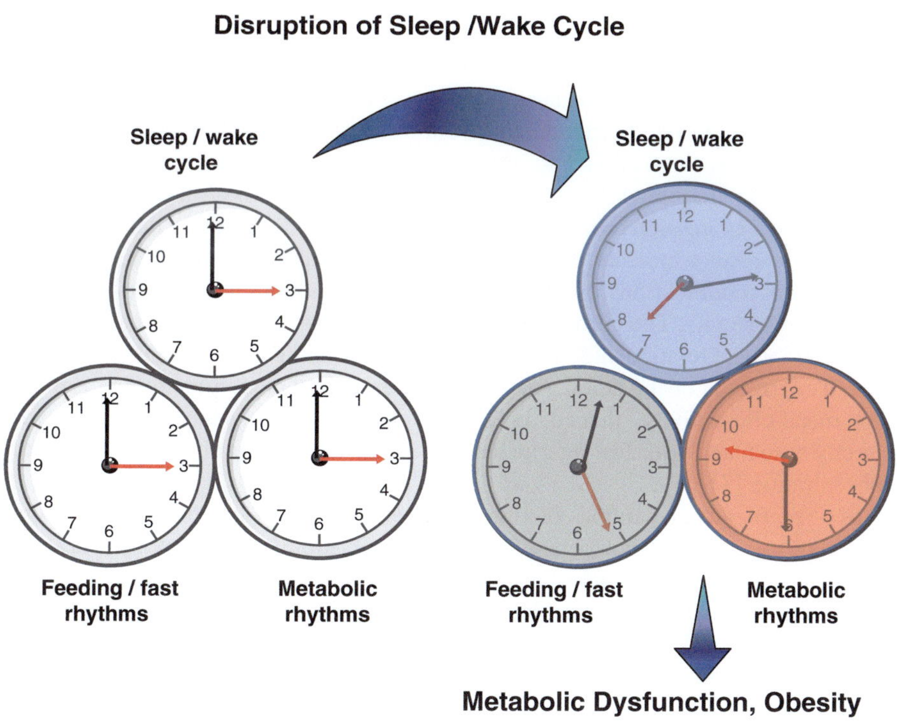

Fig. 8.13 As a disruption of the timing of bodily functions that are normally synchronized such as the sleep/wake cycle, improper timing of food intake, metabolism, and dampened clock gene expression occur. Metabolic dysfunction leads to the metabolic syndrome. Reproduced with permission from Cardinali [75]

of food administration, and that this could be accomplished without the participation of the SCN (Fig. 5.33). Taken together, the data indicate that the administration of food at inappropriate times may disrupt the metabolic profile, thus creating a desynchronized physiological state that is causally linked to the development of T2DM (Fig. 8.13).

The possible association between clock gene polymorphisms and T2DM has been explored in humans. In one study a polymorphic allele was identified in *Cry2* that correlated with T2DM [62]. Two *Bmal1* variants have been found to be associated with diabetes and hypertension in a British population [63]. In a study to examine the existence of *Per3* variants in patients with T2DM we reported that, compared with the group without diabetes, the frequency of the occurrence of the five repeat alleles of *Per3* among affected patients was greater, and that of the four repeat alleles was less [64]. Circadian clock variants have also been found to correlate with body mass index [65], and with weight loss, sleep duration, and total plasma cholesterol in obese Caucasian individuals [66]. *Clock* and *Cry1* polymorphisms are involved in individual susceptibility to abdominal obesity in a Chinese Han population [67]. Single nucleotide polymorphisms in *Bmal2* gene have been associated with a high risk of developing T2DM in obese patients [68]. Cross-sectional studies have reported associations between the clock gene polymorphisms and the prevalence of obesity, plasma glucose levels, hypertension, and T2DM. The association of the *Clock* polymorphism and stroke in T2DM was reported, indicating that core clock genes significantly contribute to increased cardiovascular risk in T2DM [69].

A high-fat diet that contributes to insulin resistance, impaired glucose metabolism, and obesity, can feedback to influence the biological clock. Rats on a 35% fat diet exhibited a disrupted 24-h rhythmicity of *Per1*, *Per2*, *Cry1*, and *Cry2* expression, the *Per2* expression profile being almost inverted by the high-fat diet (Fig. 8.14) [70]. These results indicate that the inherent transcription, translation, and post-translational modifications that give the clock its own natural rhythmicity are disrupted in obese rats. The current evidence thus suggests that these processes might have a reciprocal relationship, as an abnormal functioning of metabolism promotes altered expression of the clock genes, whereas impaired clock functioning can disrupt metabolic activity. Indeed, in T2DM patients, clock gene expression was directly associated with fasting glucose levels and with insulin mRNA and protein concentration [71].

In a recent survey, samples of subcutaneous adipose tissue from 50 overweight subjects were collected before and after an 8-week administration of a hypocaloric diet. The expression of core clock genes, *Per2* and *NR1D1*, increased after the weight loss, and their levels correlated with the expression of several genes involved in fat metabolism [72]. Another study demonstrated that a weight loss intervention based on an energy-controlled Mediterranean diet influences the methylation levels of three clock genes, *Bmal1*, *Clock*, and *NR1D1*, being an association between the methylation levels and the diet-induced serum lipid profile [73]. Thus, an abnormal functioning of metabolism promotes the low expression of clock genes, whereas impaired clock functioning can disrupt metabolic activity. Figure 8.15 summarizes the factors and their output that may be responsible for T2DM. The findings thus

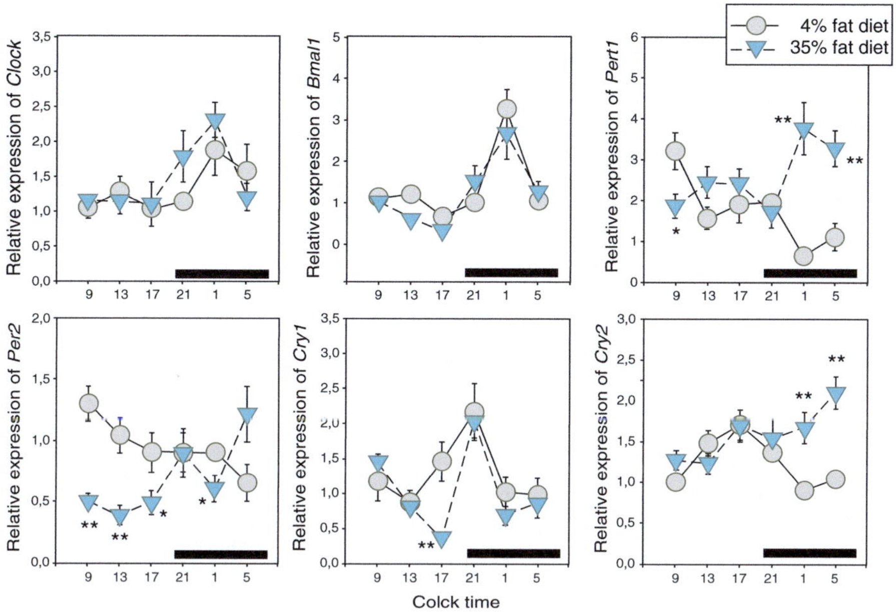

Fig. 8.14 Circadian oscillating expression of clock genes is sensitive to a high fat diet. In this experiment, rats were fed a 4% (control) or a 35% fat diet for 11 weeks. Wistar male rats ($n = 6–8$ per group) were killed by decapitation at six different time intervals throughout a 24-h cycle. Adenohypophysis was collected and RNA extracted. The means ± SEM of mRNA expression by real-time PCR (**$p < 0.01$, *$p < 0.05$ compared with control rats in a Student's t test). Results from Cardinali et al. [70]. Reproduced with permission from Cardinali [75]

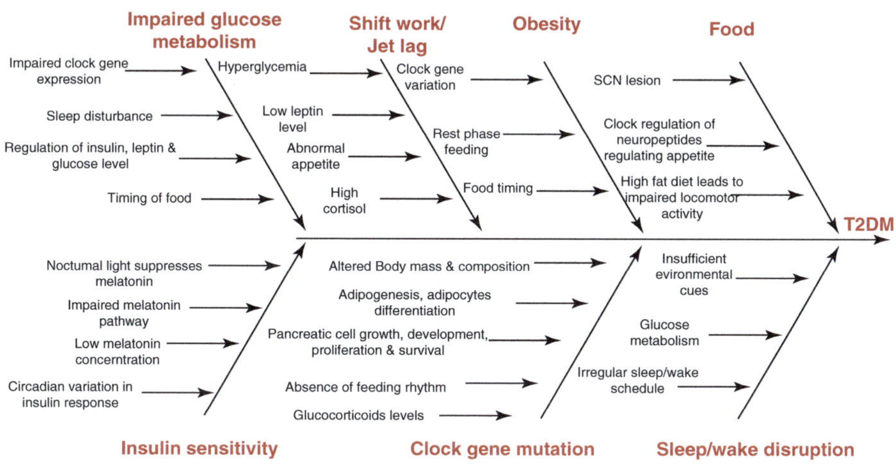

Fig. 8.15 Adapted fishbone diagram including the factors and their output responsible for type 2 diabetes mellitus

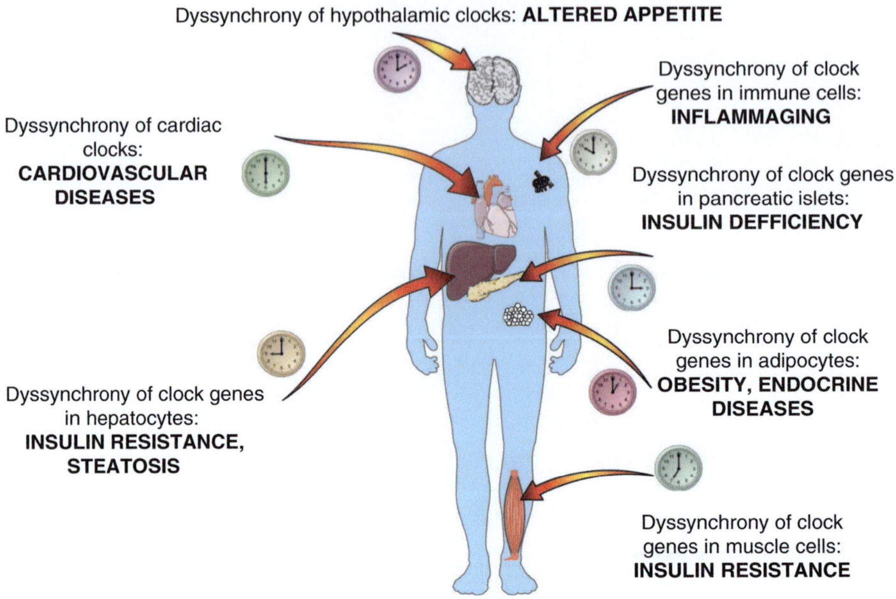

Fig. 8.16 A hypothetical distribution of circadian alterations in organs and tissues in the metabolic syndrome. Data are extrapolated from studies in transgenic mice that have provided evidence of the association between circadian clock disruptions and metabolic and behavioral events in metabolic syndrome. Reproduced with permission from Cardinali [75]

underscore the importance of circadian regulation for normal metabolic functioning, and further, that clock gene expression appears to be disturbed in T2DM (Fig. 8.16).

The possibility that a relationship might exist between melatonin and T2DM is supported by findings that insulin secretion is inversely proportional to plasma melatonin concentration. It is known that melatonin can alter insulin function and that, compared with normal healthy individuals, T2DM patients have lower circulating levels of melatonin. In vitro co-incubation of pancreatic cells with melatonin has been shown to inhibit the glucose-mediated release of insulin, additionally supporting the conclusion that melatonin activity plays a role in the function of insulin. Further, circadian disruption has been shown to alter melatonin secretion and to cause dysfunctions in pancreatic β cells. Genetic association studies that have shown that mutations in the melatonin receptor gene correlate with an increased susceptibility to T2DM [74].

It can be postulated that the alteration of phase, amplitude, and synthesis of melatonin could lead to impaired glucose metabolism in circadian disrupted individuals. Suppression of melatonin secretion by nocturnal light exposure could be a crucial factor for T2DM development [56]. In this respect, the successful management of T2DM may require an ideal drug that besides antagonizing the triggers of T2DM, also corrects the disturbed sleep–wake rhythm.

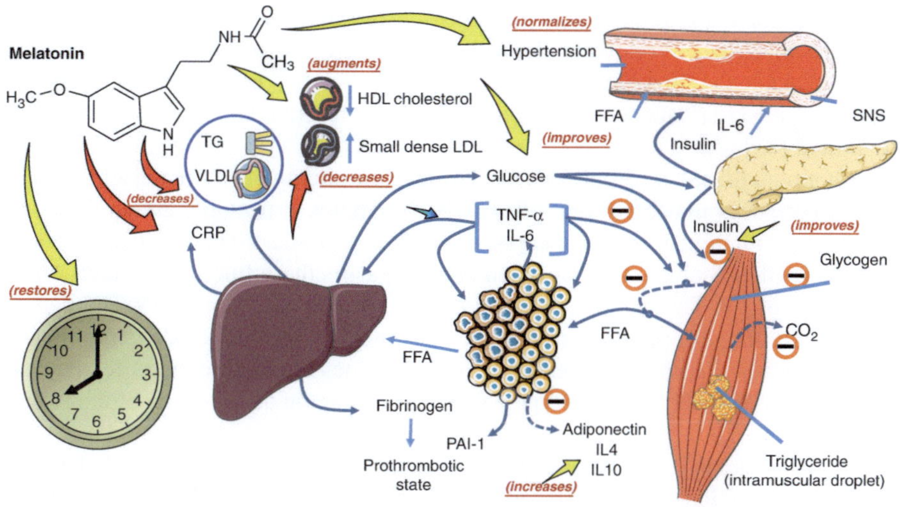

Fig. 8.17 Effects of melatonin in the metabolic syndrome. Melatonin normalizes high blood pressure and circulating indexes of inflammation. It improves insulin sensitivity and restores disrupted circadian rhythms. Reproduced with permission from Cardinali [75]

Melatonin, together with morning light, is an interesting chronotherapeutic option that can reset the phase and amplitude of circadian rhythms in T2DM patients. The combination of the specific and chronobiological treatment is needed for full recovery (Figs. 8.7 and 8.8). As has been shown in animal models of diabetes and obesity, melatonin also possesses cytoprotective properties that may prevent a number of unwanted effects in T2DM [75]. The efficacy of melatonin treatment for treating metabolic and cardiovascular comorbidities in T2DM has been reported (Fig. 8.17) [76–81]. However, melatonin treatment was reported to impair insulin release in humans who carry MTNR1B, the risk allele of the MT_2 receptor gene [82]. To what extent insulin resistance is ameliorated by melatonin in these patients remains to be defined. At an early stage of T2DM treatment, nonpharmacological approaches such as lifestyle modification, low-fat diet, and exercise are recommended [52].

Mental Illnesses

The human species is very vulnerable to psychiatric diseases: 1% of the population suffers from schizophrenia, 4% from severe uni- or bipolar depression, and 15% of some form of reactive emotional illness in isolated episodes. There are epidemiological elements to suspect a genetic predisposition in the appearance of several of these diseases.

Emil Kraepelin, at the beginning of the twentieth century, was the first to develop a modern classification of mental illness, differentiating “precocious dementia" (schizophrenia) from the more benign and recurrent forms of illness ("manic disease"). Following Kraepelin, cognitive alterations ("disorders of thought" such as schizophrenia) and affective disorders ("emotional disorders" or "affective diseases" such as depression, manic depression, and anxiety) have been identified [83].

Emotional illnesses are characterized by an abnormality of emotional and expressive experiences that constitute affectivity. Normally, a person's affectivity is in neutrality, with mild episodes of euphoria (called "happiness") and mild episodes of sadness (called "unhappiness"). When one trespasses these limits, one enters emotional pathology. The pathological end of euphoria is called "mania"; The pathological end of unhappiness is called "depression." When individuals express both abnormal ends in a single episode or at different times throughout their life, they are diagnosed with bipolar disease type 1 or 2 (depending on the intensity of the manic phase).

Mania is defined by the presence of an abnormally expansive and irritable affectivity, coupled with several related symptoms. They include decreased need for sleep, excessive loquacity, changing thoughts, grandiosity, easy distraction, and increased activities with pleasurable objectives. This has important consequences: non-normal sexual behavior, risky economic ventures, etc. A maniac patient is often enthusiastic and positive, but tolerates little frustration. Because mania is not an unpleasant experience, patients rarely admit to being treated. However, relatives or close friends recognize this situation as dangerous for the patient. Ultimately, the persistence of the patient in the manic state may be accompanied by serious affective or economic breakdown for the patients and their families.

At the opposite end of the mania is depression. The depressed patient presents a variety of disorders, including sleep disturbances, greatly diminished interest or pleasure, changes in appetite and body weight, psychomotor agitation or inhibition, decreased energy, guilt, impairment, and recurrent thoughts of death or suicide. Approximately 15% of depressed patients complete a suicide attempt. It is of interest that these alterations are linked to a high degree of creativity. For example, Van Gogh or Tolstoy were famous depressive patients.

There is a strong familial component in these diseases. The risk of becoming ill is twice that of the general population in individuals with a sick first-degree relative. However, this evidence does not necessarily indicate a genetic transmission, as educational and social factors may also explain the picture.

The link between psychiatric pathological conditions and the ANS and with brainstem monoaminergic neurons discussed in Chap. 4 is indicated by the therapeutic efficacy of various drugs that interfere with these neuronal mechanisms. It is not known whether the modification of these diffuse monoaminergic systems is the cause or effect of the psychiatric alteration that originates in other zones of the ANS (e.g., the limbic system). The introduction of methodologies such as fMRI, PET or magnetoencephalography has allowed monitoring of alterations in neuronal activity, whose correction is accompanied by improvement of psychiatric symptoms, to be systematized in different brain areas. The symptoms of mental illness, such as hearing voices in solitude (hallucination) or feeling persecuted (delirium) are the result of alterations in brain circuits.

Mental illnesses range from relatively mild to severe forms. The mildest of these are personality alterations (e.g., obsessive–compulsive disorders, psychopathic personality). Although seen as primarily "psychological," these alterations bring about changes in brain activity, as revealed by PET or fMRI, such as increased blood perfusion of the prefrontal cortex and basal ganglia. The psychopathic personality with antisocial behavior has a genetic component.

There is little controversy about the organic substrate of more severe mental illnesses, such as schizophrenia, manic–depressive (bipolar) illness, major depression, or anxiety disorders such as panic attacks or post-traumatic stress disorder. Circadian rhythm abnormalities, as shown by sleep/wake cycle disturbances, constitute one the most prevalent signs of mental illness. A substantial proportion of patients with a sleep complaint have a psychiatric illness and a significant number of patients with a psychiatric illness have a sleep complaint [83].

Mood Disorders

Mood disorders constitute a family of complex multifactorial illnesses that are characterized by disruptions of several physiological, neuroendocrine, and behavioral processes. According to the World Health Organization (WHO) reports, these disorders are the fourth leading cause of global burden of disease and by the year 2020, they are expected to be the second highest cause of morbidity. An interpretation of the public health impact by prevalence statistics, however, must take the following two factors into consideration. First, the use of antidepressants has increased considerably over the last 15 years, and concurrently the prevalence of mood disorders is also increasing. Second, despite this fact, only one third of patients with mood disorders are treated effectively with medication. This would suggest that either some unrecognized subtypes of mood disorders may exist that are resistant to current pharmacological treatments, or that present conceptualizations of the underlying causes of mood disorders need to be reconsidered. There is a growing amount of evidence that supports the latter suggestion and the accumulating evidence points to the possibility that mood disorders may be an overt symptom of what is basically a circadian rhythm disorder (Fig. 8.18). An internal desynchronization of circadian oscillators with a strong oscillator being linked to phase advances was postulated as a hypothesis for various subtypes of affective disorders.

Circadian rhythm abnormalities, as shown by sleep/wake cycle disturbances, constitute one the most prevalent signs of mood disorders, advances or delays in the circadian phase being documented in patients with major depressive disorder, bipolar disorder or seasonal affective disorder [84]. Changes in the sleep–wake cycle structure in mood disorders often precede changes in a patient's ongoing clinical state. During a depressive episode, approximately 80% of patients complain of symptoms of insomnia (frequent awakenings, early morning awakening) and 20% complain of hypersomnia. Changes in sleep during a depressive episode are a long sleep latency, reduced sleep efficiency, reduced stage N3, a short REM latency, a

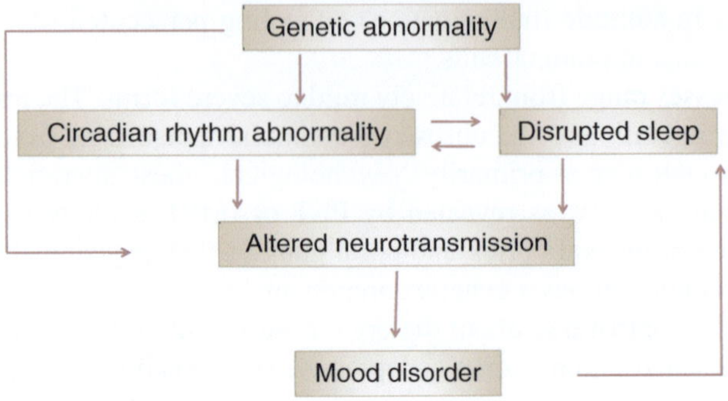

Fig. 8.18 Bidirectional relationships among circadian rhythms, sleep, and mood. It is proposed that disturbed circadian rhythm regulation has an impact on sleep rhythms to produce changes in the monoamine regulation of mood. The altered mood can then influence sleep. Reproduced with permission from Cardinali [75]

longer first REM period, and a higher REM density in the early part of the night (Chap. 2). Early morning awakening may also be present.

Insomnia can precede a major depressive episode and is often the last symptom of depression to resolve. The persistence of changes in sleep–wake cycle is a risk factor for a relapse of depression. During manic episodes, the patient reports a decreased need for sleep (feeling rested on a few hours of sleep). In turn, sleep loss can precipitate a manic episode.

Individuals suffering from mood disorders often have circadian misalignment of many physiological phenomena in addition to the sleep–wake cycle, e.g., BP, neurotransmitters, mood states, body temperature, energy balance, appetite regulation, melatonin secretion, and the levels of cortisol [85]. Variations in the sleep–wake schedule preferences and poor activity patterns have also been observed to occur more frequently than in healthy control subjects. Reduced sleep disturbances have been reported in morningness phenotypes, thus suggesting the existence of a correlation between activity pattern preference and mood regulation. Favoring a circadian rhythm hypothesis of the disease, sleep deprivation and light therapy have clinically relevant antidepressant effects in patients [86]. The combination of the specific and chronobiological treatment is usually needed for full recovery (Figs. 8.7 and 8.8).

Changes in the sleep–wake cycle structure in mood disorders often precede changes in a patient's ongoing clinical state and can even signal a relapse or predict the occurrence of suicidal behavior. In addition to an altered sleep–wake cycle, daytime mood variation and periodic recurrences are clinical findings that relate depressive states to the circadian system [87]. A significant proportion of patients have regular changes in the intensity of their depressive mood during the day, with parallel changes in anxiety symptoms, attention capacity, and psychomotor symptoms that frequently accompany depression. Depressive patients with melancholic characteristics typically have an early morning awakening, worsening their mood state, which additionally correlates with elevations in cortisol secretion. Both symptoms are part of the clinical diagnostic criteria of the melancholic depressive subtype [87].

We have discussed in Chap. 2 how circadian processes interact with homeostatic mechanisms to regulate 24-h oscillations in sleep propensity. Intensity of light and its duration ensure the proper functioning of the circadian clock. Hence, the strength of the entrained circadian pacemaker greatly influences the sleep, cognitive, and emotional functioning. Additionally, the interactive dialogue between circadian system and homeostatic mechanism affects the timing of sleep onset and wake onset in individuals (Chap. 2).

It has been proposed that sleep and mood have a bi-directional and interactive relationship ("vicious circle") in which disturbances in sleep perpetuate disturbances in mood throughout the day (Fig. 8.19) [88]. Various studies have shown that prolonged sleep disruption results in abnormal mood states. Conversely, alterations in mood can affect sleep: it has been shown that when patients with bipolar disease switch from depression to mania, they experience one or more successive rest–activity cycles [89]. Sleep profiles of both manic and depressive patients are similar in showing continuous sleep disturbance and more time spent in N1 sleep. Reduction in the tendency to sleep and the quality of sleep, disturbed and increased REM sleep, early morning awakening, and short sleep duration are observed in patients with bipolar disease. Supersensitivity to light was proposed as a trait marker in

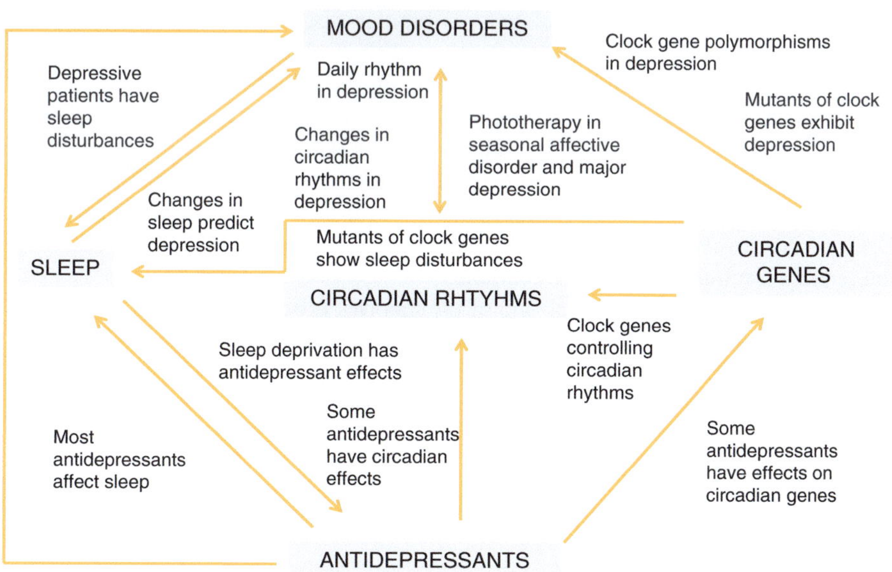

Fig. 8.19 The circadian clock influences multiple systems and pathways that are thought to underlie mood disorders. In most cases, there are reciprocal interactions that, in turn, regulate circadian rhythms. Circadian gene mutations may make an individual more vulnerable to mood changes exacerbated by environmental deviations in the daily schedule. Reproduced with permission from Cardinali [75]

bipolar disease patients after showing that melatonin levels in these patients fell twice as much as the levels of normal subjects following exposure to light during the night [90]. Phase advances in the rhythm of melatonin secretion have been documented in numerous studies of patients with major depressive disorder [91].

Bipolar patients suffer from abnormal circadian rhythms in their sleep–wake cycles, with regard to BP, hormone levels, neurotransmitters, mood states, body temperature, energy balance, appetite regulation, melatonin secretion, and the levels of cortisol [92]. Variations in sleep–wake schedule preferences and poor activity patterns have also been observed more frequently in bipolar disease patients than in healthy control subjects. Reduced sleep disturbances were reported in morningness phenotypes and this suggests a correlation between activity pattern preference and mood regulation.

Globally, places receiving a reduced amount of sunlight over the year have a higher incidence of depression. This is of importance considering that bipolar disease patients are more sensitive to light. Indeed, winter increases the chance of depressive episodes, whereas the summer period promotes manic states in bipolar individuals [92]. The loss of strength in coupling the circadian oscillator and environmental cues contributes to the deterioration of mood. Hence, perturbations in the environmental cues such as irregular sleep–wake cycle and social schedule, receiving a limited amount of sunlight and exposure to artificial light at unusual times, travelling between two different photoperiods or shift work are the factors responsible for the misalignment of the clock that may contribute to mood disorders.

A hypothesis has been postulated for bipolar disease, interlinking sleep and wakefulness in such a way that disturbed mood states during the daytime can affect the sleep process. Similarly, sleep deprivation during nighttime influences daytime mood [93]. Dysregulation of neurotransmitters is one of the major manifestations of mood disorders. An interesting hypothesis is that a high level of DA induces mania and that low levels are associated with a depressive episode of bipolar disease. Many medications prescribed to treat patients with mood disorder influence 5-HT, DA, and NE levels and abnormal emotional states could be averted by modulating the neural communication of monoamines.

The social zeitgeber theory is one of the theories proposed to explain the origin of bipolar disorder and is based on the principles of biological rhythms. It states that stressful life events influence the regular sleep–wake cycle, resulting in the disruption of the circadian clock, inducing mood disorder in vulnerable people. In addition, social zeitgebers possess a critical function in the regulation of normal mood states, and social rhythm therapy is effective for mood regulation in bipolar disease patients.

The most commonly used drug to treat maniac episodes in bipolar illness is lithium. It acts on numerous steps in intracellular signaling and generally lengthens free running rhythms [94]. Present-day psychiatric treatment of chronobiological aspects of mood disorder comprises medication and nonpharmacological treatments, including phase advance, light therapy, and sleep deprivation. The combination of specific and chronobiological treatments is often employed (Figs. 8.7 and 8.8).

The timing of melatonin secretion is important for the regulation of mood. Phase advances of at least 1 h were observed in the nocturnal melatonin peak during the manic phase of bipolar disease, whereas phase delays in circulating melatonin occurred in bipolar disease type 1 patients [92]. Patients suffering from seasonal affective disorder exhibit delayed circadian rhythms, with a delayed offset of melatonin secretion of about 2 h. Presumably, the symptoms of hypersomnia and late awakening seen in seasonal affective disorder are due to the delayed phase and long duration of melatonin secretion [92].

Elucidating how biological clocks regulate the timing of physiological processes could ultimately be helpful for understanding and treating mood disorders, which may be the result of dampened circadian oscillators. Some initial efforts have already been made to use sleep scheduling as one component of psychiatric therapy. The benefit of this approach has been demonstrated in studies showing that the close maintenance of sleep–wake timing in patients with mood disorders can produce enhancements in patients' mood profiles. Additionally, extended bedrest and optimal sleep duration have been shown to substantially reduce manic episodes in bipolar disease patients. Psychiatric treatment programs are increasingly incorporating recommendations that chronotherapies be adopted as adjunctive strategies for mood disorders (Figs. 8.7 and 8.8) [95, 96].

Clinical studies involving chronobiological manipulations, such as exposure to bright light in the morning and melatonin administration in the evening, have been found to be useful for reducing phase abnormalities and depressive symptomatology [97]. Indeed, this is the basis for postulating the association of an adjuvant chronobiological treatment to the specific treatment of the psychiatric disease to obtain full recovery of patients (Figs. 8.7 and 8.8). It must be noted that from clinical standpoints the abnormalities in amplitude of the sleep–wake cycle ("poor sleep together with poor vigilance") is a paramount sign of the disease and that its correction increases substantially the quality of life of the patient suffering an affective disorder, regardless of the uncontrolled influence of external (light/dark cycle) or internal (sleep–wake cycle) masking phenomena.

We discussed in Chap. 2 the intricacies of the cellular mechanisms of circadian oscillation. These clock genes influence mood disorders. Genes that can tolerate environmental disruption are not associated with abnormal mood swings, whereas mutations in circadian clock genes are related to them. Moreover, clock gene allelic variations can worsen mood symptoms and contribute to differential effects of psychiatric medications. Mutations of circadian genes in animal models resemble mood alterations.

In humans, the relationship between mood regulation and clock genes was first identified in seasonal affective disorder [98]. A positive association for *Npas2* and *Per3* and sleep–wake time preference was found in patients. Genetic variants of *Bmal1*, *Npas2*, and *Per2* have been found to increase susceptibility toward the development of seasonal affective disorder. *Cry2* alleles and their mRNA levels have been directly associated with a depressive mood profile and *Clock* alleles were associated with hyperactivity in bipolar disease patients, who exhibited significant delayed sleep phase and reduction in sleep duration [99]. Significant reductions in insomnia and antidepressant treatment have correlated with *Clock* variants in bipolar disease and major depressive disorder patients. The post-mortem analysis of major depressive disorder patients indicated a dampened clock gene expression in several brain regions compared with healthy controls [100]. In a collaborative study with Indian and Canadian colleagues [101], we found in a South Indian population an increased prevalence of five repeat homozygotes of Per3 in bipolar disease patients, which was particularly notable among female patients. No significant association was observed in the allele frequencies of four and five repeat alleles in schizophrenia patients. Therefore, the occurrence of the five-repeat allele of *Per3* can be a risk factor for type 1 bipolar disease onset.

Panic disorder is a complex anxiety disorder characterized by recurrent panic attacks. It is a poorly understood psychiatric condition that is associated with significant morbidity and an increased risk of suicide attempts and completed suicide. Accumulating evidence suggests that ANS mechanisms controlling acidosis constitute a contributing factor in the induction of panic [102]. Challenge studies in patients reveal that panic attacks are provoked by agents that lead to acid–base imbalance, such as CO_2 inhalation or sodium lactate infusion. Chemosensory mechanisms that translate pH into panic-relevant fear, autonomic, and respiratory responses include regions such as the subfornical organ and medullary raphe that can directly detect pH fluctuations in the internal milieu. The hypothalamus, amygdala, and periaqueductal gray, in addition to their chemosensory potential, also represent key nodes in the processing of external threats and sensory stimuli (Chaps. 4 and 5). Acidosis sensed by chemosensory mechanisms is translated to autonomic, behavioral, and respiratory symptoms of a panic attack. The amygdala, periaqueductal grey and the hypothalamus regulate behavioral and autonomic symptoms of panic, whereas respiratory symptoms are regulated by brainstem regions such as the medullary raphe and the PBN via inputs from the hypothalamus and indirectly from the subfornical organ through the OVLT [102]. Many of these structures connect via thalamic nuclei with the insula, a region relevant for interoceptive sensing and shown to be dysfunctional in panic disorder patients.

Patients with panic disorder are at an increased risk of myocardial infarction and sudden death, which is evident in epidemiological surveys. Cardiac arrhythmias and coronary artery spasm have been demonstrated during panic attacks. A constellation of brain and sympathetic nervous system abnormalities, both in the quiescent phase and during panic attacks, have been documented in panic disorder sufferers; these are believed to underlie the increased cardiovascular risk. Abnormalities demonstrable at rest are very high brain turnover of 5-HT, perhaps the CNS substrate for panic disorder, E cotransmission in sympathetic nerves, and impairment of neuronal reuptake of the NE released by sympathetic nerves. During panic attacks, there is a surge of E secretion, accompanied by high-level activation of central sympathetic outflow, including to the heart, and release of neuropeptide Y by cardiac sympathetic nerves.

Melatonin has shown efficacy for treating mood disorders and the data indirectly suggest that disruption in melatonin pathways could alter circadian clock mechanisms, leading to the disruption of physiological processes, including sleep and mood behavior. In addition to sleep promotion, MT_1 and MT_2 receptors appear to be involved in sedative and antiexcitatory effects of melatonergic drugs. This has been mainly studied in relation to anticonvulsant actions that have been linked to a facilitatory role of melatonin on GABA neurotransmission. In mammals, the antiexcitatory actions of melatonin may be also related to additional anxiolytic, antihyperalgesic, and antinociceptive effects [75]. Because melatonin impairs contextual fear conditioning, a hippocampus-dependent task, and facilitates the extinction of conditional cued fear without affecting its acquisition or expression [103], it may serve as an agent for the treatment of posttraumatic stress disorder. Clinical evidence supports such a possibility [14].

To improve the efficacy of the sleep-promoting effects of melatonin, several analogs of melatonin have been developed for treating circadian rhythm sleep disorders or insomnia. Among these, agomelatine has been licensed by the European Medicines Agency (EMA) for the treatment of major depressive disorder in adults. Agomelatine has a unique pharmacological profile, as it is both a MT_1/MT_2 melatonin receptor agonist and an antagonist of $5\text{-}HT_{2C}$ receptors [104].

For decades, the treatment of depression had revolved around drugs that increase synaptic amounts of monoamine neurotransmitters (5-HT, NE, etc.). As the first melatonergic antidepressant on the market, agomelatine displays a nonmonoaminergic mechanism that addresses sleep disturbances and depressive symptoms together. Agomelatine has an early onset of action even in a severely depressed population and may stand unique among antidepressants for effective management of a major depressive disorder. In several studies, agomelatine has not only produced a cure of depressive symptoms, but also the patients return to a normal social and occupational functioning [104, 105].

Increases in alanine aminotransferase and/or aspartate aminotransferase (three times the upper limit of normal) have been noted in patients treated with agomelatine and this has become of great concern. In early 2015, the European health authorities chose to keep agomelatine on the market despite its serious adverse hepatic effects. It is recommended that liver function tests should be performed in all patients: at initiation of treatment and then periodically after around 6 weeks, after around 12 and 24 weeks, and thereafter when clinically indicated.

Schizophrenia

Schizophrenia is a complex psychiatric disorder comprising both positive and negative symptoms, including hallucinations, delusions, poor social behavior, and low motivation [106]. The diagnosis of schizophrenia is currently based on the presence of specific symptoms in affected individuals. About 1% of the global population suffer from schizophrenia. Many genes have been identified to play a potential role in the pathogenesis of schizophrenia, but none are established as causative. To date, there has been no biological marker for schizophrenia.

Disruption of rest/activity cycles have been observed in schizophrenia patients and may be generated by abnormal functioning of the circadian clock [106]. Schizophrenic, schizoaffective and bipolar patients have greater eveningness scores than control subjects. Hence, there may be vulnerability for evening chronotype individuals to develop these disorders [107]. A recent study of 100,420 individuals from the UK Biobank confirmed this relationship between schizophrenia and a evening chronotype [108].

Schizophrenic patients exhibit altered circadian rhythms of serum concentrations of tryptophan, serotonin, prolactin, and cortisol. Aberrant rest/activity behaviors have been documented in several studies of schizophrenic patients, including phase shifts, abrupt and segmented sleep cycles, and waking during sleep [106]. The most common sleep disturbances in schizophrenia are prolonged sleep onset latency and problems of sleep maintenance [109]. Sleep disruption has been found to occur in schizophrenic patients during NREM and REM sleep stages, although no significant changes have been observed in REM sleep duration. As REM sleep is regulated by the cholinergic system, an altered cholinergic activity may have a connection to the development of hallucinations.

Proper entrainment by zeitgebers has been shown to stabilize affected schizophrenic individuals [86]. Because of an abnormal activity pattern, schizophrenia patients often have an unusual length of light exposure and this in turn, has an effect on the patient's sleep. Insufficient exposure and inappropriate timing of environmental cues can lead to phase misalignments between internal and external timing factors [110]. These in turn may have an impact on mental and neurological functioning. A combination of weak zeitgebers, delayed sleep schedule, absence of morning light, and receiving more light during the evening can generate a differential phase between internal and external time keepers. Thus, the circadian system may play a significant role in schizophrenic symptoms. A combination of the specific and chronobiological treatment may be needed for full recovery from the crisis (Figs. 8.7 and 8.8).

DA brain systems play a crucial role in schizophrenia pathology. Both the first- and second-generation antipsychotic medications, the prime drugs used to treat schizophrenia, block D_2 receptors. They are particularly effective against the positive symptoms of the disorder. The term second generation is usually reserved for agents modified to reduce side effects or to decrease negative symptoms [106]. Disruption of glutamate, GABA and orexin/hypocretin systems are also thought to be involved in schizophrenic pathology [111].

Disruption of the melatonin profile in schizophrenia has been described in several studies, often correlating with sleep–wake disruption [106, 112]. In some instances, the circadian period of melatonin in schizophrenia patients was longer than 24 h; thus, is phase shifted, in others it is phase advanced, and in others still it may be arrhythmic. Schizophrenia patients often exhibit a unique difference in the circadian phase angle, i.e., there exists a difference among the timing of melatonin secretion, sleep–wake cycle, and the zeitgeber. Moreover, this phase angle difference has been associated with the symptoms observed in schizophrenia.

Neurodegeneration is often found in schizophrenia [113]. As discussed later in this chapter, melatonin acts as a potent antioxidant and cytoprotective agent, which suggests that the reduced melatonin levels seen in schizophrenia patients may lead to or aggravate an excitatory hyperoxidative status [114]. Melatonin treatment ameliorated the metabolic syndrome caused by second-generation antipsychotics in schizophrenic patients [80, 81].

Postmortem analysis of the temporal cortex region in schizophrenia patients revealed dampened expression of the clock gene *Per1* [115]. Signaling associated with the DA D_2-receptor enhanced the transcription of clock components, whereas DA D_2-receptor mutant mice displayed impaired expression of *Per1*. In another study, single nucleotide polymorphisms identified on the locus of *Per3* and *Timeless* were associated with schizophrenia/schizoaffective disorder [107]. However, no correlation for *Per3* length variants was observed in other studies. Our report about a South Indian population was compatible with the view that there is no correlation between *Per3* polymorphism and schizophrenia [101].

In neuropsychiatric disorders, clock gene polymorphisms may have a direct impact on the regulation of neural communication. Secretion of cortisol is influenced by the clock genes and increased cortisol concentration has been observed in schizophrenia patients [116], thus suggesting that clock genes might be involved in pathogenesis. Understanding the processes relating to the internal time-keeping system and neurobehavioral processes may shed new light on and help to develop new approaches to the treatment of schizophrenia.

Brain Aging

The term longevity includes two different concepts. Average longevity is defined as the average life expectancy at birth for individuals of a given species. Maximal longevity is the maximum age that an individual of a given species can reach. The average longevity of the human species has increased considerably throughout history owing to the decline in infant mortality, the discovery of antibiotics, vaccines and, more generally, improvement in the control of infectious diseases, in addition to more balanced nutrition, better sanitary conditions, and advances in the treatment of diseases such as cancer or diabetes. In contrast, maximal longevity has remained unmodified. In ancient times, people reaching the ages of 80–90 years or more were also found, although the percentage was much lower than today.

The increase in average longevity is seen in the growing segment of the population between 60 and 100 years (Fig. 8.20). Those over 80 now constitute 1.6% of the world population and by 2050 they will constitute 4.3% (about 400 million people). During the period from 2008 to 2040, the population of individuals aged 65 years and above is projected to increase by 160%, whereas that of individuals aged 80 years and older will be over 230% (Fig. 8.20) [117]. Considering that the estimation for the number of patients with Alzheimer's disease for 2050 is 150 million people, the goal of "successful aging" has become very important in avoiding the consequences of neurodegenerative diseases, cancer, or arteriosclerosis, i.e., those most likely to affect this elderly group. These figures represent an increase of 56% in high-income and of 239% in low-income countries. They underlie the extreme importance of maintaining healthy aging for public health policies in years to come. The desired goal is to minimize chronodisruption.

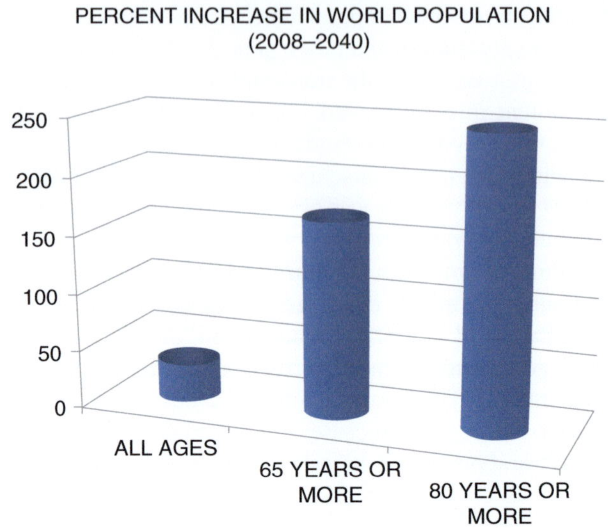

Fig. 8.20 Projected increase in the total world population contrasts with that of the elderly population (>65 years or >80 years)

Virtually all physiological functions become less efficient with age (Fig. 8.21). As discussed in Chap. 2, slow-wave sleep decreases exponentially with aging and often disappears after 60 years of age. Many elderly individuals complain of interrupted sleep and of daytime sleepiness [118]. Other common complaints are early awakenings and a poor capacity to maintain alertness in the evenings. Both are indexes of the aging of the circadian apparatus: a decrease in the amplitude and phase advance of circadian rhythms [119]. Thus, the relative distribution of wakefulness, NREM, and REM body configurations changes significantly in aged individuals (Fig. 8.2).

Aging entails a homeostatic loss of the capacity to maintain the stability of the internal environment of the individual to combat environmental disturbances. An example of this is the reduced ability of the elderly to withstand extreme temperatures, trauma, infections, and stress in general. With aging, most vital organs suffer atrophy or degeneration phenomena. This is most noticeable in differentiated cells such as neurons, myocardial cells, muscle cells or the renal parenchyma.

It is generally accepted that cell disruption due to the oxidative stress is a major physiopathological event in aging. In addition, other current hypotheses suggest a direct relationship among aging, genetic programs, and telomere loss that occurs after each cell division, leading to apoptotic cell death. Although all these processes and mechanisms are probably involved in diseases associated with aging, their role in normal aging has not yet been clarified.

Studies aimed at evaluating the impact of modifiable risk factors on pathological brain aging are justified by their potential for prevention at a population level. The risk factors with strongest evidence for possible causal associations with brain

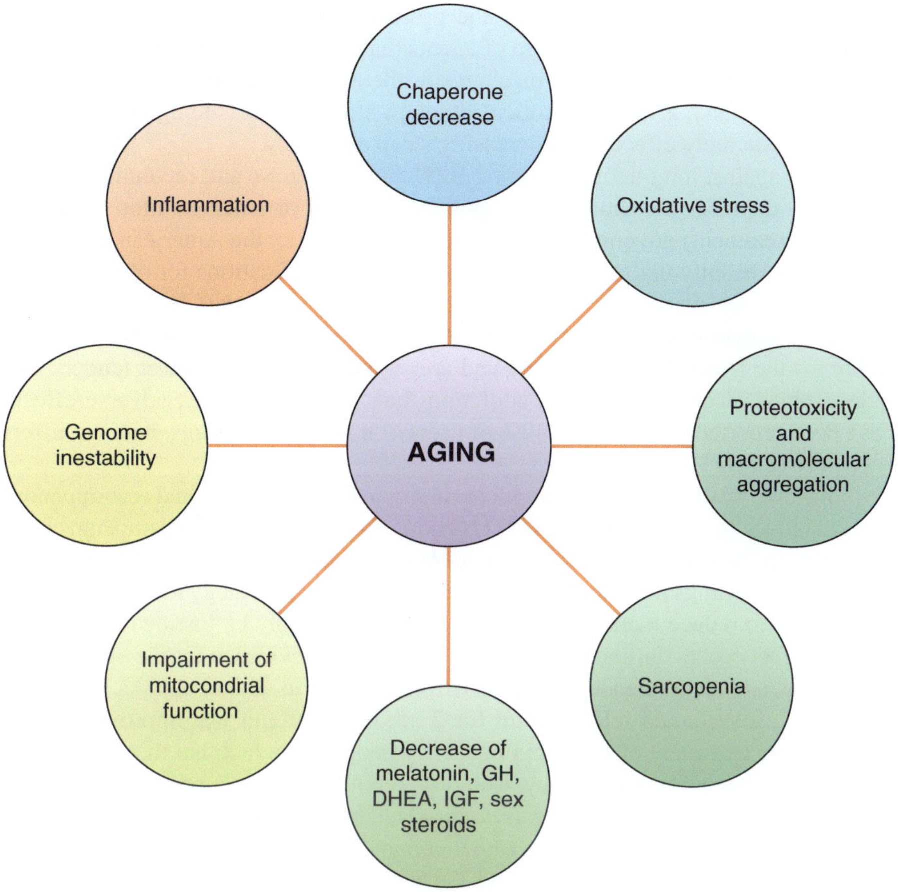

Fig. 8.21 Some consequences of aging

pathology, e.g., dementia, are poor sleep, T2DM throughout life, hypertension in midlife, smoking, and a low level of education in early life.

Concerning sleep loss in aging, it typically altered the three body configurations toward a sympathetic predominance in the face of poor NREM sleep (Fig. 8.2). Indeed, the prevalence of insomnia is up to 25–30% in the elderly, which leads to a concomitant increase in the use of hypnotics. The benzodiazepines (BZDs) and BZD receptor agonists (Z drugs: zolpidem, zaleplon, zopiclone) are the most commonly prescribed drugs for the treatment of insomnia in the elderly. The BZDs are a group of compounds that exert their therapeutic effect on sleep through the allosteric modulation of $GABA_A$ receptor complex. BZDs have broad inhibitory effects on brain functions, including promoting sleep, anxiolysis, anticonvulsant effects, cognitive and motor impairment, and reinforcing effects. In addition, significant adverse effects, such as cognitive and psychomotor impairment, next day hangover, rebound insomnia, anterograde amnesia, and dependence have been documented, thus rendering the use of BZDs for the prolonged treatment of insomnia highly controversial. “Z drugs” are a group of agents that are not part of the chemical class BZD, but act through the same mechanism, the allosteric modulation of $GABA_A$ receptor. Generally, Z drug hypnotics, although effective at reducing sleep latency, are only moderately effective at increasing sleep efficiency.

Advice against long-acting hypnotic BZD and Z drug use and recommendations to employ them for the shortest time possible in older patients (no more than 2–3 weeks of treatment) are common nowadays. For example, the American Geriatrics Society has recently updated its list of inappropriate medications for older patients and advised physicians to “avoid benzodiazepines (any type) for treatment of insomnia, agitation, or delirium” [120]. Z drugs are used, unlike the BZDs, exclusively for the treatment of insomnia and are supposed to have a lower tendency to induce physical dependence and addiction than BZDs. However, adverse effects have been reported in more than 40% of users of both types of drugs, with no difference between BZDs and Z drugs.

In Europe, health authorities are increasingly initiating policies and recommendations to reduce the consumption of BZDs and Z drugs. However, the campaigns have not generally been successful, despite national guidelines and recommendations, and the use of these drugs has continued to increase. The clearer strategy to reduce chronic BZD use is to reduce medication; abrupt cessation can only be justified if a serious adverse effect occurs during treatment. There is no clear evidence for the optimal rate of tapering, and recommended times vary from 4 weeks to several months.

In 2007, a sustained release form for 2 mg of melatonin was approved by the EMA for the treatment of insomnia in elderly people. The fact that melatonin does not show evidence for dependency, isolation, rebound insomnia or a negative influence on alertness during the day was emphasized by the EMA for melatonin and by the FDA for the melatonin analogs ramelteon and tasimelteon.

Melatonin competes with BZPs and Z drugs at their site of action. A facilitatory role of melatonin on GABA neurotransmission was first demonstrated in the author’s laboratory and explains the anxiolytic, antihyperalgesic, and antinociceptive effects of melatonergic agents [75]. Several clinical studies now support the efficacy of melatonin in reducing BZP use in chronically treated patients [121, 122]. In a pharmacoepidemiological study aimed at evaluating the impact of anti-BZD/Z drugs campaigns and the availability of alternative pharmacotherapy (melatonin) in the consumption of BZDs and Z drugs in several European countries, the results indicated that the campaigns failed unless they were associated with the availability of melatonin in the market [123]. Several consensus statements support the role of melatonin in the elderly. For example, the British Association for Psychopharmacology consensus statement on evidence-based treatment of insomnia, parasomnias, and circadian rhythm disorders concluded that “…melatonin should be the choice hypnotic for insomniacs over 55 years of age” [45].

To maintain normal sleep is crucial in the elderly. As discussed in Chap. 2, the repair function of sleep is, among others, the consequence of the increased activity of the glymphatic system with elimination of potentially neurotoxic waste products accumulated in the CNS during wakefulness. Age-associated glymphatic dysfunction has been reported, with decreased and delayed CSF penetration along paravascular pathways and the pial surface [124]. The mechanics and importance of the glymphatic system in several cerebrovascular disorders are still being unraveled and investigations of therapeutic strategies that can protect or restore its integrity are warranted [125].

Glymphatic dysfunction has been reported in neurological disease states such as stroke, traumatic brain injury, and AD. The pathological signature of AD includes extracellular senile plaques, formed mainly by amyloid β (Aβ) deposits, and intracellular neurofibrillary tangles, resulting mainly from abnormally hyperphosphorylated microtubule-associated tau protein (Fig. 8.22). Aβ is composed of 39–43 amino acid

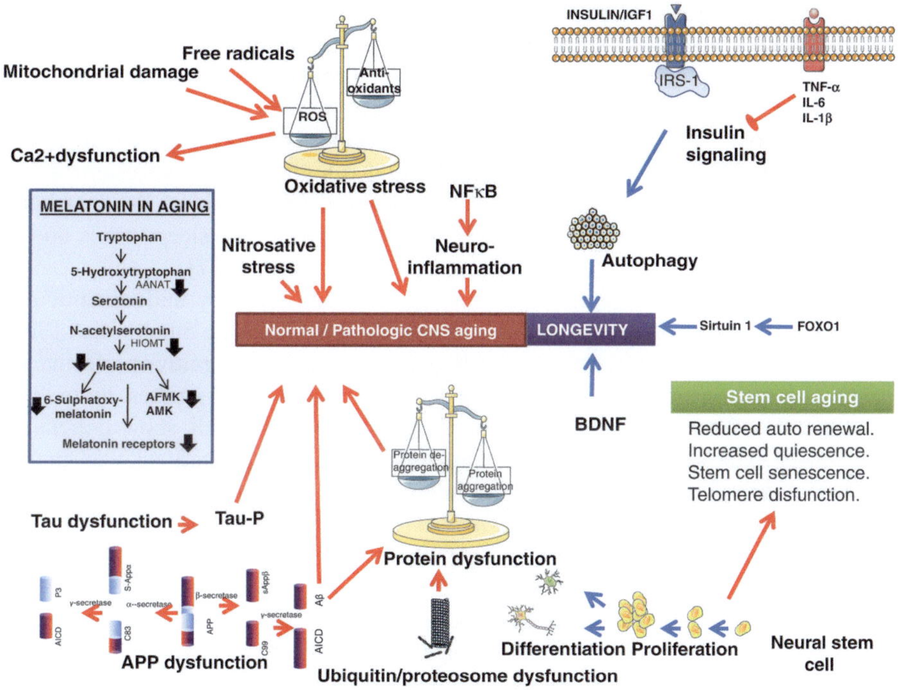

Fig. 8.22 Mechanisms promoting normal (*blue arrows*) and pathological (*red arrows*) central nervous system aging. Free radicals and mitochondrial damage promote oxidative stress and nervous system aging when the level of reactive oxygen species (ROS) production is higher than that of antioxidants. Oxidative stress leads to cell damage and calcium dysfunction. Melatonin, its metabolites, and receptors decrease in aging. Insulin, insulin-like growth factor 1 (IGF1), and insulin-like growth factor 2 (IGF2) act via insulin receptor substrate 1 (IRS-1) to trigger the insulin signaling pathway stimulating, for example, autophagy, a recycling pathway that maintains protein and organelle quality control. Inflammaging and the low-degree inflammation seen in obesity-related metabolic disorders lead to overproduction of proinflammatory cytokines and activate microglia. This interferes with the ability of IRS-1 to engage in insulin signaling and blocks the intracellular actions of insulin. Protein misfolding, aggregation, and degradation impairment are the main characteristic features of age-related neurodegenerative diseases. In the stem cell niche, depletion of the neural stem cell pool or decreased potential to produce progenitor cells leads to neural stem cell aging via impairments in self-renewal, stem cell senescence, increased quiescence, and neural stem cell fate changes

residues derived from its precursor, the amyloid precursor protein (APP). APP is proteolytically processed by α- or β-secretases in different pathways. The α-non-amyloidogenic pathway involves cleavage of APP by α-secretase to release a fragment of APP N-terminal, which, after cleavage by γ-secretase, precludes the formation of Aβ. The β-amyloidogenic pathway includes β-secretase, which results in the formation of intact Aβ peptide and is mediated by the sequential cleavage of β-secretase and γ-secretase at the N- and C-terminal of the Aβ molecule [126].

In AD, glymphatic impairment has emerged as a piece of the disease pathology puzzle. The Aβ peptide, which typically accumulates for years preceding AD dementia, is also produced by the normal brain and is present in the circulating blood and CSF. However, unlike the healthy brain, which is able to clear Aβ via glymphatic drainage, in AD there is a gradual build-up of Aβ in the brain parenchyma and vascular structures, leading to neurovascular uncoupling, including cerebral blood flow decrease, blood–brain barrier disruption, and impairment of vasculature [125]. The release of Aβ from neurons is dependent on synaptic activity and before the accumulation of Aβ plaques, brain Aβ levels fluctuate with the sleep–wake cycle in a pattern in which the Aβ concentration of interstitial fluid is higher during wakefulness and lower during sleep. Chronic sleep deprivation increases Aβ deposition. Recent results indicate that blood–brain barrier disruption induced by chronic sleep loss is due to the low-grade inflammation that the sleep deprivation entails [127].

A decline in cognitive capacities, including reasoning, memory, and semantic fluency, characterizes normal aging and is detectable as early as the 5th decade of life. There is evidence for a preclinical stage in dementia in which cognitive performance is borderline compared with normal aging. In community-based studies, up to 30% of a sample of healthy, community-dwelling, elderly individuals show deficits in performance that were not explained by age-related changes, education levels, mood, or health status. This strongly suggests the existence of early pathological changes, which is a transitional state taking place between normal aging and early AD [128].

Cross-sectional studies reveal that sleep disturbances are associated with memory and cognitive impairment. A severe disruption of the circadian timing system occurs in AD, as indicated by alterations in numerous overt rhythms such as body temperature, glucocorticoids, and/or plasma melatonin. The internal desynchronization of rhythms is significant in AD patients. One emerging symptom is "sundowning," a chronobiological phenomenon observed in AD patients in conjunction with sleep–wake disturbances [129]. Sundowning includes symptoms such as disorganized thinking, a reduced ability to maintain attention to external stimuli, agitation, wandering, and perceptual and emotional disturbances, all appearing in late afternoon or early evening. Chronotherapeutic interventions such as exposure to bright light and timed administration of melatonin in selected circadian phases alleviated sundowning symptoms and improved the sleep–wake patterns of AD patients (Fig. 8.7) [130].

Levels of melatonin in the CSF decrease, even in preclinical stages of AD when the patients do not manifest any cognitive impairment, suggesting that the reduction in CSF melatonin might be an early trigger and marker for AD. Although it is not known whether the relative melatonin deficiency is either a consequence or a cause of neurodegeneration, it seems clear that the loss in melatonin aggravates the disease and that early circadian disruption can be an important deficit to be considered [75]. Significant differences were observed in melatonin levels between mild cognitive impairment (MCI) and AD patients, with a negative correlation between the neuropsychological assessment of dementia and melatonin levels [131].

Mild cognitive impairment is diagnosed in those who have an objective and measurable deficit in cognitive functions, but with a preservation of daily activities. The estimates of annual conversion rates to dementia vary across studies, but may be as high 10–15%, as MCI represents a clinically important stage for identifying and treating individuals at risk [132]. Indeed, the degenerative process in AD brain starts 20–30 years before the clinical onset of the disease. During this phase, plaque and tangle loads increase and at a certain threshold, the first symptom appears.

An analysis of published data with melatonin in the early stages of cognitive decline consistently showed that administration of melatonin every night improves the quality of sleep and cognitive performance in this phase of the disease. Therefore, melatonin treatment can be effective in the early stages of neurodegenerative disease [75].

The mechanisms accounting for the therapeutic effect of melatonin in MCI patients remain to be defined. Melatonin treatment mainly promotes slow-wave sleep in the elderly and can be beneficial in MCI as it augments the restorative phases of sleep, including the secretion of GH and neurotrophins and the functioning of the glymphatic system. The antioxidant, mitochondrial, and anti-amyloidogenic effects of melatonin may possibly interfere with the onset of the disease (Fig. 8.23). Therefore, the point at which melatonin treatment begins can be decisive for the final response.

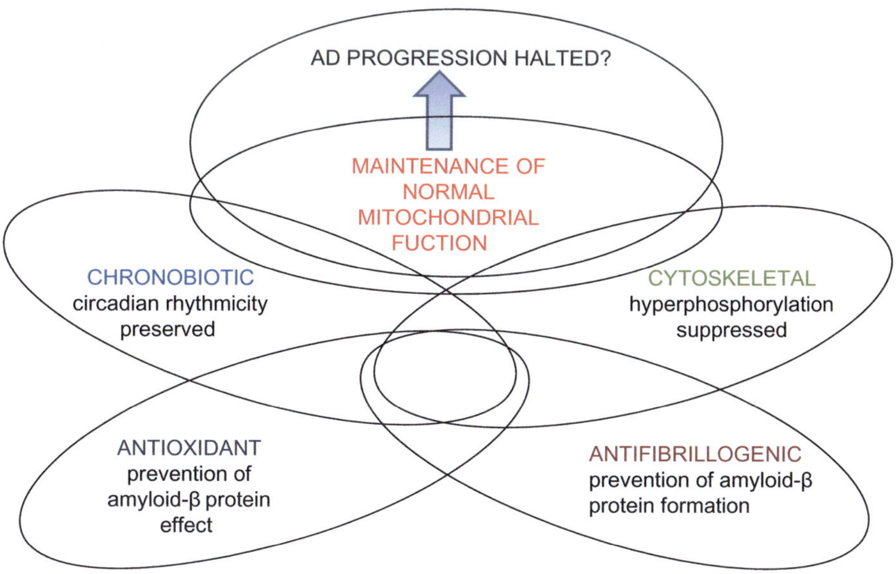

Fig. 8.23 Melatonin and Alzheimer's disease. The multiple effects of melatonin and the different degree of overlap (interrelations and mutual influences) are indicated by their respective intersections. Reproduced with permission from Cardinali [75]

Double-blind multicenter studies are needed to further explore and investigate the potential and usefulness of melatonin as an antidementia drug in the early stages of the disease. Acetylcholinesterase inhibitors, the first-line drugs used today do not prevent and treat AD. So far, over 90 phase 3 trials of AD have been unsuccessful, with a 99.0% failure rate [133]. Owing to the multifactorial nature of AD pathogenesis, polypharmacy with drugs that target heterogeneous pathophysiological pathways, must be considered. A novel pharmacological treatment paradigm (the "M" drugs) was proposed, involving the use of melatonin, minocycline, modafinil, and memantine [133]. Minocycline is neuroprotective, it reduces neuroinflammation and CNS pathology, and prevents cell death. Modafinil, a wake-promoting agent, improves global mental status, hippocampal neurogenesis, attention, and cognition. Memantine is a NMDA receptor antagonist and is approved for the management of moderate-to-severe AD. This sort of strategy could provide a comprehensive and pragmatic means of combating multiple pathological targets and ameliorating cognitive dysfunction in AD.

Other major risk factors for dementia are hypertension and T2DM. They occur in middle-aged and older adults and are strongly influenced by the dysregulation of insulin signaling, which starts with insulin resistance and is followed by hyperinsulinemia, metabolic syndrome and finally T2DM. Consumption of energy-dense diets, high in saturated fat and sugar, is associated not only with weight gain and metabolic syndrome, but also with impaired hippocampal-dependent memory and the emergence of pathological hippocampal conditions [134]. The hippocampus is one of the first brain regions in AD to show Aβ deposition, possibly associated with cognitive impairment. It has been demonstrated that greater cognitive and affective decline occurs in AD patients with metabolic syndrome than in those without, suggesting that insulin resistance and vascular endothelial dysfunction might be strongly correlated with AD before pathological changes of the brain can be observed. In this context, the hypothesis that metabolic syndrome might operate as a "second hit" is suggestive as a potential trigger of AD progression (Fig. 8.24) [134].

Inflammaging as a central phenomenon in senescence and its role in the development of age-associated brain diseases are now recognized. Genetic predispositions such as an immune risk profile, which comprises an increased tendency toward inflammatory responses, may set limits to health and lifespan, whereas an "inverted immune risk profile" found in centenarians may be the basis of successful aging. This is indicated by enhanced inflammation in neurodegenerative diseases and the effects of inflammatory signals and free radicals released by immune cells on neurons and astrocytes [10].

The immune system is remodeled with aging. One of the contributing factors is a progressive thymic involution, which leads to losses in the number of both CD4+ and CD8+ T lymphocytes, in addition to a disturbed balance among naïve, memory, and effector T cells. For example, the subset of CD8+ cells is largely replaced by the clonal expansion of cells of lower proliferative activity. Type 1 cytokines are elevated, such as IL-2, IFN-γ, and TNF-α, and, to a lesser extent, type 2 cytokines, such as IL-4, IL-6, and IL-10. This contributes to a higher inflammatory state;

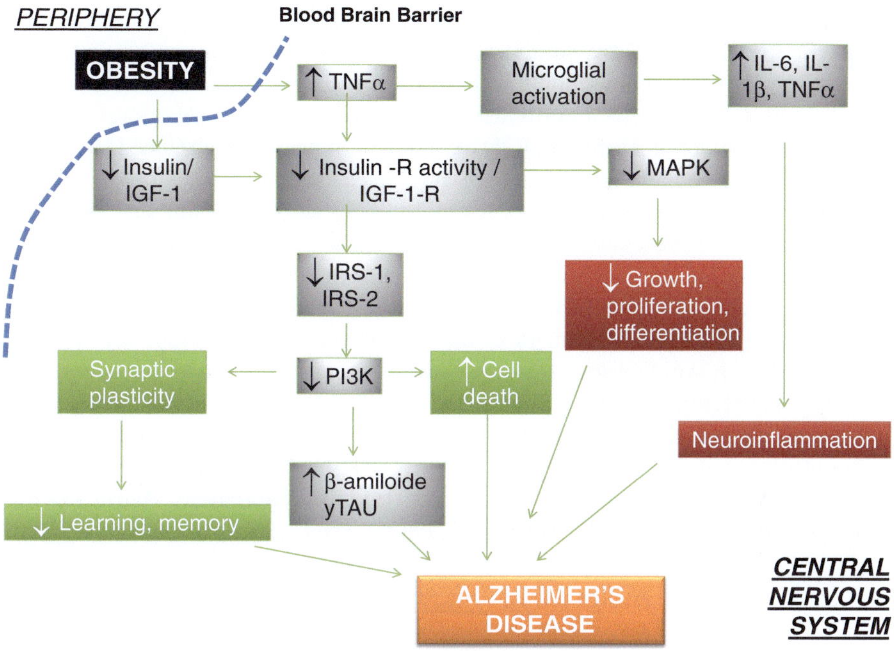

Fig. 8.24 Possible links between obesity and neurodegeneration. Obesity promotes chronic low-grade peripheral inflammation and insulin and IGF-1 resistance. Cerebrovascular dysfunction, together with increased blood–brain barrier permeability, allows macrophage/cytokine entry and reduced transport of trophic factors. Chronic inflammation, coupled with insulin and IGF-1 resistance, promotes neurodegenerative pathological conditions

however, with large interindividual differences. Age-dependent changes occur for CD4+ T lymphocytes, mostly in the activated/memory cells, among which TNFα-positive cells are decreased, but IL-4-positive cells increased, reflecting a shift from type 1 to type 2 cytokines. B cell immunosenescence results in decreases of IgM and IgD production, whereas increased levels of IgG1 have been reported. Concerning the innate immune system, centenarians have a relatively higher number of NK cells, especially NKT cells bearing the γδ-T cell receptor, in contrast to less successfully aging subjects. Moreover, these cells had a stronger response to activation in centenarians by releasing IFN-γ, underlying the importance of a robust innate immune system in longevity [135].

In addition to the gradual functional losses, immunosenescence has two other undesired consequences, a higher incidence of autoimmune responses and increased levels of inflammatory mediators. This latter observation is also made in centenarians. However, in these successfully aged subjects, the higher quantities of proinflammatory factors are associated with augmentations of anti-inflammatory cytokines and a protective genotype [136].

Inflammaging differs from acute or chronic inflammation elsewhere in the body. It is of a slowly progressing, lingering type, with moderate microglia activation that is sustained by the degenerative processes and oxidative stress resulting from the release of ROS and RNS by immune cells, astrocytes, and neurons that is further enhanced by damage to mitochondria (Fig. 8.25). Inflammaging, especially in neurodegenerative diseases, is intertwined with other potentially deteriorating processes, among which mitochondrial dysfunction is of prime importance [10].

Inflammaging entails a senescence-associated secretory phenotype, which depends on DNA damage, and becomes more likely with increasing age, accumulating over time. Senescent cells carrying damaged DNA are usually mitotically arrested, a mechanism that keeps them alive and metabolically active. However, these cells display a chronic DNA damage response leading to the release of proinflammatory cytokines, the hallmark of the senescence-associated secretory phenotype (Fig. 8.25). Aging astrocytes also express senescence-associated secretory phenotype characteristics [10].

Neuronal overexcitation and inflammatory responses are intimately associated. Although glutamate excitotoxicity can lead to microglia activation, primary immune responses involving the microglia may, in turn, lead to excitotoxicity. Astrocytes are frequently coactivated with microglial cells. They also contribute to both

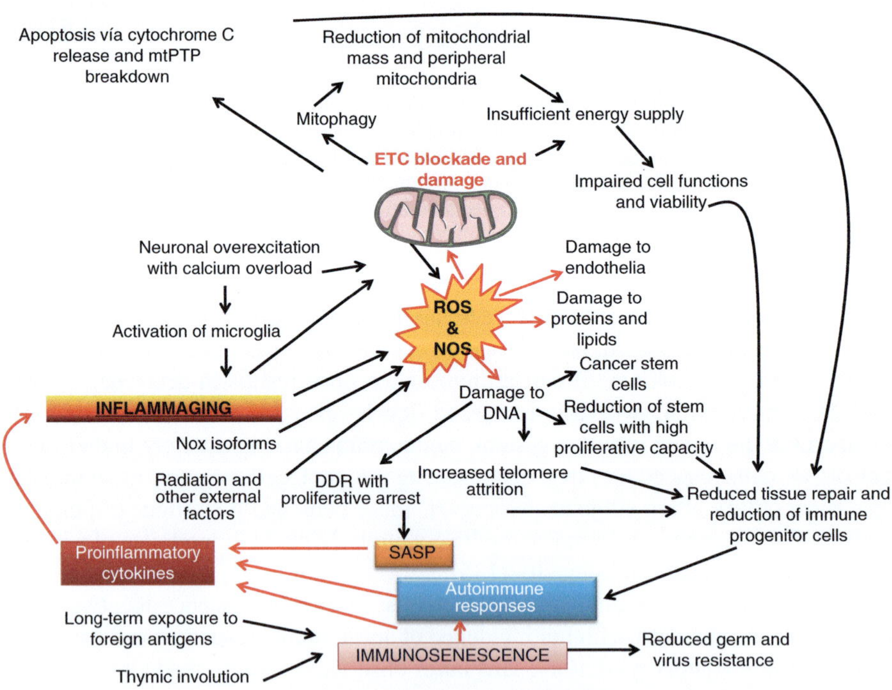

Fig. 8.25 The imbalance between inflammatory and anti-inflammatory signals is a hallmark of aging and contributes to its progression. An age-related proinflammatory tendency (inflammaging) is mostly unavoidable because of thymic involution and extended germ exposure. Mitochondrial dysfunction ensues with disruption of the electron transport chain and ROS and reactive nitrogen species generation. Reproduced with permission from Cardinali [75]

excitotoxicity and microglia activation, by mechanisms that involve impaired glutamate uptake, inflammatory signals, and/or oxidative/nitrosative stress.

Recent results indicate that blood–brain barrier disruption induced by chronic sleep loss is due to the low-grade inflammation that the sleep deprivation entails [127]. Figure 8.24 summarizes the possible links between obesity and neurodegeneration in AD. Obesity promotes chronic low-grade peripheral inflammation and insulin and insulin growth factor (IGF)-1 resistance. Cerebrovascular dysfunction, together with increased blood–brain barrier permeability, allows macrophage/cytokine entry and reduced transport of trophic factors. Chronic inflammation, coupled with insulin and IGF-1 resistance, promotes neurodegenerative pathological conditions. Potential signaling pathways further lead to AD-like molecular and cognitive changes via increased Aβ and tau phosphorylation (Fig. 8.24).

Neurodegeneration is facilitated by stimulation of proinflammatory processes in different cell types, that implies the possibility of positive feedback loops between neurons, astrocytes, and microglia, and that expand the grade and area of inflammation recruiting other immune cells. Activation of microglia can lead to different phenotypes, from neurodestructive and phagocytic cells to others that are primarily neuroprotective or promote growth.

Inflammaging is specifically accelerated and aggravated under neuropathological conditions. Sustained, progressive, primarily low-grade inflammation is observed in presumably all neurodegenerative disorders, with differences in details, pathways, and affected cells. Neuroinflammation has been demonstrated in Huntington's disease, amyotrophic lateral sclerosis, Friedreich's ataxia, Parkinson's disease, frontotemporal lobe degeneration, and AD [10].

As the early phases of neurodegenerative disorders are poorly distinguishable from normal inflammaging or other forms of low-grade inflammation, these pathological conditions display important mechanisms of aggravation, in terms of vicious cycles. These comprise the accumulation of toxic products, such as Aβ peptides, oligomers, and plaques in AD, microglial proliferation and phagocytosis after their activation, and the progressive formation of proinflammatory signal molecules. Elevated levels of IL-6 and TNFα are demonstrable in the CNS and in the circulation, reflecting the dysregulation of inflammatory pathways, and are meanwhile regarded as markers of frailty. Interrupting the proinflammatory vicious cycles represents a main challenge in combating neurodegenerative diseases [135, 136].

A particular aspect of brain inflammaging concerns the impaired function of SCN, which is observed during normal aging and, even more frequently, in neurodegenerative disorders. This change may be of high relevance, because it systemically affects a host of other functions in the body (Chap. 2). Several alterations can be involved in SCN insufficiency, from reduction of blue light perception to losses in signal transmission to the SCN, but a major factor is that of SCN neural degeneration. Because of the crucial role of the SCN in the control of the mammalian pineal gland, dysfunction of the hypothalamic master clock strongly contributes to aging- or disease-related reductions of nocturnal melatonin secretion and changes in the secretory patterns and phasing of melatonin [131]. Although SCN deterioration is more strongly pronounced in AD and other forms of dementia, it seems that the relevance of this phenomenon is already high in normal aging. In part, this may be related to changes in melatonin, but additional impairments of other rhythmic functions likely arise when a master clock decays. The replacement of the poorly functional SCN of a senescent hamster by transplantation of a juvenile SCN not only restored the previously decomposed circadian rhythmicity, but caused a rejuvenation in terms of physical appearance and extended the lifespan of the recipient [137]. These findings impressively show how important the SCN and a well-operating circadian system are for preventing aging-related impairments. Because of the reduction in neurons in the SCN the sequence and spacing of the maximum values of daily rhythms are progressively decreased. Correction of chronodisruption by chronobiological treatment is thus needed for maintaining health in the elderly (Fig. 8.26).

A role of melatonin in attenuating inflammaging and its progression has been especially discussed with regard to treatment options under conditions of reduced endogenous melatonin levels [10]. Melatonin declines during aging and, even more so in several age-related diseases, changes that have been documented in humans. Interindividual variations observed among elderly persons may be explained, to a certain extent, by differences in the acquisition of melatonin-depressing diseases and disorders.

Reversion of inflammaging by melatonin occurs at various levels (Fig. 8.27). One level concerns the correction of metabolic dysregulation, including the prevention of insulin resistance, an inflammation-promoting change and hallmark of metabolic syndrome. Notably, and as already mentioned, melatonin was effective at suppressing insulin resistance. Melatonin reversed the blockade of a key step in

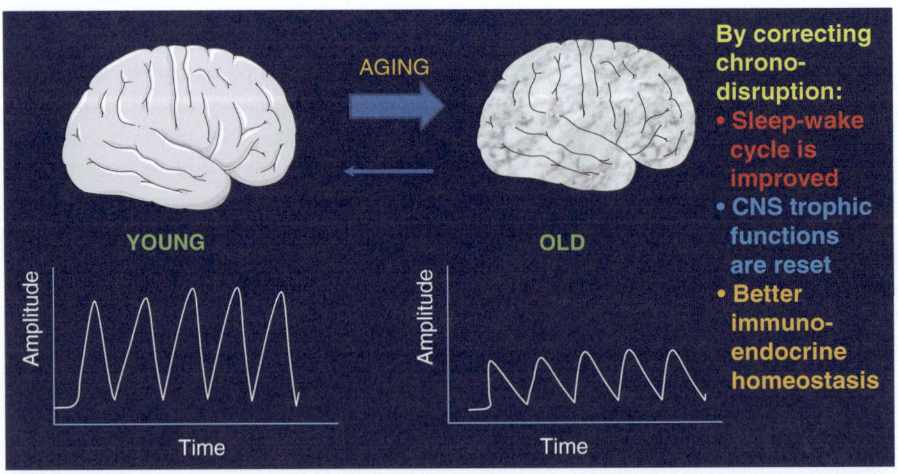

Fig. 8.26 The aging of the circadian apparatus results in a decrease in amplitude and phase-advance of circadian rhythms. By restoring chronodisruption, the sleep–wake cycle improves, central nervous system trophic functions are reset and better immunoendocrine homeostasis occurs. Reproduced with permission from Cardinali [75]

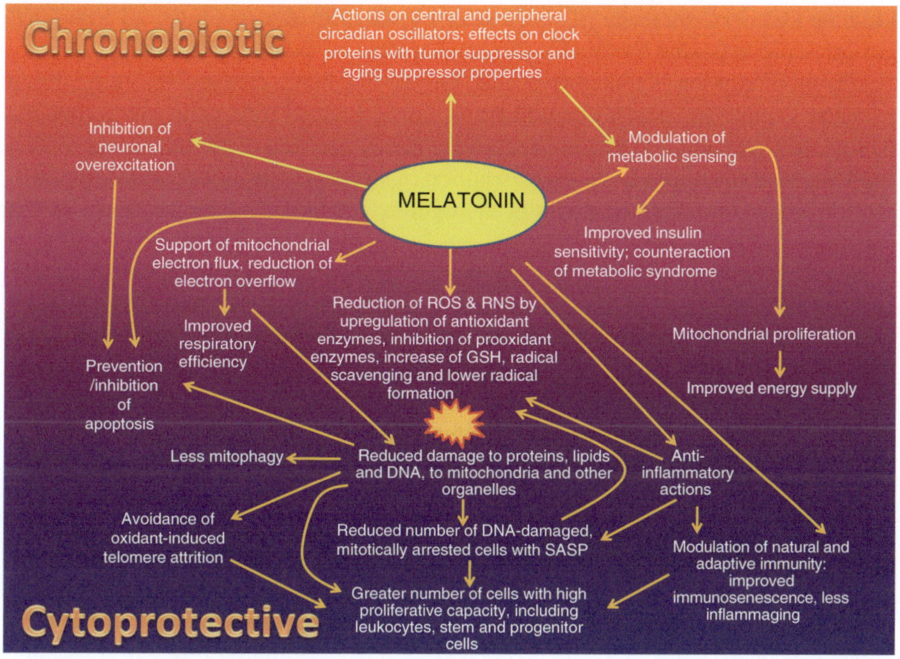

Fig. 8.27 Overview of the multiple actions of melatonin that antagonize brain inflammaging. Reproduced with permission from Cardinali [75]

insulin signal transduction, i.e., the reduced phosphorylation of IRS-1 (insulin receptor substrate 1), which is generally accompanied by an upregulation of IRS-1 expression [75].

A further level of action of melatonin concerns the suppression of processes that favor or lead to inflammation. This comprises calcium overload, excessive NO release that results in the formation of peroxynitrite, peroxynitrite-derived free radicals and tyrosine nitration, and mitochondrial dysfunction (Fig. 8.27).

The immunological effects of melatonin represent a third area relevant to inflammaging. The role of melatonin as an immunemodulatory agent comprises both pro-inflammatory and anti-inflammatory actions, which, consequently, leads to an either pro-oxidant or antioxidant balance. In several conditions concerning senescence, the anti-inflammatory side of melatonin seems to prevail [138].

Cancer

Cancer is responsible for about 25% of all deaths in the Western world. Most diagnoses of cancer occur in people over 55 years of age, with breast and prostate cancer being the leading types of cancer followed by lung cancer. More than half of new cases of cancer are breast, prostate, lung or gastrointestinal cancer [139]. Many etiological factors including genetic, environmental, dietary, hormonal and aging, the immune status or presence of medical or psychiatric illnesses have all been suggested as predisposing factors for the development of cancer in humans.

Chronodisruption is linked to cancer. On the one hand, cancer itself (e.g., tumor invasion symptoms, pain), chemotherapy, corticosteroid treatment, environmental factors, or psychological distress are among the factors that contribute to chronodisruption, immunosuppression, and disruption of sleep. On the other hand, chronodisruption given by exposure to light at night is considered a major cause of the increased risk of cancer [140].

Working in nondaylight hours are associated with an increased risk of cancer, and the International Agency for Research on Cancer, WHO, classified shift work with chronodisruption as a probable human carcinogen (group 2A carcinogen) [141]. The reduction of melatonin levels following repeated nocturnal exposure to light, as this occurs in women engaged in night-shift work, can result in higher rates of breast cancer development and proliferation [142]. Breast cancer accounts for 25% of all cancers and caused 522,000 deaths worldwide in 2012. A meta-analysis of 16 investigations, involving 2,020,641 participants, 10,004 incident breast cancer cases, 7185 cancer-related deaths, 4820 cardiovascular endpoints, and 2480 all-cause mortalities indicated that night work increased the risk of breast cancer morbidity by: 1.9% for 5 years, 2.5% for 5–10 years, 7.4% for 10–20 years, and 8.8% for >20 years of work [143]. Additionally, night work enhanced the morbidity of breast cancer by 8.9% and was associated with a 2.7% increase in cardiovascular death.

Circadian genes are essential in the regulation of the cell cycle, a finely regulated process from a cell that can generate multiple cells through a series of cell divisions, including four critical and successive steps named G1 phase (growth phase 1), S phase (synthesis), G2 (growth phase 2), and M phase (mitosis). Knock-down of the circadian gene *Bmal1* in the carcinoma cells of the colon, fibroblast cells, and intestine epithelial cells produces cellular proliferation in vitro and increments the size of tumor cells injected subcutaneously, via, among others, the inhibition of apoptosis and the reduction of the time transition between G2/M. Moreover, the knock-out for clock genes *Bmal1* and *Per2* in mice previously exposed to radiation caused hyperplastic growth and development of lymphoma, hepatic carcinomas, ovary tumors, and osteosarcomas [144].

In humans, the association between clock gene polymorphisms and cancer is common. An incremented risk of breast cancer, colorectal carcinoma, prostate cancer, lung cancer, non-Hodgkin's lymphoma, glioma, and primary hepatocellular carcinoma were associated with alleles of circadian genes. Therefore, clock genes may be modulators of the cellular cycle that modulates the risk of cancer [144].

In addition, the oncogenic alteration of circadian rhythms has also been reported to occur. Mutations in circadian clock genes, including promoter methylation, coding region mutation, deletion, or rare amplification, have been documented across many different types of cancer [145]. Given that these mutations disrupt normal oscillation, it has been suggested that the clock might be tumor-suppressive. Many proto-oncogenes and tumor suppressors are normally under circadian control; thus, disruption of oscillation could potentially release these proteins to be constitutively overexpressed or suppressed. Given the frequency of mutations, it can be speculated that many cancers with oncogenic mutations have altered or disrupted circadian rhythm and altered oscillation of gene expression and metabolism [145].

With few exceptions, melatonin levels were found to be low in cancer patients [146]. This may be an indication for the ongoing chronodisruption, as circulating melatonin levels are an index of SCN function (Chap. 2). In addition, another prevalent idea is that the reduced melatonin secretion plays a role in the occurrence of cancer in chronodisruption.

The light–melatonin–cancer hypothesis has been supported by experimental studies conducted on athymic rats in which human breast or prostate cancer tissue was transplanted. Rats that had been exposed to a constant light environment exhibited a sevenfold increase in tumor growth compared with rats that had remained in normal light/dark cycles [147]. There was an augmented uptake of linoleic acid and of its metabolism in rats kept under constant illumination. This accelerated rate of linoleic acid metabolism was attributed to the suppression of the circadian melatonin signal, which normally inhibits linoleic acid uptake at nighttime. Compared with tumors perfused with melatonin-deficient human blood collected during the daytime, human breast cancer xenografts and rat hepatomas perfused in situ with nocturnal, physiologically melatonin-rich blood collected during the night, exhibited markedly suppressed proliferative activity and linoleic acid uptake/metabolism [148]. In the case of prostate cancer xenografts, the amplification of nighttime melatonin levels by exposing nude rats to blue light during the daytime significantly reduces the human prostate cancer metabolic, signaling, and proliferative activities [149]. These studies are relevant for explaining the risk of increased breast and prostate cancer risk in night-shift workers [143] and that of colon cancer [150]. Additionally, melatonin disruption increases the risk of lung cancer [151]. The oncostatic properties of melatonin are supported by several experimental and clinical studies [146].

Sleep disturbance is a major outcome of chronodisruption among cancer patients [152]: 30–60% of cancer patients report insomnia symptoms, whereas approximately 20% of them meet the diagnostic criteria for an insomnia disorder, which is more than twice as frequent as in the general population. However, they generally remained underdiagnosed and poorly treated [153–156].

Prevalence rates of sleep disturbances have also been found to be up to three times higher in patients undergoing chemotherapy compared with the general population. Sleep difficulties may occur before, during, and may persist long after the end of cancer treatment [157]. An 18-month longitudinal study revealed that such persistence was even more likely to occur in patients with an insomnia syndrome [158]. Moreover, about 20% of patients who experienced an insomnia remission had a relapse later during the study.

Sleep–wake disturbance is among the most severe and common symptoms reported by primary brain tumor patients, particularly those undergoing radiation therapy. As with other cancers and neurological illness, sleep–wake disturbance may also be clustered or related to additional symptoms such as fatigue, depression, and cognitive impairment [159].

It must be noted that an imprecise conceptualization of chronodisruption in cancer has led to narrowly focused interventions being diffusely targeted to symptoms, rather than focused and specific to one or more chronobiological disorders underlying those symptoms. The consequences of chronodisruption in cancer are numerous and can negatively affect both psychological and physical functioning. Compared with the consequences of cancer itself, those related to sleep disruption are often overlooked both by the patients and by the health care providers.

The most common consequences reported by the patients are symptoms of fatigue, psychological distress, impaired daytime functioning, and disrupted cognitive functioning. Chronodisruption is also associated with an increased risk of subsequently developing a psychiatric disorder (e.g., anxiety and depressive disorder), exacerbation of pain, impaired immune functioning, and increased risk of infections. As stated in Figs. 8.7 and 8.8, the concomitant chronodisruption must be adequately treated to obtain full recovery, i.e., the specific treatments must be combined with a chronobiological adjuvant treatment.

Pharmacotherapy is currently the most common treatment employed to counteract sleep difficulties in the general population, and in cancer patients. Surveys that have documented the use of sedative and/or hypnotic medications among cancer patients reached utilization rates close to 25%. Among the factors associated with the increased use of hypnotic medication, being older, experiencing more stressful life events during the past 6 months, suffering higher levels of anxiety or past or current chemotherapy treatment were quoted [160]. Almost 80% of participants who were taking drugs were prescribed BZDs (mostly lorazepam and oxazepam), followed by zopiclone (9%). As discussed earlier in this chapter, the side effects and risks associated with the usage of BZPs and Z drugs include drowsiness, dizziness, headache, cognitive impairments, loss of motor coordination, and a risk of tolerance and dependence when the medication is used daily to treat chronic difficulties.

Depression is a frequent and serious comorbid condition affecting the quality of life. Such comorbidity reduces compliance with treatment and aggravates the physical consequences of the disease. Although there are studies showing that about 40% of tumor patients need professional psycho-oncological support, less than 10% of patients are referred for psychosocial intervention in daily clinical practice. Studies of effective pharmacotherapy are relatively scarce in cancer patients with depression and they are biased by a high number of dropouts because of the side effects relating to the use of antidepressants compared with placebo [161]. It is therefore difficult to determine with clarity what is the best pharmacological treatment for major depression in cancer patients.

Circadian rhythm abnormalities, as shown by the sleep–wake cycle disturbances, constitute one the most prevalent signs of depression. The disturbances in the amplitude and rhythm of melatonin secretion that occur in patients with depression resemble those seen in subjects with chronobiological disorders, thus suggesting a link between disturbance of melatonin secretion and depressed mood. As melatonin is involved in the regulation of both circadian rhythms and sleep, any antidepressant drug with effects on melatonin receptors could be an advantage in treatment. Melatonin has been found to be effective at treating circadian rhythm disorders and has antidepressant activity [75]. Among the analogs developed to improve the efficacy of the effects of melatonin, agomelatine has been licensed by the EMA for the treatment of major depression disorder in adults (Chap. 6).

Melatonergic receptors, particularly MT_1, are also involved in the sedatory and anxiolytic effects of melatonin that have been linked to a facilitatory role in GABA transmission. This anti-excitatory action of melatonin underlies the anxiolytic, antihyperalgesic, and antinociceptive effects of melatonergic agents, all of them having potential application in cancer patients [75].

Increasing importance is given to chronotherapy, or the timed administration of treatment to patients, based on circadian rhythm, to increase the efficacy and reduce the toxicity of drugs or radiation. Several traditional cancer therapeutics, including the anti-metabolite folate pathway antagonist methotrexate, have known circadian-dependent toxicity. In addition, new research indicates that several targeted therapies currently in clinical use have strongly circadian-dependent efficacy depending on the time of day given, including erlotinib (inhibits epidermal growth factor receptor, EGFR, used in lung cancer), lapatinib (inhibits HER/Neu and EGFR, used in breast cancer), and evirolimus (inhibits mTOR, used in some breast cancers and pancreatic neuroendocrine tumors). In fact, there are several chronotherapy dosing schedules under clinical trial [162]. An appropriate adjuvant chronobiological treatment (Fig. 8.8) contributes to keeping the circadian system normal, with obvious advantages for chronotherapy.

References

1. Chang AM, Aeschbach D, Duffy JF, Czeisler CA. Evening use of light-emitting eReaders negatively affects sleep, circadian timing, and next-morning alertness. Proc Natl Acad Sci U S A. 2015;112:1232–7.
2. Lavie P. Sleep-wake as a biological rhythm. Annu Rev Psychol. 2001;52:277–303.
3. Gringras P, Middleton B, Skene DJ, Revell VL. Bigger, brighter, bluer-better? Current light-emitting devices – adverse sleep properties and preventative strategies. Front Public Health. 2015;3:233.
4. Covassin N, Singh P. Sleep duration and cardiovascular disease risk: epidemiologic and experimental evidence. Sleep Med Clin. 2016;11:81–9.
5. Blanco M, Kriguer N, Pérez Lloret S, Cardinali DP. Attitudes towards treatment among patients suffering from sleep disorders. A Latin American survey. BMC Fam Pract. 2003;4:17.
6. Wright KP Jr, Bogan RK, Wyatt JK. Shift work and the assessment and management of shift work disorder (SWD). Sleep Med Rev. 2013;17:41–54.
7. Singh A, Yeh CJ, Verma N, Das AK. Overview of attention deficit hyperactivity disorder in young children. Health Psychol Res. 2015;3:2115.
8. Smolensky MH, Hermida RC, Portaluppi F. Circadian mechanisms of 24-hour blood pressure regulation and patterning. Sleep Med Rev. 2017;33:4–16.
9. Cardinali DP. Qué es el Sueño. Paidós: Buenos Aires; 2014.
10. Hardeland R, Cardinali DP, Brown GM, Pandi-Perumal SR. Melatonin and brain inflammaging. Prog Neurobiol. 2015;127-128:46–63.
11. Anothaisintawee T, Reutrakul S, Van CE, Thakkinstian A. Sleep disturbances compared to traditional risk factors for diabetes development: systematic review and meta-analysis. Sleep Med Rev. 2016;30:11–24.
12. Kaur J. A comprehensive review on metabolic syndrome. Cardiol Res Pract. 2014;2014:943162.

13. Reid KJ, Abbott SM. Jet Lag and shift work disorder. Sleep Med Clin. 2015;10:523–35.
14. Agorastos A, Linthorst AC. Potential pleiotropic beneficial effects of adjuvant melatonergic treatment in posttraumatic stress disorder. J Pineal Res. 2016;61:3–26.
15. Laermans J, Depoortere I. Chronobesity: role of the circadian system in the obesity epidemic. Obes Rev. 2016;17:108–25.
16. Oldham MA, Lee HB, Desan PH. Circadian rhythm disruption in the critically ill: an opportunity for improving outcomes. Crit Care Med. 2016;44:207–17.
17. Madrid-Navarro CJ, Sanchez-Galvez R, Martinez-Nicolas A, Marina R, Garcia JA, Madrid JA, Rol MA. Disruption of circadian rhythms and delirium, sleep impairment and sepsis in critically ill patients. potential therapeutic implications for increased light-dark contrast and melatonin therapy in an ICU environment. Curr Pharm Des. 2015;21:3453–68.
18. Waterhouse J, Reilly T, Atkinson G, Edwards B. Jet lag: trends and coping strategies. Lancet. 2007;369:1117–29.
19. Cajochen C, Munch M, Knoblauch V, Blatter K, Wirz-Justice A. Age-related changes in the circadian and homeostatic regulation of human sleep. Chronobiol Int. 2006;23:461–74.
20. Srinivasan V, Singh J, Pandi-Perumal SR, Brown GM, Spence DW, Cardinali DP. Jet lag, circadian rhythm sleep disturbances, and depression: the role of melatonin and its analogs. Adv Ther. 2010;27:796–813.
21. Golombek DA, Casiraghi LP, Agostino PV, Paladino N, Duhart JM, Plano SA, Chiesa JJ. The times they're a-changing: effects of circadian desynchronization on physiology and disease. J Physiol Paris. 2013;107:310–22.
22. Ariznavarreta C, Cardinali DP, Villanua M, Granados B, Martin M, Chiesa JJ, Golombek DA, Tresguerres JAF. Circadian rhythms in airline pilots submitted to long-haul transmeridian flights. Aviat Space Environ Med. 2002;73:445–55.
23. Tresguerres J, Ariznavarreta C, Granados B, Martín M, Villanua M, Golombek D, Cardinali DP. Circadian urinary 6-sulphatoxymelatonin, cortisol excretion and locomotor activity in airline pilots during transmeridian flights. J Pineal Res. 2001;31:16–22.
24. van DA, Boot CR, Hlobil H, Twisk JW, Smid T, van der Beek AJ. Evaluation of an mHealth intervention aiming to improve health-related behavior and sleep and reduce fatigue among airline pilots. Scand J Work Environ Health. 2014;40:557–68.
25. McEwen BS. Sleep deprivation as a neurobiologic and physiologic stressor: allostasis and allostatic load. Metabolism. 2006;55:S20–3.
26. Coutinho JF, Goncalves OF, Maia L, Fernandes VC, Perrone-McGovern K, Simon-Dack S, Hernandez K, Oliveira-Silva P, Mesquita AR, Sampaio A. Differential activation of the default mode network in jet lagged individuals. Chronobiol Int. 2015;32:143–9.
27. Pinheiro SP, Schernhammer ES, Tworoger SS, Michels KB. A prospective study on habitual duration of sleep and incidence of breast cancer in a large cohort of women. Cancer Res. 2006;66:5521–5.
28. Megdal SP, Kroenke CH, Laden F, Pukkala E, Schernhammer ES. Night work and breast cancer risk: a systematic review and meta-analysis. Eur J Cancer. 2005;41:2023–32.
29. Dawson D, Reid K. Fatigue, alcohol and performance impairment. Nature. 1997;388:235.
30. Diez JJ, Vigo DE, Pérez Lloret S, Ritgers S, Rolé N, Cardinali DP, Pérez-Chada D. Sleep habits, alertness, cortisol levels and cardiac autonomic activity in short distance bus drivers. Differences between morning and afternoon shifts. J Occup Environ Med. 2011;36:806–11.
31. Diez JJ, Vigo DE, Cardinali DP, Pérez-Chada D. Sleep habits, daytime sleepiness and working conditions in short-distance bus drivers. Int J Workplace Health Manag. 2014;7:202–12.
32. Eckel RH, Depner CM, Perreault L, Markwald RR, Smith MR, McHill AW, Higgins J, Melanson EL, Wright KP Jr. Morning circadian misalignment during short sleep duration impacts insulin sensitivity. Curr Biol. 2015;25:3004–10.
33. Hardeland R. Melatonin and circadian oscillators in aging--a dynamic approach to the multiply connected players. Interdiscip Top Gerontol. 2015;40:128–40.
34. Reid KJ, Burgess HJ. Circadian rhythm sleep disorders. Prim Care. 2005;32:449–73.
35. Shanahan TL, Czeisler CA. Physiological effects of light on the human circadian pacemaker. Semin Perinatol. 2000;24:299–320.

36. Dawson D, Armstrong SM. Chronobiotics--drugs that shift rhythms. Pharmacol Ther. 1996;69:15–36.
37. Pandi-Perumal SR, Srinivasan V, Maestroni GJM, Cardinali DP, Poeggeler B, Hardeland R. Melatonin: nature's most versatile biological signal? FEBS J. 2006;273(13):2813–38.
38. Herxheimer A, Petrie KJ. Melatonin for the prevention and treatment of jet lag. Cochrane Database Syst Rev. 2002;CD001520.
39. Czeisler CA, Kronauer RE, Allan JS, Duffy JF, Jewett ME, Brown EN, Ronda JM. Bright light induction of strong (type 0) resetting of the human circadian pacemaker. Science. 1989;244:1328–33.
40. Cardinali DP, Bortman GP, Liotta G, Perez LS, Albornoz LE, Cutrera RA, Batista J, Ortega GP. A multifactorial approach employing melatonin to accelerate resynchronization of sleep-wake cycle after a 12 time-zone westerly transmeridian flight in elite soccer athletes. J Pineal Res. 2002;32:41–6.
41. Cardinali DP, Furio AM, Reyes MP, Brusco LI. The use of chronobiotics in the resynchronization of the sleep-wake cycle. Cancer Causes Control. 2006;17:601–9.
42. Smolensky MH. Circadian rhythms in medicine. CNS Spectr. 2001;6:467–82.
43. Crowley SJ, Eastman CI. Phase advancing human circadian rhythms with morning bright light, afternoon melatonin, and gradually shifted sleep: can we reduce morning bright-light duration? Sleep Med. 2015;16:288–97.
44. Ferracioli-Oda E, Qawasmi A, Bloch MH. Meta-analysis: melatonin for the treatment of primary sleep disorders. PLoS One. 2013;8:e63773.
45. Wilson SJ, Nutt DJ, Alford C, Argyropoulos SV, Baldwin DS, Bateson AN, Britton TC, Crowe C, Dijk DJ, Espie CA, Gringras P, Hajak G, Idzikowski C, Krystal AD, Nash JR, Selsick H, Sharpley AL, Wade AG. British Association for Psychopharmacology consensus statement on evidence-based treatment of insomnia, parasomnias and circadian rhythm disorders. J Psychopharmacol. 2010;24:1577–601.
46. Dinges DF. The state of sleep deprivation: from functional biology to functional consequences. Sleep Med Rev. 2006;10:303–5.
47. O'Neill S, O'Driscoll L. Metabolic syndrome: a closer look at the growing epidemic and its associated pathologies. Obes Rev. 2015;16:1–12.
48. Makki K, Froguel P, Wolowczuk I. Adipose tissue in obesity-related inflammation and insulin resistance: cells, cytokines, and chemokines. ISRN Inflamm. 2013:139239.
49. Cardinali DP, Hardeland R. Inflammaging, metabolic syndrome and melatonin: a call for treatment studies. Neuroendocrinology. 2017;104:382–97.
50. Tchernof A, Despres JP. Pathophysiology of human visceral obesity: an update. Physiol Rev. 2013;93:359–404.
51. Lemche E, Chaban OS, Lemche AV. Neuroendocrine and epigetic mechanisms subserving autonomic imbalance and HPA dysfunction in the metabolic syndrome. Front Neurosci. 2016;10:142.
52. Pagano ES, Spinedi E, Gagliardino JJ. White adipose tissue and circadian rhythm dysfunctions in obesity pathogenesis and available therapies. Neuroendocrinology. 2017;104:347–63.
53. Arora T, Chen MZ, Cooper AR, Andrews RC, Taheri S. The impact of sleep debt on excess adiposity and insulin sensitivity in patients with early type 2 diabetes mellitus. J Clin Sleep Med. 2016;12:673–80.
54. Arora T, Chen MZ, Omar OM, Cooper AR, Andrews RC, Taheri S. An investigation of the associations among sleep duration and quality, body mass index and insulin resistance in newly diagnosed type 2 diabetes mellitus patients. Ther Adv Endocrinol Metab. 2016;7:3–11.
55. McFadden E, Jones ME, Schoemaker MJ, Ashworth A, Swerdlow AJ. The relationship between obesity and exposure to light at night: cross-sectional analyses of over 100,000 women in the Breakthrough Generations Study. Am J Epidemiol. 2014;180:245–50.
56. Fonken LK, Nelson RJ. The effects of light at night on circadian clocks and metabolism. Endocr Rev. 2014;35:648–670.
57. Karthikeyan R, Marimuthu G, Spence DW, Pandi-Perumal SR, Bahammam AS, Brown GM, Cardinali DP. Should we listen to our clock to prevent type 2 diabetes mellitus? Diabetes Res Clin Pract. 2014;106:182–90.

58. Marcheva B, Ramsey KM, Buhr ED, Kobayashi Y, Su H, Ko CH, Ivanova G, Omura C, Mo S, Vitaterna MH, Lopez JP, Philipson LH, Bradfield CA, Crosby SD, JeBailey L, Wang X, Takahashi JS, Bass J. Disruption of the clock components CLOCK and BMAL1 leads to hypoinsulinaemia and diabetes. Nature. 2010;466:627–31.
59. Shi SQ, Ansari TS, McGuinness OP, Wasserman DH, Johnson CH. Circadian disruption leads to insulin resistance and obesity. Curr Biol. 2013;23:372–81.
60. Gale JE, Cox HI, Qian J, Block GD, Colwell CS, Matveyenko AV. Disruption of circadian rhythms accelerates development of diabetes through pancreatic beta-cell loss and dysfunction. J Biol Rhythms. 2011;26:423–33.
61. Scheer FA, Hilton MF, Mantzoros CS, Shea SA. Adverse metabolic and cardiovascular consequences of circadian misalignment. Proc Natl Acad Sci U S A. 2009;106:4453–8.
62. Kelly MA, Rees SD, Hydrie MZ, Shera AS, Bellary S, O'Hare JP, Kumar S, Taheri S, Basit A, Barnett AH. Circadian gene variants and susceptibility to type 2 diabetes: a pilot study. PLoS One. 2012;7:e32670.
63. Woon PY, Kaisaki PJ, Braganca J, Bihoreau MT, Levy JC, Farrall M, Gauguier D. Aryl hydrocarbon receptor nuclear translocator-like (BMAL1) is associated with susceptibility to hypertension and type 2 diabetes. Proc Natl Acad Sci U S A. 2007;104:14412–7.
64. Karthikeyan R, Marimuthu G, Sooriyakumar M, BaHammam AS, Spence DW, Pandi-Perumal SR, Brown GM, Cardinali DP. Per3 length polymorphism in patients with type 2 diabetes mellitus. Horm Mol Biol Clin Invest. 2014;18:145–9.
65. Monteleone P, Tortorella A, Docimo L, Maldonato MN, Canestrelli B, De LL, Maj M. Investigation of 3111T/C polymorphism of the CLOCK gene in obese individuals with or without binge eating disorder: association with higher body mass index. Neurosci Lett. 2008;435:30–3.
66. Garaulet M, Corbalan MD, Madrid JA, Morales E, Baraza JC, Lee YC, Ordovas JM. CLOCK gene is implicated in weight reduction in obese patients participating in a dietary programme based on the Mediterranean diet. Int J Obes (Lond). 2010;34:516–23.
67. Ye D, Cai S, Jiang X, Ding Y, Chen K, Fan C, Jin M. Associations of polymorphisms in circadian genes with abdominal obesity in Chinese adult population. Obes Res Clin Pract. 2016;10(Suppl 1):S133–41.
68. Yamaguchi M, Uemura H, Arisawa K, Katsuura-Kamano S, Hamajima N, Hishida A, Suma S, Oze I, Nakamura K, Takashima N, Suzuki S, Ibusuki R, Mikami H, Ohnaka K, Kuriyama N, Kubo M, Tanaka H. Association between brain-muscle-ARNT-like protein-2 (BMAL2) gene polymorphism and type 2 diabetes mellitus in obese Japanese individuals: a cross-sectional analysis of the Japan Multi-institutional Collaborative Cohort Study. Diabetes Res Clin Pract. 2015;110:301–8.
69. Corella D, Asensio EM, Coltell O, Sorli JV, Estruch R, Martinez-Gonzalez MA, Salas-Salvado J, Castaner O, Aros F, Lapetra J, Serra-Majem L, Gomez-Gracia E, Ortega-Azorin C, Fiol M, Espino JD, Diaz-Lopez A, Fito M, Ros E, Ordovas JM. CLOCK gene variation is associated with incidence of type-2 diabetes and cardiovascular diseases in type-2 diabetic subjects: dietary modulation in the PREDIMED randomized trial. Cardiovasc Diabetol. 2016;15:4.
70. Cardinali DP, Cano P, Jimenez-Ortega V, Esquifino AI. Melatonin and the metabolic syndrome: physiopathologic and therapeutical implications. Neuroendocrinology. 2011;93:133–42.
71. Stamenkovic JA, Olsson AH, Nagorny CL, Malmgren S, Dekker-Nitert M, Ling C, Mulder H. Regulation of core clock genes in human islets. Metabolism. 2012;61:978–85.
72. Pivovarova O, Gogebakan O, Sucher S, Groth J, Murahovschi V, Kessler K, Osterhoff M, Rudovich N, Kramer A, Pfeiffer AF. Regulation of the clock gene expression in human adipose tissue by weight loss. Int J Obes (Lond). 2016;40:899–906.
73. Samblas M, Milagro FI, Gomez-Abellan P, Martinez JA, Garaulet M. Methylation on the circadian gene BMAL1 is associated with the effects of a weight loss intervention on serum lipid levels. J Biol Rhythms. 2016;31:308–17.
74. Hardeland R. Melatonin and the pathologies of weakened or dysregulated circadian oscillators. J Pineal Res. 2017;62.

75. Cardinali DP. Ma Vie en Noir. Fifty years with melatonin and the stone of madness. Switzerland: Springer; 2016.
76. Kozirog M, Poliwczak AR, Duchnowicz P, Koter-Michalak M, Sikora J, Broncel M. Melatonin treatment improves blood pressure, lipid profile, and parameters of oxidative stress in patients with metabolic syndrome. J Pineal Res. 2011;50:261–6.
77. Gonciarz M, Bielanski W, Partyka R, Brzozowski T, Konturek PC, Eszyk J, Celinski K, Reiter RJ, Konturek SJ. Plasma insulin, leptin, adiponectin, resistin, ghrelin, and melatonin in nonalcoholic steatohepatitis patients treated with melatonin. J Pineal Res. 2013;54:154–61.
78. Gonciarz M, Gonciarz Z, Bielanski W, Mularczyk A, Konturek PC, Brzozowski T, Konturek SJ. The effects of long-term melatonin treatment on plasma liver enzymes levels and plasma concentrations of lipids and melatonin in patients with nonalcoholic steatohepatitis: a pilot study. J Physiol Pharmacol. 2012;63:35–40.
79. Gonciarz M, Gonciarz Z, Bielanski W, Mularczyk A, Konturek PC, Brzozowski T, Konturek SJ. The pilot study of 3-month course of melatonin treatment of patients with nonalcoholic steatohepatitis: effect on plasma levels of liver enzymes, lipids and melatonin. J Physiol Pharmacol. 2010;61:705–10.
80. Romo-Nava F, Alvarez-Icaza GD, Fresan-Orellana A, Saracco AR, Becerra-Palars C, Moreno J, Ontiveros Uribe MP, Berlanga C, Heinze G, Buijs RM. Melatonin attenuates antipsychotic metabolic effects: an eight-week randomized, double-blind, parallel-group, placebo-controlled clinical trial. Bipolar Disord. 2014;16:410–21.
81. Modabbernia A, Heidari P, Soleimani R, Sobhani A, Roshan ZA, Taslimi S, Ashrafi M, Modabbernia MJ. Melatonin for prevention of metabolic side-effects of olanzapine in patients with first-episode schizophrenia: randomized double-blind placebo-controlled study. J Psychiatr Res. 2014;53:133–40.
82. Tuomi T, Nagorny CL, Singh P, Bennet H, Yu Q, Alenkvist I, Isomaa B, Ostman B, Soderstrom J, Pesonen AK, Martikainen S, Raikkonen K, Forsen T, Hakaste L, Almgren P, Storm P, Asplund O, Shcherbina L, Fex M, Fadista J, Tengholm A, Wierup N, Groop L, Mulder H. Increased melatonin signaling is a risk factor for type 2 diabetes. Cell Metab. 2016;23:1067–77.
83. American Psychiatric Association. Diagnostic and statistical manual of mental disorders. 5th ed. 2015.
84. Campos CI, Nogueira CH, Fernandes L. Aging, circadian rhythms and depressive disorders: a review. Am J Neurodegener Dis. 2013;2:228–46.
85. Milhiet V, Boudebesse C, Bellivier F, Drouot X, Henry C, Leboyer M, Etain B. Circadian abnormalities as markers of susceptibility in bipolar disorders. Front Biosci (Schol Ed). 2014;6:120–37.
86. Wirz-Justice A. Diurnal variation of depressive symptoms. Dialogues Clin Neurosci. 2008;10:337–43.
87. Jagannath A, Peirson SN, Foster RG. Sleep and circadian rhythm disruption in neuropsychiatric illness. Curr Opin Neurobiol. 2013;23:888–94.
88. Bersani FS, Iannitelli A, Pacitti F, Bersani G. Sleep and biorythm disturbances in schizophrenia, mood and anxiety disorders: a review. Riv Psichiatr. 2012;47:365–75.
89. Lewy AJ, Kern HA, Rosenthal NE, Wehr TA. Bright artificial light treatment of a manic-depressive patient with a seasonal mood cycle. Am J Psychiatry. 1982;139:1496–8.
90. Lewy AJ, Nurnberger JI Jr, Wehr TA, Pack D, Becker LE, Powell RL, Newsome DA. Supersensitivity to light: possible trait marker for manic-depressive illness. Am J Psychiatry. 1985;142:725–7.
91. Cardinali DP, Pandi-Perumal SR, Brown GM. Sleep and circadian dysregulation in depressive illness. Pharmacological implications. Clin Neuropsychiatry. 2011;8:321–38.
92. Soreca I. Circadian rhythms and sleep in bipolar disorder: implications for pathophysiology and treatment. Curr Opin Psychiatry. 2014;27:467–71.
93. Ng TH, Chung KF, Ho FY, Yeung WF, Yung KP, Lam TH. Sleep-wake disturbance in interepisode bipolar disorder and high-risk individuals: a systematic review and meta-analysis. Sleep Med Rev. 2015;20:46–58.

94. McCarthy MJ, Welsh DK. Cellular circadian clocks in mood disorders. J Biol Rhythms. 2012;27:339–52.
95. Dallaspezia S, Benedetti F. Chronobiological therapy for mood disorders. Expert Rev Neurother. 2011;11:961–70.
96. Coogan AN, Thome J. Chronotherapeutics and psychiatry: setting the clock to relieve the symptoms. World J Biol Psychiatry. 2011;12(Suppl 1):40–3.
97. Lewy AJ, Emens J, Jackman A, Yuhas K. Circadian uses of melatonin in humans. Chronobiol Int. 2006;23:403–12.
98. Johansson C, Willeit M, Smedh C, Ekholm J, Paunio T, Kieseppa T, Lichtermann D, Praschak-Rieder N, Neumeister A, Nilsson LG, Kasper S, Peltonen L, Adolfsson R, Schalling M, Partonen T. Circadian clock-related polymorphisms in seasonal affective disorder and their relevance to diurnal preference. Neuropsychopharmacology. 2003;28:734–9.
99. Benedetti F, Dallaspezia S, Colombo C, Pirovano A, Marino E, Smeraldi E. A length polymorphism in the circadian clock gene Per3 influences age at onset of bipolar disorder. Neurosci Lett. 2008;445:184–7.
100. Li SX, Liu LJ, LZ X, Gao L, Wang XF, Zhang JT, Lu L. Diurnal alterations in circadian genes and peptides in major depressive disorder before and after escitalopram treatment. Psychoneuroendocrinology. 2013;38:2789–99.
101. Karthikeyan R, Marimuthu G, Ramasubramanian C, Arunachal G, BaHammam AS, Spence DW, Cardinali DP, Brown GM, Pandi-Perumal SR. Association of *Per3* length polymorphism with bipolar I disorder and schizophrenia. Neuropsychiatr Dis Treat. 2014;110:2025–30.
102. Vollmer LL, Strawn JR, Sah R. Acid-base dysregulation and chemosensory mechanisms in panic disorder: a translational update. Transl Psychiatry. 2015;5:e572.
103. Huang F, Yang Z, Li CQ. The melatonergic system in anxiety disorders and the role of melatonin in conditional fear. Vitam Horm. 2017;103:281–94.
104. Cardinali DP, Vidal MF, Vigo DE. Agomelatine: Its role in the management of major depressive disorder. Clin Med Insights Psychiatry. 2012;4:1–23.
105. Liu J, Clough SJ, Hutchinson AJ, Adamah-Biassi EB, Popovska-Gorevski M, Dubocovich ML. MT1 and MT2 melatonin receptors: a therapeutic perspective. Annu Rev Pharmacol Toxicol. 2016;56:361–83.
106. Monti JM, BaHammam AS, Pandi-Perumal SR, Bromundt V, Spence DW, Cardinali DP, Brown GM. Sleep and circadian rhythm dysregulation in schizophrenia. Prog Neuropsychopharmacol Biol Psychiatry. 2013;43:209–16.
107. Mansour HA, Wood J, Logue T, Chowdari KV, Dayal M, Kupfer DJ, Monk TH, Devlin B, Nimgaonkar VL. Association study of eight circadian genes with bipolar I disorder, schizoaffective disorder and schizophrenia. Genes Brain Behav. 2006;5:150–7.
108. Lane JM, Vlasac I, Anderson SG, Kyle SD, Dixon WG, Bechtold DA, Gill S, Little MA, Luik A, Loudon A, Emsley R, Scheer FA, Lawlor DA, Redline S, Ray DW, Rutter MK, Saxena R. Genome-wide association analysis identifies novel loci for chronotype in 100,420 individuals from the UK Biobank. Nat Commun. 2016;7:10889.
109. Benson KL. Sleep in schizophrenia: pathology and treatment. Sleep Med Clin. 2015;10:49–55.
110. Gerbaldo H, Demisch L, Cardinali DP. Light exposure patterns in schizophrenia. Acta Psychiatr Scand. 1992;85:94–5.
111. Ju P, Cui D. The involvement of N-methyl-D-aspartate receptor (NMDAR) subunit NR1 in the pathophysiology of schizophrenia. Acta Biochim Biophys Sin (Shanghai). 2016;48:209–19.
112. Wulff K, Dijk DJ, Middleton B, Foster RG, Joyce EM. Sleep and circadian rhythm disruption in schizophrenia. Br J Psychiatry. 2012;200:308–16.
113. Pasternak O, Westin CF, Dahlben B, Bouix S, Kubicki M. The extent of diffusion MRI markers of neuroinflammation and white matter deterioration in chronic schizophrenia. Schizophr Res. 2015;161:113–8.
114. Morera-Fumero AL, Abreu-Gonzalez P. Role of melatonin in schizophrenia. Int J Mol Sci. 2013;14:9037–50.
115. Aston C, Jiang L, Sokolov BP. Microarray analysis of postmortem temporal cortex from patients with schizophrenia. J Neurosci Res. 2004;77:858–66.

116. Zorn JV, Schur RR, Boks MP, Kahn RS, Joels M, Vinkers CH. Cortisol stress reactivity across psychiatric disorders: a systematic review and meta-analysis. Psychoneuroendocrinology. 2016;77:25–36.
117. Tang Q, Song P, Xu L. The government's role in regulating, coordinating, and standardizing the response to Alzheimer's disease: anticipated international cooperation in the area of intractable and rare diseases. Intractable Rare Dis Res. 2016;5:238–43.
118. Zdanys KF, Steffens DC. Sleep disturbances in the elderly. Psychiatr Clin North Am. 2015;38:723–41.
119. Duffy JF, Zitting KM, Chinoy ED. Aging and circadian rhythms. Sleep Med Clin. 2015;10:423–34.
120. American Geriatrics Society 2015. Updated Beers criteria for potentially inappropriate medication use in older adults. J Am Geriatr Soc. 2015;63:2227–46.
121. Cardinali DP, Golombek DA, Rosenstein RE, Brusco LI, Vigo DE. Assessing the efficacy of melatonin to curtail benzodiazepine/Z drug abuse. Pharmacol Res. 2016;109:12–23.
122. Baandrup L, Fasmer OB, Glenthoj BY, Jennum PJ. Circadian rest-activity rhythms during benzodiazepine tapering covered by melatonin versus placebo add-on: data derived from a randomized clinical trial. BMC Psychiatry. 2016;16:348.
123. Clay E, Falissard B, Moore N, Toumi M. Contribution of prolonged-release melatonin and anti-benzodiazepine campaigns to the reduction of benzodiazepine and Z-drugs consumption in nine European countries. Eur J Clin Pharmacol. 2013;69:1–10.
124. Zeppenfeld DM, Simon M, Haswell JD, D'Abreo D, Murchison C, Quinn JF, Grafe MR, Woltjer RL, Kaye J, Iliff JJ. Association of perivascular localization of aquaporin-4 with cognition and Alzheimer disease in aging brains. JAMA Neurol. 2017;74:91–9.
125. Venkat P, Chopp M, Chen J. New insights into coupling and uncoupling of cerebral blood flow and metabolism in the brain. Croat Med J. 2016;57:223–8.
126. MacLeod R, Hillert EK, Cameron RT, Baillie GS. The role and therapeutic targeting of alpha-, beta- and gamma-secretase in Alzheimer's disease. Future Sci OA. 2015;1:FSO11.
127. Hurtado-Alvarado G, Dominguez-Salazar E, Pavon L, Velazquez-Moctezuma J, Gomez-Gonzalez B. Blood-brain barrier disruption induced by chronic sleep loss: low-grade inflammation may be the link. J Immunol Res. 2016;2016:4576012.
128. Collins O, Dillon S, Finucane C, Lawlor B, Kenny RA. Parasympathetic autonomic dysfunction is common in mild cognitive impairment. Neurobiol Aging. 2012;33:2324–33.
129. Ooms S, Ju YE. Treatment of sleep disorders in dementia. Curr Treat Options Neurol. 2016;18:40.
130. Riemersma-van der Lek RF, Swaab DF, Twisk J, Hol EM, Hoogendijk WJ, Van Someren EJ. Effect of bright light and melatonin on cognitive and noncognitive function in elderly residents of group care facilities: a randomized controlled trial. JAMA. 2008;299:2642–55.
131. Sirin FB, Kumbul DD, Vural H, Eren I, Inanli I, Sutcu R, Delibas N. Plasma 8-isoPGF2alpha and serum melatonin levels in patients with minimal cognitive impairment and Alzheimer disease. Turk J Med Sci. 2015;45:1073–7.
132. Allan CL, Behrman S, Ebmeier KP, Valkanova V. Diagnosing early cognitive decline-When, how and for whom? Maturitas. 2017;96:103–8.
133. Daulatzai MA. Pharmacotherpy and Alzheimer's disease: the M-Drugs (melatonin, minocycline, modafinil, and memantine) approach. Curr Pharm Des. 2016;22:2411–30.
134. Martino Adami PV, Galeano P, Wallinger ML, Quijano C, Rabossi A, Pagano ES, Olivar N, Reyes Toso C, Cardinali DP, Brusco LI, DoCarmo S, Radi R, Gevorkian G, Castaño EM, Cuello AC, Morelli L. Worsening of memory deficit induced by energy-dense diet in a rat model of early-Alzheimer's disease is associated to neurotoxic Aβ species and independent of neuroinflammation. Biochim Biophys Acta. 2017;1863:731–43.
135. Monti D, Ostan R, Borelli V, Castellani G, Franceschi C. Inflammaging and omics in human longevity. Mech Ageing Dev. 2016.
136. Minciullo PL, Catalano A, Mandraffino G, Casciaro M, Crucitti A, Maltese G, Morabito N, Lasco A, Gangemi S, Basile G. Inflammaging and anti-inflammaging: the role of cytokines in extreme longevity. Arch Immunol Ther Exp (Warsz). 2016;64:111–26.

137. Ralph MR, Lehman MN. Transplantation: a new tool in the analysis of the mammalian hypothalamic circadian pacemaker. Trends Neurosci. 1991;14:362–6.
138. Carrillo-Vico A, Lardone PJ, Alvarez-Sanchez N, Rodriguez-Rodriguez A, Guerrero JM. Melatonin: buffering the immune system. Int J Mol Sci. 2013;14:8638–83.
139. Siegel R, Naishadham D, Jemal A. Cancer statistics, 2013. CA Cancer J Clin. 2013;63:11–30.
140. Stevens RG, Zhu Y. Electric light, particularly at night, disrupts human circadian rhythmicity: is that a problem? Philos Trans R Soc Lond B Biol Sci. 2015;370.
141. Straif K, Baan R, Grosse Y, Secretan B, El GF, Bouvard V, Altieri A, Benbrahim-Tallaa L, Cogliano V. Carcinogenicity of shift-work, painting, and fire-fighting. Lancet Oncol. 2007;8:1065–6.
142. Hill SM, Belancio VP, Dauchy RT, Xiang S, Brimer S, Mao L, Hauch A, Lundberg PW, Summers W, Yuan L, Frasch T, Blask DE. Melatonin: an inhibitor of breast cancer. Endocr Relat Cancer. 2015;22:R183–204.
143. Lin X, Chen W, Wei F, Ying M, Wei W, Xie X. Night-shift work increases morbidity of breast cancer and all-cause mortality: a meta-analysis of 16 prospective cohort studies. Sleep Med. 2015;16:1381–7.
144. Valenzuela FJ, Vera J, Venegas C, Munoz S, Oyarce S, Munoz K, Lagunas C. Evidences of polymorphism associated with circadian system and risk of pathologies: a review of the literature. Int J Endocrinol. 2016;2016:2746909.
145. Altman BJ. Cancer clocks out for lunch: disruption of circadian rhythm and metabolic oscillation in cancer. Front Cell Dev Biol. 2016;4:62.
146. Cardinali DP, Escames G, Acuña-Castroviejo D, Ortiz F, Fernández-Gil B, Guerra Librero A, García-López S, Shen Y, Florido J. Melatonin-induced oncostasis. Mechanisms and clinical relevance. Integr Oncol. 2016;S1:006. doi:10.4172/2329-6771.S1-006.
147. Blask DE, Dauchy RT, Sauer LA, Krause JA, Brainard GC. Growth and fatty acid metabolism of human breast cancer (MCF-7) xenografts in nude rats: impact of constant light-induced nocturnal melatonin suppression. Breast Cancer Res Treat. 2003;79:313–20.
148. Blask DE, Brainard GC, Dauchy RT, Hanifin JP, Davidson LK, Krause JA, Sauer LA, Rivera-Bermudez MA, Dubocovich ML, Jasser SA, Lynch DT, Rollag MD, Zalatan F. Melatonin-depleted blood from premenopausal women exposed to light at night stimulates growth of human breast cancer xenografts in nude rats. Cancer Res. 2005;65:11174–84.
149. Dauchy RT, Hoffman AE, Wren-Dail MA, Hanifin JP, Warfield B, Brainard GC, Xiang S, Yuan L, Hill SM, Belancio VP, Dauchy EM, Smith K, Blask DE. Daytime blue light enhances the nighttime circadian melatonin inhibition of human prostate cancer growth. Comp Med. 2015;65:473–85.
150. Wang X, Ji A, Zhu Y, Liang Z, Wu J, Li S, Meng S, Zheng X, Xie L. A meta-analysis including dose-response relationship between night shift work and the risk of colorectal cancer. Oncotarget. 2015;6:25046–60.
151. Ma Z, Yang Y, Fan C, Han J, Wang D, Di S, Hu W, Liu D, Li X, Reiter RJ, Yan X. Melatonin as a potential anticarcinogen for non-small-cell lung cancer. Oncotarget. 2016;7:46768–84.
152. Howell D, Oliver TK, Keller-Olaman S, Davidson JR, Garland S, Samuels C, Savard J, Harris C, Aubin M, Olson K, Sussman J, MacFarlane J, Taylor C. Sleep disturbance in adults with cancer: a systematic review of evidence for best practices in assessment and management for clinical practice. Ann Oncol. 2014;25:791–800.
153. Dahiya S, Ahluwalia MS, Walia HK. Sleep disturbances in cancer patients: underrecognized and undertreated. Cleve Clin J Med. 2013;80:722–32.
154. Chen WY, Giobbie-Hurder A, Gantman K, Savoie J, Scheib R, Parker LM, Schernhammer ES. A randomized, placebo-controlled trial of melatonin on breast cancer survivors: impact on sleep, mood, and hot flashes. Breast Cancer Res Treat. 2014;145:381–8.
155. Dickerson SS, Connors LM, Fayad A, Dean GE. Sleep-wake disturbances in cancer patients: narrative review of literature focusing on improving quality of life outcomes. Nat Sci Sleep. 2014;6:85–100.
156. Harris B, Ross J, Sanchez-Reilly S. Sleeping in the arms of cancer: a review of sleeping disorders among patients with cancer. Cancer J. 2014;20:299–305.

157. Palesh OG, Roscoe JA, Mustian KM, Roth T, Savard J, Ancoli-Israel S, Heckler C, Purnell JQ, Janelsins MC, Morrow GR. Prevalence, demographics, and psychological associations of sleep disruption in patients with cancer: University of Rochester Cancer Center-Community Clinical Oncology Program. J Clin Oncol. 2010;28:292–8.
158. Jia Y, Lu Y, Wu K, Lin Q, Shen W, Zhu M, Huang S, Chen J. Does night work increase the risk of breast cancer? A systematic review and meta-analysis of epidemiological studies. Cancer Epidemiol. 2013;37:197–206.
159. Armstrong TS, Shade MY, Breton G, Gilbert MR, Mahajan A, Scheurer ME, Vera E, Berger AM. Sleep-wake disturbance in patients with brain tumors. Neuro Oncol. 2017;19:323–35.
160. Casault L, Savard J, Ivers H, Savard MH, Simard S. Utilization of hypnotic medication in the context of cancer: predictors and frequency of use. Support Care Cancer. 2012;20:1203–10.
161. Thekdi SM, Trinidad A, Roth A. Psychopharmacology in cancer. Curr Psychiatry Rep. 2015;17:529.
162. Dallmann R, Okyar A, Levi F. Dosing-time makes the poison: circadian regulation and pharmacotherapy. Trends Mol Med. 2016;22:430–45.

Epilogue

The history of the ANS starts in the Western world with the ancient Greeks, who coined the idea that the body was divided into two systems (somatic and autonomic) [1]. It was Galen (second century AD) who followed the vagus into the chest and abdominal cavities, documenting its communications with the viscera. Galen identified the sympathetic trunks as crossing the ribs, connecting with the thoracic and lumbar spinal cord, and continuing further to communicate with organs within the body cavity via the plexuses. Galen believed that the nerves acted as pipes, allowing the flow of "animal spirits" to pass among the organs.

He conceived the notion of "sympathy," i.e., when a change in the condition of one organ or part of the body occurs, this in turn causes another organ or portion of the body to react or alter its function. Sympathy remained a vague notion for centuries. It explains the cooperation or coordination of organs, such as an irritation of the stomach that produces syncope or seizures, as it is transmitted by the brain and nerves to the heart. Galen also postulated a humoral sympathy through the blood vessels, such as the relationship between the pregnant uterus and the mammary glands.

Indeed, sympathy is the major inspiring idea for this book. The bio-psycho-social-ecological nature of the individual can only be embraced if the enlarged and timed view of the ANS is followed, as outlined in this book.

It is futile to approach the disease from the statistical point of view of evidence-based medicine, as the popular paradigm claims nowadays. The physician must know in depth the individual reality of the person he is attempting to cure and this part of medicine is forgotten because of the inappropriate pressure of public health administrators and pharmaceutical companies. The enlarged and timed view of the ANS holds that all body systems are dependent and affected by the actions of others in a multicellular organization.

In summary, the ANS innervates the entire human body, and is involved in the regulation of every single organ in the body. Thus, perturbations in autonomic function account for everything from abnormalities in pupillary function to gastroparesis, intestinal dysmotility, diabetic diarrhea, genitourinary dysfunction, amongst others. Know autonomic function and you will know the whole of medicine! [2].

D.P. Cardinali, *Autonomic Nervous System*, DOI 10.1007/978-3-319-57571-1

References

1. Oakes PC, Fisahn C, Iwanaga J, DiLorenzo D, Oskouian RJ, Tubbs RS. A history of the autonomic nervous system: part I: from Galen to Bichat. Childs Nerv Syst. 2016;32:2303–8.
2. Vinik AI, Erbas T. 22 – Diabetic autonomic neuropathy. In: Buijs RM, Swaab D, editors. Handbook of clinical neurology. Autonomic nervous system. New York: Elsevier; 2013, p. 279–94.

Index

D.P. Cardinali, *Autonomic Nervous System*, DOI 10.1007/978-3-319-57571-1

T

MIX
Papier aus verantwortungsvollen Quellen
Paper from responsible sources
FSC® C105338

If you have any concerns about our products,
you can contact us on
ProductSafety@springernature.com

In case Publisher is established outside the EU,
the EU authorized representative is:
Springer Nature Customer Service Center GmbH
Europaplatz 3, 69115 Heidelberg, Germany

Printed by Libri Plureos GmbH
in Hamburg, Germany